Gas Accretion onto Galaxies

Astrophysics and Space Science Library

More information about this series at http://www.springer.com/series/5664

Andrew Fox • Romeel Davé
Editors

Gas Accretion onto Galaxies

Springer

Editors
Andrew Fox
Space Telescope Science Institute
Baltimore, MD, USA

Romeel Davé
University of the Western Cape
Cape Town, Western Cape
South Africa

ISSN 0067-0057 ISSN 2214-7985 (electronic)
Astrophysics and Space Science Library
ISBN 978-3-319-52511-2 ISBN 978-3-319-52512-9 (eBook)
DOI 10.1007/978-3-319-52512-9

Library of Congress Control Number: 2017933289

Printed on acid-free paper

This Springer imprint is published by Springer Nature
The registered company is Springer International Publishing AG
The registered company address is: Gewerbestrasse 11, 6330 Cham, Switzerland

Foreword

Astrophysics is hard. This branch of physics presents a number of obvious challenges to observers and experimental scientists. The targets are at tremendous distances, signals are weak, experimental setup is difficult or impossible to control (i.e., we must analyze the data that nature provides and cannot carefully design experiments), and results are often limited by cosmic variance and telescope time. Edwin Hubble's famous characterization of observational astrophysics is apt: "...we search among ghostly errors of observations for landmarks that are scarcely more substantial." Similarly, modern astrophysics makes intense demands on theorists. Current problems are rarely tractable with analytical treatments, and computer simulations require exquisite resolution and extreme dynamic range in order to adequately capture crucial small-scale microphysical processes in a cosmological (large-scale) context. Indeed, in the area of galaxy formation and evolution, full numerical modeling of all of the relevant physics is usually impossible; many important but unresolved processes must be handled with sub-grid prescriptions, and different prescriptions for sub-grid physics can lead to profoundly different results. To make progress today, theorists must be inventive and resourceful, but they must also exercise caution about systematic uncertainties in simulation outcomes.

For these reasons, it is perhaps not surprising that just a few years ago, a highly influential paper by Kereš et al. presented a seemingly simple question: how do galaxies get their gas? This is clearly a fundamental question about galaxy evolution, and at first glance this seems like a relatively straightforward issue. After all, the (presumably pristine) intergalactic gas reservoir from which galaxies form is mostly a simple hydrogen plasma, and many of the complications that plague other topics (e.g., dust, molecules, turbulence, magnetic fields, and cosmic rays) might be negligible or at least of secondary importance. However, in reality this simple question has proved to be a recalcitrant problem, for many reasons. On the observational side, accreting intergalactic gases have very low densities, and at the expected densities, emission from the accreting gas is very difficult (often impossible) to detect. Moreover, the infalling material can be shock heated into temperature ranges (e.g., the so-called warm-hot intergalactic medium at 10^5–10^7 K) that require ultraviolet or X-ray observations with cutting-edge telescopes in space, an expensive endeavor. This seemingly simple question creates headaches for theorists as well, e.g., in addition to shock heating, infalling gas can be shredded by processes such as the Kelvin-Helmholtz instability. Conversely, some accreting material could be thermally unstable and could fragment into rather small and low-mass clouds that ultimately drop into a galaxy and fuel star formation. These processes can be difficult to accurately model in computational simulations, especially if the simulation has a large enough size to provide a proper cosmological context. In addition, gas accretion is not an isolated phenomenon; as the material descends into a galaxy, it may encounter outflowing and enriched material driven away by star formation or feedback from supermassive nuclear black holes, and the interactions between the infalling and outflowing matter can significantly change how accretion works (and the theoretical predictions to be tested with observations). Stripping of gas from satellite galaxies may play a role in addition to infalling primordial material, and of course dark matter cannot be ignored. In the end, understanding how galaxies get their gas turns out to be a very difficult question.

However, there are reasons to feel optimistic about the likelihood of progress on understanding galactic gas accretion. Absorption spectroscopy can detect low-density gas and is orders of magnitude more sensitive than emission studies, and access to high-resolution spectroscopy in the rest-frame UV and X-ray bands provides detailed information on all of the likely phases in circumgalactic and intergalactic media from $z = 0$ to $z > 5$, including the elusive warm-hot gas. The deployment of the Cosmic Origins Spectrograph (COS) on the Hubble Space Telescope has been particularly transformative. By providing coverage of UV resonance lines from a wide variety of elements and ionization stages at low and intermediate redshifts, COS has enabled statistically useful studies of absorption lines from circumgalactic/intergalactic plasmas in a variety of contexts, and programs such as the COS-Halos survey have led to rapid progress on low-density and highly ionized gas in galaxy halos. On the theoretical side, Moore's Law continues to hold, and advances in computational power support increasingly sophisticated simulations. Theoretical modeling is improving by leaps and bounds.

For these reasons, this is an ideal time for a set of detailed reviews of recent observational and theoretical work on the topic of how galaxies acquire their gas. The chapters in this book present a set of reviews that span many of the key observations of circumgalactic material ranging from cool and neutral matter to hot and highly ionized plasma over a wide range of redshifts. The book also presents excellent discussions of theoretical motivations and progress on several aspects of accretion and galactic feedback. I expect that this publication will provide a valuable tool for pundits and highly experienced researchers as well as students that are just beginning to come up to speed on galaxy evolution. I am sure that I will often reach for this set of reviews, and I commend the authors and editors for assembling an excellent compendium on a crucial aspect of galaxy evolution.

Amherst, MA, USA Todd Tripp
November 2016

Preface

From majestic spirals to behemoth ellipticals to disordered dwarfs, the richness and diversity of galaxies has been a subject of study since the time of Hubble. A common feature among all galaxies is that their growth is driven by accretion of material from a vast reservoir of surrounding intergalactic gas, which provides fuel for forming new stars and growing supermassive black holes. Yet this ubiquitously predicted accretion has been notoriously difficult to detect directly. Until recently accretion was only seen around our own Milky Way, but advancing facilities have now enabled astronomers to obtain tantalizing evidence of accretion out to much earlier epochs, back to when galaxies were in their heyday of growth. Meanwhile, supercomputer simulations have highlighted that simple gravitationally driven accretion is only one aspect of a vast and complex story for how galaxies obtain their fuel, a story that includes energetic processes such as supernova-driven winds and black hole accretion. This edited volume presents the current state of accretion studies from both observational and theoretical perspectives, and charts our progress towards answering the fundamental yet elusive question, *"how do galaxies get their gas?"*

Understanding how galaxies form and evolve has been a central focus in astronomy for over a century. These studies have accelerated in the new millennium, driven by two key advances: the establishment of a firm concordance cosmological model that provides the backbone on which galaxies form and grow, and the recognition that galaxies grow not in isolation but within a "cosmic ecosystem" that includes the vast reservoir of gas filling intergalactic space. This latter aspect in which galaxies continually exchange matter with the intergalactic medium via inflows and outflows has been dubbed the "baryon cycle", and is featured as one of the central questions in the 2010 Astronomy Decadal Survey (*New Worlds, New Horizons*). The topic of our book is directly related to the baryon cycle, in particular its least well-constrained aspect, namely gas accretion.

Accretion is a rare area of astrophysics in which the basic theoretical predictions are established, but the observations have been as yet unable to verify the expectations. Accretion has long been seen around the Milky Way in so-called High Velocity Clouds, but the inferred accretion rates are uncertain. Detecting accretion even around nearby galaxies has proved challenging; its multiphase nature requires sensitive observations across the electromagnetic spectrum for full characterization. Theory also strongly predicts that accretion is much more rapid in the early universe, so much effort has gone into developing new ways to detect accretion in distant, unresolved galaxies. A promising approach involves looking for kinematic signatures, but accretion signatures are often confused with internal motions within galaxies. Meanwhile, theorists have realized that accretion left unchecked would lead to galaxies that look nothing like observed galaxies. Hence accretion must somehow be a self-regulating process. Understanding the physical origin of this delicate balance of the baryon cycle that leads to galaxies as we see them has proved to be an immense challenge, requiring the most advanced supercomputer simulations to model properly. Accretion studies therefore touch a wide range of astrophysical processes, and hence a wide cross section of the astronomical community.

An edited volume on this topic is timely for a number of reasons. Observational facilities are finally able to access the wavelength ranges and depths at which accretion processes may be manifest. Because inflowing gas is diffuse and does not glow like stars, the best hope for direct detection generally lies in absorption-line spectroscopy. It turns out that the ultraviolet waveband contains the most interesting lines for this purpose, which has made the Cosmic Origins Spectrograph on Hubble a game changer for baryon cycle observations. Meanwhile, the emergence of multi-object spectroscopy on 10m-class ground-based telescopes such as Keck and VLT has likewise revolutionized our understanding of baryon cycle processes at intermediate redshifts, where the UV lines are redshifted into the more accessible optical band. These baryon cycle studies represent a key line of investigation for upcoming 30m-class facilities and the proposed next-generation UV/optical space telescope (LUVOIR), which may even be sensitive enough to map UV line emission from accreting gas. At the same time, radio investigations at low redshift continue to unravel the properties of the neutral gas around galaxies in high spatial resolution. Hence the time is right to survey these multiple lines of investigation and determine the state of the field in accretion studies of the baryon cycle.

Acknowledgments We are grateful to Nora Rawn, Sheik Mohideen, and the Springer staff for their advice and support in the preparation of this volume. Andrew Fox thanks Dr. Fred Lo for hosting a workshop on gas accretion at NRAO that helped to formulate the idea for this book. During the preparation of the final manuscript, we learned that Fred had sadly passed away. We acknowledge his key role in bringing this volume together.

Baltimore, MD, USA
Cape Town, South Africa
December 2016

Andrew Fox
Romeel Davé

Contents

Contributors

Nicolas F. Bouché IRAP, 9 Av. Colonel Roche, F-31400 Toulouse, France

Sebastiano Cantalupo Institute for Astronomy, ETH-Zürich, Zürich, Switzerland

Hsiao-Wen Chen Department of Astronomy & Astrophysics and Kavli Institute for Cosmological Physics, The University of Chicago, Chicago, IL, USA

Claude-André Faucher-Giguère Department of Physics and Astronomy and Center for Interdisciplinary Exploration and Research in Astrophysics (CIERA), Northwestern University, Evanston, IL, USA

Kristian Finlator New Mexico State University, Las Cruces, NM, USA

Filippo Fraternali Department of Physics and Astronomy, University of Bologna, Bologna, Italy

Kapteyn Astronomical Institute, University of Groningen, The Netherlands

Glenn G. Kacprzak Center for Astrophysics & Supercomputing, Swinburne University of Technology, Hawthorn, VIC, Australia

Nicolas Lehner Center for Astrophysics, Department of Physics, University of Notre Dame, Notre Dame, IN, USA

Felix J. Lockman Green Bank Observatory, Green Bank, WV, USA

Mary E. Putman Department of Astronomy, Columbia University, New York City, NY, USA

Philipp Richter Institut für Physik und Astronomie, University of Potsdam, Potsdam, Germany

Kate H.R. Rubin Department of Astronomy, San Diego State University, San Diego, CA, USA

Jorge Sanchez Almeida Instituto de Astrofísica de Canarias, Canary Islands, Spain

Departamento de Astrofísica, Universidad de La Laguna, Tenerife, Spain

Kyle R. Stewart Department of Mathematical Sciences, California Baptist University, Riverside, CA, USA

Freeke van de Voort Theoretical Astrophysics, Heidelberg Institute for Theoretical Studies, Heidelberg, Germany

Academia Sinica Institute of Astronomy and Astrophysics, Taipei, Taiwan

Heidelberg Institute for Theoretical Studies, Heidelberg, Germany

Astronomy Department, Yale University, New Haven, CT, USA

Acronyms

Gas Accretion onto Galaxies, Fox and Davé, eds.

ACS	Advanced Camera for Surveys
AGN	Active galactic nuclei
ALMA	Atacama Large Millimeter/Submillimeter Array
AMR	Adaptive mesh refinement
AO	Adaptive optics
ASKAP	Australian Square Kilometer Array Pathfinder
BAL	Broad Absorption Line
BASIC	Bimodal Absorption System Imaging Campaign
BAT	Burst Alert Telescope
BCD	Blue compact dwarf
BOSS	Baryon Oscillation Spectroscopic Survey
CGM	Circumgalactic medium
CL	Confidence limit
CLUES	Constrained Local UniversE Simulations
COS	Cosmic Origins Spectrograph
DIG	Diffuse ionized gas
DLA	Damped Lyman-alpha (system)
DM	Dark Matter
EAGLE	Evolution and Assembly of Galaxies and their Environments
EBHIS	Effelsberg-Bonn HI Survey
ESI	Echelette Spectrograph and Imager (instrument on Keck)
FIRE	Feedback in Realistic Environments (simulations)
FMR	Fundamental metallicity relation
FWHM	Full width at half maximum
GASS	Galactic All-Sky Survey
GBT	Green Bank Telescope
GMM	Gaussian mixture modeling
GMOS	Gemini Multi-Object Spectrograph

HALOGAS	Hydrogen Accretion in LOcal GAlaxieS (survey)
HST	Hubble Space Telescope
HVC	High-velocity cloud
HzRG	High-z radio galaxy
ICM	Intracluster medium
IFU	Integral field unit
IGM	Intergalactic medium
IMF	Initial mass function
ISM	Interstellar medium
IVC	Intermediate-velocity cloud
JVLA	Jansky Very Large Array
JWST	James Webb Space Telescope
KCWI	Keck Cosmic Web Imager
KMOSS	K-Band Multi-Object Spectrograph
KODIAQ	Keck Observatory Database of Ionized Absorbers toward Quasars
LAB	Leiden Argentine Bonn (survey)
LAB	Lyman-alpha blob
LAE	Lyman-alpha emitter
LBG	Lyman-break galaxy
LBT	Large Binocular Telescope
LINER	Low-ionization nuclear emission line regions
LLS	Lyman limit system
LMC	Large Magellanic Cloud
LRG	Luminous red galaxy
LRIS	Low Resolution Imaging Spectrometer
LUVOIR	Large UltraViolet/Optical/InfraRed (mission concept)
LVC	Low-velocity cloud
ΛCDM	Lambda Cold Dark Matter (cosmology/model)
MANGA	Mapping Nearby Galaxies at APO (survey)
MEGAFLOW	MusE GAs FLOw and Wind (survey)
MOSFIRE	Multi-Object Spectrometer For Infra-Red Exploration
MQN	MUSE Quasar Nebulae
MS	Magellanic Stream
MUSE	Multi-Unit Spectroscopic Explorer (instrument on VLT)
MZR	Mass-metallicity relation
QBCD	Quiescent blue compact dwarf
QSO	Quasi-stellar object
PCWI	Palomar Cosmic Web Imager
PDF	Probability distribution function
PLLS	Partial Lyman limit system
PSF	Point spread function
SDSS	Sloan Digital Sky Survey
SFG	Star forming galaxy
SFR	Star formation rate
SIMPLE	SInfoni Mg II Program for Line Emitters

SINFONI	Spectrograph for INtegral Field Observations in the Near Infrared
SKA	Square Kilometer Array
SLLS	Super Lyman limit system
SN	Supernovae
SPH	Smoothed particle hydrodynamics
Sub-DLA	Sub-damped Lyman-alpha (system)
SXRB	Soft X-ray background
TMT	Thirty-Meter Telescope
TPCF	Two-point correlation function
TTT	Tidal torque theory
UV	Ultraviolet
UVB	Ultraviolet Background
VLT	Very Large Telescope
VMP	Very metal poor (gas or absorbers)
WHAM	Wisconsin H-alpha Mapper
WSRT	Westerbork Synthesis Radio Telescope
XMP	Extremely metal poor (galaxy)

An Introduction to Gas Accretion onto Galaxies

Mary E. Putman

1 Introduction

The idea of gas accretion onto galaxies first came ∼50 years ago. In the 1960s and 1970s, observations of high velocity hydrogen clouds were made and it was proposed that they represent infalling Galactic fuel (Muller et al. 1963; Hulsbosch 1968; Dieter 1971; Oort 1970). This infall was soon understood to have consequences on the Milky Way's star formation and distribution of stellar metallicities (Larson 1972a,b; van den Bergh 1962). Since the original detection of Galactic cold hydrogen halo clouds, observations have been made of halo gas in a variety of phases for galaxies throughout the universe. The observations have made it clear that there are abundant baryons surrounding galaxies and that some of these baryons will accrete and fuel future star formation.

There are numerous observations of galaxies and the intergalactic medium (IGM) that infer gas accretion is needed throughout cosmic time. There are galaxies at all redshifts observable with star formation rates that indicate they will run out of fuel within a few Gyrs without replenishment. The metallicities of their stars also suggest galaxy evolution models without accretion will not work. At the level of the census of baryons in the universe, the decrease in the mass density of cold hydrogen with time does not closely track the steady increase in the mass density of stars, and this requires the ionized IGM to cool and accrete onto galaxies. The first section of this introduction provides an overview of this type of indirect observational evidence for gas accretion.

Theoretically, ongoing gas accretion is required to produce realistic galaxies and it largely occurs through the accretion of the intergalactic medium and satellites.

M.E. Putman (✉)
Department of Astronomy, Columbia University, New York, NY 10027, USA
e-mail: mputman@astro.columbia.edu

A. Fox, R. Davé (eds.), *Gas Accretion onto Galaxies*, Astrophysics and Space Science Library 430, DOI 10.1007/978-3-319-52512-9_1

Feedback from galaxies is also a key component of galaxy formation models and the interplay between accretion and feedback is important to understand. There is some observational support for the IGM, satellite gas, and feedback material all being sources of future star formation fuel. The second section of this chapter discusses these sources of gas accretion and the different modes with which the gas may ultimately reach the star-forming core of a galaxy.

Direct evidence for gas accretion, as in the kinematic signature of gas falling directly onto the stellar component of a galaxy, is relatively rare and the third section of this chapter summarizes the direct observational evidence currently available. Some of the numerous additional observational claims for gas accretion onto individual systems are also discussed. I conclude this chapter with a brief summary and thoughts on directions for the future.

2 The Need for Accretion Through Cosmic Time

The need for ongoing gas accretion is evident in observations of galaxies and the IGM at all redshifts. In this section, the broader indirect pieces of observational evidence for gas accretion are discussed. This includes the star formation rates of populations of galaxies, the state of the baryons in the universe, and the metal enrichment history of galaxies.

Star formation rates that will exhaust a galaxy's gas supply on a relatively rapid timescale are commonly derived from observations. At high redshift ($z > 1$), there are measured gas depletion times of less than a Gyr (e.g., Genzel et al. 2010; Daddi et al. 2010; Tacconi et al. 2013). The high redshift observations are limited to those galaxies with accessible gas and star formation tracers (Shapley 2011), but this result is found for a wide variety of tracers. At lower redshift, where there is a more complete census of the gas content and star formation rate of galaxies, the measured gas depletion times are longer, but still typically less than a few Gyrs (Kennicutt and Evans 2012; Bigiel et al. 2011; Leroy et al. 2013; Schiminovich et al. 2010). The primary exception is the gas-rich dwarf galaxies that can have depletion times closer to a Hubble time (van Zee 2001; Hunt et al. 2015; Huang et al. 2012). There is evidence that galaxies are undergoing a more efficient mode of star formation at higher redshift, leading to the shorter depletion times and the need for a greater rate of accretion at early times (Scoville et al. 2016; Santini et al. 2014). This is also evident in the evolution of the specific star formation rate (SFR/M_*) that gradually decreases towards lower redshifts at a given stellar mass (Madau and Dickinson 2014; Karim et al. 2011). The accretion process may be key to regulating a galaxy's SFR and without it the depletion times at all redshifts suggest a large fraction of galaxies are not far from being red and dead.

Beyond individual galaxies, the cold gas content of the universe as a whole should decrease as more stars are formed. If it does not correspondingly decrease there must be continuous cooling of the ionized gas in the IGM and halos that harbor the majority of the baryons in the universe (Shull et al. 2012; Bregman 2007).

This can be investigated with a comparison of the mass density of atomic hydrogen to the mass density of stars through cosmic time, as shown in Fig. 1. This figure shows that the HI mass density has an overall decrease from $z = 3$ to $z = 0$ ($\sim$11 Gyr ago to today), but this evolution is mild compared to the increase in stellar mass density with time. Where the decrease in H I mass density specifically happens depends on the measurements adopted (Rao et al. 2006; Lah et al. 2007; Prochaska et al. 2005; Zafar et al. 2013), but the values at $z = 3$ derived from damped Lyman-α absorbers (DLAs) are clearly higher than the H I emission measurements at $z = 0$ (Zwaan et al. 2005; Hoppmann et al. 2015). A recent study by Neeleman et al. (2016) at $z \sim 0.6$ ($\sim$6 Gyr ago) is consistent with a gradual decline from $z = 2$ to today. This measurement was achieved with a blind DLA survey, an improvement over the higher point at approximately this redshift/time that is based on Mg II surveys (see Prochaska and Wolfe 2009). In any case, the evolution of the H I does not appear to closely follow the evolution of the stellar mass density or star formation rate density of the universe (Madau and Dickinson 2014; Putman et al. 2009; Hopkins et al. 2008). Naively a direct correlation would be expected if there is no accretion and cooling of new H I gas.

There is clearly more work to be done to understand the evolution of baryons across time. For the mass density of atomic hydrogen, we are currently limited to using absorption line studies at every redshift but $z = 0$, where H I emission measurements are available. This will improve with H I emission surveys that can reach $z = 0.5$ ($\sim$5 Gyr ago) in progress with the JVLA (Fernández et al. 2016), and planned with SKA precursor telescopes (ASKAP; Duffy et al. 2012). The MeerKAT survey LADUMA is designed to detect H I in emission out to at least $z = 1$ (over half the age of the universe), and this will significantly add to our knowledge of the evolution of the H I mass density. CO surveys with ALMA are also an important component to our future understanding of the evolution of the cold gas content of the universe. The CO has a direct correlation with star formation (i.e., a short consumption time; Bigiel et al. 2011; Leroy et al. 2008), and correspondingly the molecular (traced by CO) to stellar mass ratio already shows indications of a clear decline with redshift (Carilli et al. 2013; Bauermeister et al. 2013). The molecular gas depletion rate does still require continuous replenishment via gas inflow (Bauermeister et al. 2010), but it is not as directly apparent as with the H I evolution.

The final piece of indirect observational evidence for accretion discussed in this section is the metallicity distribution of the stars in galaxies. In the local universe, the metallicity distribution of the long-lived stars supports galaxy evolution models with a continuous inflow of relatively low metallicity gas (Chiappini 2009; Fenner and Gibson 2003; Larson 1972b). Closed box galaxy evolution models produce a wider distribution of stellar metallicities than is observed and this has traditionally been referred to as the G-dwarf problem. This evidence for accretion is strongest in the local universe where the metallicity of individual long-lived stars can be measured (Holmberg et al. 2007; Kirby et al. 2013; Grillmair et al. 1996), but it is also consistent with the metallicities derived from the integrated stellar light observations of galaxies (e.g., Henry and Worhey et al. 1999; Bressan et al. 1994; Stott et al. 2014;

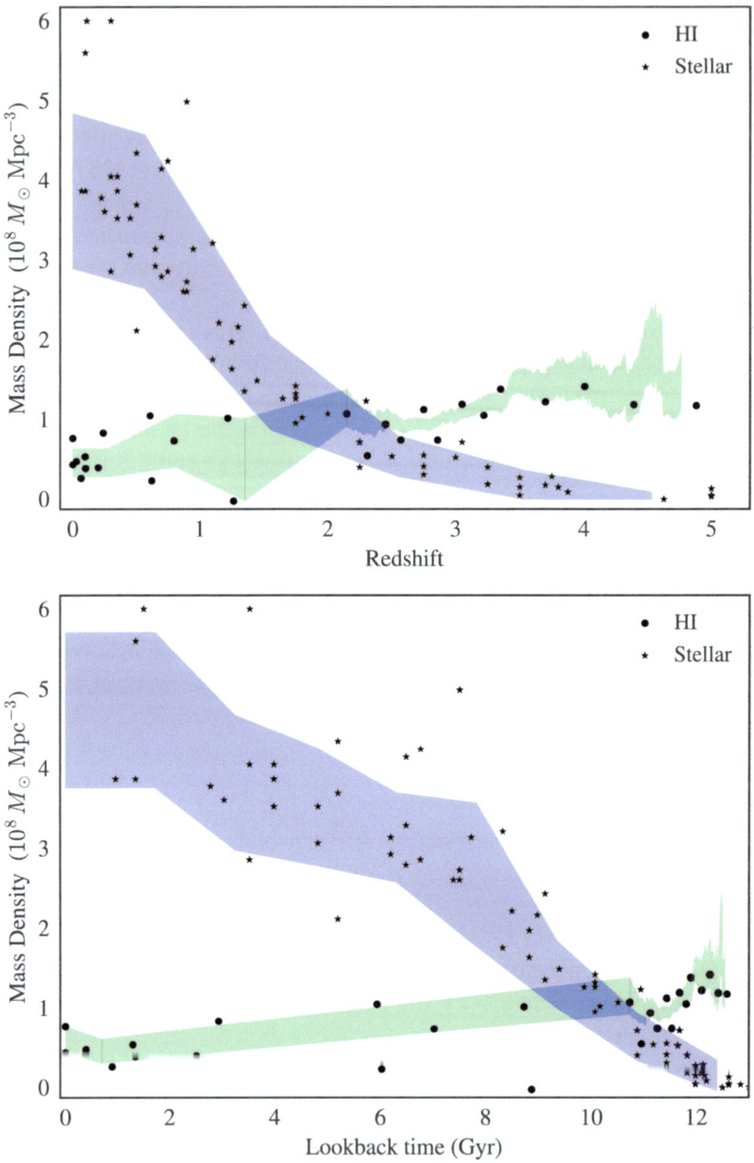

Fig. 1 A comparison of the evolution of the stellar mass density (*stars*/*purple*) and atomic gas (H I) mass density (*points*/*green*) with redshift (*top*) and lookback time (*bottom*). The H I data are from the compilation by Neeleman et al. (2016) and Sánchez-Ramírez et al. (2016), and the stellar mass densities are taken from the Madau and Dickinson (2014) compilation. The H I measurements are largely based on damped Lyman-α absorption measurements except at $z \sim 0$. The *purple* and *green shades* represent the running median and scatter of the data with the shaded H I incorporating the $z > 1.6$ results of Sánchez-Ramírez et al. (2016)

Gallazzi et al. 2005). The metallicity distribution of planetary nebulae has also been used as evidence for gas accretion (e.g., Magrini et al. 2007). These metallicity results are for a variety of galaxy types and the fact that the accreting gas needs to be relatively low metallicity is considered support for the IGM being a major source of accretion.

3 Expected Modes of Accretion

Gas accretion onto the stellar component of a galaxy can proceed in several ways and from multiple sources. Most of the gas in the halos of galaxies is ionized, and since the ionized gas mass in a galaxy's disk is smaller than the mass in cold gas ($<10^4$ K; e.g., Ferrière et al. 2001), the halo gas must rapidly cool as it accretes. The gas may come in as large cool clouds, dribble onto the disk from smaller warm clouds, enter preferentially at the edges of the galaxy, or the gas may accrete with a combination of these methods. The dominant mode of accretion at the star-forming component of a galaxy remains to be determined. The interplay between enriched outflowing gas with gas coming in may be key in this process (Fraternali and Binney 2008; Putman et al. 2012; Marinacci et al. 2010; Voit et al. 2015).

The major sources of the accreting material are thought to be the IGM, satellites, and recycled feedback gas (see Fig. 2). Theoretically, all three of these sources are expected and observationally there is evidence for all three, although much of the evidence is indirect. For instance, the continuous distribution of gas from a galaxy through its halo to the IGM, as found with absorption line experiments (Tumlinson et al. 2013; Prochaska et al. 2011; Penton et al. 2002; Wakker and Savage 2009; Chen et al. 2001), is consistent with the IGM as an important fuel source, though not direct evidence of its accretion. Theoretically, the inflowing filaments of IGM are expected to be the largest source of ongoing accretion for a galaxy (Joung et al. 2012a; Kereš et al. 2005; Brooks et al. 2009). Depending on the mass of the galaxy halo, some percentage of the inflowing IGM is heated to high temperatures in the simulations. Figure 3 shows the state of the accreting case for an L_* galaxy and how the filaments of low metallicity IGM (top panel) are partially heated as they move through the halo and also cool in the central regions as they approach the disk. How the gas is able to ultimately cool to below 10^4 K and feed the star formation in the disk may be related to density enhancements in the filaments and the mixing with satellite and feedback material (Joung et al. 2012b; Fraternali and Binney 2008). The simulations also find that much of the ongoing IGM accretion occurs towards the edges of the galaxy to avoid the dominant feedback from the central regions (Stewart et al. 2011; Fernández et al. 2012).

It is clear that satellite gas is stripped within a galaxy halo as observations have captured this directly (e.g., Fig. 4), and satellites closer to galaxies typically do not have gas and are redder (Grcevich and Putman 2009; Spekkens et al. 2014; Geha et al. 2012). Ram pressure stripping by the CGM of the host galaxy is thought to be the dominant stripping mechanism, but other forces can be important, for instance,

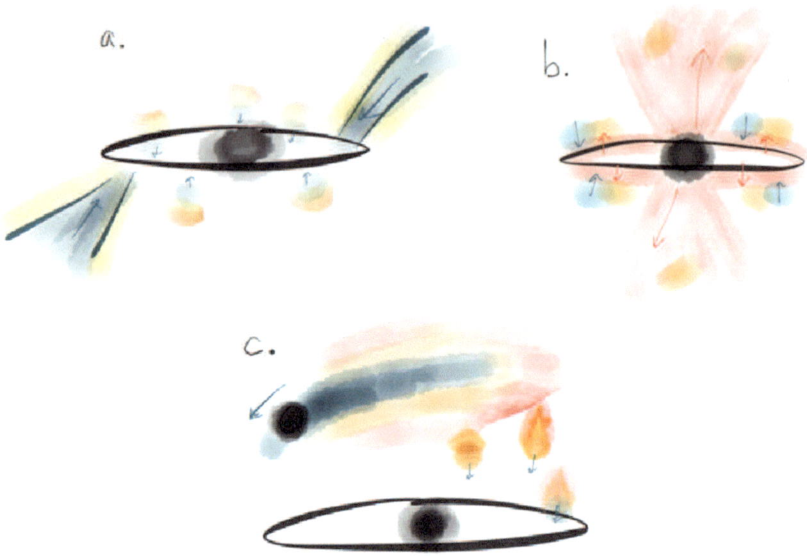

Fig. 2 A representation of the three expected sources of accretion with red indicating hot gas and blue cooler gas. (**a**) (*top-left*) Accretion from the IGM along filaments where the outer parts are heated and the inner parts are able to cool. The hot IGM-originated halo gas cooling near the disk as it mixes with denser gas is also indicated. (**b**) (*top-right*) Feedback material can accrete as part of a fountain flow close to the disk with hot gas from stellar feedback rising and then cooling and falling back down. Gas from a central outflow will mix with existing halo density enhancements and this may also result in cool clumps that eventually accrete. (**c**) (*bottom*) Satellites are stripped of their gas as they move through the diffuse halo medium and this gas will fall to the disk as warm clouds. As the gas slows and mixes with denser feedback material it can potentially re-cool close to the disk

when satellites come in as an interacting pair (Pearson et al. 2016; Marasco et al. 2016). The gas is largely heated when it is ram pressure stripped (Tepper-García et al. 2015; Gatto et al. 2013; Fox et al. 2014), and (again) it is not completely clear how it ultimately cools to feed the galaxy's star formation. It may sink to the disk as density enhancements in the halo and ultimately cool closer to the disk as it slows and encounters a denser surrounding medium (Heitsch and Putman 2009; Joung et al. 2012b; Bland-Hawthorn et al. 2007). The numerous small satellites found in the Local Group and predicted by simulations do not provide a significant amount of gas, but larger satellites, such as the Magellanic Clouds for the Milky Way, can provide gigayears worth of star formation fuel to a galaxy.

There are numerous observed kinematic signatures of feedback mechanisms putting gas into the halos of galaxies (Rubin et al. 2014; Shapley et al. 2003; Weiner et al. 2009; Chen et al. 2010; Heckman et al. 2000). While simulations require many of the metals created by a galaxy are ejected from the system (see panel c of Fig. 3), there is a large mass of metals detected in galaxy halos that remains bound

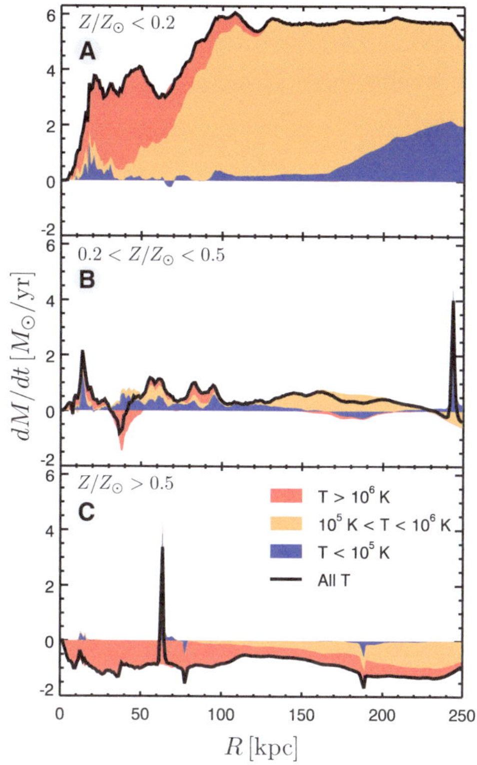

Fig. 3 The mass accretion rate of different temperature and metallicity gas at $z = 0$ in an AMR simulation of a Milky Way mass galaxy (Joung et al. 2012a). Most of the accreting gas is low metallicity material from the IGM as shown in panel (**a**). The highest metallicity gas (panel (**c**)) has a net outflow at all radii. The central panel (**b**) shows the intermediate metallicity gas inflow with the spikes corresponding to satellites. The change in temperature of the accreting gas is largely due to the heating of the inflowing gas as it interacts with existing halo gas and the cooling in the central regions of the filaments

(Tumlinson et al. 2011; Werk et al. 2014). The results therefore indicate there is abundant future star formation fuel that has already cycled through the galaxy. As discussed at the end of Sect. 2, the metallicities of the stars in galaxies indicate feedback should not dominate as the fuel source. Mixing feedback material with IGM and satellite gas is key to balance this out. This is consistent with the results of simulations that can produce the large amount of detected ions in galaxy halos while remaining consistent with the observed mass-metallicity relation for galaxies (Oppenheimer et al. 2016; Muratov et al. 2016).

4 Direct Observational Evidence for Accretion

Direct unambiguous kinematic evidence of gas falling onto a galaxy is relatively rare. The simple approximation of the mass accretion rate is $\dot{\mathrm{M}}$ ($M_{\odot}$/yr) $= Mv/z$, where M is the mass of the accreting material, v is its velocity, and z is the height the material is falling from. Each of these parameters usually has significant uncertainties in the observations depending on the gas phase probed and the geometry of the system. In all cases we are only capable of measuring one

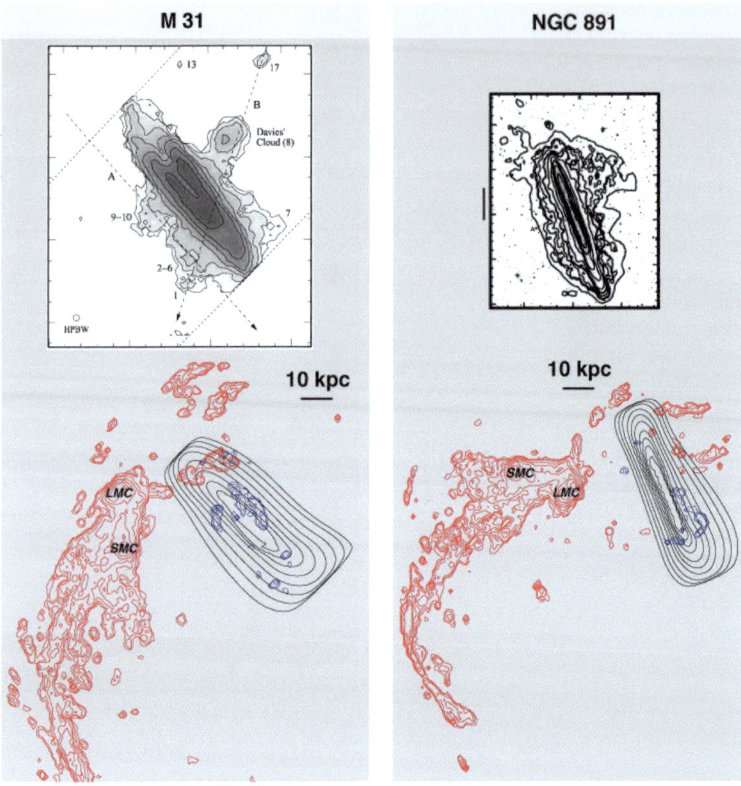

Fig. 4 Evidence for accretion from H I observations for the Milky Way (*bottom*), M31 (*top left*; Westmeier et al. 2007), and NGC 891 (*top right*; Oosterloo et al. 2007). The Milky Way observations are shown from an external point of view with the viewing angle and distance to M31 (*bottom left*) and NGC 891 (*bottom right*). The accreting Milky Way H I halo gas at <10 kpc (*blue contours*) is difficult to discern from the disk gas (*black contours*) with an external view. The *red contours* show the accretion of satellite H I from the Magellanic System (Putman et al. 2003). The origin of the extraplanar H I for M31 and NGC 891 is unknown, but some of the larger H I features are potentially linked to satellite accretion. The extraplanar H I shown is not a substantial amount of mass, but some of it may represent the cooling of the large reservoir of ionized halo gas. This figure is from Putman et al. (2012)

component of the gas velocity, and accretion can be more accurately assessed when the gas is known to be close to the disk and unmeasured tangential velocities cannot easily dominate.

The Milky Way is an example of gas observed at 1–15 kpc above the disk that is clearly infalling. The actual rate of accretion depends on the 3D motions of the gas and the full extent of the accreting layer, but the rates calculated are 0.1–0.4 $M_{\odot}$/yr for the coldest gas (Putman et al. 2012), and closer to 1 $M_{\odot}$/yr when the ionized gas is included (Lehner and Howk 2011). Most of the gas thought to be in the halo of the Milky Way has unknown distances. The only halo gas known to be at large radii is that associated with the Magellanic System (Putman et al. 2003; Fox et al. 2014). It is difficult to say when the gas of Magellanic origin will accrete as it is likely to have a large tangential velocity component and will slow and be heated as it falls. An extended, diffuse halo medium is inferred to exist for the Milky Way from the nature of the Magellanic System and stripped satellites (Salem et al. 2015; Emerick et al. 2016; Grcevich and Putman 2009). The motion of this diffuse halo medium is largely unknown, and much of it is thought to be hot and not easily observed. There is at least a consistency between the radial velocities of the ions that probe the warm-hot halo gas and the simulated extended halo medium represented in Fig. 3 (Zheng et al. 2015).

There is limited direct evidence for accretion beyond the Milky Way. Absorption line experiments that use background QSOs do not know the location and motion of the gas relative to the galaxy's stars. Experiments that use objects within the galaxy itself do not have the uncertainty of the gas being on the near or far side and the corresponding ambiguity of the velocity potentially representing inflow or outflow. The vast majority of the observations using the galaxy itself show significant outflows, with only a few examples of detected inflow. Rubin et al. (2012) and Martin et al. (2012) detected Mg II and Fe II absorption for 100+ star-forming galaxies at $z = 0.4$–1.3 and found only 4–6% show cool gas inflow with velocities $<200\,\mathrm{km\,s^{-1}}$. The low detection rate may be related to the covering fraction of the cold gas inflow, the low velocities of the inflow relative to their velocity resolution, and/or significant inflow could be an intermittent process. Rubin et al. (2012) found all but one of the galaxies with inflow have an inclination $>60°$; and this may have helped to differentiate the inflowing material from the ubiquitous outflows. Many other claims of inflow that use the galaxy itself are tenuous given the velocity resolution of the observations and the difficulty in separating the inflow component from the galaxy in the spectrum (Sato et al. 2009; Giavalisco et al. 2011).

One of the strongest examples of the direct detection of gas accretion beyond the Milky Way is for the small spiral galaxy M33 in the Local Group. Zheng et al. (2017) used UV-bright stars in M33's disk as background probes and found the kinematic signature of inflow across the star-forming disk in the Si IV absorption lines (Fig. 5). An accreting layer of gas at the disk-halo interface is the most consistent model with the distribution and velocities of the inflowing gas. A layer close to the disk is consonant with the difficulty in detecting the inflow in other systems. The accretion rate obtained ($\sim 2.9\,M_{\odot}$/yr) is relatively large for this small galaxy, and may be further evidence for the infall of fuel being intermittent in nature.

With the numerous indirect methods of detecting gas accretion, it is difficult to provide a complete census of the results. This paragraph gives an overview of some of these methods. As mentioned previously, absorption line experiments that use distant background probes and model the likely location of the gas relative to the galaxy are often used to claim accretion (Bouché et al. 2016; Bowen et al. 2016). The actual location of the gas is not known, but cases where the absorbing gas is close in position-velocity space to the galaxy are more likely to be capturing the accretion process. When particularly low metallicity gas is detected in a galaxy halo it is often claimed to be the accretion of the IGM filaments mentioned in

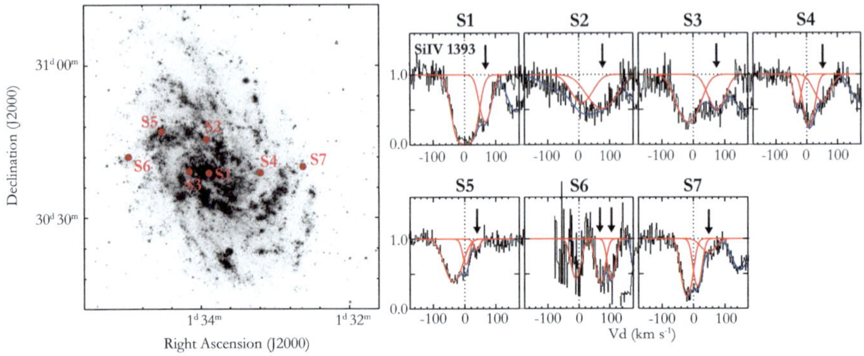

Fig. 5 The direct detection of accretion for the Local Group dwarf spiral galaxy M33 (Zheng et al. 2016). The *left* is a UV image of M33's star formation from *GALEX* with the stars targeted with the Cosmic Origins Spectrograph on *HST* labeled as S1–S7. The *right* shows the Si IV (λ1393) absorption lines detected for each sightline. The $0\,\mathrm{km\,s^{-1}}$ *dotted line* is the observed systemic velocity of the rotating H I disk, the *red* represents the fits to the individual lines and the *blue* is the composite fit. The *arrows* indicate gas that is inflowing with respect to M33's disk

Sect. 3 (e.g., Lehner et al. 2013; Cooper et al. 2015; Crighton et al. 2013). It would be difficult to explain this gas as anything else, but since the exact location and kinematics of the gas is unknown, it may or may not be accreting. It is also difficult to separate an IGM origin from satellite material at low redshift using metallicity alone (Muratov et al. 2016). Filamentary extensions of Lyman-α emission have been taken to be the detection of accretion at high redshift. In cases when multiple datasets are combined, the evidence is particularly strong (e.g., Rauch et al. 2016; Fumagalli et al. 2016; Martin et al. 2016). Finally, at low redshift, extensions of H I emission have been published as evidence for accretion (e.g., Kreckel et al. 2012; Putman et al. 2009; Sancisi et al. 2008, top panels of Fig. 4). The gas detected will certainly eventually accrete onto the nearby galaxy, but the origin of the gas and the direction the gas is currently moving are usually uncertain.

5 Summary

As outlined in this chapter and throughout this book, it is clear that gas accretion onto galaxies is occurring. We now know that the accreting cold hydrogen clouds originally found for the Milky Way are a small component of a process found throughout the universe. There are clear kinematic signatures of infalling gas for multiple galaxies and numerous other observations of individual systems that are consistent with gas accretion (Sect. 4). Beyond the evidence in individual systems, galaxies throughout time have star formation rates and metallicities that require ongoing accretion and the evolution of the H I mass density with time suggests it needs a constant source of replenishment (Sect. 2). The three main sources of accretion seen in simulations (IGM, satellites, feedback material) are consistent with what is found in observations of galaxy halos (Sect. 3).

Though we have made tremendous progress from when the idea of gas accretion originated, there remain many open questions as to exactly how gas accretion proceeds and how frequently. In particular, how the abundant ionized baryons within galaxy halos ultimately become star formation fuel is unclear, as is whether the infall occurrence is preferentially linked to satellite accretion or a feedback cycle. Future observations and simulations should further investigate how accretion and feedback are consistent with each other. In simulations the two mechanisms often preferentially occur along different axes, but observations have yet to solidly confirm this. It is also possible that accretion turns on when a galaxy is not in as dominant of a feedback stage. We are well-poised with the future direction of observations and simulations to address these questions. Sensitive, high velocity resolution absorption line experiments that use the galaxy itself combined with spatially and kinematically resolved observations of galaxies in emission will reveal additional clear cases of gas accretion and begin to place the origin of the gas. Sophisticated, high resolution simulations that include realistic feedback prescriptions and gas mixing can be compared to observations and will serve as a useful guide to our understanding.

Acknowledgements I thank J. Xavier Prochaska for his help making Fig. 1 pretty, Marcel Neeleman for sharing his data, Yong Zheng for providing the M33 figure, Jim Putman for useful plotting advice, Dan and Elaine Putman for daycare support, and NSF support from grants AST-1410800 and AST-1312888. Thanks to the University of Colorado for hosting me while I wrote this article, in particular conversations with Ben Oppenheimer and Julie Comerford.

References

Bauermeister, A., Blitz, L., Bolatto, A., et al. 2013, ApJ, 768, 132
Bauermeister, A., Blitz, L., & Ma, C.-P. 2010, ApJ, 717, 323
Bigiel, F., Leroy, A. K., Walter, F., et al. 2011, ApJL, 730, L13
Bland-Hawthorn, J., Sutherland, R., Agertz, O., & Moore, B. 2007, ApJL, 670, L109
Bouché, N., Finley, H., Schroetter, I., et al. 2016, ApJ, 820, 121
Bowen, D. V., Chelouche, D., Jenkins, E. B., et al. 2016, ApJ, 826, 50
Bregman, J. N. 2007, ARA&A, 45, 221
Bressan, A., Chiosi, C., & Fagotto, F. 1994, ApJS, 94, 63
Brooks, A. M., Governato, F., Quinn, T., Brook, C. B., & Wadsley, J. 2009, ApJ, 694, 396
Carilli, C. L., & Walter, F. 2013, ARA&A, 51, 105
Chen, H.-W., Lanzetta, K. M., & Webb, J. K. 2001, ApJ, 556, 158
Chen, H.-W., Wild, V., Tinker, J. L., et al. 2010, ApJL, 724, L176
Chiappini, C. 2009, in IAU Symposium Vol. 254, ed. J. Andersen, J. Bland-Hawthorn, & B. Nordström, 191
Cooper, T. J., Simcoe, R. A., Cooksey, K. L., O'Meara, J. M., & Torrey, P. 2015, ApJ, 812, 58
Crighton, N. H. M., Hennawi, J. F., & Prochaska, J. X. 2013, ApJL, 776, L18
Daddi, E., Bournaud, F., Walter, F., et al. 2010, ApJ, 713, 686
Dieter, N. H. 1971, A&A, 12, 59
Duffy, A. R., Meyer, M. J., Staveley-Smith, L., et al. 2012, MNRAS, 426, 3385

Emerick, A., Mac Low, M.-M., Grcevich, J., & Gatto, A. 2016, ApJ, 826, 148
Fenner, Y., & Gibson, B. K. 2003, PASA, 20, 189
Fernández, X., Gim, H. B., van Gorkom, J. H., et al. 2016, ApJL, 824, L1
Fernández, X., Joung, M. R., & Putman, M. E. 2012, ApJ, 749, 181
Ferrière, K. M. 2001, Reviews of Modern Physics, 73, 1031
Fox, A. J., Wakker, B. P., Barger, K. A., et al. 2014, ApJ, 787, 147
Fraternali, F., & Binney, J. J. 2008, MNRAS, 386, 935
Fumagalli, M., Cantalupo, S., Dekel, A., et al. 2016, MNRAS, 462, 1978
Gallazzi, A., Charlot, S., Brinchmann, J., White, S. D. M., & Tremonti, C. A. 2005, MNRAS, 362, 41
Gatto, A., Fraternali, F., Read, J. I., et al. 2013, MNRAS, 433, 2749
Geha, M., Blanton, M. R., Yan, R., & Tinker, J. L. 2012, ApJ, 757, 85
Genzel, R., Tacconi, L. J., Gracia-Carpio, J., et al. 2010, MNRAS, 407, 2091
Giavalisco, M., Vanzella, E., Salimbeni, S., et al. 2011, ApJ, 743, 95
Grcevich, J., & Putman, M. E. 2009, ApJ, 696, 385
Grillmair, C. J., Lauer, T. R., Worthey, G., et al. 1996, AJ , 112, 1975
Heckman, T. M., Lehnert, M. D., Strickland, D. K., & Armus, L. 2000, ApJS, 129, 493
Heitsch, F., & Putman, M. E. 2009, ApJ, 698, 1485
Henry, R. B. C., & Worthey, G. 1999, PASA, 111, 919
Holmberg, J., Nordström, B., & Andersen, J. 2007, A&A, 475, 519
Hopkins, A. M., McClure-Griffiths, N. M., & Gaensler, B. M. 2008, ApJL, 682, L13
Hoppmann, L., Staveley-Smith, L., Freudling, W., et al. 2015, MNRAS, 452, 3726
Huang, S., Haynes, M. P., Giovanelli, R., & Brinchmann, J. 2012, ApJ, 756, 113
Hulsbosch, A. N. M. 1968, Bull. Astron. Inst. Netherlands, 20, 33
Hunt, L. K., García-Burillo, S., Casasola, V., et al. 2015, A&A, 583, A114
Joung, M., Putman, M., Bryan, G., Fernandez, X., & Peek, J. 2012a, ApJ
Joung, M. R., Bryan, G. L., & Putman, M. E. 2012b, ApJ, 745, 148
Karim, A., Schinnerer, E., Martínez-Sansigre, A., et al. 2011, ApJ, 730, 61
Kennicutt, R. C., & Evans, N. J. 2012, ARA&A, 50, 531
Kereš, D., Katz, N., Weinberg, D. H., & Davé, R. 2005, MNRAS, 363, 2
Kirby, E. N., Cohen, J. G., Guhathakurta, P., et al. 2013, ApJ, 779, 102
Kreckel, K., Platen, E., Aragón-Calvo, M. A., et al. 2012, AJ, 144, 16
Lah, P., Chengalur, J. N., Briggs, F. H., et al. 2007, MNRAS, 376, 1357
Larson, R. B. 1972a, Nature Physical Science, 236, 7
Larson, R. B. 1972b, Nature , 236, 21
Lehner, N., & Howk, J. C. 2011, Science, 334, 955
Lehner, N., Howk, J. C., Tripp, T. M., et al. 2013, ApJ, 770, 138
Leroy, A. K., Walter, F., Brinks, E., et al. 2008, AJ , 136, 2782
Leroy, A. K., Walter, F., Sandstrom, K., et al. 2013, AJ , 146, 19
Madau, P., & Dickinson, M. 2014, ARA&A, 52, 415
Magrini, L., Corbelli, E., & Galli, D. 2007, A&A, 470, 843
Marasco, A., Crain, R. A., Schaye, J., et al. 2016, MNRAS, 461, 2630
Marinacci, F., Binney, J., Fraternali, F., et al. 2010, MNRAS, 404, 1464
Martin, C. L., Shapley, A. E., Coil, A. L., et al. 2012, ApJ, 760, 127
Martin, D. C., Matuszewski, M., Morrissey, P., et al. 2016, ApJL, 824, L5
Muller, C. A., Oort, J. H., & Raimond, E. 1963, *Academie des Sciences Paris Comptes Rendus*, 257, 1661
Muratov, A. L., Keres, D., Faucher-Giguère, C.-A., et al. 2016, arXiv:1606.09252
Neeleman, M., Prochaska, J. X., Ribaudo, J., et al. 2016, ApJ, 818, 113
Oort, J. H. 1970, A&A, 7, 381
Oosterloo, T., Fraternali, F., & Sancisi, R. 2007, AJ , 134, 1019
Oppenheimer, B. D., Crain, R. A., Schaye, J., et al. 2016, MNRAS
Pearson, S., Besla, G., Putman, M. E., et al. 2016, MNRAS, 459, 1827
Penton, S. V., Stocke, J. T., & Shull, J. M. 2002, ApJ, 565, 720

Prochaska, J. X., Herbert-Fort, S., & Wolfe, A. M. 2005, ApJ, 635, 123
Prochaska, J. X., Weiner, B., Chen, H.-W., Mulchaey, J., & Cooksey, K. 2011, ApJ, 740, 91
Prochaska, J. X., & Wolfe, A. M. 2009, ApJ, 696, 1543
Putman, M. E., Henning, P., Bolatto, A., et al. 2009, *Astro2010: The Astronomy and Astrophysics Decadal Survey*, 241 (arXiv:0902.4717)
Putman, M. E., Peek, J. E. G., & Joung, M. R. 2012, ARA&A, 50, 491
Putman, M. E., Staveley-Smith, L., Freeman, K. C., Gibson, B. K., & Barnes, D. G. 2003, ApJ, 586, 170
Rao, S. M., Turnshek, D. A., & Nestor, D. B. 2006, ApJ, 636, 610
Rauch, M., Becker, G. D., & Haehnelt, M. G. 2016, MNRAS, 455, 3991
Rubin, K. H. R., Prochaska, J. X., Koo, D. C., & Phillips, A. C. 2012, ApJL, 747, L26
Rubin, K. H. R., Prochaska, J. X., Koo, D. C., et al. 2014, ApJ, 794, 156
Salem, M., Besla, G., Bryan, G., et al. 2015, ApJ, 815, 77
Sánchez-Ramírez, R., Ellison, S. L., Prochaska, J. X., et al. 2016, MNRAS, 456, 4488
Sancisi, R., Fraternali, F., Oosterloo, T., & van der Hulst, T. 2008, Astron. Astrophys. Rev., 15, 189
Santini, P., Maiolino, R., Magnelli, B., 2014, A&A, 562, A30
Sato, T., Martin, C. L., Noeske, K. G., et al. 2009, ApJ, 696, 214
Schiminovich, D., Catinella, B., Kauffmann, G. et al. 2010, MNRAS, 408, 919
Scoville, N., Sheth, K., Aussel, H., et al. 2016, ApJ, 820, 83
Shapley, A. E. 2011, ARA&A, 49, 525
Shapley, A. E., Steidel, C. C., Pettini, M., & Adelberger, K. L. 2003, ApJ, 588, 65
Shull, J. M., Smith, B. D., & Danforth, C. W. 2012, ApJ, 759, 23
Spekkens, K., Urbancic, N., Mason, B. S., Willman, B., & Aguirre, J. E. 2014, ApJL, 795, L5
Stewart, K. R., Kaufmann, T., Bullock, J. S., et al. 2011, ApJ, 738, 39
Stott, J. P., Sobral, D., Swinbank, A. M., et al. 2014, MNRAS, 443, 2695
Tacconi, L. J., Neri, R., Genzel, R., et al. 2013, ApJ, 768, 74
Tepper-García, T., Bland-Hawthorn, J., & Sutherland, R. S. 2015, ApJ, 813, 94
Tumlinson, J., Thom, C., Werk, J. K., et al. 2013, ApJ, 777, 59
Tumlinson, J., Thom, C., Werk, J. K., et al. 2011, Science, 334, 948
van den Bergh, S. 1962, AJ , 67, 486
van Zee, L. 2001, AJ , 121, 2003
Voit, G. M., Bryan, G. L., O'Shea, B. W., & Donahue, M. 2015, ApJL, 808, L30
Wakker, B. P., & Savage, B. D. 2009, ApJS, 182, 378
Weiner, B. J., Coil, A. L., Prochaska, J. X., et al. 2009, ApJ, 692, 187
Werk, J. K., Prochaska, J. X., Tumlinson, J., et al. 2014, ApJ, 792, 8
Westmeier, T., Braun, R., Brüns, C., Kerp, J., & Thilker, D. A. 2007, New Astron. Rev., 51, 108
Zafar, T., Péroux, C., Popping, A., et al. 2013, A&A, 556, A141
Zheng, Y., Putman, M. E., Peek, J. E. G., & Joung, M. R. 2015, ApJ, 807, 103
Zheng, Y., Peek, J., Werk, J., & Putman, M. 2017, ApJ, 834, 179
Zwaan, M. A., Meyer, M. J., Staveley-Smith, L., & Webster, R. L. 2005, MNRAS, 359, L30

Gas Accretion onto the Milky Way

Philipp Richter

1 Introduction

We start this chapter on Galactic accretion with an introduction, in which we first summarize early measurements of circumgalactic gas and the star-formation history of the Milky Way from a historical perspective. We then shortly discuss the role of gas-accretion processes in Milky Way-type galaxies in a cosmological context and highlight their importance for galaxy evolution in general. The introduction section ends with a proper definition of the gas-accretion rate, $\mathrm{d}M_{\mathrm{gas}}/\mathrm{d}t$, and an assessment of the physical parameters that need to be constrained from observations and simulations to estimate $\mathrm{d}M_{\mathrm{gas}}/\mathrm{d}t$ for the Milky Way.

1.1 Historical Remarks

The presence of gas above/below the Milky Way disk has been established already more than 60 years ago, when absorption-line measurements demonstrated the presence of gas clouds at high galactic latitudes that exhibit relatively high radial velocities (Adams 1949; Münch 1952; Münch and Zirin 1961). In 1956, Lyman Spitzer argued that these structures, if located in the halo, must be surrounded by a hot, gaseous medium (he named this medium the *Galactic Corona*) that provides the necessary pressure-confinement of these clouds, otherwise they should disperse on relatively short timescales (Spitzer 1956).

P. Richter (✉)
Institut für Physik und Astronomie, Universität Potsdam, Karl-Liebknecht-Str. 24/25, Golm, 14476 Potsdam, Germany
e-mail: prichter@astro.physik.uni-potsdam.de

A. Fox, R. Davé (eds.), *Gas Accretion onto Galaxies*, Astrophysics and Space Science Library 430, DOI 10.1007/978-3-319-52512-9_2

With the new receiver technologies and the resulting improved sensitivity of radio telescopes in the 1960s, high-velocity H I 21 cm emission at high galactic latitudes was found by Muller et al. (1963), Smith (1963), Dieter (1964), Blaauw and Tolbert (1966), Hulsbosch and Raimond (1966), and Mathewson (1967), supporting the conclusions from the earlier absorption-line measurements. The observed distribution of radial velocities of the 21 cm emission features lead to the definition of two classes of Galactic halo clouds: as "high-velocity clouds" (HVCs) those halo structures were labeled that have radial velocities, $|v_{\rm LSR}| > 100\,{\rm km\,s^{-1}}$, while features with somewhat smaller radial velocities ($|v_{\rm LSR}| \approx 30$–$100\,{\rm km\,s^{-1}}$) were given the name "intermediate-velocity clouds" (IVCs). As will be discussed later, there also might exist a population of halo clouds with very low LSR velocities ("low-velocity clouds" (LVCs)), similar to those in the disk.

Several scenarios for the origin of the 21 cm neutral halo clouds were discussed by Jan Oort in 1966, among which the infall of intergalactic gas, the accretion of gas from satellite galaxies, the condensation of neutral gas patches from cooling coronal gas, and the ejection of gaseous material from the Milky Way disk were regarded as plausible scenarios (Oort 1966). Indeed, as we will discuss in this review, these scenarios are still up-to-date.

The need for feeding the Milky Way disk with low-metallicity gas also comes from early studies of the Galaxy's star-formation activity and stellar content. Already in the 1970s it was realized that star formation in the Milky Way would have come to a halt early on, if the Galaxy was not fed with fresh material from outside. This is because the gas-consumption timescale (even for a moderate star-formation rate of $\sim 1\,M_{\odot}{\rm yr}^{-1}$) is short compared to the age of the Milky Way (Larson 1972). Another argument for gas accretion comes from the observed metallicity distribution of low-mass stars in the solar neighborhood, which cannot be reproduced by closed-box models of the chemical evolution of the stellar disk. To match the observations, such models *require* the continuous accretion of metal-poor gas (van den Bergh 1962; Chiappini et al. 2001). These findings provide additional strong arguments that the Milky Way has accreted (and is still doing so) large amounts of low-metallicity gas to continuously built up its stellar content as observed today.

In conclusion, the observed presence of large amounts of gas above/below the disk, the past and present star-formation rate of the Milky Way, and the metallicity distribution of low-mass stars in the solar neighborhood demonstrate that gas accretion represents an important process that has strong implications for the past, present, and future evolution of our Galaxy.

1.2 Cosmological Context

In the overall context of galaxy formation in the Universe, gas accretion and feedback nowadays are regarded as the main processes that regulate the star-formation activity in galaxies. Many of the cosmological aspects of gas accretion in galaxies will be discussed in detail in other chapters of this book. Still, for our discussion on gas accretion in the Milky Way, the main aspects need to be summarized here.

In the conventional sketch of galaxy formation, gas is falling into a dark matter (DM) halo and then is shock-heated to approximately the halo virial temperature (a few 10^6 K, typically), residing in quasi-hydrostatic equilibrium with the DM potential well (Rees and Ostriker 1977). The gas then cools and sinks into the center of the potential where it is transformed into stars. This model is often referred to as the "hot mode" of gas accretion. It has been argued that for smaller DM potential wells the infalling gas may reach the disk *directly* at much shorter timescales, without being shock-heated to the virial temperature ("cold mode" of gas accretion; e.g., White and Rees 1978; Kereš et al. 2005). In the latter case, the star-formation rate of the central galaxy would be directly coupled to its gas-accretion rate (White and Frenk 1991). In these simple pictures, the dominating gas-accretion mode depends on the mass and the redshift of the galaxy (e.g., Birnboim and Dekel 2003; Kereš et al. 2005), where at low redshift the critical halo mass that separates the hot mode from the cold mode is $\sim 10^{12}\,M_\odot$ (van de Voort et al. 2011). However, the underlying physics that describes the large-scale flows of multi-phase gas from the outer to the inner regions of a dynamically evolving galaxy is highly complicated (e.g., Mo and Miralda-Escude 1996; Maller and Bullock 2004). To understand these processes, high-resolution hydrodynamical simulations with well-defined initial conditions (e.g., Bauermeister et al. 2010; Fumagalli et al. 2011; van de Voort and Schaye 2012; Vogelsberger et al. 2012; Shen et al. 2013) are required. Yet, even the most advanced hydrodynamical simulations do not provide tight constraints for the gas-accretion rate of *individual* galaxies without knowing their exact halo masses, their cosmological environment, and the initial circumgalactic gas distribution (Nuza et al. 2014; hereafter referred to as N14).

Next to the feeding of the halos of Milky Way-type galaxies through intergalactic gas, galactic-fountain type processes (from supernova (SN) feedback; Fraternali and Binney 2008) and mergers with satellite galaxies (Di Teodoro and Fraternali 2014) need to be considered. The vast amounts of neutral and ionized gas carried by the Magellanic Stream underline the importance of merger processes for the Milky Way's gas-accretion rate (D'Onghia and Fox 2016; see Sect. 2). The preconditions under which such gas clouds are generated and falling towards the disk are different from those for clouds being accreted from the intergalactic medium (IGM; e.g., Peek 2009) and thus they need to be explored separately by both observations and simulations. Finally, it is important to keep in mind that the Milky Way is not an isolated galaxy, but is embedded in the Local Group, being close to another galaxy of similar mass, M31. The original distribution of gas that is entering the virial radius of the Milky Way from outside thus depends on the spatial distribution of satellite galaxies and the distribution of intragroup gas in the cosmological filament that builds the Local Group. In Fig. 1 we sketch the local galaxy distribution in the Local Group and the gas distribution around the Milky Way and M31 from a hypothetical external vantage point.

Turning back to the Milky Way, we know that since $z = 1$ the Milky Way has produced $\sim 8 \times 10^9 M_\odot$ of stars, while the current star-formation rate of the Milky Way is ~ 0.7–$2.3\,M_\odot\,\mathrm{yr}^{-1}$ (Levine et al. 2006; Robitaille and Whitney 2010; Chomiuk and Povich 2011; see also Peek 2009). To relate this stellar mass and

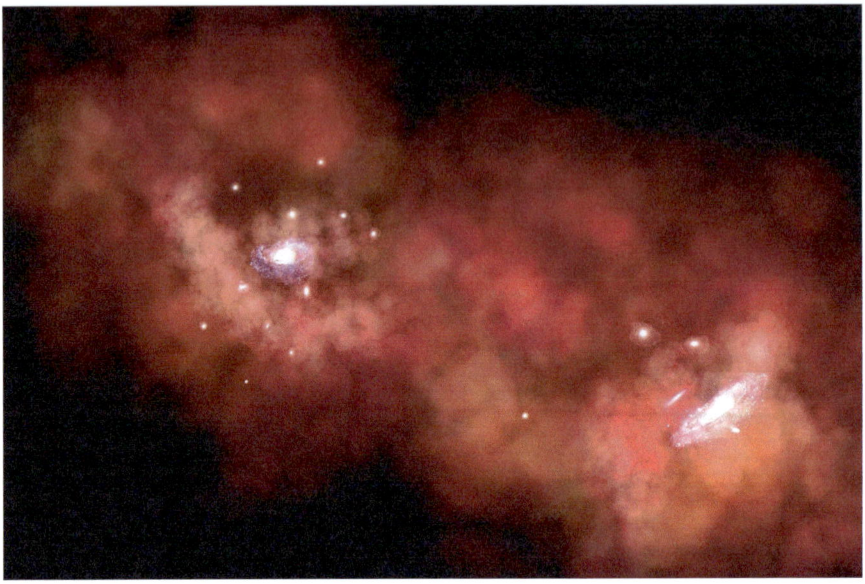

Fig. 1 Sketch of the expected distribution of galaxies and multi-phase gas in the Local Group from an external vantage point. The Milky Way (*left big galaxy*) and M31 (*right big galaxy*) are surrounded by their populations of satellite galaxies and by large amounts of multi-phase gas. Both galaxies are interconnected by a gaseous bridge, which spatially falls together with the Local Group barycenter (see, e.g., N14). Figure produced by the author for this review article

star-formation rate to the gas-accretion rate it is important to remember that as much as 50% of the initial material from which a generation of stars is formed will be returned back to the ISM and will be recycled in later stellar generations (Rana 1991). This means that for one accreted mass unit of gas, *two* mass units of *evolved* stars will have emerged after several star-formation cycles. For the Milky Way, this implies that it has accreted a gas mass of $\geq 4 \times 10^9 M_\odot$ during the last 8 Gyr.

1.3 Parameterization of Gas Accretion

Before we start to discuss in detail the various observational and theoretical aspects of gas-accretion processes in the Milky Way, we need a proper definition of the most important parameters involved, in particular the accretion *rate*. From a cosmological perspective (e.g., in studies using cosmological simulations), the growth and evolution of galaxies through cosmic times is governed by the gain of gas mass within their gravitational sphere of influence (i.e., within their virial radius, $R_{\rm vir}$) through mergers and gas infall. In this context, one can simply define the gas accretion rate of a galaxy as the net mass inflow of gas through an imaginary sphere with radius $R_{\rm vir}$. From a galaxy-evolution perspective, in contrast, the only relevant accretion rate is that of the disk, where the infalling gas is being transformed into stars, while the total amount of gas cycling within the galaxy's extended halo is relatively unimportant. One of the most burning questions in gas-accretion research therefore is, how much of the gas entering the virial radius of a Milky Way-type galaxy actually makes it to the disk and what are the typical timescales for this process?

In the following, we address these conceptional issues in two steps. First, we define the overall *current-day gas-accretion rate* of the Milky Way simply by relating the total mass of infalling gas, $M_{\rm gas}$, with its infall velocity, $v_{\rm infall}$, and its galactocentric distance, d, so that

$$\frac{{\rm d}M_{\rm gas,halo}}{{\rm d}t} = \frac{M_{\rm gas}\, v_{\rm infall}}{d}. \tag{1}$$

Because of the cloud's passage through the halo and the interaction with the ambient hot coronal gas, only a fraction of this initial gas mass will end up in the disk to power star formation therein. The *future* disk gas-accretion rate thus can be defined as

$$\frac{{\rm d}M_{\rm gas,disk}}{{\rm d}t} = \eta\, \frac{{\rm d}M_{\rm gas,halo}}{{\rm d}t}, \tag{2}$$

where $\eta \leq 1$ represents the *fueling parameter* that modulates the disk's gas-accretion and star-formation rate at the time of impact. The accretion time of each infalling gas cloud from its initial position seen today to the disk is $t_{\rm acc} = d/\langle v_{\rm infall}\rangle$. Here, $\langle v_{\rm infall}\rangle$ represents the average infall velocity along the cloud's passage towards the disk. In conclusion, it is the today's 3D distribution and space motion of gas around the Milky Way (parameterized by the current-day halo gas-accretion rate) that governs the *future* star-formation activity in the Milky Way disk.

From Eqs. (1) and (2) it becomes immediately clear, which parameters need to be constrained by observations to get an insight into the gas-accretion rate of the Milky Way.

- The **total mass of halo gas** that is potentially flowing towards the Milky Way disk needs to be constrained from observations. Such observations need to take into account the huge span in physical conditions such gas might have: from cold neutral gas (that can be observed in H I 21 cm emission) to million-degree shock-heated gas (potentially visible in X-ray emission and absorption) all relevant circumgalactic gas phases need to be considered. Next to sensitivity issues, disentangling distant halo gas from foreground disk gas is difficult for material that has low radial velocities similar to those expected for the rotating interstellar gas disk.
- Also the **distance** of extra-planar/circumgalactic gas needs to be determined by observational means. Reliable distance estimates for gas complexes in the Milky Way halo are particularly challenging. The so-called bracketing method requires the presence of foreground and background stars with known distances in the

general direction of a halo gas cloud to pinpoint a distance range for it (Prata and Wallerstein 1967). If high-velocity absorption of a given ion is observed in the spectrum of a star with distance $d_{\rm star}$, it is clear that the absorbing gas is in front of the star at $d_{\rm gas} < d_{\rm star}$. The *significant* absence of absorption (for instance, taking into account the absorption strength predicted from the 21 cm emission spectrum) instead implies that the gas lies behind the star. This bracketing method is, however, limited to large, spatially extended gas complexes and requires a substantial observational effort. Other (indirect) distance estimates for extra-planar/circumgalactic gas features (e.g., from modeling the observed Hα fluxes) are afflicted with systematic uncertainties.

- The **infall velocity** of Milky Way halo gas cannot be observed directly, but must be estimated from theoretical considerations in combination with observational constraints. This is because the observed radial velocities, $v_{\rm rad}$, of extra-planar/circumgalactic gas features do not reflect their 3D space velocities. Also, the effect of galactic rotation adds another velocity component that needs to be taken into account for the interpretation of $v_{\rm rad}$. As infalling gas is expected to interact with its ambient medium, the infall velocity is likely to depend on the overall halo-gas properties and the position of the infalling gas cloud in the Milky Way potential well, i.e., on its distance to the disk.
- The **fueling parameter** η can only be estimated from full-fledged hydrodynamical models that realistically describe the passage of neutral and ionized gas through the hot coronal gas of the Milky Way. Such models need to take into account the initial gas distribution of the infalling and ambient medium and all relevant physical processes (e.g., instabilities, turbulent mixing layers, conductive interfaces, and many others). In addition, possibly existing *outflowing* gas components (e.g., from SN feedback) are likely to be important.

In the following sections, we will discuss our understanding of these individual parameters in detail, summarizing past and recent observational and theoretical studies.

2 The Observed Distribution of Gas Around the Milky Way

To pinpoint the 3D distribution of gas around the Milky Way and to estimate its total mass, the different gas *phases* (in the density-temperature phase-space) need to be considered. In the following, we separate the Milky Way halo gas in three such phases: (1) neutral/molecular gas, (2) warm ionized gas with temperatures $T \leq 10^5$ K, and (3) hot ionized gas with $T > 10^5$ K. First, we discuss the angular distribution of these phases, their galactocentric distances, and their total mass considering recent observational results. Then, we provide estimates on the gas-accretion rate of the Milky Way based on these observations. Note that many of these aspects have also been discussed in previous reviews (Wakker and van Woerden 1997; Richter 2006; Putman et al. 2012) and in the book on HVCs (van Woerden et al. 2004).

2.1 Neutral Gas

As discussed earlier, radio observations in the H I 21 cm line have become the most powerful method to study the distribution and internal structure of neutral halo clouds (Bajaja et al. 1985; Hulsbosch and Wakker 1988; Wakker and van Woerden 1991; Hartmann and Burton 1997; Morras et al. 2000; Kalberla et al. 2005; McClure-Griffiths et al. 2009; Winkel et al. 2010). Absorption-line measurements against bright extragalactic background sources (e.g., quasars) provide additional information on the chemical composition of the neutral gas and its connection to the other circumgalactic (ionized) gas phases.

In Fig. 2 we show the sky distribution of the 21 cm IVCs (upper panel) and HVCs (lower panel) plotted in galactic coordinates in an Aitoff projection centered on $l = 180°$. Because of their different kinematics and intrinsic properties, it is useful to discuss the properties of 21 cm IVCs and HVCs individually.

2.1.1 IVCs

The sky distribution of IVCs indicates that intermediate-velocity gas has predominantly negative radial velocities. The most prominent IVC features (see labels in Fig. 2) are in the northern sky the IV Arch and its low-latitude extension (LLIV Arch), IV Spur, the low-velocity part of Complex L, Complex K, and the low-velocity part of the Outer Arm. In the southern sky, there is the Anticenter (AC) shell and the Pegasus-Pisces (PP) Arch (see Kuntz and Danly 1996; Wakker 2001). The total sky covering fraction of neutral IVC gas is $f_c \approx 0.30$ for column densities $N(\mathrm{H\,I}) \geq 10^{19}\,\mathrm{cm}^{-2}$ and deviation velocities $|v_{\mathrm{dev}}| = 30$–$90\,\mathrm{km\,s}^{-1}$ (Wakker 2004; see caption of Fig. 2). Many of the IVC features are spatially and kinematically connected with 21 cm disk gas. The typical H I column densities in IVCs in the 21 cm surveys lie in the range $N(\mathrm{H\,I}) = 10^{19} - 10^{20}\,\mathrm{cm}^{-2}$, but some cores in the IV Arch and the Outer Arm exhibit column densities above $10^{20}\,\mathrm{cm}^{-2}$ (Fig. 2). At positive radial velocities, there is the low-velocity extension of HVC Complex WA in the north and Complex gp in the south. The positive velocity IVCs are much smaller in angular size and column density and they exhibit a very complex internal structure.

Direct distance measurements (mostly using the bracketing technique) place the IVCs relatively close to the Milky Way disk with z heights <2.5 kpc, typically. The most massive IVC complexes, for example, have distances of 0.8–1.8 kpc (IV Arch), $\sim$0.9 kpc (LLIV Arch), and 0.3–2.1 kpc (IV Spur; see IVC compilations by Wakker 2001, 2004 and the references therein). Most of the IVCs thus belong to the disk-halo interface, which possibly represents a crucial component for gas-accretion processes in the Galaxy, as will be discussed below. For the most prominent, large-scale IVC complexes, the 21 cm data indicate individual H I masses on the order of 1–$8 \times 10^5 M_{\odot}$ (e.g., IV Arch, LLIV Arch, IV Spur; Wakker 2001). The total neutral gas mass of the large-scale IVCs thus can estimated to be $\sim 10^6 M_{\odot}$.

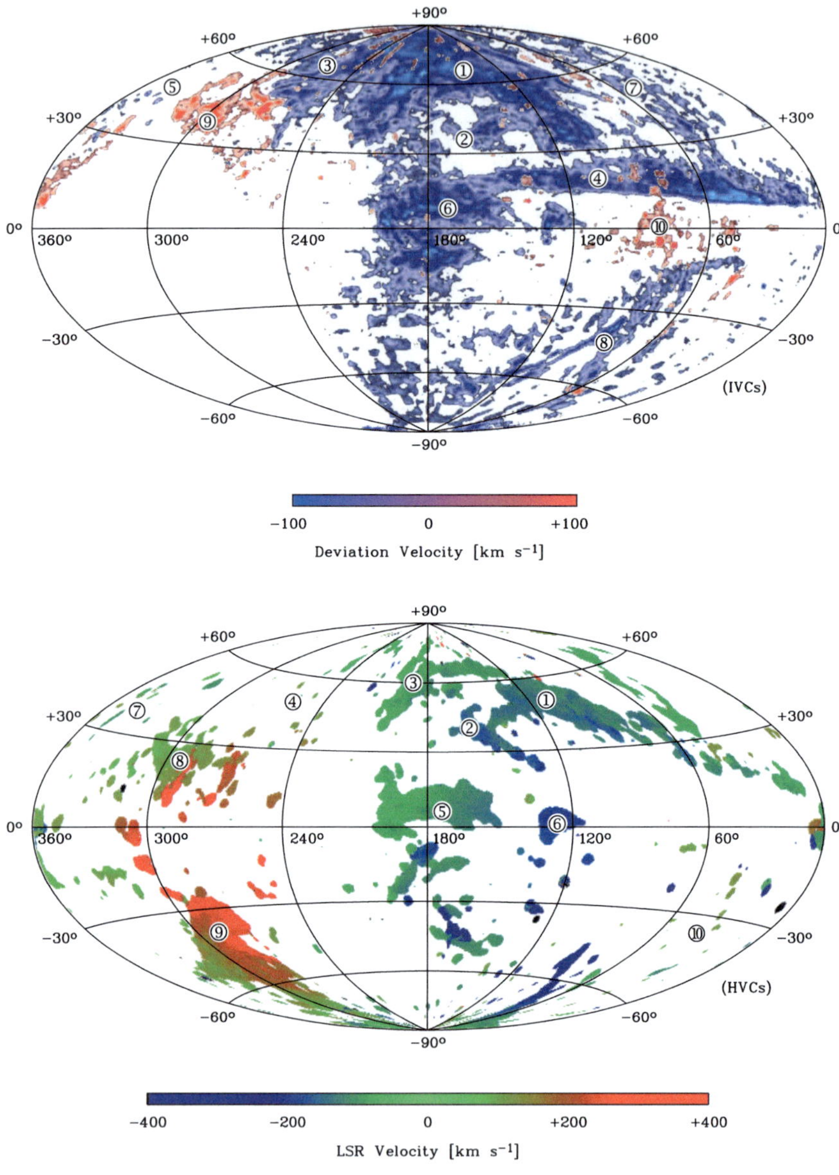

Fig. 2 All-sky maps of 21 cm emission for intermediate-velocity gas (*upper panel*) and high-velocity gas (*lower panel*) in an Aitoff projection centered on $l = 180°$. The maps show the sky distribution of neutral gas in the Galactic halo. The maps have been generated from different data sets described in Wakker (2004), Kalberla et al. (2005), and by Tobias Westmeier (priv. comm.). For the IVCs we show (*color-coded*) the deviation velocity of the gas from a simple model of Galactic rotation (see Wakker 2004) in the range $|v_{\rm dev}| = 30$–$90\,{\rm km\,s^{-1}}$. For the HVCs, we display the color-coded LSR velocity ($|v_{\rm LSR}| = 100$–$500\,{\rm km\,s^{-1}}$). Individual neutral IVC and HVC complexes (see Wakker 2001, 2004) are labeled with numbers. For IVCs: (1) IV Arch, (2) LLIV Arch, (3) IV Spur, (4) Outer Arm, (5) Complex L, (6) AC Shell, (7) Complex K, (8) PP Arch, (9) IV-WA, (10) Complex gp. For HVCs: (1) Complex C, (2) Complex A, (3) Complex M, (4) Complex WA, (5) AC Cloud, (6) Complex H, (7) Complex L, (8) Leading Arm of MS, (9) Magellanic Stream, (10) Complex GCN

Absorption-line measurements indicate that the metallicity of most of the large northern IVCs is relatively high with typical values between 0.5 and 1.0 solar (e.g., Wakker 2001; Richter et al. 2001a,b). For the Outer Arm, Tripp and Song (2012) derive a lower metallicity of 0.2–0.5 solar, suggesting that this gas (albeit being close to the disk) might have an extragalactic origin, such as many HVCs (see below). Another interesting feature is the core IV 21, which has a metallicity of just 0.4 solar (Hernandez et al. 2013) at a z height of $\sim$300 pc above the disk, thus also pointing towards an extragalactic origin. Dust is also present in IVCs, as is evident from the observed depletion patterns of heavy elements in intermediate-velocity gas (Richter et al. 2001a,b; Wakker 2001; see also Savage and Sembach 1996) and from the observed excess of extra-planar infrared emission in the direction of IVCs (Desert et al. 1988, 1990; Weiss et al. 1999). Also molecules (H_2, CO) have been detected in several IVCs (Richter et al. 2001b, 2003b; Wakker 2006; Gillmon et al. 2006; Hernandez et al. 2013; Röhser et al. 2016), but the molecular gas fraction is very small, so that the molecular phase does not contribute significantly to the overall IVC masses. Yet, the presence of molecular gas implies the presence of substantial small-scale structure in the gas down to AU scales (Richter et al. 2003a). Small-scale structure is also evident from high-resolution 21 cm data, which show, next to the coherent large-scale 21 cm IVC complexes, a population of several thousand compact H I clumps at $z = 1$–2 kpc (e.g., Lockman 2002; Kalberla and Kerp 2009; Saul et al. 2012). These clumps, that have very small masses of only 10^1–$10^4 M_\odot$, may represent cloudlets that have condensed out of the ambient multi-phase medium. They are raining down to the disk, thus fueling it. Because of their small cross section for absorption spectroscopy, the metal and dust content of these clumps remains unknown so far.

In view of the measured overall chemical composition of the large IVCs and their location in the disk-halo interface, the favored scenario for the origin of near solar-metallicity IVCs is that these structures represent the back-flow of cooled gas from the galactic-fountain process (Shapiro and Field 1976; Houck and Bregman 1990); i.e., they originate from metal-enriched gas that has been ejected from the disk by supernova explosions that is now cycling back to the disk due to gravitational forces.

2.1.2 HVCs

The sky distribution of the Milky Way HVCs is far more complex than that of the IVCs (Fig. 2, lower panel); HVCs span a huge LSR velocity range of $\sim$700 km s^{-1}.

Among the most prominent Milky HVCs is Complex C, which covers $\sim$1500 square-degree on the northern sky in 21 cm emission, which is about 4% of the entire sky (Wakker 2004). Complex C is a cloud that presumably is being accreted from the IGM or from a satellite galaxy (e.g., Sembach et al. 2004). Another prominent 21 cm HVC is the Magellanic Stream (MS) in the south. The MS also covers an area of $\sim$1500 square-degree in 21 cm, but it spreads over the entire southern sky, forming a coherent stream of neutral gas (D'Onghia and Fox 2016). The MS represents a tidal feature expelled from the Magellanic Clouds as they approach the Milky Way halo (e.g., Gardiner and Noguchi 1996; Connors et al. 2006; Besla et al. 2010, 2012; Nidever et al. 2010; Diaz and Bekki 2011, 2012). Over its entire body the MS spans a distance range of $d = 50$–100 kpc (or even further) from the Galactic disk, suggesting that it extends over several hundred kpc in the outer halo of the Milky Way (Putman et al. 1998, 2003; Stanimirović et al. 2002, 2008; Brüns et al. 2005). Other prominent Galactic HVCs are Complex A, Complex H, the Anti-Center Cloud, and Complexes WA–WE. Their positions are indicated in Fig. 2. The individual properties of all HVC complexes are discussed in detail in Wakker (2001).

The entire HVC population of the Milky Way shown in Fig. 2 has a total sky covering fraction in 21 cm of $f_c \approx 0.35$ for neutral gas column densities $N(\mathrm{H\,I}) \geq 7 \times 10^{17}\,\mathrm{cm}^{-2}$ (Murphy et al. 1995; Wakker 2004 and the references therein). The covering fraction reduces to $f_c \approx 0.15$ for larger column densities $N(\mathrm{H\,I}) \geq 2 \times 10^{18}\,\mathrm{cm}^{-2}$. The H I column densities in HVCs follow a well-defined column density distribution function of the form $f(N_{\mathrm{HI}}) \propto N_{\mathrm{HI}}^{-\beta}$ with $\beta = 1.42$ for $\log N(\mathrm{H\,I}) \geq 18$ (Lockman et al. 2002). With the exception of the MS, all HVCs for which direct distance information from the bracketing method is available are located within 20 kpc. For instance, the Complexes A and C have distances $d \approx 10$ kpc (van Woerden et al. 1999; Wakker et al. 2007; Smoker et al. 2011; Thom et al. 2008), the Cohen Stream and Complex GCP are at $d = 5$–15 kpc (Wakker et al. 2008), and the HVC towards the LMC has $d \approx 9$ kpc (Richter et al. 2015). For these structures in the inner Milky Way halo the distances estimated indirectly from the measured Hα fluxes agree well with the values derived from the bracketing method (e.g., Bland-Hawthorn and Putman 2001; Putman et al. 2003; Tufte et al. 2002). Combining the 21 cm emission in Milky Way HVCs and in M31 halo clouds, Richter (2012) predicts that the H I covering fraction in HVCs around Milky Way-type galaxies declines exponentially with galactocentric distance with $f_c < 0.01$ for $d > 70$ kpc. From deep 21 cm observations of the M31 halo (Westmeier et al. 2007) further follows that also the so-called compact high-velocity clouds (CHVCs), isolated high-velocity 21 cm gas clumps with very small angular sizes of $<2°$ (Braun and Burton 1999; de Heij et al. 2002), are located at $d < 50$ kpc, disproving a previous scenario in which CHVCs are regarded as gas-filled dark matter halos residing in the Local Group (Blitz et al. 1999; Braun and Burton 1999).

From the 21 cm data and the available distance information it follows that the total neutral HVC gas mass is $M_{\mathrm{HI,HVC}} \approx 2.5 \times 10^{8}\,M_{\odot}$ (Wakker 2004; Brüns et al. 2005). The MS contributes with more than 60% to this mass ($M_{\mathrm{HI,MS}} \approx 1.6 \times 10^{8}\,M_{\odot}$ for $d = 55$ kpc), while the other HVCs have substantially smaller masses (e.g., $M_{\mathrm{HI,Complex\,C}} \approx 5 \times 10^{6}\,M_{\odot}$; Wakker 2004). Since the two Magellanic Clouds are located at only $\sim$80 kpc distance, their interstellar gas content adds to the gas mass being accreted by the Milky Way. Therefore, if we add the neutral gas mass of the Magellanic Clouds and their gaseous interconnection, the Magellanic Bridge, the total H I budget in the Milky Way halo sums up to a value of $M_{\mathrm{HI,MWhalo}} \approx 1.3 \times 10^{9}\,M_{\odot}$, which is $\sim$20% of the neutral gas mass in the Milky Way disk ($M_{\mathrm{HI,MWdisk}} \approx 7 \times 10^{9}\,M_{\odot}$; Ferriere 2001).

Other tracers of predominantly neutral gas in the Milky Way halo are absorption lines of neutral and low ions, such as O I, N I, Ar I, S II with transitions in the UV, and Ca II and Na I in the optical regime. From a survey of high-velocity Ca II/Na I absorption in the Milky Way halo against several hundred extragalactic background sources, Ben Bekhti et al. (2008, 2012) derive a covering fraction of $f_c \approx 0.50$ for log N(Ca II) > 11.4, which is $\sim$2 times higher than the 21 cm covering fraction. These results further indicate the widespread presence of cold, neutral gas structures away from the large 21 cm complexes. Such structures possibly are too small to be seen in all-sky 21 cm survey because of the limited angular resolution of these surveys in combination with beam-smearing effects.

While most of the IVCs have near-solar metallicities, the metal abundance in many HVCs is substantially lower (by a factor 5–10, typically). For Complex C, the (mean) metallicity has been constrained to 0.15 solar (Wakker et al. 1999; Richter et al. 2001a; Tripp et al. 2003; Collins et al. 2003; Sembach et al. 2004). The main body of the Magellanic Stream also has a metallicity of only 0.1 solar (Fox et al. 2010, 2013), but the MS contains a filament that is more metal rich (0.3 solar; Richter et al. 2013; Gibson et al. 2001). Similarly, Complex A most likely has metallicity of only $\sim$0.1 solar (Wakker 2004). Such low metallicities are in line with the idea that HVCs represent gas infalling from the pre-enriched intergalactic medium (or intragroup gas), but the clouds may also trace material stripped from satellite dwarf galaxies as they are being accreted by the Galaxy. Moreover, it cannot be ruled out that some of the gas has been part of the Milky Way a long time ago, then was ejected (at relatively low metallicity) by a former Galactic outflow or wind, and now is raining back towards the disk ("intergalactic fountain"). An example for this latter scenario is the Smith Cloud (also called Complex GCP), which has a metallicity of 0.5 solar and is believed to originate in the outer Galactic disk (Lockman et al. 2008; Hill et al. 2009; Fox et al. 2016).

There appears to be only little dust in neutral and ionized HVCs (e.g., Wakker and Boulanger 1986; Bates et al. 1988; Tripp et al. 2003; Richter et al. 2001a, 2009; Williams et al. 2012). The few tentative detections of far-IR emission in some HVCs (e.g., Miville-Deschênes et al. 2005; Peek et al. 2009; Planck Collaboration 2011) remain inconclusive with respect to their dust abundance. Diffuse molecular gas is present only in high-column density regions of the Magellanic Stream (Sembach et al. 2001; Richter et al. 2001c, 2013), the Magellanic Bridge (Lehner 2002; Murray et al. 2015), and in a dense clump of an HVC projected onto the LMC (Richter et al. 1999). However, the molecular component is not of importance for the total mass of HVCs.

Finally, it is worth noting that some of the above mentioned neutral halo clouds (e.g., Complex L) exhibit a radial velocity range that extends from values below 100 km s^{-1} to values above this threshold, i.e., these complexes can be regarded as *both* IVCs and HVCs, although they each represent a single, kinematically coherent structure. The question arises, whether the separation of IVCs and HVCs as different halo-cloud populations is justified, or whether they just represent the same population of objects with just different radial velocities. As discussed above, distance measurements place the IVCs within 2.5 kpc of the Milky Way disk, while most of the HVCs are located much further away. This, together with the on average higher metallicity of IVCs compared to HVCs, indeed indicates that low-velocity halo clouds (with LSR velocities *typically* $<100\,\mathrm{km\,s^{-1}}$) predominantly reside in the disk-halo interface, while high-velocity halo clouds (with LSR velocities *typically* $\geq 100\,\mathrm{km\,s^{-1}}$) predominantly trace gas at $d = 2$–100 kpc.

2.2 *Warm Ionized Gas*

Warm ionized gas in the Milky Way halo is even more widespread than the neutral gas traced by 21 cm emission. We here define warm ionized gas as gas that is predominantly ionized (i.e., with small neutral gas fractions) and has a temperature $< 10^5$ K. Circumgalactic gas at such temperatures is expected to be photoionized by the combined ionizing radiation from stars in the Milky Way disk and the ambient extragalactic UV background at $z = 0$ (see model by Fox et al. 2005). Warm ionized gas in the Milky Way halo can be detected either in emission in recombination lines such as Hα or in high-velocity UV absorption of the so-called intermediate ions that have lower ionization boundaries in the range 15–40 eV (e.g., C III, N III; Si III, Si IV, Fe III; Morton 2003; R17).

In Fig. 3, upper panel, we show the sky distribution of Hα emission in the range 0.3–100 Rayleigh, based on data obtained from the Wisconsin Hα Mapper (WHAM; Haffner et al. 2003, 2016). As can be seen, Hα emission is widespread at latitudes $b < 30°$, situated in distinct coherent spatial structures such as lobes and arches. These features reflect the complex motions of diffuse ionized gas (DIG) in the disk-halo interface (DHI) that is believed to be shaped by the on-going star-formation in the Milky Way disk. Based on the derived gas densities and volume-filling factors (Reynolds et al. 2012; Haffner et al. 2003), the total mass of ionized gas in the DHI of the Milky Way can be estimated to be $\sim 10^8 M_{\odot}$, in line with estimates for extra-planar ionized gas in other low-redshift disk galaxies (e.g., NGC 891; Dettmar 1990).

Many of the 21 cm HVCs at much larger distances from the disk are also detected in Hα emission (e.g., Weiner and Williams 1996; Tufte et al. 1998; Bland-Hawthorn et al. 1998), indicating that the neutral gas clouds are surrounded by envelopes of ionized gas, whose masses are comparable with or even larger than the neutral gas body (e.g., Fox et al. 2004). Also the Magellanic Stream at $d = 50$–100 kpc is detected in Hα (e.g., Putman et al. 2003; Fox et al. 2014), proving that the MS is surrounded by substantial amounts of warm H II. From the models of Bland-Hawthorn et al. (2007) it follows that the mass-weighted H II column density in the MS is $> 10^{20}\,\mathrm{cm^{-2}}$, thus larger than the mass-weighted H I column density. A similar conclusion was drawn by Fox et al. (2014), who determine an ionized-to-neutral hydrogen mass ratio of ~ 3 based on the absorption strength of intermediate and high ions associated with the 21 cm body of the Stream (Fig. 4).

The most sensitive ions to trace the warm ionized gas in the halo are C III and Si III with strong transitions in the UV at 977.02 Å (C III) and 1206.50 Å (Si III; see

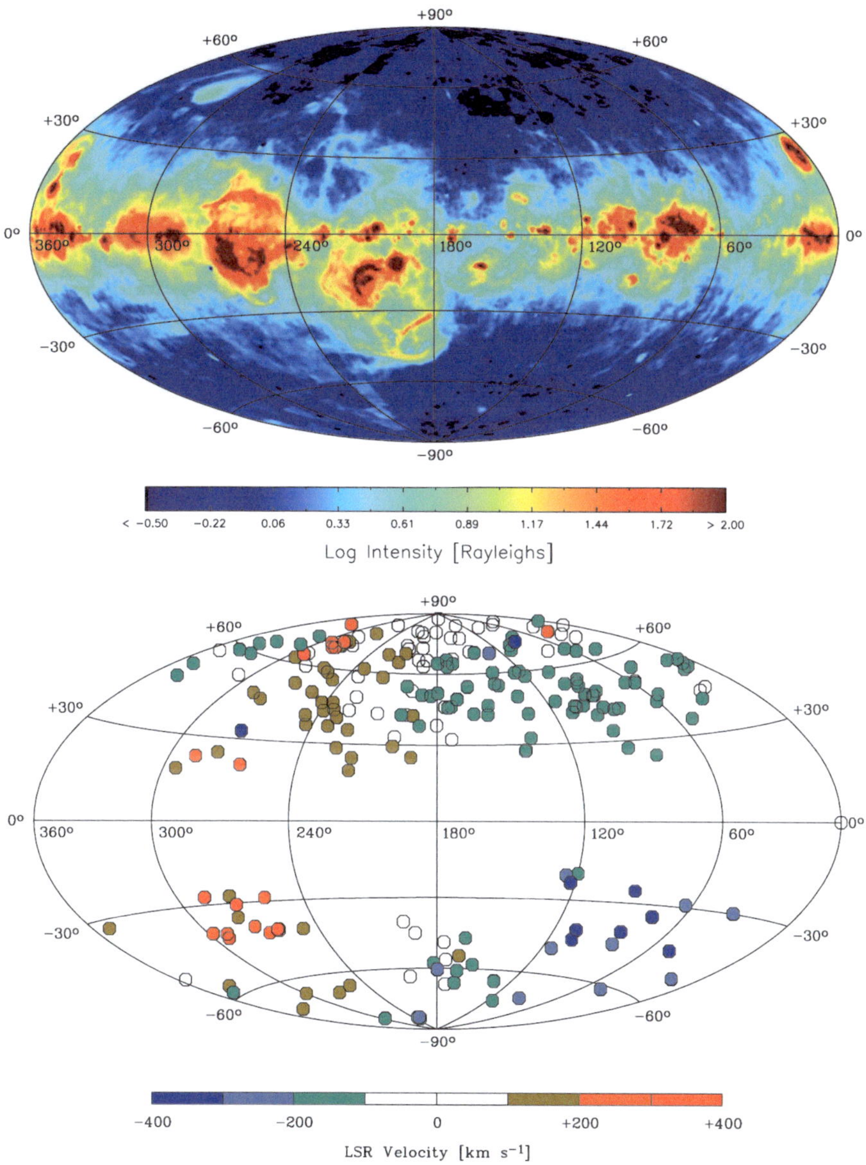

Fig. 3 All-sky maps of Hα emission in the Milky Way (*upper panel*) and high-velocity UV absorption of Si III absorption (*lower panel*) in an Aitoff projection centered on $l = 180°$ The maps show the sky distribution of diffuse ionized gas (DIG) in the lower and upper Galactic halo, respectively. The Hα map (kindly provided by Matt Haffner) has been compiled from data of the Wisconsin Hα mapper (WHAM; Haffner et al. 2003, 2016). It shows emission from relatively dense gas that predominantly resides in the inner halo and disk-halo interface of the Milky Way (DIG layer; Reynolds 1991). The Si III data stem from the high-velocity ($|v_{\rm LSR}| = 100$–$500\,{\rm km\,s^{-1}}$) UV absorption survey from Richter et al. (2017; hereafter referred to as R17) using 265 HST/COS spectra of extragalactic background sources. Diffuse ionized gas has a substantially larger sky covering fraction (∼2–3 times higher) than the neutral gas (Fig. 2)

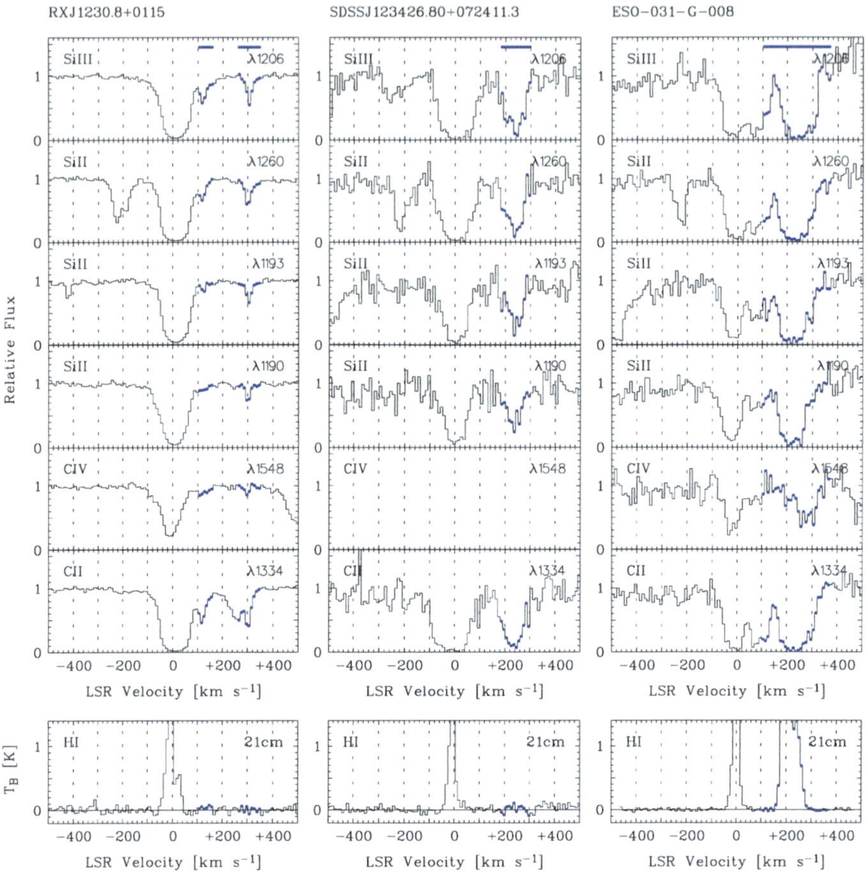

Fig. 4 Velocity profiles of high-velocity UV absorbers in the Galactic halo along three lines of sight (from R17), based on HST/COS data. Absorption profiles of various transitions from Si III, Si II, C IV, and C II are shown, where the high-velocity absorption at $|v_{\rm LSR}| \geq 100\,{\rm km\,s^{-1}}$ is indicated in *blue*. We also show the H I 21 cm emission profiles (from GASS/EBHIS data) for the same sightlines in the lowest panels. The individual high-velocity absorption features towards RXJ1230.8+0115 and SDSSJ123426.80+072411.3 trace gas streams in the halo that are predominantly ionized (without H I 21 cm counterpart), while the high-velocity absorption towards ESO−031−G−008 traces neutral *and* ionized halo gas related to the MS (UV absorption plus H I 21 cm emission)

Richter et al. 2016). In their recent legacy survey of high-velocity UV absorption in the Milky Way halo, R17 found that warm ionized halo gas, as traced by Si III at velocities $|v_{\rm LSR}| \geq 100\,{\rm km\,s^{-1}}$ and column densities log N(Si III) ≥ 12.1, has a covering fraction as high as $f_{\rm c} = 0.74$, confirming earlier results based on much smaller samples (Collins et al. 2009; Shull et al. 2009; Lehner et al. 2012; Herenz et al. 2013). This covering fraction is more than twice the value obtained for the neutral HVCs from the 21 cm observations (see above). In the lower panel of

Fig. 3 we show the sky distribution of high-velocity Si III absorption, which can be directly compared with the 21 cm HVC map (Fig. 2, lower panel). Like the Hα emission, high-velocity Si III absorption is often associated in radial velocity with the 21 cm HVC features, even if located several degrees away from the 21 cm contours. This further implies that HVCs represent coherent multi-phase gas streams (with a neutral gas body surrounded by an ionized gas layer) that move through the Milky Way halo (Lehner et al. 2012). Figures 2 and 3 also indicate that there are several regions in the high-velocity sky that exhibit pronounced Si III absorption but do not show significant large-scale H I. In contrast to the 21 cm HVCs and their ionized envelopes, gas in these regions can be regarded as *coherent ionized gas streams* in which patchy condensations of cooler, neutral gas clumps are embedded. Particularly interesting are the regions $l > 200^\circ, b > 0^\circ$ and $l < 120^\circ, b < 0^\circ$, which form a velocity dipole on the sky in UV absorption (Fig. 2; R17; Collins et al. 2003) in a direction that forms the major axis of the Local Group cosmological filament (N14). The observed kinematically distinct absorption features at high positive and high negative radial velocities possibly indicate that the Milky Way is ramming into ionized intragroup gas because it follows the general flow of galaxies in the direction of the Local Group barycenter (Peebles et al. 2001, 2011; Whiting 2014), while it is moving away from Local Group gas in the opposite direction along the filament (R17). Therefore, the Milky Way's accretion of warm ionized gas might be strongly influenced by the local galaxy environment and cosmological structure formation in the Local Group (N14). This important aspect will be further discussed in Sect. 3.2.

The ionized gas components that are associated with the 21 cm features obviously have the same distances as the neutral halo clouds. They also have comparable metallicities, if the gas has not yet been mixed with the ambient hot coronal gas. This implies that the majority of the diffuse ionized halo clouds that are *not* associated with the MS are located at $d < 20$ kpc, while the ionized envelope of the Stream is at $d = 50$–100 kpc. The exact angular extent of the Stream's ionized gas component is unknown, but it may well cover 30–50% of the entire sky (R17; Fox et al. 2014). From their survey of high-velocity UV absorption towards Galactic halo stars with known distances ($d < 15$ kpc) and extragalactic background sources Lehner and Howk (2011) and Lehner et al. (2012) find that the sky covering fraction of high-velocity UV absorption increases only marginally from the halo-star sample to the QSO sample, if velocities $|v_{\rm LSR}| \leq 170\,{\rm km\,s^{-1}}$ are considered. This indicates that HVCs in this velocity range are predominantly located at $d < 15$ kpc. In contrast, HVCs with absolute LSR velocities larger than $170\,{\rm km\,s^{-1}}$ (e.g., the MS) are only seen against extragalactic background sources, demonstrating that the gas is located at $d > 15$ kpc, and being in line with the distance constraints for the neutral gas (see above). The halo-star sample of Lehner and Howk (2011) and Lehner et al. (2012) covers only a limited fraction of the sky, however, so that the possible presence of more distant ionized gas structures even at low velocities (in particular in the directions of a possible Local Group filament) cannot be ruled out with these data.

The total mass of diffuse ionized high-velocity gas in the Galactic halo is dominated by the extended envelope of the MS (Fox et al. 2014; R17). Assuming

$d = 55$ kpc and calculating the amount of H II from the observed ion abundances in combination with an ionization model, both studies obtain a gas mass of the ionized component of the MS of $M_{\rm MS} \approx 1$–$3 \times 10^9\, M_\odot$. This mass would be substantially higher, if some of the gas from the MS was located at larger distances. For instance, if the distance of MS would lie in the range $d = 100$–150 kpc (see Besla et al. 2012; Jin and Lynden-Bell 2008; Bland-Hawthorn et al. 2013), the mass of the ionized component of the MS would be as large as $\sim$3–$7 \times 10^9\, M_\odot$, thus very close to the total ISM gas mass in the Galactic disk (Ferriere 2001). The contribution of high-velocity absorbers at $d < 20$ kpc to the ionized gas mass in the halo is small instead; their gas mass sums up to a total value of no more than $M_{\rm HVCs, \mathit{d}<20\,kpc} = 2 \times 10^7\, M_\odot$ (R17). This value still is comparable to or even higher than the mass of the neutral gas in the same distance range.

2.3 Hot Ionized Gas

Ever since the prediction of Lyman Spitzer in 1956 on the existence of a Galactic Corona (see Sect. 1), the search for a low-density, high-temperature ($T > 10^5$ K) gaseous medium that surrounds the Milky Way has been of high priority for astrophysicists, as the Corona links the observed properties of the Galaxy to cosmological structure formation (see, e.g., Oort 1966). From more recent theoretical work (e.g., Maller and Bullock 2004), it is indeed expected that all MW-type galaxies are surrounded by massive, hot gaseous halos of typical mass of $10^{11} M_\odot$ and temperature $T \sim 10^6$ K (the virial temperature of the galaxy's DM halo). If some fraction of the gas was able to cool, it would sink towards the disk, feeding the galaxy with fuel for future star formation. Therefore, hot coronal gas may serve as a huge baryon reservoir from which MW-type galaxies gain their gas. In addition, the hot Milky Way halo might be further fed with gas from a possible large-scale outflow from the Galactic center region (Fox et al. 2015; Lehner et al. 2012; Zech et al. 2008; Bland-Hawthorn and Cohen 2003; Su et al. 2010).

Despite the obvious importance of the hot, ionized circumgalactic gas phase for galaxy evolution, our knowledge about the properties and spatial extent of hot coronal gas in the Milky Way still is very limited. This is because it is very difficult to detect such coronal gas that is expected to have very low densities ($n_{\rm H} < 10^{-3}\,{\rm cm}^{-3}$), in particular in the outer regions of the halo. Observational evidence for the existence of a hot Milky Way Corona comes from observations in the X-ray regime, where the gas can be observed in emission or in absorption against extragalactic X-ray point sources. Using ROSAT data, Kerp et al. (1999) systematically searched for X-ray emission spatially associated with neutral HVCs and reported several positive detections, e.g., in the direction of Complex GCN, Complex C, and Complex D. In Fig. 5, we show as an example the ROSAT emission map of hot halo gas in the direction of Complex GCN from the study of Kerp et al. Other observations of the soft X-ray background support the existence of hot coronal gas in the Milky Way (e.g., Kuntz and Snowden 2000). X-ray absorption of O VII

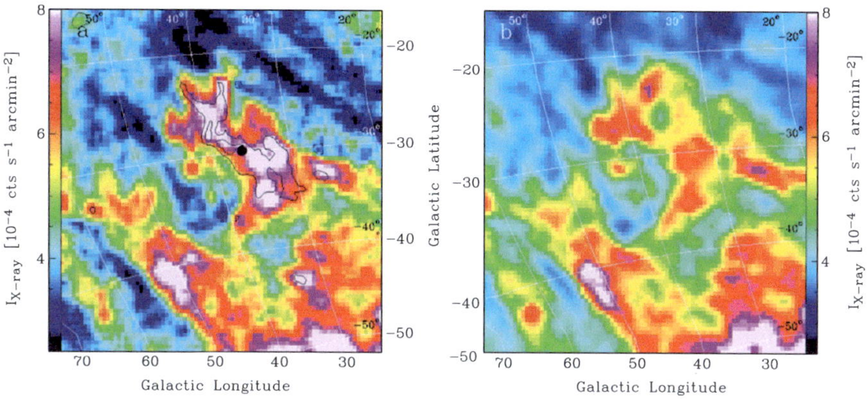

Fig. 5 *Left panel*: X-ray emission map of hot gas in the direction of HVC complex GCN, based on ROSAT 0.25 keV soft-X ray background (SXRB) data (see Kerp et al. 1999 for details). The position of the background quasar Mrk 509 is indicated with the *black dot*. In this direction, hot halo gas has been detected in high-velocity O VI absorption using FUV data (Sembach et al. 2003; see also Winkel et al. 2011). *Right panel*: model of the SXRB emission in the same direction, based on 21 cm H I data from foreground gas (causing photoelectric absorption of the background X-ray photons) and assuming a homogenous distribution of the X-ray background (Kerp et al. 1999). Both maps indicate the presence of hot coronal gas in the Galactic halo in this area of the sky. Maps kindly provided by Jürgen Kerp

and O VIII in the Galactic halo has been reported by several groups (Wang et al. 2005; Fang et al. 2002, 2003, 2006; Mathur et al. 2003; Bregman 2007; McKernan et al. 2004; Williams et al. 2005; Gupta et al. 2012; see also Miller et al. 2016), but the interpretation of these low-resolution spectra is afflicted with systematic uncertainties (see Richter et al. 2008). Also pulsar dispersion measures have been used to constrain the properties of hot coronal gas in the Milky Way halo (e.g., Gaensler et al. 2008). All these observations are biased towards the regions with the highest gas densities in the Corona, however, so that the bulk of the hot gas detected in this manner presumably resides in the inner halo at $d < 20$ kpc (e.g., Rasmussen et al. 2003). It therefore remains unknown whether the coronal gas really extends to the virial radius of the Milky Way ($R_{\rm vir} \sim 260$ kpc; e.g., Tepper-García et al. 2015) and how it connects to gas gravitationally bound to the Local Group. From the UV observations of O VI in the thick disk of the Milky Way (Widmann et al. 1998; Savage et al. 2003; Wakker et al. 2003; Savage and Wakker 2009) indeed follows that there are large amounts of extra-planar warm-hot gas at vertical heights $z < 5$ kpc.

Indirect evidence for the widespread presence of hot gas in the Milky Way halo comes from the many UV absorption-line detections of *high-velocity* O VI (Sembach et al. 2003; Wakker et al. 2003), which has a sky covering fraction as large as $f_c = 0.60$–0.85. O VI arises in warm-hot gas at $T \sim 3 \times 10^5$ K and is believed to trace the interface regions between the hot coronal gas and cooler halo clouds embedded therein (i.e., the neutral and ionized HVCs; see above). Isolated, strong O VI absorption at high velocities is also seen in the direction of the Local Group barycenter and, in particular, towards the quasar Mrk 509 in HVC Complex GCN (Fig. 5), where enhanced X-ray emission and high-velocity Si III absorption is observed (Collins et al. 2005; see above). The coincidence of UV absorption and X-ray emission further indicates that there is an excess of hot gas in this direction that possibly is related to highly ionized intragroup gas near the Local Group barycenter, towards which the Milky Way is moving (see Sect. 2.2).

To estimate the baryon content of the Milky Way's coronal gas, its radial density profile needs to be constrained. Due to on-going accretion of gas and stars from satellite galaxies and the IGM, substantial deviations from a simple hydrostatic density distribution are likely. Even for a hot halo that is not perfectly hydrostatic, however, the average gas density in the coronal gas is expected to decrease for increasing distances to the disk. From X-ray absorption, spectra Bregman and Lloyd-Davies (2007) estimate a gas density of $n_{\rm H} \approx 8 \times 10^{-4}\,{\rm cm}^{-3}$ for the inner halo ($d < 20\,{\rm kpc}$). From pulsar dispersion measures instead follows that the average coronal gas density must be smaller, $n_{\rm H} < 8 \times 10^{-4}\,{\rm cm}^{-3}$, in line with studies that estimate $n_{\rm H}$ indirectly from considering the interaction between the cool HVCs and the ambient hot medium ($n_{\rm H} \approx 2 \times 10^{-4}\,{\rm cm}^{-3}$; Grcevich and Putman 2009; Peek et al. 2007; Tepper-García et al. 2015). Because of the unknown extent and the unknown gas properties at the virial radius, the total mass of the Milky Way's hot coronal gas is very uncertain. For $d < 250\,{\rm kpc}$ the total mass is estimated to be $M_{\rm Corona} \approx 10^{10} - 10^{11} M_{\odot}$ (Anderson and Bregman 2010; Yao et al. 2008; Gupta et al. 2012; Miller and Bregman 2013, 2015; Fang et al. 2013; Salem et al. 2015), in line with the idea, the Milky Way's hot Corona represents a huge baryon reservoir. However, the hot coronal gas must cool and condense into streams of denser gas to be able to sink to the Milky Way disk, i.e., it must transit through the diffuse ionized and/or neutral phase to contribute to the gas-accretion rate. To understand the details of this important phase transition, hydrodynamical simulations are required, which will be discussed in Sect. 3.

2.4 Gas-Accretion Rates from Observations

The observations presented in the previous sections suggest that HVCs (and IVCs) represent coherent entities of multi-phase gas that move within the Milky Way halo (Lehner et al. 2012; R17). Independent of their individual origin (tidal interactions with satellite galaxies, infall from the IGM or intragroup medium, condensations from the hot coronal gas, the back-flow of material expelled previously from the disk), these streams of gas trigger the Galaxy's present-day gas-accretion rate as defined in Eq. (1) and we will discuss their contribution to ${\rm d}M_{\rm gas,halo}/{\rm d}t$ in the following.

While the 3D distribution of neutral and diffuse ionized gas around the Galaxy (i.e., $M_{\rm gas,halo}$, d) is constrained by observations, the space *motion* of the gas is not, as only the radial component of the velocity can be observed. In lack of further information, the mean infall velocity often is assumed to be constant for all halo clouds, although it is likely that $v_{\rm infall}$ spans a large range for the IVCs and HVCs and systematically depends on d (see Sect. 1). For the MS, recent studies assume $\langle v_{\rm infall}\rangle = 100\,{\rm km\,s^{-1}}$ (e.g., Fox et al. 2014), a value that also has been used for other HVCs (e.g., Complex C; Wakker et al. 1999, 2008). A more complex model for $v_{\rm infall}$ is presented by Putman et al. (2012), where they try to separate for the HVC population the azimuthal velocity component from the accretion velocity in the Galactic center (GC) direction and derive $\langle v_{\rm infall,GC}\rangle \approx -50\,{\rm km\,s^{-1}}$. This value is consistent with the *mean* radial velocity of HVCs of $\langle v_{\rm rad}\rangle \approx -50\,{\rm km\,s^{-1}}$, but the most distant halo structures, such as the MS, might have somewhat larger accretion velocities (Mathewson et al. 1974).

Because of its large mass, the MS dominates by far the total present-day gas accretion rate of the Milky Way. From their UV absorption-line survey Fox et al. (2014) derive a mass inflow rate of neutral and diffuse ionized gas from the Stream of $dM_{\rm MS}/dt \approx 2\,M_{\odot}\,{\rm yr^{-1}}$ for a fixed distance of the MS of $d = 55$ kpc and $\langle v_{\rm infall}\rangle = 100\,{\rm km\,s^{-1}}$. At this distance and infall velocity, the gas (or better said, a fraction η of it; see Eq. (2)) would reach the disk in $\sim$540 Myr. If, instead, the distance of the Stream was $d = 100$ kpc, then it would take $\sim$1 Gyr for the gas to reach the disk and the accretion rate would be higher by a factor of $\sim$2, because the estimate of $M_{\rm MS}$ depends on its assumed distance. Taking $d = 55$ kpc as a conservative lower limit and adding the gas mass associated with the Magellanic Clouds and the Magellanic Bridge (with all these components together forming the *Magellanic System*), the total accretion rate from all these components sums up to a value of $dM_{\rm MSys}/dt \geq 3.7\,M_{\odot}\,{\rm yr^{-1}}$; Fox et al. (2014).

Based on the observed properties discussed above, the contribution of the other *individual*, 21 cm-selected HVC Complexes at $d < 20$ kpc to the mass-inflow rate (including both neutral and ionized gas) is expected to be rather small (e.g., Complex C: 0.1–0.2 $M_{\odot}\,{\rm yr^{-1}}$; Complex A: 0.05 $M_{\odot}\,{\rm yr^{-1}}$; Cohen Stream: 0.01 $M_{\odot}\,{\rm yr^{-1}}$; see Wakker et al. 1999, 2007, 2008; Thom et al. 2006, 2008; Putman et al. 2012; R17). Putman et al. (2012) derive a maximum accretion rate of 0.4 $M_{\odot}\,{\rm yr^{-1}}$ for all HVCs *except* the MS. If we consider only the neutral gas mass in the Galactic HVC population, the total H I gas-accretion rate from all 21 cm HVCs (including the MS) comes out to $dM_{\rm HI}/dt = 0.7\,M_{\odot}\,{\rm yr^{-1}}$ (Richter 2012), again assuming that the MS is at $d = 55$ kpc.

The general distribution of UV-absorbing gas in the halo (and its total mass) yet implies that the *ionized* component (independent of whether it is associated with H I emission or not) dominates the gas accretion not only for the MS, but also for the nearby HVCs at $d \leq 20$ kpc. From their absorption-line survey towards halo stars and extragalactic background sources, Lehner and Howk (2011) determine an accretion rate of predominantly ionized high-velocity gas at $d < 15$ kpc of 0.45–1.40 $M_{\odot}\,{\rm yr^{-1}}$. Putting it all together, R17 estimate from their UV absorption-line survey of 265 sightlines a total gas-accretion rate of neutral and ionized high-velocity gas in the halo (including gas from the Magellanic System) of $dM_{\rm HVC}/dt \geq 5\,M_{\odot}\,{\rm yr^{-1}}$. This limit is higher by a factor of >2 than the *current* star-formation rate of the Milky Way ($\sim$0.7–2.3$M_{\odot}\,{\rm yr^{-1}}$; e.g., Robitaille and Whitney 2010; Chomiuk and Povich 2011).

The contribution of the 21 cm IVCs and the DIG in the disk-halo interface (DHI) at z-heights <2.5 kpc to the Milky Way's gas-accretion rate is significantly smaller than that of the neutral HVCs. Considering the IVC distances and neutral-gas masses discussed in Sect. 2.1.1, the neutral gas-accretion rate from IVCs is only $\sim$0.01–0.05$M_{\odot}$ yr^{-1}. A much higher gas-accretion rate can be determined considering the ionized gas reservoir in the DIG (Sect. 2.2). In principle, the overall gas flow in the DIG of Milky Way type is expected to be strongly influenced by the various feedback processes from the disk (e.g., radiative and mechanical feedback from supernovae, stellar winds, and AGN; e.g., Bland-Hawthorn and Maloney 2002; MacLow and Klessen 2004; Springel and Hernquist 2005; Marasco et al. 2013), but the exact role of these processes in the gas-circulation cycle of the Milky Way's DHI is uncertain. Assuming that at least half of the diffuse ionized gas in the disk-halo interface is currently being accreted (the rest being related to outflowing gas), the observational constraints imply $\mathrm{d}M_{\mathrm{DHI}}/\mathrm{d}t = 1$–$2\,M_{\odot}\,\mathrm{yr}^{-1}$ for $v_{\mathrm{infall}} \leq 20\,\mathrm{km\,s}^{-1}$, thus in line with the current star-formation rate (see Fraternali et al. 2013). A net-infall of ionized gas is in line with the observed Hα kinematics (Haffner et al. 2003).

Some interesting conclusions can possibly be drawn from these numbers. Obviously, the amount of large-scale *neutral* gas in the disk-halo interface is much smaller than the amount of large-scale neutral gas at larger distances; it is also much smaller than the amount required to keep up the current star-formation rate in the disk. Infalling neutral gas structures thus might be disrupted and ionized when entering the disk-halo interface, where it might re-cool and condense again before it enters the disk. This re-processing of infalling gas in the disk-halo interface is sometimes referred to as "quiet accretion." The galactic-fountain flow is believed to play a crucial role in the the gas cooling and condensation processes (Marinacci et al. 2010; Armilotta et al. 2016). The infalling gas thus might enter the disk in the form of low-velocity, mildly ionized gas clumps or tiny 21 cm drops (Lockman 2002; Begum et al. 2010; Ford et al. 2010; see Sect. 2.1), thus in a form that is difficult to identify observationally.

Another possible reason for the apparent discrepancy between the neutral gas budget in the DHI and that at larger distances might be the existence of neutral DHI gas that is "hidden" to us: clouds that have low radial velocities (LVCs; see Sect. 1.1), similar to those in the disk, but that reside in the (lower) Galactic halo (e.g., Zheng et al. 2015; Peek et al. 2009). Finally, also gas from the outer disk might contribute to the fueling of star formation in the inner regions of the Milky Way disk through a *radial* inflow of gas (e.g., Elson et al. 2011; Sellwood and Binney 2002; see also Putman et al. 2012 and the references therein).

Obviously, observations alone cannot provide a full insight into the complex processes that govern the past, present, and future gas-accretion rate of the Milky Way. Hydrodynamical simulations represent an important toolkit to further study the dynamics and physical properties of gas falling towards Milky Way-type galaxies and to pinpoint the overall mass-inflow rate to the disk. The most relevant of these aspects will be discussed in the following section.

3 Simulations of Milky Way Gas Accretion

In this section, we briefly discuss results from theoretical studies of accretion processes of Milky Way-type galaxies based on hydrodynamical simulations. Although these studies will also be discussed in a more general context in other chapters in the second part of this book, they are particularly important for our understanding of gas accretion in the Milky Way and thus need to be considered in this review.

3.1 Hydrodynamical Simulations of Gas Infall

Whatever the initial conditions of an infalling gas structure at a given distance to the Galaxy might be, the gas will interact with the ambient hot coronal medium that fills the galaxy's dark matter potential. Modeling these interaction processes are of fundamental importance to understand origin and fate of accreted material and to explain the observations discussed in the previous sections.

As mentioned earlier, one plausible scenario for gas accretion in Milky Way size gaseous halos with $M_{\rm halo} \sim 10^{12}\,M_{\odot}$ and $R_{\rm vir} \sim 250\,{\rm kpc}$ is the cooling and fragmentation of hot ($T \sim 10^6$ K) halo gas that falls towards the Galactic disk in the form of overdense clouds (Maller and Bullock 2004). Support for this scenario comes from early SPH simulations (Kaufmann et al. 2006, 2009; Sommer-Larsen 2006). In these simulations, cool pockets of gas condense out from the hot coronal gas from thermal instabilities, mimicking the properties of the Galactic 21 cm HVCs that have $d < 20\,{\rm kpc}$ (Peek et al. 2008). More recent studies indicate, however, that these early studies may draw a too simplistic picture of the HVC condensation process. One problem lies in the buoyancy of thermally unstable gas, which is expected to disrupt condensing seed structures before they can cool efficiently (Burkert and Lin 2000). Thus, linear isobaric perturbations in a homogeneous, hot coronal gas are expected to be inefficient to develop cool gas patches in the halo, unless the entropy gradient is very small. An alternative scenario is offered by Joung et al. (2012a), who study the condensation process in Galactic coronal gas from non-linear perturbations that might be generated from either infalling intergalactic gas (e.g., Kereš and Hernquist 2009; Kereš et al. 2009; Brooks et al. 2009) or from the gaseous leftovers of satellite accretion (Bland-Hawthorn et al. 2007; Grcevich and Putman 2009; Nichols and Bland-Hawthorn 2011; see below). Joung et al. (2012a) find from their high-resolution adaptive mesh refinement (AMR) hydrodynamical simulations that the efficiency for condensing out cool gas patches from non-linear perturbations depends critically on the ratio between the cooling time and the acceleration time (to reach the sound speed) in the gas. If cooling is efficient (such as in clumpy gas and/or in gas with a high-metallicity), cool patches can condense out in a Milky Way-type halo before being disrupted and may show up as 21 cm HVCs. The mixing of outflowing, metal-enriched fountain material, and infalling coronal material possibly represents a key process that determines the net accretion through the disk-halo interface, as it regulates the cooling efficiency of the gas (Armilotta et al. 2016; Marinacci et al. 2010).

Cosmologically "cold" ($T < T_{\rm vir}$) streams of intergalactic gas are expected to feed the halos of Milky Way-type galaxies (e.g., van de Voort et al. 2011) and such streams may penetrate deep into the hot halo before they are being disrupted. From their AMR simulations, Fernandez et al. (2012) find that cold streams from the IGM may continuously bring up to $\sim 10^8 M_\odot$ of H I into the Milky Way halo and this amount of neutral gas appears to be fairly constant over time (at least for the last 5 Gyr, or $z = 0$–0.45). The H I gas accretion rate for a Milky Way-type galaxy in these simulations comes out to $0.2\,M_\odot\,{\rm yr}^{-1}$, thus very similar to what has been derived for the Galactic HVC population at $d < 20$ kpc, i.e., *without* the contribution of the MS (see Sect. 2.4).

Next to the cloud-condensation scheme and the cold streams, the interaction of Milky Way-type galaxies with dwarf satellites can transport large amounts of relatively cool gas in the halos of MW-type galaxies *directly*. From their study of a Milky Way-type galaxy and its satellite-galaxy environment Fernandez et al. (2012) derive a median mass-loss rate of H I of $\sim 3 \times 10^{-3}\,M_\odot\,{\rm yr}^{-1}$, suggesting that satellites add $\sim 3 \times 10^6\,M_\odot$ of neutral gas to the host galaxy within a Gyr. Based on their overall results, this H I mass flow represents a significant albeit not dominating contribution to the neutral gas-accretion rate of the host galaxy. In the light of these results, the observed neutral gas supply from Magellanic Stream for the Milky Way ($M_{\rm HI,MS} \sim 2.5 \times 10^8\,M_\odot$ within 0.5–1.0 Gyr; see Sect. 2.1) is huge, underlining that the MS represents a rather extreme (and atypical) example for gas accretion by satellite interactions.

As suggested by Peek (2009), the specific angular momentum of the gas might be a key parameter to discriminate between the various scenarios for the origin of the HVCs. Gas that is accreted from dwarf satellites by either ram-pressure stripping or tidal interactions enters the halo with a large initial angular momentum ($L \sim 3 \times 10^4$ km s^{-1} kpc), while the angular momentum of the cooling halo clouds is one order of magnitude less (see, e.g., Peek et al. 2008; Kaufmann et al. 2009). Thus, gas accretion from mergers with Milky Way satellites takes place predominantly in the outer regions ($R > 15$ kpc) of the Galactic disk. In fact, observations indicate that part of the MS (i.e., the Leading Arm) might already be close to the outer disk of the Milky Way, where it possibly already interacts with the underlying interstellar gas (McClure-Griffiths et al. 2008; Casetti-Dinescu et al. 2014), thus providing support for this scenario. Following Peek (2009), the HVCs at small galactocentric distances ($R < 15$ kpc), in contrast, are more likely to be produced by the condensed halo clouds and it is this process that appears to dominate the feeding of the inner-disk regions where most of the star formation takes place.

Independently of the origin of the infalling gas, how much of it makes it into the disk? To answer this crucial question for the Milky Way and other galaxies of similar mass and type, the initial conditions for the gas infall (e.g., infall velocities, individual cloud masses, density profile of the ambient coronal gas) turn out to be particularly important, as we will discuss in the following.

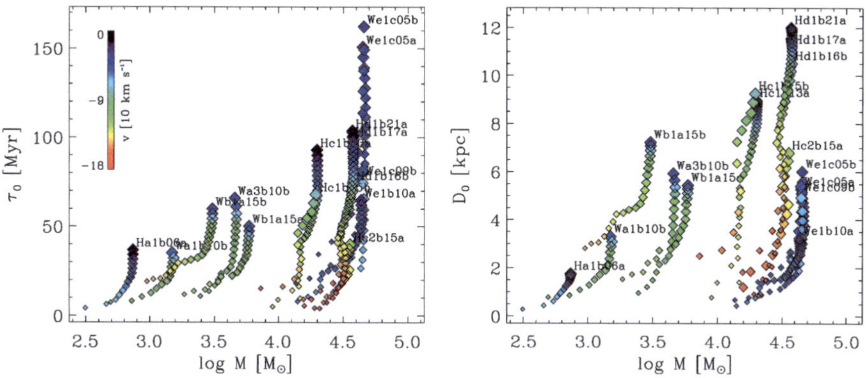

Fig. 6 Time sequences of cloud survival-times (*left panel*) and disruption length-scale (*right panel*) of neutral halo clouds of different mass that move through hot coronal gas in a Milky Way-type galaxy (Heitsch and Putman 2009). In these models, H I clouds with masses $< 10^{4.5} M_{\odot}$ do not survive their passage through the halo when infalling from $d = 10$–12 kpc. Figure adopted from Heitsch and Putman (2009)

At $T = 10^6$ K in the coronal gas, the infall velocity of the HVCs is close to the sound speed in ambient hot medium ($v_s \sim 150\,\mathrm{km\,s^{-1}}$), so that infalling material moves either in the subsonic, or transonic, or supersonic regime, leading to different ablation scenarios of infalling gas structures (Kwak et al. 2011). Using a grid-based hydrodynamical 3D code, Heitsch and Putman (2009) simulated the fate of H I clouds moving through the hot coronal gas to explore the characteristic morphologies of the infalling structures, such as head-tail structures, infall velocities, disruption path lengths, and timescales. In Fig. 6 we highlight some of their results. From their study it follows that H I clouds with relatively low initial masses of $M_{\rm cloud} < 10^{4.5} M_{\odot}$ lose basically all their neutral gas content when infalling from $d = 10$–12 kpc. Thus, wherever such cloudlets might be formed in the halo, i.e., either from condensing out of the coronal gas or from dissolving infalling tidal or cosmological gas streams at much larger distances, their gas content will not reach the disk in *neutral* form. However, if the density contrast between the break-up material and the surrounding medium is large enough, the cloud remnants might reach the disk as warm ionized material (e.g., Shull et al. 2009; Heitsch and Putman 2009; Bland-Hawthorn 2009; Joung et al. 2012b) or will otherwise fuel the Galactic Corona.

At the high-mass end, Kwak et al. (2011) predict that for cloud masses $M_{\rm cloud} > 10^5 M_{\odot}$ up to 70% of the H I mass remains cool at $T < 10^4$, even after a possible break-up of the initial infalling cloud. Such massive structures may even survive the trip from larger distances, as they can move several hundred Myr through the hot halo before being destroyed completely. A similar conclusion was drawn from Joung et al. (2012b) from their high-resolution AMR simulations (see above). This result is relevant for the Magellanic Stream at $d = 50$–100 kpc. Although observations suggest that the MS is further braking up into smaller gas clumps as it moves towards the Milky Way disk (e.g., Stanimirović et al. 2002, 2008; Westmeier and Koribalski 2008; see also Tepper-García et al. 2015), some dense cores with masses $M_{\rm cloud} > 10^5 M_{\odot}$ may survive and these clumps could enter the disk in the form of neutral gas clouds.

Note that a certain fraction of the halo clouds that condense out of the coronal gas may also reach the disk *before* the gas cools efficiently, i.e., in the form of warm, ionized gas that never becomes neutral and visible in 21 cm emission. Those structures possibly explain some of the isolated high-velocity UV absorbers that have no 21 cm counterparts (Collins et al. 2003, 2009; Shull et al. 2009; Richter et al. 2009; Fraternali et al. 2013; Marasco et al. 2013; R17).

Even with the most advanced hydrodynamical simulations at hand, the exact role of many of the involved physical processes in the time evolution of accreted gas remains unclear. For instance, thermal conduction can suppress shear instabilities and thus stabilize clouds from evaporation by smoothing out steep density/temperature gradients between the cool infalling gas and the hot ambient medium (e.g., Vieser and Hensler 2007). However, such steep gradients appear to be unimportant for the cool/warm gas clouds in the Milky Way halo (e.g., Kwak et al. 2011) and also the diffusion by thermal conduction possibly is small compared to turbulent diffusion processes. Not all of the possibly relevant processes can be included simultaneously in the simulations. In particular, the role of magnetic fields (and resulting magnetohydrodynamical effects) have been mostly ignored so far. Also, the spatial (or mass) resolution of many of the simulations discussed above still is very limited. Since early SPH simulation codes, for instance, had problems in resolving Kelvin–Helmholtz instabilities, cool circumgalactic gas clumps are artificially stabilized by SPH particle effects (Agertz et al. 2007), probably leading to misleading results.

In summary, hydrodynamical simulations predict that the life-time, the morphology, and the mass distribution of infalling gas clouds in the Milky Way halo depend strongly on the local boundary conditions under which the gas is generated and moving through the hot halo. Whether or not an infalling cool/warm gas cloud reaches the Milky Way disk in a region where it supplements star formation depends on its initial distance to the disk, its mass and density, its infall velocity, its angular momentum, and other parameters.

3.2 Cosmological Hydrodynamical Simulations

Given the complexity of the physics of gas accretion outlined above and the strong dependence of the gas-accretion rate on the local boundary conditions, additional large-scale simulations are desired that also consider a realistic *cosmological* environment of the Milky Way. In particular, the role of a second nearby spiral galaxy (M31), the gas motions within the super-ordinate galaxy group environment (Local Group), as well as the streaming of intergalactic gas within the local cosmic web that connects the Local Group with its surrounding large-scale structure (e.g., the Virgo cluster) should be studied in such a context.

In a recent paper, N14 have studied the large-scale distribution and overall physical properties of gas in the Local Group and around Milky Way and M31 based on simulation data from the Constrained Local UniversE Simulations (CLUES) project (www.clues-project.org). In their study, the authors separate the circumgalactic and intragroup gas into three different phases: neutral gas, cold/warm ionized gas with $T < 10^5$ K, and hot gas with $T \geq 10^5$ K, similar as done here. The total neutral gas mass in the simulated Milky Way at galactocentric distances $d < 50$ kpc comes out to $M_{\rm HI} \approx 3 \times 10^8 M_{\odot}$, thus in excellent agreement with the observations (Sect. 2.1). The total mass of the cold/warm ionized gas component is as large as $M_{\rm HII} \approx 3 \times 10^{10} M_{\odot}$ for the entire halo out to the virial radius, but reduces to $M_{\rm HII} \approx 2 \times 10^8 M_{\odot}$ for $d \leq 10$ kpc, where the bulk of UV absorption of warm ionized gas is observed in the Milky Way (Lehner and Howk 2011; Sect. 2.2). The mass of the Milky Way's hot coronal gas in the simulation is $M_{\rm Corona} \approx 4 \times 10^{10} M_{\odot}$ for $d \leq R_{\rm vir}$ and $\approx 10^{10} M_{\odot}$ for $d \leq 100$ kpc, the latter value being consistent with the estimates from X-ray observations (Sect. 2.3). The Milky Way's neutral gas-accretion rate from gas at $d \leq 50$ kpc is estimated as $M_{\rm HI} \approx 0.3\, M_{\odot}\,{\rm yr}^{-1}$, which is only $\sim$40% of the value derived from the 21 cm observations (Richter 2012), but more in line with what is expected for the neutral HVCs without the Magellanic Stream (Sect. 2.4). For larger distances to the disk, the neutral gas-accretion rate quickly falls below $10^{-2}\, M_{\odot}\,{\rm yr}^{-1}$ in the simulations (N14; their Fig. 14). The accretion rate of cold/warm gas instead is fairly independent of the distance for $d > 15$ kpc at a level of $M_{\rm HII} \approx 5\, M_{\odot}\,{\rm yr}^{-1}$, thus in line with the estimate from the UV observations (Fox et al. 2014; R17). Finally, the simulations imply that only for very large distances $d > 100$ kpc the accretion rate of hot ($T > 10^6$ K) gas dominates the mass inflow of gas for the Milky Way.

The influence of the Local Group environment in the simulations is reflected particularly in the anisotropic distribution of gas near the viral radius of the MW, because the gas follows the large-scale matter distribution in the elongated cosmological filament that forms the Local Group (N14). To visualize this, we show in Fig. 7 the gas distribution and gas kinematics around the simulated Milky Way and M31 galaxies from the CLUES simulations. The two galaxies move towards the LG barycenter while the ambient gas is circulating around MW and M31 within the elongated filament in a complex pattern of infall and outflow channels. There is a significant gas excess between the two galaxies, as compared to any other direction, resulting from the overlap of their gaseous halos. Because of the Milky Way's flow towards the LG barycenter, a velocity dipole pattern for high-ion absorption from LG gas/M31 halo gas is expected from a perspective within the Milky Way (R17). Such a dipole pattern is indeed observed in UV absorption in the direction of M31 and its antipode on the sky (Sect. 2.2). This possibly implies that warm/hot LG gas/M31 halo gas is pushed into the Milky Way halo due to the large-scale motion of both galaxies in their group environment. If true, this effect might cause a major boost in the Milky Way's future gas-accretion rate.

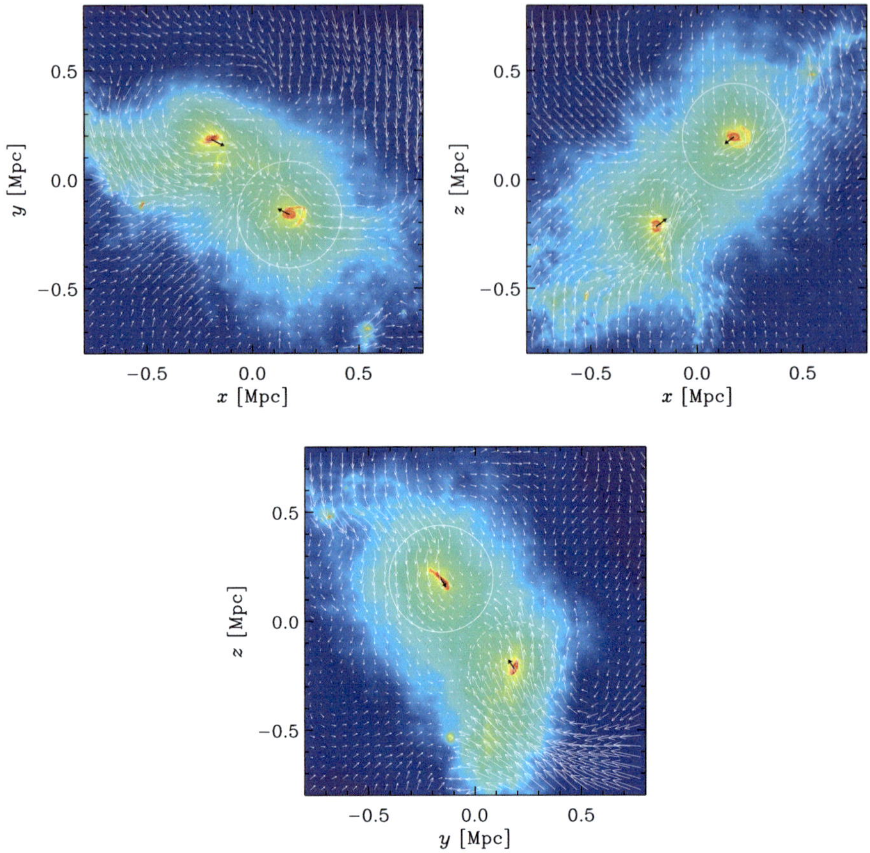

Fig. 7 Projected distribution and bulk motions of gas in and around the Milky Way (*dashed circle*, indicating the virial radius of the MW) and M31 (*solid circle*, indicating the virial radius of M31) based on constrained cosmological simulations from the CLUES project (N14). Coordinates and velocities of the gas are given with respect to the Local Group barycenter in the x/y, x/z, and y/z planes, respectively. The *white arrows* show the velocity field of the gas, with the *longest arrows* representing a velocity of 130 km s^{-1}. The *black arrows* indicate the velocities of the MW and M31 galaxies. Their absolute space velocities are 67 and 76 km s^{-1}, respectively. Maps kindly provided by Sebastian Nuza

3.3 Comparison with Observations

For a better understanding of the gas-accretion processes in the Milky Way, the comparison between observational data and predictions from simulations are essential. Next to the gas masses in the individual phases and the accretion rates (see above), the spatial distribution of the various gas phases in the simulations and their kinematics can be compared (in a statistical sense) with the observational constraints from 21 cm data and UV absorption spectra.

Combining 21 cm data from the Milky Way and M31, Richter (2012) predicted that the volume-filling factor of neutral gas in the halo of Milky Way/M31 type galaxies declines exponentially with radius, leading to an exponential decline of the observed (projected) H I covering fraction. The study suggests that the covering fraction f_c(H I) drops below 0.05 for $d > 50$ kpc. A similar trend indeed has also been found in the recent CGM simulations of Milky Way-type galaxies (Fernandez et al. 2012; N14), indicating that basically all neutral gas in MW-type galaxies is concentrated in the inner halo region. This conclusion is supported by the observed cosmological cross section of neutral gas around low-redshift galaxies (Zwaan et al. 2005; Richter et al. 2011).

Also the apparent interaction between the MS and the surrounding coronal gas as well as the Hα emission from the Stream have been investigated in simulations to reproduce the observational results. Bland-Hawthorn et al. (2007) and Tepper-García et al. (2015) modeled the Hα emission from the MS based on a shock-cascade model. Tepper-García et al. (2015) conclude that the Hα emission from the Stream can only be reproduced if the density of the ambient medium is $n_{\rm H} = 2$–$4 \times 10^{-4}\,{\rm cm}^{-3}$, indicating that H$\alpha$ emitting regions in the MS must be within $d \leq 75$ kpc from the Galactic center.

Next to these examples, there are several other studies that have addressed these and other aspects by comparing observational results with simulations. Describing all of these unfortunately is beyond the scope of this review. Clearly, with future, more detailed simulations and additional constraints from multi-wavelength observations the systematic combination of simulations and observations will provide crucial new insights into the properties of the Milky Way's CGM.

4 Concluding Remarks

As for many other aspects of galaxy evolution, the Milky Way and its gaseous environment represent an excellent laboratory to study the details of gas-accretion processes of L* galaxies in the local Universe. Although it remains a challenging task to reconstruct the 3D distribution and galactocentric kinematics of the Milky Way's circumgalactic medium from an internal vantage point in the rotating disk, the combination of multi-wavelength observations of the gas in all its phases, the measurement and modeling of the stellar composition of the Milky Way disk and its star-formation rate and history, the numerical modeling of the hydrodynamic processes that shape the properties of the Galaxy's CGM, and the deep observations of the Milky Way and its satellite galaxies in a cosmological context together provide a particularly rich database that cannot be achieved for any other galaxy in the Universe.

The above discussed observations and simulations imply that the combination of gas infall, outflows, and mergers generates a multi-phase gaseous halo that is characterized by a highly complex spatial distribution of gas structures of different age and origin. The cycle of processes that is believed to lead to the continuous feeding of the Milky Way disk with fresh gas to supplement subsequent star-formation therein can be summarized as follows.

Cold and warm gas from the intergalactic medium and from satellite galaxies enters the Milky Way halo at its virial radius and its then processed by the ambient hot coronal gas. Fragments of the originally infalling gas may reach the disk in the form of warm or cold gas streams, while the remaining gas fraction is being incorporated into the hot Galactic Corona. In this way, the Corona is continuously fed with fresh gas from outside, while it is further stirred up and heated by gas outflowing from the star-forming disk and (eventually) from the Galactic center region (i.e., it is influenced by feedback processes). From the hot Corona, warm ionized and/or warm neutral gas patches may condense out through cooling processes, and these structures will sink down to the disk through the disk-halo interface, further contributing to the overall accretion rate of the Milky Way disk.

The accretion of cold and warm gas from the Magellanic Stream is a direct result of the interaction between the Milky Way and its population of satellite galaxies (here: the Magellanic Clouds). The MS adds more than one billion solar masses of gaseous material to the Milky Way halo, material that either directly or indirectly feeds the MW disk to supplement star formation therein. Thus, there is sufficient cold and warm gaseous material present in the outer halo to maintain the Milky Way's star-formation rate *in the far future* at its *current* level, although it remains unclear, how much of the material from the MS will finally end up in the disk (and at what timescale). The amount of gas that is currently being accreted through the disk-halo interface, and that will determine the star-formation rate *in the near future*, remains uncertain, however.

Our understanding of the gas-accretion processes in the Milky Way is far from being complete. On the observational side, additional constraints on distances and 3D velocities of the HVCs, the role of low-velocity halo gas, and the mass and spatial extent of the hot coronal gas based on multi-wavelength observations are highly desired on the long way towards a complete census of the Milky Way's circumgalactic gas. On the theoretical side, more advanced hydrodynamical simulations of the Milky Way's gaseous halo, that include all relevant physical processes in a realistic cosmological environment, will be of great importance to study the dynamics of gas flows around the Galaxy. Finally, a systematic comparison between gas-accretion processes in the Milky Way and similar processes in other low- and high-redshift galaxies in the general context of galaxy evolution (with particular focus on environmental issues, feedback effects, and other important aspects) will provide crucial information that will help to better understand the cycling of gas on galactic and super-galactic scales. Many of these aspects will also be discussed in the following chapters.

Acknowledgements The author would like to thank Andy Fox, Matt Haffner, Fabian Heitsch, Jürgen Kerp, and Sebastián Nuza for providing helpful comments and supplementary material for the figures.

References

Adams, W.S. 1949, ApJ, 109, 354
Agertz, O., Moore, B., Stadel, J., et al. 2007, MNRAS, 380, 963
Anderson, M.E., & Bregman, J.N. 2010, ApJ, 714, 320
Armilotta, L., Fraternali, F., & Marinacci, F. 2016, MNRAS, 462, 4157
Bajaja, E., Cappa de Nicolau, C.E., Cersosimo, J.C. et al. 1985, ApJS, 58, 143
Bates, B., Catney, M.G., & Keenan, F.P. 1988, ASS 146, 195
Bauermeister, A., Blitz, L., & Ma, C.-P. 2010, ApJ, 717, 323
Begum, A., Stanimirovic, S., Peek, J.E.G., et al. 2010, ApJ, 722, 395
Ben Bekhti, N., Richter, P., Westmeier, T., & Murphy, M.T. 2008, A&A, 487, 583
Ben Bekhti, N., Winkel, B., Richter, P., et al. 2012, A&A, 542, A110
Besla, G., Kallivayalil, N., Hernquist, L., et al. 2010, ApJL, 721, L97
Besla, G., Kallivayalil, N., Hernquist, L., et al. 2012, MNRAS, 421, 210
Birnboim, Y., & Dekel, A. 2003, MNRAS, 345, 349
Blaauw, A. & Tolbert, C.R. 1966, BAN, 18, 405
Bland-Hawthorn, J., Veilleux, S., Cecil, G. N., et al. 1998, MNRAS, 299, 611
Bland-Hawthorn, J., & Putman, M. E. 2001, in ASP Conf. Ser. 240, *Gas and Galaxy Evolution*, ed. J. E. Hibbard, M. Rupen, J.H. van Gorkom, p. 369
Bland-Hawthorn, J.,& Maloney, P. R. 2002, in ASP Conf. Ser. 254, *Extragalactic Gas at Low Redshift*, ed. J. S. Mulchaey & J. T. Stocke (San Francisco, CA: PASP), p. 267
Bland-Hawthorn, J., & Cohen, M. 2003, ApJ, 582,246
Bland-Hawthorn, J., Sutherland, R., Agertz, O., & Moore, B. 2007, ApJL, 670, L109
Bland-Hawthorn, J. 2009, IAUS, 254, 241
Bland-Hawthorn, J., Maloney, P., Sutherland, R.S., & Madsen, G.J. 2013, ApJ,778, 58
Blitz, L., Spergel, D.N., Teuben, P.J., Hartmann, D., & Burton, W.B. 1999, ApJ, 514, 818
Bregman, J.N. & Lloyd-Davies, E.J. 2007, ApJ, 669, 990
Bregman, J.N. 2007, ARA&A, 45,221
Braun, R. & Burton, W.B. 1999, A&A, 341, 437
Brooks, A.M, Governato, F., Quinn, T,, Brook, C.B., & Wadsley, J. 2009, ApJ, 694,396
Brüns, C, Kerp, J., Staveley Smith, L., et al. 2005, A&A, 432, 45
Burkert, A., & Lin, D.N.C. 2000, ApJ, 537, 270
Casetti-Dinescu, D.I., Moni Bidin, C., Girard, T.M., et al. 2014, ApJL, 784, L37
Chiappini, C., Matteucci, F., & Romano, D. 2001, ApJ, 554, 1044
Chomiuk, L. & Povich, M.S. 2011, AJ, 142, 197
Collins, J.A., Shull, J.M., & Giroux, M.L. 2003, ApJ, 585, 336
Collins, J.A., Shull, J.M., & Giroux, M.L. 2005, ApJ, 623, 196
Collins, J.A., Shull, J.M., & Giroux, M.L. 2009, ApJ, 705, 962
Connors, T.W., Kawata, D., & Gibson, B.K. 2006, MNRAS, 371, 108
de Heij, V., Braun, R. & Burton, W.B. 2002, A&A, 391, 67
Desert, F.-X., Bazell, D. & Boulanger, F. 1988, ApJ, 334, 815
Desert, F.-X., Bazell, D. & Blitz, L. 1990, ApJ, 355, L51
Dettmar, R.-J. 1990, A&A, 232, L15
Diaz, J D., & Bekki, K. 2011, ApJ, 413, 2015
Diaz, J D., & Bekki, K. 2012, ApJ, 750, 36
Dieter, N.H. 1964, AJ, 79, 288
Di Teodoro, E.M. & Fraternali, F. 2014, A&A, 567, 68
D'Onghia, E. & Fox, A.J. 2016, ARA&A, 54, 363
Elson, E.C., de Blok, W.J.G. & Kraan-Korteweg, R.C. 2011, MNRAS, 411, 200
Fang, T., Marshall, H.L., Lee, J.C., Davis, D.S., & Canizares, C.R. 2002, ApJ, 572, L127
Fang, T., Sembach, K.R., & Canizares, C.R. 2003, ApJ, 586, L49
Fang, T., McKee C.F., Canizares, C.R., & Wolfire, M. 2006, ApJ, 644,174
Fang, T., Bullock, J. & Boylan-Kolchin, M. 2013, ApJ, 762, 20

Fernandez, X., Joung, M. R., & Putman, M. E. 2012, ApJ, 749, 181
Ferriere, K. 2001, RvMP, 73, 103
Ford, H.A., Lockman, F.J, & McClure-Griffiths, N.M. 2010, 722, 367
Fox, A.J., Savage, B.D., Wakker, B.P., et al. 2004, ApJ, 602, 738
Fox, A.J., Wakker, B.P., Savage, B.D., et al. 2005, ApJ, 630, 332
Fox, A.J., Wakker, B.P., Smoker, J.V., et al. 2010, ApJ, 718, 1046
Fox, A.J., Richter, P., Wakker, B.P., et al. 2013, ApJ, 772, 110
Fox, A.J., Wakker, B.P., Barger, A., et al. 2014, ApJ, 787, 147
Fox, A.J., Bordoloi, R., Savage, B.D., et al. 2015, ApJ, 799, L7
Fox, A.J., Lehner, N., Lockman, F.J. et al. 2016, ApJ, 816, L11
Fraternali, F. & Binney, J.J. 2008, MNRAS, 386, 935
Fraternali, F., Marasco, A., Marinacci, F., & Binney, J.J. 2013, ApJL, 764, L21
Fumagalli, M., Prochaska, J.X., Kasen, D., et al. 2011, MNRAS, 418, 1796
Gardiner, L.T. & Noguchi, M. 1996, MNRAS, 278, 191
Gaensler, B.M., Madsen, G.J., Chatterjee, S. & Mao, S.A. 2008, PASA, 25,184
Gibson, B.K., Giroux, M.L., Penton, S.V., et al. 2001, AJ, 122, 3280
Gillmon, K., Shull, J.M., Tumlinson, J., & Danforth, C. 2006, ApJ, 636, 891
Grcevich, J., & Putman, M. E. 2009, ApJ, 696, 385
Gupta, A., Mathur, S., Krongold, Y., Nicastro, F., & Galeazzi, M. 2012, ApJ, 756, L8
Haffner, L.M., Reynolds, R.J., Tufte, S.L., et al. 2003, ApJS, 149, 405
Haffner, L.M., Reynolds, R.J., Babler, B.L., et al. 2016, Proc. AAS, 227, id347
Hartmann, D., & Burton, W. B. 1997, *Atlas of Galactic Neutral Hydrogen*, (Cambridge: Cambridge Univ. Press)
Herenz, P., Richter, P., Charlton, J.C., & Masiero, J.R. 2013, A&A, 550, A87
Heitsch, F., & Putman, M. E. 2009, ApJ, 698, 1485
Hernandez, A.K., Wakker, B.P., Benjamin, R.A., et al. 2013, ApJ, 777, 19
Hill, A.S., Haffner, L.M., & Reynolds, R.J. 2009, ApJ, 703, 1832
Houck, J.C. & Bregman, J.N. 1990, ApJ, 352, 506
Hulsbosch, A.N.M. & Wakker, B.P. 1988, A&AS, 75, 191
Hulsbosch, A.N.M. & Raimond, R. 1966, BAN, 18, 413
Jin, S. & Lynden-Bell, D. 2008, MNRAS, 383, 1686
Joung, M.R., Bryan, G.L., & Putman, M.E. 2012a, ApJ, 745, 148
Joung, M.R., Putman, M.E., Bryan, G.L., Fernandez, X., & Peek, J.E.G. 2012b, ApJ, 759, 137
Kalberla, P.M.W., Burton, W.B., Hartmann, D., et al. 2005, A&A, 440, 775
Kalberla, P.M.W. & Kerp, J. 2009, ARA&A, 47, 27
Kaufmann, T., Mayer, L., Wadsley, J., Stadel, J., & Moore, B. 2006, MNRAS, 370, 1612
Kaufmann, T., Bullock, J.S., Maller, A.H., Fang, T., & Wadsley, J. 2009, MNRAS, 396, 191
Kereš, D., Katz, N., Weinberg, D.H,, & Dave, R. 2005, MNRAS, 363,2
Kereš, D., Katz, N., Fardal, M., Davé, R., & Weinberg, D.H. 2009, MNRAS, 395, 160
Kereš, D. & Hernquist, L. 2009, ApJL, 700, L1
Kerp, J., Burton, W.B., Egger, R., et al. 1999, A&A, 342,213
Kuntz, K.D. & Danly, L. 1996, ApJ, 457, 703
Kuntz, K.D. & Snowden, S.L. 2000, ApJ, 543, 195
Kwak, K., Henley, D.B. & Shelton, R.L. 2011, ApJ, 739,30
Larson, R.B. 1972, Nature, 236,21
Lehner N. 2002, ApJ, 578,126
Lehner, N., & Howk, J.C. 2011, Sci, 334, 955
Lehner, N., Howk, J.C., Thom, C., et al. 2012, MNRAS, 424, 2896
Levine, E.S., Blitz, L. & Heiles, C. 2006, ApJ, 643, 881
Lockman, F.J. 2002, ApJL, 580, L47
Lockman, F.J., Murphy, E.M., Petty-Powell, S., & Urick, V.J. 2002, ApJS, 140, 331
Lockman, F.J., Benjamin, R.A., Heroux, A.J., & Langston, G.I. 2008, ApJL, 679, L21
MacLow, M. & Klessen, R. 2004, RvMP, 76, 125
Maller, A.H. & Bullock, J,S. 2004, MNRAS, 355,694

Marasco, A., Marinacci, F., & Fraternali, F. 2013, MNRAS, 433, 1634
Marinacci, F., Binney, J.J., Fraternali, F., et al. 2010, MNRAS, 404, 1464
Mathewson, D.S. 1967, PASA, 1, 21
Mathewson, D.S., Cleary, M.N., & Murray, J.D. 1974, ApJ, 190, 291
Mathur, S., Weinberg, D. H., & Chen, X. 2003, ApJ, 582, 82
McClure-Griffiths, N.M., Staveley-Smith, L, Lockman, F.J., et al. 2008, ApJL, 673, L143
McClure-Griffiths, N.M., Pisano, D.J., Calabretta, M.R., et al. 2009, ApJS, 181, 398
McKernan, B., Yaqoob, T., & Reynolds, C.S. 2004, ApJ, 617, 232
Miller, M.J., & Bregman, J.N. 2013, ApJ, 770, 118
Miller, M.J., & Bregman, J.N. 2015, ApJ, 800, 14
Miller, M.J., Kluck, E.J. & Bregman, J.N. 2016, ApJ, 818, 112
Miville-Deschênes, M.A., Boulanger, F., Reach, W.T. & Noriega-Crespo, A. 2005, ApJ, 631, L57
Mo, H.J., Miralda-Escude, J. 1996, ApJ, 469, 589
Morras, R., Bajaja, E., Arnal, E.M. & Pöppel, W.G.L. 2000, A&AS, 142, 25
Morton, D.C. 2003, ApJS, 149, 205
Muller, C.A., Oort, J.H., & Raimond, E. 1963, C.R. Acad. Sci. Paris, 257, 1661
Münch, G. 1952, PASP, 64, 312
Münch, G. & Zirin, H. 1961, ApJ, 133, 11
Murphy, E.M., Lockman, F.J., & Savage, B.D. 1995, ApJ, 447, 642
Murray, C.E., Stanimirovic, S., McClure-Griffiths, N.M. et al. 2015, ApJ, 808, 41
Nichols, M. & Bland-Hawthorn, J. 2011, ApJ, 732, 17
Nidever, D.L., Majewski, S.R., Butler Burton, W., & Nigra, L. 2010, ApJ, 723, 1618
Nuza, S.E., Parisi, F., & Scannapieco, C., et al. 2014, MNRAS, 441, 2593
Oort, J.H. 1966, BAN, 18, 421
Peebles, P.J.E., Phelps, S.D., Shaya, E.J., & Tully, R.B. 2001, ApJ, 554, 104
Peebles, P.J.E., Tully, R.B., & Shaya, E.J. 2011, preprint (arXiv:1105.5596)
Peek, J.E.G., Putman, M.E., McKee, C.F., Heiles, C., & Stanimirovic, S. 2007, ApJ, 656,907
Peek, J.E.G., Putman, M.E., & Sommer-Larsen, J. 2008, ApJ, 674,227
Peek, J.E.G. 2009, ApJ, 698, 1429
Peek, J.E.G., Heiles, C., Putman, M.E., & Douglas, K. 2009, ApJ, 692, 827
Planck Collaboration.: Abergel, A., Ade, P.A.R., Aghanim, N. et al. 2011, A&A, 536, A24
Prata, S., & Wallerstein, G. 1967, PASP, 79, 202
Putman, M.E., Gibson, B.K., Staveley-Smith, L., et al. 1998, Nature, 394, 752
Putman, M.E., Staveley-Smith, L., Freeman, K.C., Gibson, B.K., & Barnes, D.G. 2003, ApJ, 586, 170
Putman, M.E., Peek, J.E.G., & Joung, M.R. 2012, ARA&A, 50, 491
Rasmussen, A., Kahn, S. M., & Paerels, F. 2003, in ASSL Conf. Proc. 281, *The IGM/Galaxy Connection. The Distribution of Baryons at* $z = 0$, ed. J.L. Rosenberg & M.E. Putman (Dordrecht: Kluwer), 109
Rana, N. 1991, ARA&A, 29, 129
Rees, M.J. & Ostriker, J.P. 1977, MNRAS, 179, 541
Reynolds, R.J. 1991, ApJ, 372, L17
Reynolds, R.J., Haffner, L.M., Madsen, G.J., Wood, K.,& Hill, A.S. 2012, EAS, 56, 213
Richter, P., de Boer, K.S., Bomans, D.J., et al. 1999, Nature, 402, 386
Richter, P., Sembach, K.R., Wakker, B.P., et al. 2001a, ApJ, 559, 318
Richter, P., Sembach, K.R., Wakker, B.P., & Savage, B.D. 2001c, ApJ, 562, L181
Richter, P., Savage, B.D., Wakker, B.P., Sembach, K.R., Kalberla, P.M.W. 2001b, ApJ, 549, 281
Richter, P., Sembach, K.R., & Howk, J.C. 2003a, A&A, 405, 1013
Richter, P., Wakker, B.P., Savage, B.D., & Sembach, K.R. 2003b, ApJ, 586, 230
Richter, P. 2006, RvMA 19, 31
Richter, P., Paerels, F.B.S., & Kaastra, J.S. 2008, SSRv, 134, 25
Richter, P., Charlton, J.C., Fangano, A.P.M., Ben Bekhti, N.,& Masiero, J.R. 2009, ApJ, 695, 1631
Richter, P., Krause, F., Fechner, C., Charlton, J.C., & Murphy, M.T. 2011, A&A, 528, A12
Richter, P. 2012, ApJ, 750, 165

Richter, P., Fox, A. J., Wakker, B. P., et al. 2013, ApJ, 772, 111
Richter, P., de Boer, K.S., Werner, K., & Rauch, T. 2015, A&A, 584, L6
Richter, P., Wakker, B.P., Fechner, C., et al. 2016, A&A, 590, A68
Richter, P., Ben Bekhti, N., Nuza, S.E., et al. 2017, A&A, submitted (R17)
Robitaille, T.P., & Whitney, B.A. 2010, ApJL, 710, L11
Röhser, T., Kerp, J., Lenz, D., & Winkel, B. 2016, A&A, 596, A94
Salem, M., Besla, G., Bryan, G., et al. 2015, ApJ, 815, 77
Saul, D.R., Peek, J.E.G., Grcevich, J., et al. 2012, ApJ, 758, 44
Savage, B.D. & Sembach, K.R. 1996, ARA&A, 34, 279
Savage, B.D., Sembach, K.R., Wakker, B.P., et al. 2003, ApJS, 146, 125
Savage, B.D. & Wakker, B.P. 2009, ApJ, 702, 1472
Sellwood, J.A. & Binney, J.J. 2002, MNRAS, 336,785
Sembach, K.R., Howk, J.C., Savage, B.D., & Shull, J.M. 2001, AJ, 121, 992
Sembach, K.R., Wakker, B.P., Savage, B.D., Richter, P. et al. 2003, ApJS, 146, 165
Sembach, K.R., Wakker, B.P., Tripp, T.M. 2004, ApJS, 150, 387
Shapiro, P.R & Field, G.B. 1976, ApJ, 205, 762
Shen, S., Madau, P., Guedes, J., et al. 2013, ApJ, 765, 89
Shull, J.M., Jones, J.R., Danforth, C.W., & Collins, J.A. 2009, 699, 754
Smith, G.P. 1963, BAN, 17, 203
Smoker, J.V., Fox, A.J., & Keenan, F.P. 2011, MNRAS, 415, 1105
Sommer-Larsen, J. 2006, ApJL, 644, L1
Spitzer, L. 1956, ApJ, 124, 40
Springel, V. & Hernquist, L. 2005, ApJ, 622, L9
Stanimirović, S., Dickey, J.M., Krco, M., & Brooks, A.M. 2002, ApJ, 576, 773
Stanimirović, S., Hoffman, S., Heiles, C., et al. 2008, ApJ, 680, 276
Su, M., Slatyer, T.R., & Finkbeiner, D.P. 2010, ApJ, 724, 1044
Tepper-García, T., Bland-Hawthorn, J., & Sutherland, R.S. 2015, ApJ, 813, 94
Thom, C., Putman, M. E., Gibson, B. K., et al. 2006, ApJ, 638, L97
Thom, C., Peek, J. E. G., Putman, M. E., et al. 2008, ApJ, 684, 364
Tripp, T.M., Wakker, B.P., Jenkins, E.B., et al. 2003, AJ, 125, 3122
Tripp, T. M., & Song, L. 2012, ApJ, 746, 173
Tufte, S.L., Reynolds, R.J., & Haffner, L.M. 1998, ApJ, 504, 773
Tufte, S.L., Wilson, J.D., Madsen, G.J., Haffner, L.M., & Reynolds, R.J. 2002, ApJL, 572, L153
van den Bergh, S. 1962, AJ, 67, 486
van de Voort, F., Schaye, J., Booth, C.M., Haas, M.R., & Dalla Vecchia, C. 2011, MNRAS, 414, 2458
van de Voort, F., & Schaye, J. 2012, MNRAS, 423, 2991
van Woerden, H., Schwarz, U.J., Peletier, R.F., Wakker, B.P., & Kalberla, P.M.W. 1999, Nature, 400, 138
van Woerden, H., Wakker, B.P., Schwarz, U.J., & de Boer, K.S. 2004, *High Velocity Clouds*, (Dordrecht: Kluwer)
Vieser, W. & Hensler, G. 2007, A&A, 475, 251
Vogelsberger, M., Sijacki, D., Keres, D., Springel, V., & Hernquist, L. 2012, MNRAS, 425, 3024
Wakker, B.P., & Boulanger, F. 1986, A&A, 170, 84
Wakker, B.P. & van Woerden, H. 1991, A&A, 250, 509
Wakker, B.P. & van Woerden, H. 1997, ARA&A, 35, 217
Wakker, B.P., Howk, J.C., Savage, B.D. 1999, Nature, 402, 388
Wakker, B.P. 2001, ApJS, 136, 463
Wakker, B.P., Savage, B.D., Sembach, K.R., et al. 2003, ApJS, 146, 1
Wakker, B.P. 2004, in *High Velocity Clouds*, ed. H. van Woerden, B.P.Wakker, U.J. Schwarz, & K.S. de Boer (Dordrecht: Kluwer), 25
Wakker, B.P. 2006, ApJS, 163, 282
Wakker, B.P., York, D.G., Howk, J.C., et al. 2007, ApJ, 670, L113
Wakker, B.P., York, D.G., Wilhelm, R., et al. 2008, ApJ, 672, 298

Wang Q.D., Yao, Y., Tripp, T.M., et al. 2005, ApJ, 635, 386
Weiner, B., & Williams, T. 1996, AJ, 111, 1156
Weiss, A., Heithausen, A., Herbstmeier, U., & Mebold, U. 1999, A&A, 344, 955
Westmeier, T., Braun, R., Brüns, C., Kerp, J., & Thilker, D.A. 2007, New. Astron. Rev., 51, 108
Westmeier, T. & Koribalski, B.S. 2008, MNRAS, 388, L29
White, S.D.M. & Frenk, C.S. 1991, ApJ, 379, 52
White, S.D.M. & Rees, M.J. 1978, MNRAS, 183, 341
Whiting, A.B. 2014, ApJ, 793, 63
Widmann, H., de Boer, K.S., Richter, P., et al. 1998, A&A, 338, L1
Williams, R.J., Mathur, S., Nicastro, F., et al. 2005, ApJ, 631, 856
Williams, R.J., Mathur, S., Poindexter, S., Elvis, M., & Nicastro, F., 2012, AJ, 143, 82
Winkel, B., Kalberla, P.M.W., Kerp, J., Flöer, L. 2010, ApJS, 188, 488
Winkel, B., Ben Bekhti, N., Darmstädter, V, et al. 2011, A&A, 533, A105
Yao, Y., Nowak, M.A., Wang, Q.D., Schulz, N.S., & Canizares, C.R. 2008, ApJ, 672, L21
Zech, W., Lehner, N., Howk, J.C., Dixon, W., & Brown, T. 2008, ApJ, 679, 460
Zheng, Y., Putman, M.E., Peek, J.E.G., & Joung, M.R. 2015, ApJ, 807, 103
Zwaan, M., van der Hulst, J.M., Briggs, F.H., Verheijen, M.A., & Ryan-Weber, E.V. 2005, MNRAS, 364, 1467

Neutral Gas Accretion onto Nearby Galaxies

Felix J. Lockman

1 Galaxies Then and Now

> The evidence at present available points strongly to the conclusion that the spirals are individual galaxies, or island universes, comparable with our own galaxy in dimensions and in number of component units. – H.D. Curtis

> …the extraplanar gas seems to consist of two parts: a large one from galactic fountains and a smaller part accreted from intergalactic space. There is direct (HVCs in our galaxy and filaments in external galaxies) and indirect (rotational velocity gradients) evidence for the accretion from outside. — Sancisi et al. (2008)

> Galaxies are like people. Every time you get to know one well, it turns out to be a little peculiar. – Sidney van den Bergh

Unlike the days of the *Island Universe*, when galaxies floated in solitary splendor on Hubble's Tuning Fork (Hubble 1958), today's galaxies are a mess (Fig. 1). Evidence for the growth and evolution of galaxies by the capture of stellar systems is everywhere and there are arguments for continued accretion of gas as well. Certainly some gas will arrive with the small stellar systems that large galaxies devour, but how does it get to the disk, and is there neutral gas accreting from other sources? In this article I consider the evidence for accretion of neutral gas onto nearby galaxies, especially gas that is not associated with stars. This is not a comprehensive review, but instead explores the connection between gas likely to be accreting onto galaxies in the Local Group, where we can examine it with high sensitivity and

The Green Bank Observatory is a facility of the National Science Foundation, operated under a cooperative agreement by Associated Universities, Inc.

F.J. Lockman (✉)
Green Bank Observatory, 155 Observatory Rd., P.O. Box 2, Green Bank, WV, USA
e-mail: jlockman@nrao.edu

A. Fox, R. Davé (eds.), *Gas Accretion onto Galaxies*, Astrophysics and Space Science Library 430, DOI 10.1007/978-3-319-52512-9_3

Fig. 1 Examples of stellar accretion in two galaxies from Carlin et al. (2016). Images on the left from the Sloan Digital Sky Survey of the nearby galaxies NGC 4013 (*top*) and M 63 (*bottom*) show regular disks while the much deeper images on the right reveal streams of stars accreted from smaller galaxies

linear resolution, and that seen in more distant systems with a better vantage but considerably less sensitivity. Is there local accretion of starless gas, and what form does it take?

Two comprehensive reviews are directly relevant to this topic. Sancisi et al. (2008) consider the evidence for accretion of neutral gas by analyzing high-resolution H I maps of several dozen nearby galaxies. They find ample evidence of kinematic anomalies, tails and filaments, warps, lopsided disks, and interaction. They adopt the stance that any significant deviation of a galaxy's H I from symmetry is evidence for interaction, and that interaction implies accretion. Accretion also has a prominent place in the recent review of gaseous galaxy halos by Putman et al. (2012) which takes a thorough look at circumgalactic gas in all phases. I will refer to these reviews throughout this article for more complete discussion of some topics.

The volume around galactic disks contains material in many forms. Both the Milky Way (MW) and M31 have a hot 10^{6-7} K circumgalactic medium (CGM) of enormous mass and extent, possibly dwarfing the baryon content of the disks themselves (Anderson and Bregman 2011; Hodges-Kluck et al. 2016; Lehner et al. 2015). This gas may cool and condense, feeding the disk. There may also be cosmological accretion—cold flows—which may have to traverse the CGM to reach the disk and in that passage may be disrupted entirely (Putman et al. 2012). Then there is gas that has been stripped from one galaxy through interaction and ultimately ends up in the disk of another. Add to that the ejection of gas from a galaxy's disk from supernovae or a nuclear wind, and the processes can become quite difficult to disentangle. Considerable insight into the CGM of galaxies has come through studies of Lyα and MgII absorption lines (Kacprzak et al. 2013; Wakker and Savage 2009), but those data will not be discussed here. Instead, we will concentrate on 21 cm H I observations, and work from the inside out, from the interstellar medium (ISM) disk-halo interface, to high velocity clouds (HVCs), to the products of interaction between galaxies, and finally to H I "clouds" that don't easily fit into any of these categories.

But first a note on the detectability of H I. Filled-aperture (single dish) radio telescopes can easily detect H I column densities of $N_{\rm HI} = 10^{18}\,{\rm cm}^{-2}$, and with a little work, $10^{17}\,{\rm cm}^{-2}$. This comes at the expense of angular resolution, which for modern instruments is in the range $3'$–$10'$. Hydrogen clouds with $M_{\rm HI} \geq 10^5$ $M_{\odot}$ can thus be detected anywhere in the Local Group at a linear resolution of 1–3 pc $d_{\rm kpc}$. Aperture synthesis instruments provide much higher angular resolution of $<1'$ but at the cost of reduced sensitivity. The smallest object in the table of plumes, wings, and other peculiar H I structures of Sancisi et al. (2008) has $10^8\,M_{\odot}$. The deepest aperture synthesis H I observations of galaxies so far have come from the HALOGAS survey Heald et al. (2011) which reaches a limiting sensitivity of $N_{\rm HI} = 10^{18.5\text{–}19.0}\,{\rm cm}^{-2}$.

This sensitivity gap is not unbridgeable, but we should keep it in mind when comparing local with distant objects.

2 The Disk-Halo Interface: Clouds and Shells

The Milky Way gives us a close view of process like the galactic fountain, which lifts gas several kpc away from a spiral galaxy's disk (Shapiro and Field 1976; Bregman 1980). Figure 2 shows Milky Way H I along a cut $\sim 20^{\circ}$ in longitude through the Galactic plane approximately along the tangent points. It is easy to find H I loops, filaments, clouds, and a diffuse component extending many degrees away from the plane. In this figure each degree of latitude corresponds to a displacement ≈ 135 pc from the midplane, so the H I emission at $b = \pm 4^{\circ}$ arises in neutral gas at a height $z \sim 0.5$ kpc, several H I scale heights above the main H I layer (Dickey and Lockman 1990). In the inner Milky Way about 10% of the H I can be found more than 0.5 kpc from the disk (Lockman 1984). Some of this H I is contained in discrete clouds that can be identified to $z \approx 2$ kpc (Lockman 2002; Ford et al. 2010). Some of the extraplanar H I is also organized as large "supershells," whose tops can reach $|z| > 3$ kpc. They can be ~ 1 kpc in size and contain $3 \times 10^4\,M_{\odot}$ of neutral gas (Heiles 1979; McClure-Griffiths et al. 2002; Pidopryhora et al. 2007).

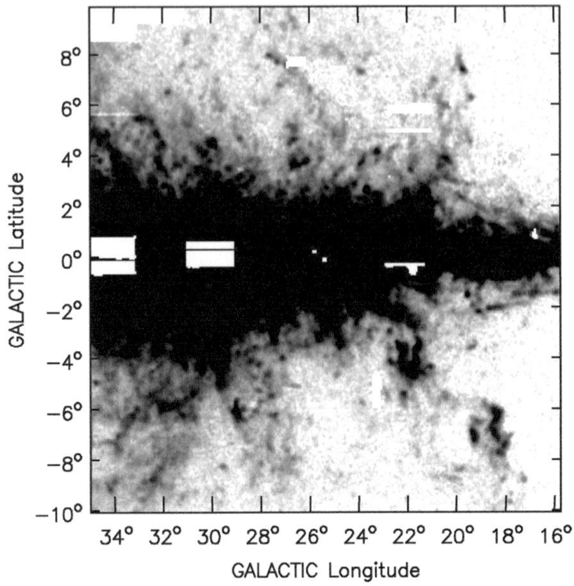

Fig. 2 Cut through the Galactic plane at velocities approximately along the tangent point showing the vertical extent of Galactic HI. In this projection a latitude of 8° corresponds to a distance ~1 kpc from the Galactic plane. These observations can resolve structures down to a few 10s of pc in size

Supershells certainly have their origin in supernovae and stellar winds (Tomisaka and Ikeuchi 1986), and can often be linked directly to sites of star formation (Pidopryhora et al. 2007; Kaltcheva et al. 2014). The cloud-like component of extraplanar H I has a less certain origin, though it may result from the breakup of shells. Often referred to as "disk-halo" clouds, they appear to be related both spatially and kinematically to the disk, though the population extends well into the lower halo (Lockman 2002; Ford et al. 2010). Some of the larger Milky Way disk-halo clouds with $M_{HI} \approx 500\,M_{\odot}$ have been imaged at 2–3 pc resolution and show a variety of shapes with some sharp boundaries and evidence of a two-component thermodynamic structure (Pidopryhora et al. 2015). Note that from the above discussion on sensitivity, these disk-halo clouds could not be detected beyond the Milky Way as individual objects. Within the Milky Way there is a general correlation between the amount and vertical extent of disk-halo clouds and the spiral arms (Ford et al. 2010) although there is no detailed correlation with individual star-forming regions.

A similar extraplanar H I layer is seen in a number of galaxies, often containing 10% of the disk H I mass, though the percentage has large variations from galaxy to galaxy (Sancisi et al. 2008). While it is not possible to isolate individual disk-halo clouds in other galaxies, the better vantage afforded in observations of extragalactic systems allows the kinematics of this component to be analyzed more completely.

The disk-halo (or extraplanar) gas often shows evidence of a vertical lag in rotational velocity of 10–20 $km\,s^{-1}\,kpc^{-1}$ but with a large range; sometimes an inflow toward the center of a galaxy is also inferred at the level of 10–20 $km\,s^{-1}$ (Fraternali et al. 2002; Sancisi et al. 2008; Zschaechner et al. 2015). This extraplanar gas does not show large deviations from prograde galactic rotation, and constitutes a dynamic neutral galactic atmosphere.

The disk-halo gas extends to heights where it may mix with material accreting through or cooling from the CGM (Putman et al. 2012). Thus the presence of neutral gas many scale heights away from the plane is not unexpected, and distinguishing between gas that is "recycled" as much of the disk-halo material must be, and gas that has never been in the disk, may not be straightforward in the absence of other information. Although the disk-halo clouds are concentrated to the disk in the sense that their numbers increase toward the disk, there is no way to know if any individual cloud had its origin in the Milky Way, or has been accreted, or is a combination of both processes (Marasco et al. 2012). It would be very interesting to have information on the elemental abundances in disk-halo clouds at different heights.

Any gas accreting onto the Milky Way has to pass through the extended disk-halo layer before it reaches the inner disk. One massive cloud passing through this layer shows evidence of disruption, and is discussed in Sect. 3.2.

3 High Velocity Clouds

It has been known for more than 50 years that there are significant amounts of neutral hydrogen around the Milky Way that do not follow Galactic rotation—the high velocity clouds (HVCs) (Muller et al. 1963; Wakker et al. 2004). For many years the lack of information on their distances has allowed speculation that some of them might be ∼1 Mpc or more away from the Milky Way and thus have $M_{HI} = 10^{7-8}\,M_{\odot}$ (Blitz et al. 1999; Braun and Burton 1999). However, the discovery of a similar population around M31, M33, and other galaxies (Thilker et al. 2004; Westmeier et al. 2008; Grossi et al. 2008; Putman et al. 2009; Keenan et al. 2016; Miller et al. 2009) suggests that the phenomenon is likely common and, as importantly, confined to the inner CGM of spirals. This does not mean that some of the more compact, more isolated HVCs might not be free-floating independent systems (Adams et al. 2013), but the total H I mass within the CGM of normal galaxies (aside from the products of tidal interaction) must be relatively small (Pisano et al. 2007, 2011). The HVC system of the Milky Way has been estimated to contain an H I mass of $\sim 3 \times 10^{7}\,M_{\odot}$ (Putman et al. 2012). The HVC population likely contains as much ionized gas as neutral gas (Shull et al. 2009, 2011; Lehner and Howk 2011).

One of the most important facts about HVCs is that they do not contain stars (Hopp et al. 2007). In this sense, compact HVCs that are discovered to be dwarf galaxies (e.g., Leo P; Rhode et al. 2013) are simply objects that have been misclassified as HVCs. Another important property is that their velocities are not really that high: none have velocities near or greater than the escape velocity of the Milky Way (Wakker et al. 2004). Putman et al. (2012) estimate that the larger Milky Way HVCs have kinematics that can be fit by a prograde rotational component about one-third the circular velocity of the Sun, and an inflow component of a few tens of km s^{-1}, though one prominent cloud almost certainly has a large retrograde motion (Lockman 2003).

3.1 High Velocity Clouds in M31 and M33

The HVC population of M31 is shown in Fig. 3. The typical M31 HVC has M_{HI} of a few $10^5\,M_{\odot}$, and would not be detectable as an individual object in most measurements of more distant galaxies. The right panel of Fig. 3 emphasizes the proximity of HVCs to that galaxy's disk, and the difficulty one would have in disentangling them from the disk-halo layer in more distant systems. The HVCs of M31 seem to be confined to within 50 kpc of that galaxy, though there are indications that some may be found to greater distances to the north (Westmeier et al. 2008; Wolfe et al. 2016). The total H I mass in the M31 HVCs is $2 \times 10^7\,M_{\odot}$; when extraplanar gas possibly associated with the disk is included, the amount may be as high as $5 \times 10^7\,M_{\odot}$ (Westmeier et al. 2008).

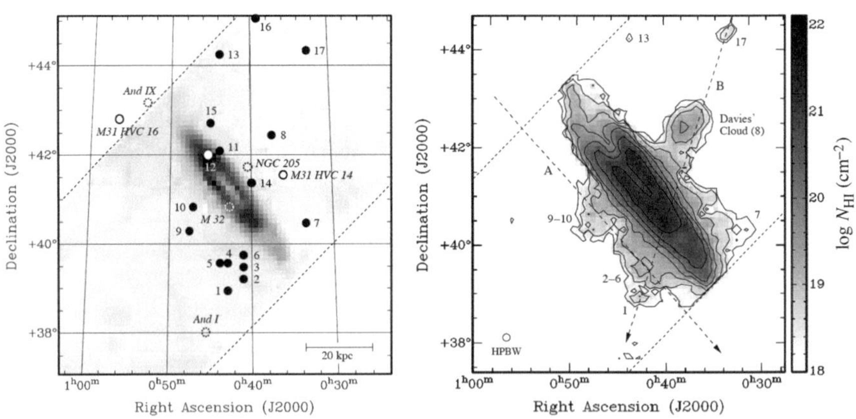

Fig. 3 From Westmeier et al. (2008). *Left panel*: the HVC system of M31 in relationship to the bright H I in the disk. *Right panel*: integrated H I over all velocities with contours at $\log_{10}(N_{HI})$ 18, 18.5, 19, 19.5, etc. Many of the HVCs seem to blend into the disk, but in maps at different velocities it is clear that they are distinct objects. This is how M31 would appear to current instruments were it ~10 Mpc distant, except that the two lowest contours would be missing. As noted by Putman et al. (2012) this image of M31 bears a strong resemblance to the H I image of the galaxy NGC 891 (Oosterloo et al. 2007) which has been observed with about the similar linear resolution as M31

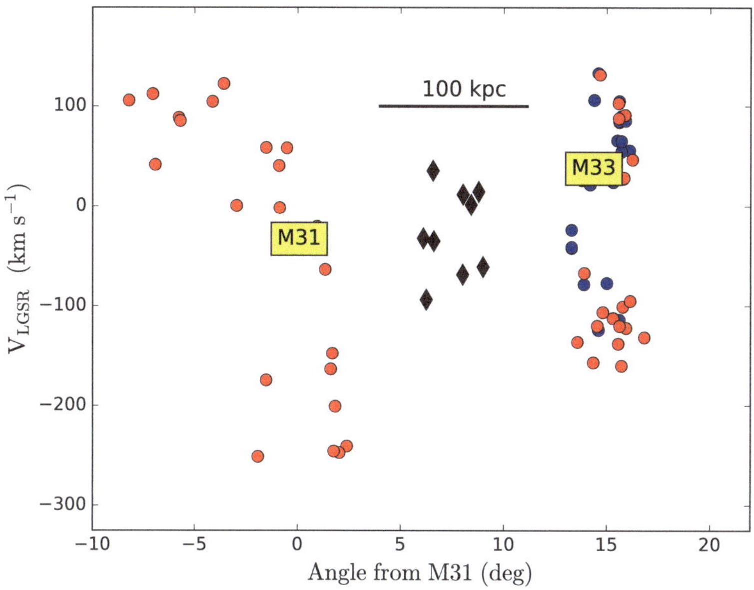

Fig. 4 The HVC systems of M31 and M33 (*red circles*), where the velocity with respect to the Local Group Standard of rest, V_{LGSR}, is plotted against the angular distance from M31 in the direction of M33. To the samples of M31 and M33 HVCs (Westmeier et al. 2008; Keenan et al. 2016) are added clouds that may be in the halo of M33 (*blue circles*) (Grossi et al. 2008; Putman et al. 2009). It can be difficult to distinguish between the two groups, which may actually blend together. The systemic velocities of M31 and M33 are indicated. *Diamonds mark* the M31–M33 clouds discussed in Sect. 5.2. They have velocities more in common with the systemic velocity of M31 and M33 than of their system of HVCs. There are no stars in any of these clouds

There are also HVCs around M33, though here confusion with that galaxy's disk and possibly unrelated extraplanar gas makes it difficult to separate the populations cleanly (Grossi et al. 2008; Putman et al. 2009; Keenan et al. 2016). The total H I mass in the M33 HVCs is $3.5 \times 10^7\,M_\odot$ using only the data from the most recent study (Keenan et al. 2016). If we include clouds that may be located in the disk-halo region (Grossi et al. 2008; Putman et al. 2009) the total M33 HVC H I mass increases to $\sim 5 \times 10^7\,M_\odot$.

Figure 4 shows the velocity and location of the HVCs of M31 and M33 with respect to M31. At the distance of these galaxies 1° on the sky corresponds to about 15 kpc. This figure includes some objects in M33 (marked in blue) that are probably part of the disk-halo interface and not true HVCs (Putman et al. 2009; Keenan et al. 2016), but are included to show that aspects of these two populations can blend and they often occupy very similar areas of position and velocity space. This figure shows another interesting aspect of HVCs. The velocity spread of those associated with M33 is considerably smaller than those associated with M31. If HVCs are condensations from the CGM this would follow naturally, as M33 is less massive than M31 and so presumably has a CGM that rotates more slowly.

A key fact that has been established through distance measurements to Milky Way HVCs and studies of M31 and M33 is that HVCs are not free-floating in the Local Group, but are concentrated around the large spirals. If the HVCs are even in approximate pressure equilibrium with a galaxy's CGM they must be orders of magnitude more dense than the hot gas around them and thus falling toward the disk. There is some evidence for the interaction of Galactic HVCs with the CGM in the form of distortions of the cloud shapes (Brüns et al. 2000), and there is rather spectacular evidence for the direct accretion of H I onto the Milky Way from one HVC: The Smith Cloud.

3.2 The Smith Cloud: Accretion in Action

In the same year that the discovery of HVCs was announced, a short paper appeared reporting observations of a peculiar H I feature that now appears to be a very important object, the Smith Cloud (Smith 1963). Figure 5 shows an H I channel map from a recent 21 cm survey made with the Green Bank Telescope (GBT). The cloud has a good distance estimate (Wakker et al. 2008), and thus a well-defined mass and size; it is moving toward the Galactic plane, which it should intersect in

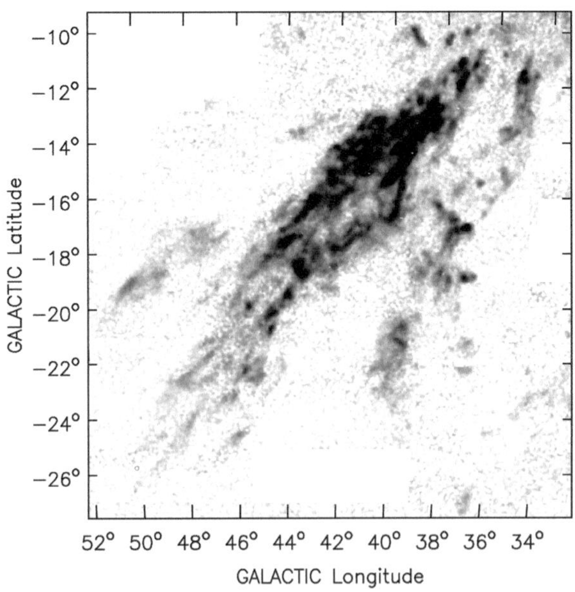

Fig. 5 Channel map of the Smith Cloud from new GBT H I data. All the emission in this figure is associated with the Cloud, which has other components extending down to latitude −40°. The part of the Cloud displayed in this image is about 4 kpc in length. The Galactic plane lies above this figure, and the Cloud is moving toward it at an angle

~30 Myr if it survives as a coherent entity (Lockman et al. 2008). Its mass in H I is $\sim 2 \times 10^6\, M_\odot$ and it has an ionized component with a similar mass detected in faint H_α emission (Hill et al. 2009). There is evidence that it has a magnetic field $\approx 8\,\mu G$ (Hill et al. 2013) and an S/H abundance that is about one-half Solar (Fox et al. 2016). Its total space velocity is below the escape velocity of the Milky Way, and the largest component is in the direction of Galactic rotation. It appears to be entering the Milky Way at a rather shallow angle (Lockman et al. 2008; Nichols and Bland-Hawthorn 2009). The Smith Cloud is thus adding angular momentum to the disk. It has no detectable stars (Stark et al. 2015).

The brightest, most compact component of the Cloud lies about 3 kpc below the Galactic plane. It is thus in the disk-halo transition, and it appears that it is encountering a clumpy medium. The Cloud has holes that are matched by small fragments at $60\,\mathrm{km\,s^{-1}}$ lower velocity, consistent with the velocity of the Milky Way's halo at that location (Fig. 6). This object shows clear signs of interaction with the Milky Way's extraplanar gas.

There are many puzzles surrounding the Smith Cloud. If it has condensed from the CGM than why does it have an S/H metallicity ratio higher than typical HVCs (Fox et al. 2016; Wakker et al. 2004)? If it originated from the Galaxy, then how did it acquire such a large peculiar motion and mass, which implies a kinetic energy

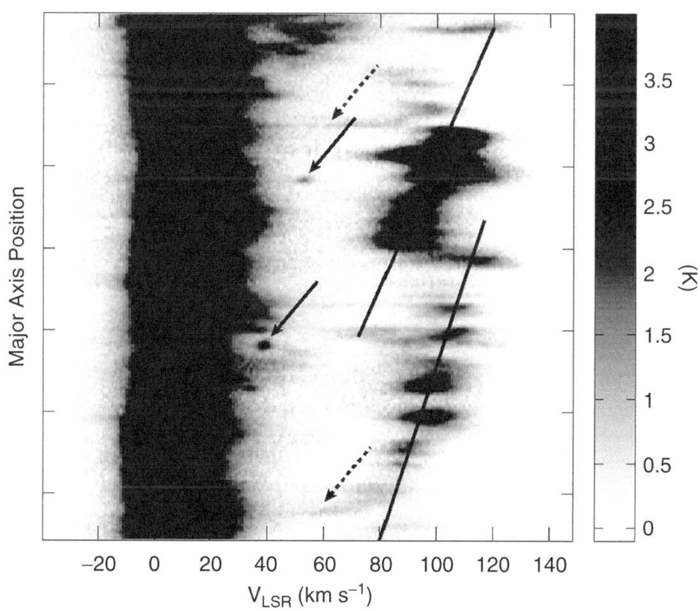

Fig. 6 A velocity-position plot down the major axis of the Smith Cloud from Lockman et al. (2008). *Arrows mark* lumps or streams of H I that have been decelerated by interaction with the Milky Way halo. Besides a general stripping and deceleration of the cloud edges (H I indicated by *dashed arrows*), it appears that the Cloud is encountering dense lumps in the Milky Way disk-halo interface that remove chunks of the Cloud (noted with *solid arrows*). Some of these correspond in detail to voids in the body of the Smith Cloud.

$\sim 5 \times 10^{53}$ ergs (Marasco and Fraternali 2017)? Does it require a significant dark matter component to maintain its stability, as suggested by some investigations (Nichols and Bland-Hawthorn 2009; Nichols et al. 2014)?

If the Smith Cloud were at the distance of M31 it would appear to the HVC surveys with a peak $N_{\rm HI}$ around $2 \times 10^{19}\,\rm cm^{-2}$, a few times higher than the M31 HVCs, but not especially anomalous. With an $M_{\rm HI}$ of $2 \times 10^6\,M_\odot$, the Smith Cloud has about the median H I mass of the M33 HVCs and lies at the upper range of those around M31. Given the observational uncertainties, the Smith Cloud would be an inconspicuous addition to either galaxy (Lockman et al. 2008; Westmeier et al. 2008; Keenan et al. 2016). However, as the brighter parts of the Smith Cloud lie only $\sim$3 kpc away from the Galactic plane, if it were at the distance of M31 or M33 it would lie projected on the disk of those galaxies, and from even the most favorable vantage would be separated by only $12'$ from their disk. It would almost certainly be considered part of the disk-halo interface and not cataloged as an HVC.

Some HVCs in M31 and M33 have linewidths $>50\,\rm km\,s^{-1}$, suggesting large internal motions or disruption (Westmeier et al. 2008; Keenan et al. 2016). If the Smith Cloud were at the distance of M31, our poorer linear resolution would blend its different components, raising its H I linewidth from the $<20\,\rm km\,s^{-1}$ that we typically measure in its brighter parts, to $40\,\rm km\,s^{-1}$.

4 H I Outside Local Group Galaxies

The Local Group contains two wonderful examples of H I without stars, both associated with the Magellanic Clouds: the Magellanic Stream (MS) and the Leading Arm (LA). These are the rather enormous streams of gas (containing about 10% of the mass of the Milky Way's ISM) that are currently being lost from the Magellanic Clouds and are entering the Milky Way's CGM. A distance of 55 kpc is assumed for both objects, though this has a large uncertainty; for a recent review, see D'Onghia and Fox (2016). The relevant H I masses are $3 \times 10^8\,M_\odot$ and $3 \times 10^7\,M_\odot$, for the Stream and Leading Arm, respectively (Brüns et al. 2005). The total mass of the Stream is $\sim 10^9\,M_\odot$, and is dominated by ionized gas (Fox et al. 2014). The Magellanic Clouds are the only stellar systems in the Local Group that are observed to be losing their gas. The Magellanic Stream is important for our purposes because it gives us a model for how we might interpret observations of other systems. Although the combined MS + LA is 200° long (Putman et al. 2003; Nidever et al. 2010) I estimate that no more than half this length would appear with $N_{\rm HI} > 10^{19}\,\rm cm^{-2}$ at a few Mpc distance, and thus be detectable in 21 cm H I emission in another galaxy group. The detectable part of the MS would have a length $\sim$100 kpc and for the Leading Arm, about 60 kpc.

Although we can see that the Stream and the Leading Arm are anchored in the stellar systems of the Magellanic Clouds, there are no stars associated with the gas of either object (D'Onghia and Fox 2016). The absence of stars in the gas mirrors the fact that none of the stellar streams around the Milky Way or M31 have been found to have a convincing association with any neutral gas (Lewis et al. 2013).

The Milky Way's CGM seems quite resistant to the passage of neutral gas clouds even when they are attached to galaxies. Dwarf spheroidals within a few 100 kpc of the Milky Way seem utterly devoid of H I: the limits on some systems with $L_* \approx 2.5 \times 10^5\ L_\odot$ are smaller than $100\,M_\odot$ (Spekkens et al. 2014). A similar deficit is found for the dwarf satellites of M31 (Beaton, private communication). Because dwarf galaxies >400 kpc from the Milky Way seem to have retained their gas, the deficit for nearer dwarfs presumably reflects their stripping by the CGM (Blitz and Robishaw 2000; Grcevich and Putman 2009). Distant dwarfs have $M_{HI}/L_* \approx 1$ (Spekkens et al. 2014), so we can estimate the H I mass that has been deposited in the CGM through this stripping. For the dwarf galaxies currently known around the Milky Way, it amounts to $M_{HI} \approx 3 \times 10^7\,M_\odot$, an value almost identical to the estimated total H I mass of HVCs in each of the HVC systems of the Milky Way, M31, and M33 (Putman et al. 2012; Westmeier et al. 2008; Keenan et al. 2016). It is tempting to dismiss this as a coincidence, especially because the inferred mass of the "stripped" H I for the Milky Way is dominated by the most massive galaxy, the Sagittarius dSph (Spekkens et al. 2014). But we do not have a good understanding of the inflow rate and location of the stripped material, which likely eventually makes its way to the disk.

5 Other Neutral Gas in the Local Group

5.1 IC 10

This blue compact dwarf galaxy, a satellite of M31, is undergoing a burst of star formation and has a complex H I structure with counter-rotating components and two tails or streamers shown in Fig. 7 (Shostak and Skillman 1989; Nidever et al. 2013; Ashley et al. 2014). Kinematic analysis of the H I suggests that the gas is inflowing rather than outflowing. Its stellar component, however, does not show evidence of streams or shells, or any kind of disturbance, though the young stellar populations are spatially offset from the older stellar population (Gerbrandt et al. 2015). The data thus suggest that IC 10 is accreting nearly starless gas and not a companion. The mass in the southern and northern H I features is $10^7\,M_\odot$ and $6 \times 10^5\,M_\odot$, respectively, out of a total M_{HI} for the entire system of $8 \times 10^7\,M_\odot$. The disturbed material is a significant fraction of the gas mass. It is interesting that the two anomalous H I features of IC 10 lie approximately on a line pointing back toward M31.

With a stellar mass of $8.6 \times 10^7\,M_\odot$ (McConnachie 2012), IC 10 has an $M_{HI}/L_V \approx 1$, similar to gas-rich Local Group dwarfs, even though it is only 250 kpc from M31 and thus is not only a satellite of M31, but lies well within the CGM of that galaxy. IC 10 and the Magellanic Clouds appear to be the only gas-rich dwarf galaxies within a few hundred kpc of the Milky Way or M31 in the Local Group, and the Magellanic Clouds are losing their gas while IC 10 seems to be accreting.

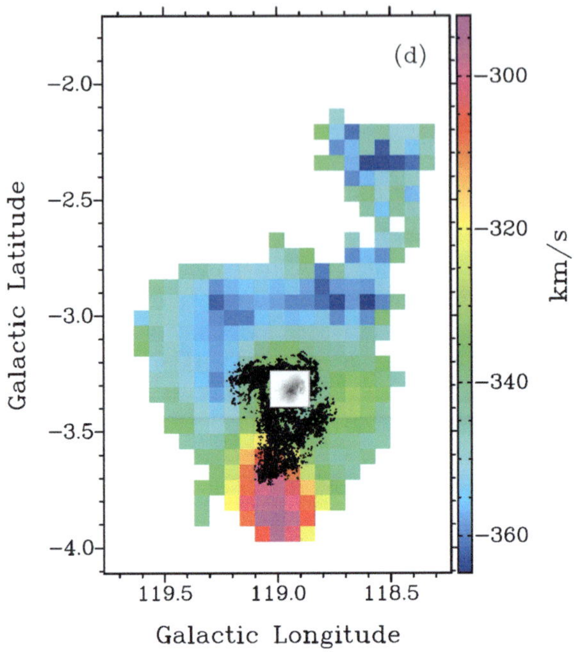

Fig. 7 Composite image of the stars and H I of IC 10 adapted from Ashley et al. (2014). Colors show the H I velocity field, highlighting the plume to the north. *Black* shows $N_{\rm HI}$ in the inner regions from a high-resolution map, showing the gas extension to the south. The *insert* shows the stars of IC 10 in the optical V-band

5.2 M31–M33 Clouds

In a map that reached a sensitivity to $N_{\rm HI}$ of a few $10^{17}\,{\rm cm}^{-2}$, Braun and Thilker (2004) detected very faint H I emission in a long plume extending from M31 in the general direction of M33. Subsequent observations with the GBT at higher angular resolution showed that most of the emission was concentrated into discrete clouds (Wolfe et al. 2013, 2016). Assuming that these clouds are at a distance of 800 kpc, between M31 and M33, they have $M_{\rm HI} = 0.4 - 3.3 \times 10^5\,M_\odot$, sizes of a few kpc, and are not associated with any stars or stellar stream. Figure 8 shows the clouds colored by their velocities. The field lies ∼100 kpc from M31. There are substantial differences in the velocities from cloud to cloud, $>100\,{\rm km\,s}^{-1}$, indicating that these are discrete objects and not simply the brighter portions of a single extended sheet. Velocity gradients across each cloud, in contrast, are only $\approx 10\,{\rm km\,s}^{-1}$. The dynamical mass of the clouds, i.e., the mass needed for them to be self-gravitating, is typically 10^3 times more than the observed mass in H I. There are indications in the data that more clouds like these exist adjacent to the area covered by the GBT observations.

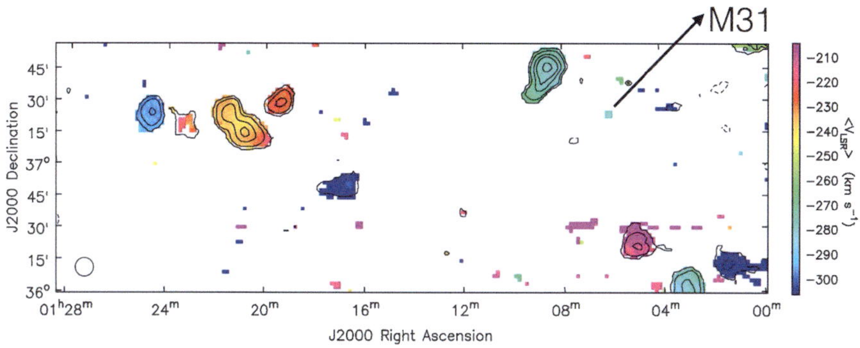

Fig. 8 Discrete H I clouds found in a region about 100 kpc away from M31 in the direction of M33 (adapted from Wolfe et al. 2016). These are also displayed as the *diamonds* in Fig. 4. Contours mark $N_{\rm HI}$ in units of $5 \times 10^{17}\,{\rm cm}^{-2}$, scaled by −1 (*dashed*), 1, 2, 4, 6, 10. The typical cloud has $M_{\rm HI} = 10^5\,M_\odot$. M31 lies to the upper right in the direction of the *arrow*; M33 to the lower left

Figure 4 shows the relationship between the M31–M33 clouds and the galaxies M31 and M33, together with their systems of HVCs. It is apparent that the M31–M33 clouds lack the large velocity spread of the HVCs, and that their kinematics has more in common with the systemic velocities of the galaxies than with their HVCs. There is no apparent connection between these clouds and the M31 system of satellite galaxies (Wolfe et al. 2016).

The M31–M33 clouds appear to be a new population in the Local Group with no known analogs. Beyond the work of Braun and Thilker (2004) and the data shown in Fig. 8, no other areas in the Local Group have been surveyed to the sensitivity necessary to detect H I emission at these faint levels. There is some indication that similar objects may exist to the north of M31 (Wolfe et al. 2016) but current data are inconclusive. Braun and Thilker (2004) initially proposed that this H I condensed from an intergalactic filament and there are models where it arises as the result of an interaction between M31 and M33 (Bekki 2008; Lewis et al. 2013). The later suggestion, while attractive, now seems unlikely given the past history of the Local Group (Shaya and Tully 2013). We note that the M31–M33 clouds, like IC 10, reside well within the CGM of M31. It is estimated that the CGM has a total column density $N_{\rm H} \approx 10^{18}\,{\rm cm}^{-2}$ at the location of Fig. 8 (Lehner et al. 2015). The average $N_{\rm HI}$ of the clouds over the entire field is only $9 \times 10^{16}\,{\rm cm}^{-2}$, so even if the ionization fraction of the CGM is >90%, there would still be enough neutral material to account for the clouds (Wolfe et al. 2016).

6 Starless H I Near and Far

This survey of H I in the Local Group has ranged from the disk-halo clouds, which are certainly confined to the stellar disks, through the high velocity clouds, which seem preferentially located around the disks of galaxies, to material that is not

Table 1 Inventory of starless H I "Clouds" in the Local Group

Object	M_{HI} ($M_\odot$)	Size (kpc)	Reference	Notes
Disk-halo clouds	7×10^2	0.06	Ford et al. (2010)	
Individual HVCs	$\sim 10^6$	2–15	Putman et al. (2012)	
All HVCs in MW, M31 or M33	2–5 $\times 10^7$		Westmeier et al. (2008); Putman et al. (2012); Keenan et al. (2016)	
H I stripped from MW dwarfs	3×10^7		Spekkens et al. (2014)	
IC 10 northern stream	6×10^5	7	Nidever et al. (2013); Ashley et al. (2014)	
IC 10 southern stream	$\sim 10^7$	7	Nidever et al. (2013); Ashley et al. (2014)	
Magellanic Stream	3×10^8	~100	Brüns et al. (2005); Nidever et al. (2010)	a
Leading Arm	3×10^7	~60	Putman et al. (2003)	a
M31–M33 clouds	$\sim 10^5$	2	Wolfe et al. (2016)	b
M31–M33 clouds all	$>1.6 \times 10^6$		Wolfe et al. (2016)	b

a: Assumed distance 55 kpc, b: Assumed distance 800 kpc

incorporated into galaxies: the Magellanic Stream and Leading Arm, the H I streams intersecting IC 10, and finally the M31–M33 clouds. The first two, at least, are associated with stellar systems, while the M31–M33 clouds are not. Their properties are summarized in Table 1. How common are objects like this in other galaxies and other groups?

The disk-halo clouds and most HVCs would be blended together in current H I observations of galaxies outside the Local Group, but similar material appears to be present in many other galaxies (see Sect. 2). In very sensitive H I observations many, if not most spiral galaxies show H I streams or features extending from the disk. This suggests that we are witnessing interaction, even though in many cases an accreting stellar companion cannot be identified (Sancisi et al. 2008). True starless H I clouds not associated with a galaxy seem rare, existing perhaps at the level of 2% of the H I objects detected in deep surveys, and even so, many of these seem to be located near large galaxies or be associated with tidal debris (Kovač et al. 2009; Haynes et al. 2011). Some of the most interesting objects, like the M31–M33 clouds and the plumes of IC 10, could be detected only out to a few Mpc from the Milky Way with current instrumentation.

The amount of H I in starless gas in Local Group galaxies is rather modest compared to the enormous structures produced by galaxy interactions in other groups. The classic example is the M81 group (Yun et al. 1994; Chynoweth et al. 2008), but evidence for interactions like this in other galaxy groups abounds. The data are too numerous to summarize here, but as merely one example, recent observations (Leisman et al. 2016) have detected extended H I features 600 kpc in length without stellar counterparts around the NGC 3590 and NGC 3227 groups (Fig. 9).

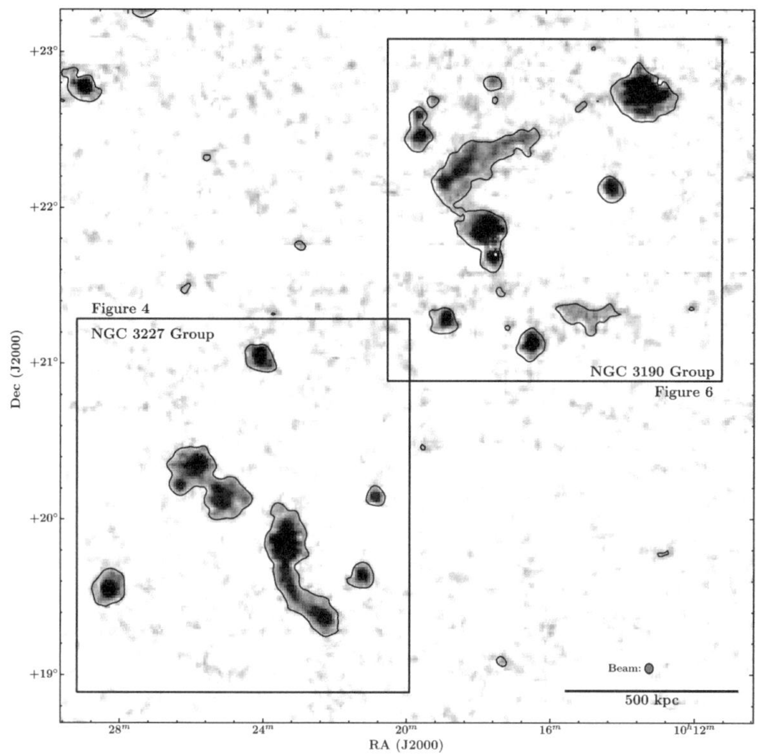

Fig. 9 Map of the H I around the NGC 3190 and NGC 3227 groups, showing extended tails and intragroup gas probably resulting from interactions between galaxies (from Leisman et al. 2016)

If our Local Group is any example, the best place to detect starless neutral gas clouds is relatively close to galaxies. As deep H I surveys of the Local Group cover more area, we will obtain better insight into the origin of starless gas, and determine if it is always located near galaxies, or is more widespread.

References

Adams, E. A. K., Giovanelli, R., & Haynes, M. P. 2013, ApJ, 768, 77
Anderson, M. E., & Bregman, J. N. 2011, ApJ, 737, 22
Ashley, T., Elmegreen, B. G., Johnson, M., et al. 2014, AJ, 148, 130
Bekki, K. 2008, MNRAS, 390, L24
Blitz, L., & Robishaw, T. 2000, ApJ, 541, 675
Blitz, L., Spergel, D. N., Teuben, P. J., Hartmann, D., & Burton, W. B. 1999, ApJ, 514, 818
Braun, R., & Thilker, D. A. 2004, A&A, 417, 421
Braun, R., & Burton, W. B. 1999, A&A, 341, 437
Bregman, J. N. 1980, ApJ, 236, 577
Brüns, C., Kerp, J., Kalberla, P. M. W., & Mebold, U. 2000, A&A, 357, 120

Brüns, C., Kerp, J., Staveley-Smith, L., et al. 2005, A&A, 432, 45
Carlin, J. L., Beaton, R. L., Martínez-Delgado, D., & Gabany, R. J. 2016, Astrophysics and Space Science Library, 420, 219
Chynoweth, K. M., Langston, G. I., Yun, M. S., et al. 2008, AJ, 135, 1983
Dickey, J. M., & Lockman, F. J. 1990, ARA&A, 28, 215
D'Onghia, E., & Fox, A. J. 2016, ARA&A, 54, 363
Ford, H. A., Lockman, F. J., & McClure-Griffiths, N. M. 2010, ApJ, 722, 367
Fox, A. J., Wakker, B. P., Barger, K. A., et al. 2014, ApJ, 787, 147
Fox, A. J., Lehner, N., Lockman, F. J., et al. 2016, ApJ, 816, L11
Fraternali, F., van Moorsel, G., Sancisi, R., & Oosterloo, T. 2002, AJ, 123, 3124
Gerbrandt, S. A. N., McConnachie, A. W., & Irwin, M. 2015, MNRAS, 454, 1000
Grcevich, J., & Putman, M. E. 2009, ApJ, 696, 385
Grossi, M., Giovanardi, C., Corbelli, E., et al. 2008, A&A, 487, 161
Haynes, M. P., Giovanelli, R., Martin, A. M., et al. 2011, AJ, 142, 170
Heald, G., Józsa, G., Serra, P., et al. 2011, A&A, 526, A118
Heiles, C., 1979, ApJ, 229, 533
Hill, A. S., Haffner, L. M., & Reynolds, R. J. 2009, ApJ, 703, 1832
Hill, A. S., Mao, S. A., Benjamin, R. A., Lockman, F. J., & McClure-Griffiths, N. M. 2013, ApJ, 777, 55
Hodges-Kluck, E. J., Miller, M. J., & Bregman, J. N. 2016, ApJ, 822, 21
Hopp, U., Schulte-Ladbeck, R. E., & Kerp, J. 2007, MNRAS, 374, 1164
Hubble, E. 1958, *The Realm of the Nebulae*, Dover Publications, Inc.
Kaltcheva, N. T., Golev, V. K., & Moran, K. 2014, A&A, 562, A69
Kacprzak, G. G., Cooke, J., Churchill, C. W., Ryan-Weber, E. V., & Nielsen, N. M. 2013, ApJ, 777, L11
Keenan, O. C., Davies, J. I., Taylor, R., & Minchin, R. F. 2016, MNRAS 456, 951
Kovač, K., Oosterloo, T. A., & van der Hulst, J. M. 2009, MNRAS, 400, 743
Lehner, N., Howk, J. C., & Wakker, B. P. 2015, ApJ, 804, 79
Lehner, N., & Howk, J. C. 2011, Science, 334, 955
Leisman, L., Haynes, M. P., Giovanelli, R., et al. 2016, MNRAS, 463, 1692
Lewis, G. F., Braun, R., McConnachie, A. W., et al. 2013, ApJ, 763, 4
Lockman, F. J. 1984, ApJ, 283, 90
Lockman, F. J. 2002, ApJ, 580, L47
Lockman, F. J. 2003, ApJ, 591, L33
Lockman, F. J., Benjamin, R. A., Heroux, A. J., & Langston, G. I. 2008, ApJ, 679, L21
Marasco, A., Fraternali, F., & Binney, J. J. 2012, MNRAS, 419, 1107
Marasco, A., & Fraternali, F. 2017, MNRAS, 464, L100
McClure-Griffiths, N. M., Dickey, J. M., Gaensler, B. M., & Green, A. J. 2002, ApJ, 578, 176
McConnachie, A. W. 2012, AJ, 144, 4
Miller, E. D., Bregman, J. N., & Wakker, B. P. 2009, ApJ, 692, 470
Muller, C. A., Oort, J. H., & Raimond, E. 1963, *Acad. Sci. Paris Comptes Rendus*, 257, 1661
Nichols, M., & Bland-Hawthorn, J. 2009, ApJ, 707, 1642
Nichols, M., Mirabel, N., Agertz, O., Lockman, F. J., & Bland-Hawthorn, J. 2014, MNRAS, 442, 2883
Nidever, D. L., Majewski, S. R., Burton, W. B., & Nigra, L. 2010, ApJ, 723, 1618
Nidever, D. L., Ashley, T., Slater, C. T., et al. 2013, ApJ, 779, L15
Oosterloo, T., Fraternali, F., & Sancisi, R. 2007, AJ, 134, 1019
Pidopryhora, Y., Lockman, F. J., & Shields, J. C. 2007, ApJ, 656, 928
Pidopryhora, Y., Lockman, F. J., Dickey, J. M., & Rupen, M. P. 2015, ApJS, 219, 16
Pisano, D. J., Barnes, D. G., Gibson, B. K., et al. 2007, ApJ, 662, 959
Pisano, D. J., Barnes, D. G., Staveley-Smith, L., et al. 2011, ApJS, 197, 28
Putman, M. E., Staveley-Smith, L., Freeman, K. C., Gibson, B. K., & Barnes, D. G. 2003, ApJ, 586, 170
Putman, M. E., Peek, J. E. G., Muratov, A., et al. 2009, ApJ, 703, 1486

Putman, M. E., Peek, J. E. G., & Joung, M. R. 2012, ARA&A, 50, 491
Rhode, K. L., Salzer, J. J., Haurberg, N. C., et al. 2013, AJ, 145, 149
Sancisi, R., Fraternali, F., Oosterloo, T., & van der Hulst, T. 2008, Astron. Astophys. Rev., 15, 189
Shapiro, P. R., & Field, G. B. 1976, ApJ, 205, 762
Shaya, E. J., & Tully, R. B. 2013, MNRAS, 436, 2096
Shull, J. M., Jones, J. R., Danforth, C. W., & Collins, J. A. 2009, ApJ, 699, 754
Shull, J. M., Stevans, M., Danforth, C., et al. ApJ, 739, 105 (2011)
Shostak, G. S., & Skillman, E. D. 1989, A&A, 214, 33
Smith, G. P. 1963, BAAN, 17, 203
Spekkens, K., Urbancic, N., Mason, B. S., Willman, B., & Aguirre, J. E. 2014, ApJ, 795, L5
Stark, D. V., Baker, A. D., & Kannappan, S. J. 2015, MNRAS, 446, 1855
Thilker, D. A., Braun, R., Walterbos, R. A. M., et al. 2004, ApJ, 601, L39
Tomisaka, K., & Ikeuchi, S. 1986, PASJ, 38, 697
Wakker, B. P. 2004, in *High Velocity Clouds*, ed. H. van Woerden, B. P. Wakker, U. J. Schwarz, & K. S. de Boer (Dordrecht: Kluwer), 25
Wakker, B. P., York, D. G., Wilhelm, R., et al. 2008, ApJ, 672, 298
Wakker, B. P., & Savage, B. D. 2009, ApJS, 182, 378
Westmeier, T., Brüns, C., & Kerp, J. 2008, MNRAS, 390, 1691
Wolfe, S. A., Pisano, D. J., Lockman, F. J., et al. 2013, Nature, 497, 224
Wolfe, S. A., Lockman, F. J., & Pisano, D. J. 2016, ApJ, 816, 81
Yun, M. S., Ho, P. T. P., & Lo, K. Y. 1994, Nature, 372, 530
Zschaechner, L. K., Rand, R. J., & Walterbos, R. 2015, ApJ, 799, 61

Gas Accretion and Star Formation Rates

Jorge Sánchez Almeida

1 Introduction: Key Physical Parameters

Cosmological numerical simulations show that model galaxies tend to reach a subtle stationary state where the gas accretion rate from the cosmic web balances the star formation rate (SFR) once outflows are taken into account (e.g., Finlator and Davé 2008; Schaye et al. 2010; Fraternali and Tomassetti 2012; Davé et al. 2012; Dekel et al. 2013; Bothwell et al. 2013; Feldmann 2013; Altay et al. 2013; Forbes et al. 2014a). The balance is set because the time-scale to transform gas into stars is significantly shorter than the Hubble time and, thus, galaxies must rely on external gas accretion to maintain star formation for a long period of time (e.g., Kennicutt 1983; Sancisi et al. 2008, and also Sect. 3.1).

Numerical simulations reveal an intimate connection between SFR and gas accretion rate, and so, provide the rationale to study the relation observationally. In order to identify the physical parameters that have to be measured, one can resort to the toy galaxy model often referred to as bathtub model or self-regulator model. It is amply described in the literature (Tinsley 1980; Edmunds 1990; Bouché et al. 2010; Peeples and Shankar 2011; Brisbin and Harwit 2012; Davé et al. 2012; Lilly et al. 2013; Dayal et al. 2013; Dekel and Mandelker 2014; Forbes et al. 2014b; Peng and Maiolino 2014; Harwit and Brisbin 2015; Rodríguez-Puebla et al. 2016; Somerville and Davé 2015; Ascasibar et al. 2015), and it provides the physical insight to understand the self-regulation of the star formation (SF) process in galaxies. In this simple model, galaxies are described as structureless entities characterized by a single mass of gas M_g, an SFR, an outflow rate $\dot{M}_{\rm out}$, and an inflow rate $\dot{M}_{\rm in}$. We take

J. Sánchez Almeida (✉)
Instituto de Astrofísica de Canarias, La Laguna, Tenerife, Spain

Departamento de Astrofísica, Universidad de La Laguna, Tenerife, Spain
e-mail: jos@iac.es

A. Fox, R. Davé (eds.), *Gas Accretion onto Galaxies*, Astrophysics and Space Science Library 430, DOI 10.1007/978-3-319-52512-9_4

the nomenclature and the equations from the particular implementation by Sánchez Almeida et al. (2014a). If the model galaxy is isolated and does not receive any external gas supply, then the initial mass of gas $M_g(0)$ drops exponentially in time t due to star formation (SF),

$$M_g(t) = M_g(0)\exp(-t/\tau_{\rm in}), \tag{1}$$

with a characteristic time-scale, $\tau_{\rm in}$, given by,

$$\tau_{\rm in} = \tau_g/(1 - R + \eta), \tag{2}$$

which depends on the so-called gas depletion time-scale τ_g,

$$\tau_g = M_g/{\rm SFR}, \tag{3}$$

and on the mass loading factor η,

$$\eta = \dot{M}_{\rm out}/{\rm SFR}, \tag{4}$$

defined to be the scale factor between the SFR and the mass outflows that the SF drives. R in Eq. (2) stands for the fraction of stellar mass that returns to the interstellar medium (ISM) rather than being locked into stars and stellar remnants. If rather than being isolated our toy galaxy is fed at a gas accretion rate $\dot{M}_{\rm in}(t)$, then after a transient that lasts $\tau_{\rm in}$, it reaches a stationary state where,

$${\rm SFR}(t) = \dot{M}_{\rm in}(t)/(1 - R + \eta). \tag{5}$$

Even if oversimplified, the above equations include all the essential ingredients giving rise to the expected relationship between gas accretion rate and SFR. Equation (5) indicates that the stationary-state SFR is set by the gas infall rate, becoming zero when the accretion rate goes to zero. Often $\eta \gg 1$, and so SFR $\ll \dot{M}_{\rm in}$ and $\tau_{\rm in} \ll \tau_g$. In this case, only a minor fraction of the accreted gas is used to form stars. The rest is returned unused to the circum-galactic medium (CGM) and inter-galactic medium (IGM). When this happens, the time-scale to consume the gas $\tau_{\rm in}$ becomes much shorter than the already short gas depletion time-scale (Sect. 3.1).

Therefore, in order to provide an observational overview of the relation between gas infall rate and SFR, it is essential to keep in mind and constrain the key parameters characterizing the relation, namely the gas depletion time-scale, the mass loading factor, and the returned mass fraction. Thus, the first section of the paper collects observational constraints on these parameters (Sect. 2). Section 3 constitutes the main body of the work, and it describes observational evidence for a relationship between SFR and metal-poor gas accretion. Unfortunately, despite the large volume of circumstantial evidence for feeding from external gas accretion, we still lack direct evidence. Several factors explain the difficulty. They are pointed out and discussed in Sects. 4 and 5, where we also mention future lines of research.

2 Characteristic Physical Parameters

Gas Depletion Time-Scale τ_g The ratio between M_g and SFR, τ_g, can be measured directly. The mass of gas and the SFR are correlated as given by the so-called Kennicutt-Schmidt relation (Schmidt 1959; Kennicutt 1998), which is usually formulated in terms of the gas surface density, Σ_g, and the SFR surface density, $\Sigma_{\rm SFR}$,

$$\Sigma_{\rm SFR} = A\,\Sigma_g^N, \tag{6}$$

so that

$$\tau_g = \Sigma_g/\Sigma_{\rm SFR} = A^{-1}\,\Sigma_g^{(1-N)}. \tag{7}$$

Kennicutt and Evans (2012) give A and N for disk-averaged galaxies, which lead to τ_g from 3 to 0.5 Gyr for gas surface densities between 10 and 1000 $M_\odot\,{\rm pc}^{-2}$, typical of star-forming galaxies. Similar τ_g are obtained when considering spatially resolved measurements with sub-kpc resolution; see Fig. 1, adapted from Bigiel et al. (2008). It shows the lines of constant τ_g, and how the observations in the range of interests are close to the $\tau_g = 1$ Gyr line. The Kennicutt-Schmidt law provides an empirical recipe linking SF with its fuel, and it has become central to the current numerical simulations of galaxy formation (e.g., Schaye and Dalla Vecchia 2008; Dalla Vecchia and Schaye 2008, 2012). Therefore, the relation has deserved much attention in the literature, rendering depletion time-scales in the range of the above values (see Filho et al. 2016, and the references therein).

Observations suggest that τ_g decreases with increasing redshift, so that it goes from 0.5 to 2 Gyr for galaxies in the redshift range between 2 and 0 (e.g., Genzel et al. 2010; Gnedin et al. 2014). This is consistent with the theoretical expectation that τ_g scales with the instantaneous Hubble time, t_H, as $\tau_g \simeq 0.17\,t_H$, so that at any time $\tau_g \ll t_H$ (Dekel et al. 2013).

Mass Loading Factor η Stellar feedback is an essential ingredient of the current galaxy formation scenario since it partly controls the predicted number of low-mass galaxies. Thus, its tuning allows us to reproduce the observed galaxy mass function (e.g., Silk and Mamon 2012; Vogelsberger et al. 2014; Schaye et al. 2015). The mass loading factor quantifies the importance of this stellar feedback [Eq. (4)]. Using the toy model, it is easy to argue that w should increase with decreasing stellar mass $M_\star$, so that for low-mass galaxies,

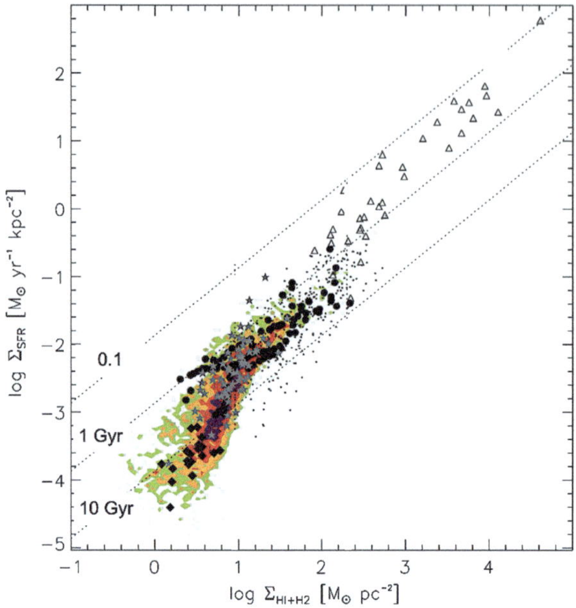

Fig. 1 Kennicutt-Schmidt relation at sub-kpc scales, i.e., SFR surface density versus gas surface density (labeled in the plot as $\Sigma_{\rm HI+H2} = \Sigma_g$). The *slanted dotted lines* represent the expected relation for constant values of τ_g of 0.1, 1, and 10 Gyr. They encompass the observed data points. Adapted from Bigiel et al. (2008, their Fig. 15)

$$\eta \geq 1. \tag{8}$$

In the stationary state, the metallicity of the gas that forms stars, Z, is independent of SFR and M_g. Z only depends on parameters characteristic of the stellar population, and on η, explicitly,

$$Z = Z_{\rm in} + y\,(1 - R)\big/(1 - R + \eta), \tag{9}$$

with $Z_{\rm in}$ the metallicity of the accreted gas, and y the stellar yield (the mass of ejected metals per unit mass locked into stars). Galaxies follow a well-known mass-metallicity relation (MZR; see, e.g., Skillman et al. 1989; Tremonti et al. 2004) which, according to Eq. (9), can only be due to the dependence of η on $M_\star$ since y and R are universals given the initial mass function (IMF). Moreover, for η to modify Z in a substantial way, Eq. (8) must be satisfied.

The fact that η varies with $M_\star$ reaching large values for low-mass galaxies is found both in numerical simulations and in observational works. In order to reproduce the MZR, Peeples and Shankar (2011), Davé et al. (2012), and Dayal et al. (2013) use η varying from 1 to 6 when $M_\star$ goes from 10^{11} to $10^{9}\,M_\odot$. The numerical simulations by Shen et al. (2012) lead to η between 1 and 10 for $M_\star$ from 2×10^{11}

to $10^9\,M_\odot$. In the cases modeled by Christensen et al. (2016), it goes from 0.5 to 50 for galaxies with virial masses from 10^{12} to $2 \times 10^9\,M_\odot$. The numerical simulations carried out by Davé et al. (2013) assume η to be proportional to $M_\star^{-1/3}$, with an even steeper dependence at low masses. The numerical simulations of giant star-forming clumps in gas-rich galaxies by Bournaud et al. (2014) render an effective η exceeding 2. Thompson and Krumholz (2016) model supersonic-turbulence driven outflows, and infer η ranging from 10^{-3} to 10. Observations of Mg II absorption around massive galaxies with $M_\star \sim 10^{11}\,M_\odot$ are used by Bordoloi et al. (2014) to set a conservative lower limit to η, which has to be larger than 0.24. Mass loading factors observed at high redshift generally refer to massive objects ($M_\star > 10^{10}\,M_\odot$), and turn out to be between 0.5 and 2 (Newman et al. 2012; Martin et al. 2012). It is not uncommon to infer factors up to 10 in local dwarfs (Martin 1999; Veilleux et al. 2005). An extreme case is presented by Olmo-García et al. (2017), where faint multiple components in the wings of Hα are interpreted as produced by SN driven outflows with η often larger than 20.

Returned Fraction *R* Once the IMF is set, the stellar mass returned to the ISM is provided by stellar evolution models. 10 Gyr after the starburst, R is typically in the range between 0.2 and 0.3 when a Salpeter IMF is adopted (e.g., Edmunds 2005; Leitner and Kravtsov 2011). It can reach 0.5 for a Kroupa IMF, that has more massive stars for the same total mass of the stellar population (e.g., Leitner and Kravtsov 2011; Segers et al. 2016).

3 Evidence for a Relationship Between the SFR and the Gas Infall Rate

3.1 The Gas-Consumption Time-Scale

The time-scale to consume the gas is too short for the observed stellar ages, therefore, a continuous gas supply is needed to explain why all types of galaxies have been forming stars for extended periods of time. This is an old idea (e.g., Roberts 1963; Kennicutt 1983) that has been reformulated in various ways along the years. An updated account is given next.

Except for the most massive ones, galaxies have been forming star during the whole life-time of the universe. Figure 2 shows the observed SFR versus redshift for typical disk galaxies forming stars at present. The galaxies have been divided into three bins according to the present stellar mass. The figure shows how their SFR has been declining slightly with time but, overall, galaxies of all masses maintain a

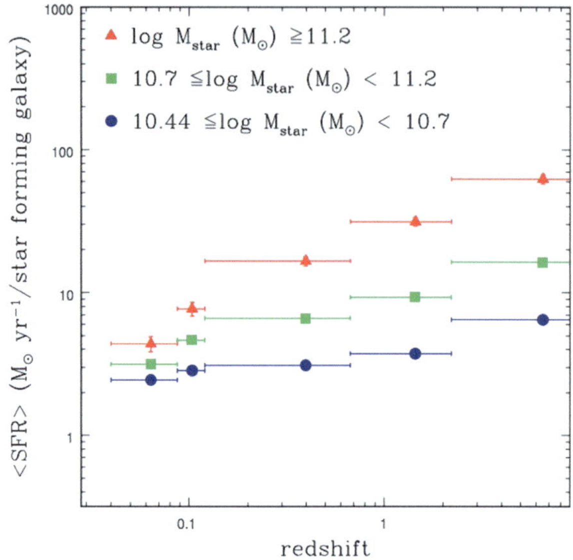

Fig. 2 Observed SFR versus redshift for typical galaxies forming stars at present. The galaxies are divided into three bins according to the present stellar mass (see the *inset*). Galaxies of all masses have been forming stars over the whole history of the Universe. Figure adapted from Guglielmo et al. (2015, their Fig. 6)

significant level of SF along the Hubble time (see also Heavens et al. 2004). If no gas supply is provided, the SF shuts off exponentially with a time-scale of

$$\tau_{\rm in} \simeq 0.75\,{\rm Gyr}, \tag{10}$$

where we have plugged into Eq. (2) the values 2 Gyr, 2, and 0.3 for τ_g, η, and R, respectively (Sect. 2). Since the SF has been active for much longer, a supply of external gas must have been feeding the SF process in all galaxies.

3.2 Relationship Between Stellar Mass, SFR, and Gas Metallicity

In two independent papers, Mannucci et al. (2010) and Lara-López et al. (2010) found that the scatter in the well-known mass-metallicity relation correlates with the SFR. Such correlation had been suggested in previous studies (Ellison et al. 2008; Peeples et al. 2009; López-Sánchez 2010). The fact that galaxies with higher SFR show lower metallicity at a given stellar mass is called fundamental metallicity relation (FMR). Figure 3 shows a recent account of the FMR by Salim et al. (2014). Galaxies are separated into two plots according to their SFR, and the difference is clear: high SFR objects (panel A) have lower metallicity than low SFR objects (panel B).

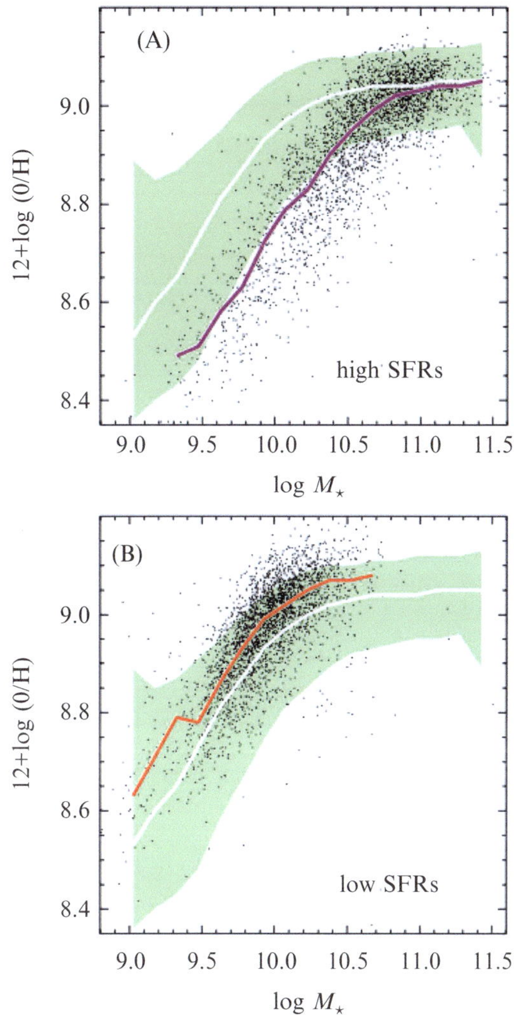

Fig. 3 Gas-phase metallicity versus stellar mass for galaxies with extreme SFR. From a sample of ∼250,000 SDSS objects, the *dots* in the *upper panel* (**a**) show the galaxies with the highest SFRs. The *lower panel* (**b**) contains those galaxies with the lowest SFRs. The *colored lines* represent the median of the distribution of points for a given stellar mass. The *green shaded region* is the same in (**a**) and (**b**) and gives the 90 percentile range of the full sample, with the white line representing the median. Note the systematically lower metallicity for the sample of higher SFR. Figure adapted from Salim et al. (2014)

Neither metal-rich outflows nor the variation of the SF efficiency with $M_\star$ explain the FMR. However, the stochastic feeding of the SF process with external metal-poor gas provides a natural explanation. The advent of external gas does not change $M_\star$, but it decreases the mean metallicity of the star-forming gas while simultaneously triggering SF. As time goes on, the SF consumes gas and increases its metallicity, until new metal-poor gas arrives and the cycle re-starts. The process is illustrated in Fig. 4. It contains the temporal variation of the gas mass and metallicity predicted by the toy model in Sect. 1, assuming the gas accretion events to be stochastic. The figure shows how the arrival of gas at a particular time increases the gas mass and thus the SFR [Eq. (3)]. This gas accretion event comes with a drop in metallicity. The gas is consumed by the SF, that rises the metallicity in a

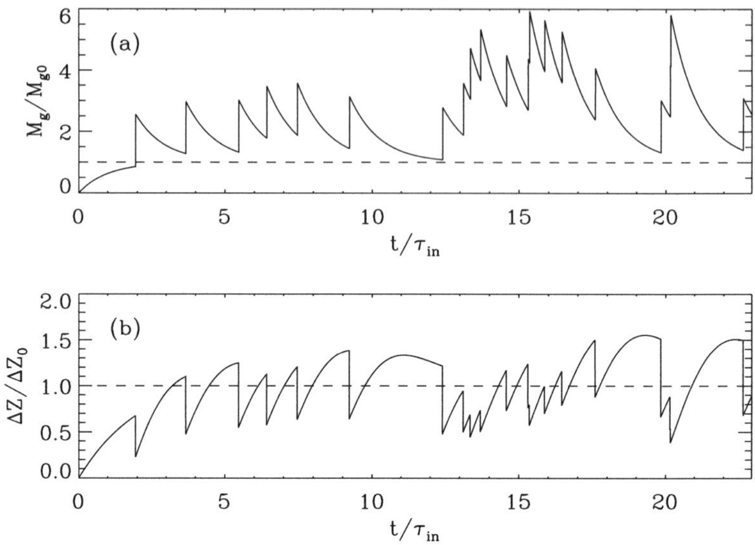

Fig. 4 Variation with time of gas mass (**a**) and metallicity (**b**) in a star-forming system where gas clumps are accreted randomly, i.e., at random times and with random masses. Time units are normalized to τ_{in}, the characteristic time-scale for the exponential fall-off of the gas content. The gas of mass and metallicity are normalized to their stationary-state value, indicated by *horizontal dashed lines* in the figures

process that in the long run yields the stationary-state value given in Eq. (9). If a collection of galaxies with similar stellar masses are observed at different phases of the cycle triggered by a gas accretion event, they will show a dispersion in metallicity which anti-correlates with the instantaneous SFR, i.e., they will show the FMR. This explanation was already advanced by Mannucci et al. (2010) and it is generally accepted today.

The FMR has received considerable attention in the literature,[1] both from an observational point of view, and from the point of view of its interpretation. Sometimes the claims seem to be contradictory, although the community is reaching a consensus in the sense that (1) the FMR is not an observational artifact, (2) it changes with redshift so that all metallicities decrease with increasing redshift, and (3) it is produced by metal-poor gas accretion triggering SF. The next paragraphs summarize the recent observational and theoretical work on the subject.

Observations of the FMR The correlation between SFR and metallicity is weaker if systematic errors are taken into account (Telford et al. 2016). The correlation may disappear depending on the strong-line ratio used to estimate metallicities (Kashino et al. 2016). The FMR remains even if different metallicity and SFR indicators are used (Salim et al. 2014; Andrews and Martini 2013).

[1]The discovery papers have about 600 citations in the ADS.

The FMR disappears when using single H II regions rather than galaxy integrated parameters (Sánchez et al. 2013). There is no FMR in the local star-forming galaxies analyzed by Izotov et al. (2014), whereas it is present in the local sample discussed by Wu et al. (2016). Arguments against the existence of a FMR in star-forming galaxies with redshift smaller than one are presented by Izotov et al. (2015). There is no obvious FMR for galaxies with redshifts between 1 and 2, according to Divoy et al. (2014). There is not relationship at redshift 0.8 (de los Reyes et al. 2015).

At redshift around 1.4, the deviation from the MZR depends on the SFR, so that galaxies with higher SFR show lower metallicities at a given $M_\star$ (Yabe et al. 2012). The FMR is still in place at redshift around 0.9, but the metallicities are systematically lower given $M_\star$ and SFR (Ly et al. 2014, 2015). Galaxies with younger and more vigorous star formation tend to be more metal poor at a given $M_\star$ (redshift between 0.3 and 0.9; Amorín et al. 2014). The FMR is in place at $z \geq 2$ (Cullen et al. 2014). Galaxies at redshift 2 show evidence that the SFR is still a second parameter in the MZR, and are consistent with a non-evolving FMR (Maier et al. 2014). The FMR is in place at redshift 1.6, but it has evolved with respect to the FMR in the local universe so that metallicities are smaller (Zahid et al. 2014). At redshift 3.4, the metallicity generally anti-correlates with the distribution of SFR and with the gas surface density, although the relation differs from the FMR in the local universe (Troncoso et al. 2014). The evolution of the FMR previously reported in the literature may be an artifact introduced by the use of the different metallicity indicators at different redshifts (Cullen et al. 2014). There is a FMR at redshift 0.7 that seems to agree with the local one (Maier et al. 2015). There is no correlation at redshift 2.3 (Sanders et al. 2015). The FMR evolves with redshift (Brown et al. 2016). There is no significant dependence of the metallicity on SFR at fixed redshift and $M_\star$ (objects with redshifts between 0.6 and 2.7; Wuyts et al. 2016).

There is also a *more fundamental* FMR where the SFR is replaced with the gas mass (Bothwell et al. 2013). The scatter of the FMR is reduced if H I mass is used instead of SFR (Jimmy et al. 2015). The central role assigned to the gas mass at the sacrifice of the SFR is also defended by Bothwell et al. (2016a,b). Moreover, Lian et al. (2015) claim that stellar age, rather than SFR or gas mass, is the third parameter in the FMR .

There is a correlation between the metallicity gradient along the radial distance in a galaxy and the SFR, in the sense that galaxies with high SFR tend to show flatter gradient (Stott et al. 2014).

Interpretations of the FMR Most of the available explanations are based on simple analytical models very much in the spirit of the one described in Sect. 1. For example, Lilly et al. (2013) present a model galaxy whose properties self-regulate due to the short gas depletion time-scale. The model galaxy is near the stationary state, but the gas reservoir available to form stars is allowed to change in time. This drives the system out of the stationary state and provides a dependence of the metallicity on the SFR and mass gas. [The metallicity does not depend on the SFR in the stationary state; see Eq. (9).] The work by Lilly et al. successfully reproduces the FMR, allowing both w and τ_g to vary with stellar mass. It reproduces the overall drop of metallicity with increasing redshift by steadily increasing the gas infall rate. Other works with this type of interpretation are those by Davé et al. (2011), Brisbin and Harwit (2012), Dayal et al. (2013), Forbes et al. (2014b), Pipino et al. (2014), and Harwit and Brisbin (2015). Mergers are also able to reproduce the FMR according to Grønnow et al. (2015).

Dekel and Mandelker (2014) use one of these simple toy models to study the redshift dependence of the FMR, finding problems to reproduce some of the observational constrains, in particular, the ratio SFR/M_g. The need to go beyond simple models because they do not reproduce the observed variation with redshift of SFR/M_g is also argued by Peng and Maiolino (2014).

Yates et al. (2012) use thousands of galaxies from dark-matter numerical simulations to interpret the FMR. Baryons that follow the dark-matter are added, generating a non-stationary clumpy gas accretion that drives the evolution of the model galaxies. The numerical simulation reproduces the main observational trends, including an apparent turnover of the mass-metallicity relationship at very high $M_\star$. The temporal evolution of the gas mass and metallicity of individual galaxies is qualitatively similar to the variations displayed in Fig. 4.

Romeo Velonà et al. (2013) present SPH-cosmological simulations of hundreds of galaxies. Surprisingly, more active galaxies in terms of SFR are also metal-richer (see their Fig. 12). The reason of this contradictory result is not properly understood.

De Rossi et al. (2015) employ hydrodynamical zoom-in cosmological simulations of 500 galaxies to study the scaling relations. The model galaxies show the trends corresponding to the observed FMR. They also find that satellite galaxies have higher metallicity for the same stellar mass, as it is indeed observed (Pasquali et al. 2012).

Kacprzak et al. (2016) find galaxies at redshift around 2 following the FMR. They show that the gas masses and metallicities required to reproduce the observed FMR are consistent with cold-accretion predictions obtained from their hydrodynamical simulations.

3.3 Relationship Between Lopsidedness and Metallicity

Surprisingly, the extremely metal-poor (XMP) galaxies of the local universe turn out to show a particular morphology consisting of a bright head and a faint tail, which is commonly referred to as *cometary* or *tadpole*. This correspondence between low metallicity and shape was first noted by Papaderos et al. (2008), and then it has been confirmed in other studies (e.g., Morales-Luis et al. 2011; Filho et al. 2013; Sánchez Almeida et al. 2016). These morphologies represent 80% of the objects in the XMP catalog used by Filho et al. (2013). The tadpole morphology is not unusual at high redshift, where galaxies tend to be clumpy and elongated (Elmegreen et al. 2005), however, it is rather uncommon in the local universe where XMPs reside. For reference, only 0.2% of the star-forming galaxies in the Kiso survey are cometary (Elmegreen et al. 2012). Figure 5 displays several of these XMP galaxies with the characteristic morphology.

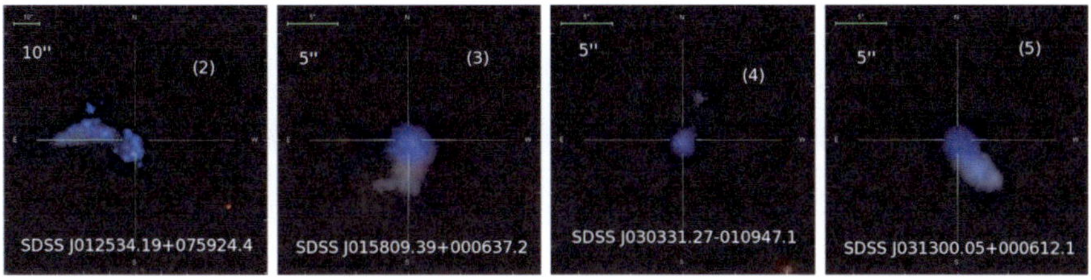

Fig. 5 Typical set of galaxies selected only because they are XMP. Surprisingly, they tend to show cometary or tadpole morphology, with a *bright blue head* and a *faint redder tail*. The images are composite color images from SDSS broad-band filters, therefore, they trace stellar light. Adapted from Fig. 5 in Morales-Luis et al. (2011)

Even though XMP galaxies are a very particular type of galaxy, the morphology-metallicity relation that they exhibit is only the extreme case of a common behavior followed by many star-forming galaxies. Reichard et al. (2009) quantify the lopsidedness of 25,000 star-forming galaxies from SDSS using the amplitude of the $m = 1$ azimuthal Fourier mode. At a fixed mass, the more metal-poor galaxies turn out to be more lopsided, extending the morphology-metallicity relation to the full population of star-forming galaxies.

This property of the XMPs and the other galaxies can be naturally understood within the *gas accretion triggering* scenario. The actual characteristics of the starbursts induced by gas accretion are far from being properly understood and modeled (e.g., Verbeke et al. 2014; Casuso et al. 2006, see also Sect. 4). However, a few general trends seem to be clear. The accreted gas is metal-poor (e.g., Dekel et al. 2009; van de Voort and Schaye 2012), and it induces off-center giant star-forming clumps that gradually migrate toward the disk centers (Ceverino et al. 2010; Mandelker et al. 2014). The giant star-forming clumps may be born in-situ or ex-situ. In the first case, the accreted gas builds up the gas reservoir in the disk to a point where disk instabilities set in and trigger SF. In the second case, already formed clumps are incorporated into the disk. They may come with stars and dark-matter, and thus, they are often indistinguishable from gas-rich minor mergers (Mandelker et al. 2014). In any case, a significant part of the SF in the disks occurs in these giant clumps. As a result of the whole process, the gas accretion produces bright off-center starbursts increasing the lopsidedness of the host disk. This increase and the decrease of metallicity come hand-to-hand together, giving rise to a relation between morphology and metallicity qualitatively similar to the observed one.

We note, however, that the same trend can also be reproduced by gas-rich metal-poor mergers (e.g., Kazantzidis et al. 2009; Pawlik et al. 2016). As we pointed out above, *gas-rich minor mergers* and *gas accretion events* are often impossible to distinguish, both observationally and from the point of view of the numerical simulations. On the one hand, it is unclear how to define *galaxy* at the low-mass end of the galaxy mass function. If the presence of stars is essential (see Forbes and Kroupa 2011), whether a gas dominated system is or is not a galaxy ultimately depends on the sensitivity of the observation (e.g., Cannon et al. 2014; Serra et al.

2015; Janowiecki et al. 2015). On the other hand, the presence or absence of stars in a particular dark-matter halo of a numerical simulation depends on details of the assumed sub-grid physics, which may or may not be adequate to describe the formation of stellar systems in objects with sizes and masses at the resolution of the simulation.

This remarkable association between SFR and lopsidedness has been observed in the H I morphology too. Lelli et al. (2014) find that dwarfs with active SF have more asymmetric outer H I envelopes than typical irregulars. Moreover, galaxies hosting an old burst (≥ 100 Myr) have more symmetric H I morphology than those with a young one (≤ 100 Myr).

3.4 Metallicity Drops in Starbursts of Local Star-Forming Galaxies

According to the conventional wisdom, the gas of the local gas-rich dwarf galaxies has uniform metallicity (Kobulnicky and Skillman 1996; Croxall et al. 2009; Pilyugin et al. 2015). However, there is mounting evidence that some particular objects do show metallicity inhomogeneities (Papaderos et al. 2006; Izotov et al. 2009, 2012; Werk et al. 2010; Levesque et al. 2011; Haurberg et al. 2013; Sánchez Almeida et al. 2013, 2014b; Thöne et al. 2014; Richards et al. 2014). They are often associated with star-forming regions, so that a drop in metallicity occurs in regions of intense SF.

Sánchez Almeida et al. (2015) carried out a systematic study of the variation of gas metallicity along the major axis of a representative sample of XMP galaxies. Metallicities were inferred using HCM (Pérez-Montero 2014), a code that compares the observed optical emission lines with photoionization models and which provides metallicity measurements in agreement with direct-method within 0.07 dex. Figure 6 contains the result for one of the galaxies. It shows a clear drop in metallicity at the peak surface SFR. The pattern is the same in 9 out of the 10 studied galaxies. The XMP star-forming clumps are immersed in a host galaxy which is several times more metal-rich. Figure 7 summarizes these results. Independent observations proof that the XMP galaxies rotate, and that the star-forming clumps of low metallicity are dynamically decoupled from the underlying disk (Olmo-García et al. 2017).

The existence of localized metallicity drops suggests a recent metal-poor gas accretion episode. The time-scale for the azimuthal mixing of the gas in turbulent disk is short, of the order of a fraction of the rotational period (e.g., de Avillez and Mac Low 2002; Yang and Krumholz 2012; Petit et al. 2015), equivalent to a few hundred Myr. Therefore, the metal-poor gas forming stars must have arrived to the disk recently, as naively expected for SF episodes driven by external metal-poor gas accretion. As we discussed in Sect. 3.3, the triggering of SF feeding from external gas is a complex process not properly understood yet (see also Sect. 4).

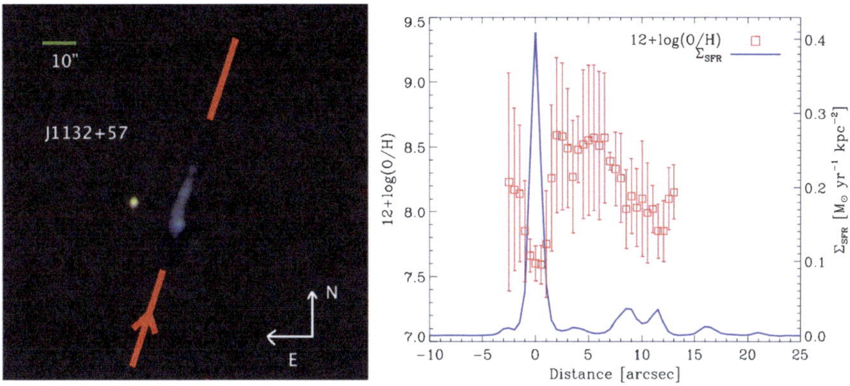

Fig. 6 *Left*: SDSS image of the XMP galaxy J1132+57, with the *red bar* indicating the position of the slit during observation. The *arrows* indicate north and east, and the scale on the top left corner corresponds to 10 arcsec. *Right*: variation of surface SFR (*blue solid line*) and oxygen abundance (*red symbols* with error bars) along the slit. Note the drop in abundance associated with the peak SFR. Figure adapted from Sánchez Almeida et al. (2015)

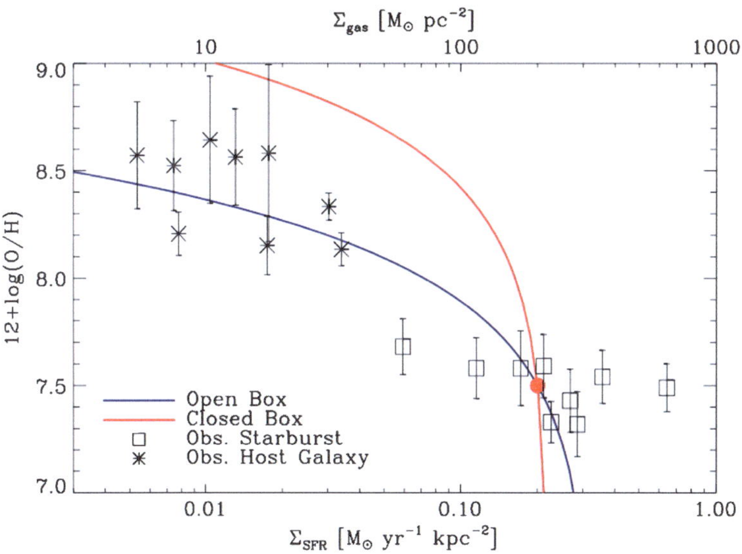

Fig. 7 Summary plot with the oxygen abundance of the starburst (*square symbols*) and the host galaxy (*asterisks*) for the XMP galaxies studied by Sánchez Almeida et al. (2015). They are represented versus the surface SFR inferred from Hα. The star-forming clumps are 0.5 dex metal-poorer than the host, and have an SFR between 10 and 20 times larger. The axis on top gives the gas surface density. The *lines* show the chemical evolution of a clump at the position of the red bullet depending on whether it evolves as a closed-box (*red line*) or as an open box (*blue line*). Taken from Sánchez Almeida et al. (2015)

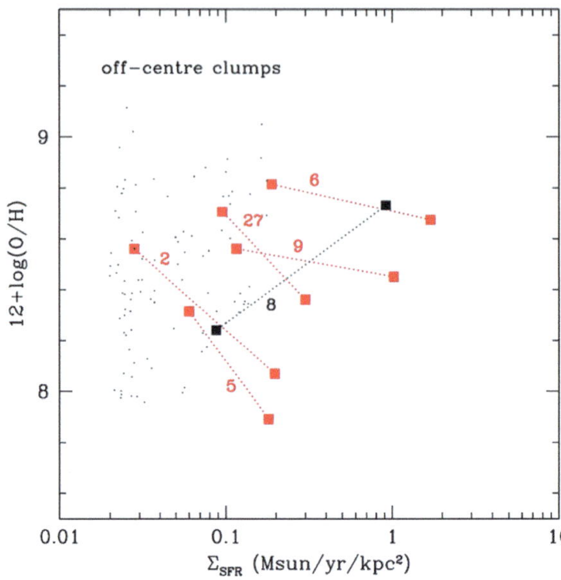

Fig. 8 Metallicity [$12 + \log(O/H)$] versus surface SFR (Σ_{SFR}) for a number of clumps and their nearby intra-clump medium in the cosmological numerical simulations of galaxies by Ceverino et al. (2016). Each pair is joined by a *dotted line*. In all but one case (the one shown in *black*), the region of large SFR (the clump) has lower metallicity than the nearby region of small SFR (the intra-clump medium). The *small black dots* represent 100 randomly chosen apertures in one of the model galaxies. Adapted from Fig. 3 in Ceverino et al. (2016)

There is a significant degree of gas mixing in the CGM (Sect. 4) and the *naive* interpretation may not be correct. However, the cosmological numerical simulations of galaxies analyzed by Ceverino et al. (2016) are reassuring. The model galaxies produce off-center star-forming clumps with a metallicity lower than the metallicity of their immediate surroundings. Figure 8, taken from Ceverino et al. (2016), shows metallicity versus surface SFR for a number of clump intra-clump pairs. Each pair is joined by a dotted line. They follow a clear pattern where the point of lower metallicity coincides with the point of larger SFR. Qualitatively, the figure resembles the behavior of the star-forming clumps in XMPs (see Fig. 7).

3.5 *The Traditional G-dwarf Problem*

This observation is included here both for historical reasons, and because the accepted interpretation is easy to understand in terms of the toy model described in Sect. 1. The so-called *G-dwarf problem* was the first clear indication that an external metal-poor gas supply was needed to explain the observed properties, a stellar population.

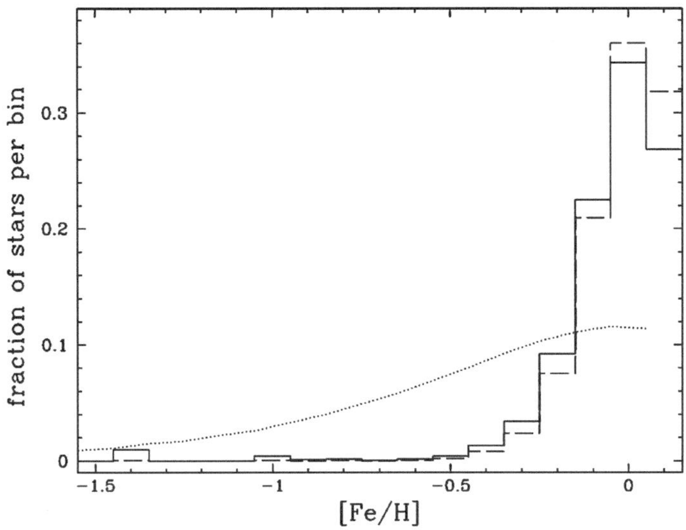

Fig. 9 Fraction of stars in each 0.1 dex metallicity bin. [Fe/H] denotes the Fe abundance referred to the solar metallicity in a decimal logarithm scale. The *dashed* and *solid lines* represent observed uncorrected and corrected data, respectively. The *dotted line* is the distribution predicted by a closed-box chemical evolution model, and it largely deviates from the observed one. The observed distribution, from Woolf and West (2012), corresponds to M dwarf stars in the solar neighborhood, but is very similar to the distribution of G dwarfs discussed in the text (see, e.g., Rocha-Pinto and Maciel 1996)

If a system of pure metal-poor gas evolves as a closed box, each new generation of stars must be less numerous and more metal-rich than the previous one. Therefore, in such a system, the number of stars is expected to decrease with increasing metallicity. However, the distribution of metallicities of the G dwarf stars in the solar neighborhood does not show the fall-off expected in a closed box. There is a deficit of sub-solar metallicity G dwarf stars in the solar neighborhood (van den Bergh 1962; Schmidt 1963; Lynden-Bell 1975)—see Fig. 9. This *problem* has deserved careful attention in the literature, with solutions going from variations of the IMF (Carigi 1996; Martinelli and Matteucci 2000) to inhomogeneous chemical evolution and star formation (Malinie et al. 1993). Among them, a continuous metal-poor gas inflow sustaining the formation of the G stars seems to be the preferred mechanism (Pagel 2001; Edmunds 2005). The explanation was first proposed by Larson (1972), who discovered that the SF maintained by constant metal-poor gas accretion reaches a constant value set only by the stellar yield (see Eq. [9]), which implies a value around the solar metallicity. In the context of this explanation, the apparent deficit of sub-solar metallicity G dwarfs is actually an excess of solar metallicity G dwarf stars formed over time out of an ISM always near equilibrium at approximately the solar metallicity.

The G dwarf problem has also been observed in K dwarfs (e.g., Casuso and Beckman 2004) and in M dwarfs (e.g., Woolf and West 2012), and it exists in other galaxies as well (e.g., Worthey et al. 1996). Current chemical evolution models resort to metal-poor gas inflow to reproduce the spatial distribution of stellar metallicities observed in the disk of spirals (e.g., Chiappini et al. 2001; Magrini et al. 2010; Mollá et al. 2016; Pezzulli and Fraternali 2016). Such gas inflow is needed for the same reasons invoked to solve the G dwarf problem.

3.6 Existence of a Minimum Metallicity for the Star-Forming Gas

XMPs are defined as galaxies where the gas that produces stars has a metallicity smaller than 10% of the solar metallicity. They turn out to be quite rare. Systematic searches, such as that carried by Sánchez Almeida et al. (2016), render a few hundred objects in catalogs containing of the order of one million galaxies (XMP represent $\ll$0.1% of the known galaxies). Interestingly, their metallicity, and therefore the metallicity of all local galaxies, seems to have a lower limit at around 2% of the solar metallicity. Figure 10 displays the distribution of metallicities found by Sánchez Almeida et al. (2016) in the search for XMP candidates in SDSS. It has a sharp cut-off at low metallicity, with no galaxy with $12 + \log(O/H) \leq 7.0$. The current record-breaking object is AGC 198691, with a metallicity around 2.1% times the solar metallicity (Hirschauer et al. 2016). It is also included in Fig. 10 for reference.

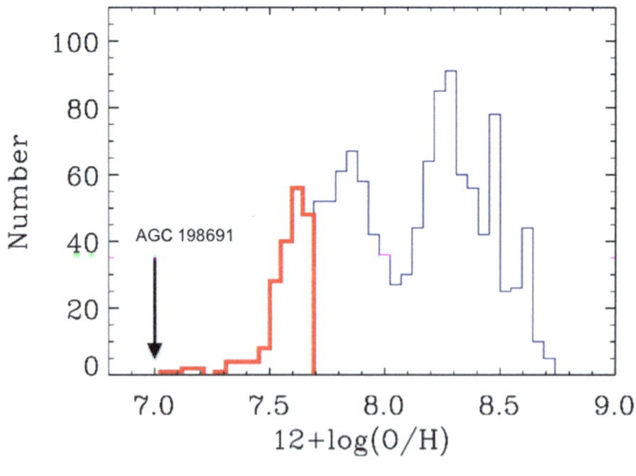

Fig. 10 Distribution of oxygen abundance $[12 + \log(O/H)]$ for all the objects found in the search for low-metallicity galaxies by Sánchez Almeida et al. (2016). The solar metallicity is set at $12 + \log(O/H)_{\odot} = 8.69$, therefore, the value $12 + \log(O/H) = 7.69$ separates the XMPs (the *thick red line*) and the failed XMP candidates (the *thin blue line*). There is a clear drop in the distribution toward low metallicity, with no object metal-poorer than 2% of the solar metallicity. AGC 198691 is the star-forming galaxy with the lowest metallicity known (Hirschauer et al. 2016), and the lower limit it sets is marked with an *arrow*. Figure adapted from Sánchez Almeida et al. (2016)

The existence of this metallicity threshold is not an artifact. Observers have been looking for record-breaking galaxies during the last 45 years, after the discovery of the prototypical XMP galaxy IZw 18 (Sargent and Searle 1970). These efforts led to enlarging the number of known XMPs (Terlevich et al. 1991; Guseva et al. 2009, 2015; Morales-Luis et al. 2011; Izotov et al. 2012; Sánchez Almeida et al. 2016), but the lower limit metallicity set by IZw 18 remains almost unchanged (the metallicity of IZw 18 is about 3% times the solar metallicity; see, e.g., Thuan and Izotov 2005).

Several explanations have been put forward to account for the existence of a minimum metallicity in the gas that forms stars. Kunth and Sargent (1986) point out the self-enrichment of the H II region used for measuring. However, the time-scale for the SN ejecta to cool down and mix is of the order of several hundred Myr (e.g., Legrand et al. 2001) and, thus, longer than the age of the H II regions. This fact eliminates the possibility of self-contamination. Self-enrichment by massive star winds seems to be negligible too (Kröger et al. 2006). Kunth and Lebouteiller (2011) suggest the pre-enrichment of the proto-galactic clouds, but it is unclear why there should be a minimum metallicity for such clouds, except perhaps the value set by the metal contamination produced by Pop III stars. Pop III star contamination has been suggested too (Audouze and Silk 1995; Thuan and Izotov 2005), but the expected level of metal enrichment is of the order of 10^{-4} times the solar metallicity (Bromm and Larson 2004; Muratov et al. 2013), and so much lower than the observed threshold.

Alternatively, if the SF is feeding from gas of the IGM, there is a minimum gas-phase metallicity set by the metallicity of the local IGM. This possibility provides a natural explanation for the long-lasting puzzle. Numerical simulations predict the local cosmic web gas to have a metallicity of the order of 1% times the solar value (e.g., van de Voort and Schaye 2012; Rahmati et al. 2016). The metal content of the IGM has been rising over time contaminated by galactic winds, and now it happens to be at the level of the observed metallicity threshold.

Other independent observations also support the existence of a minimum metallicity in the CGM of galaxies at the level of 1% times the solar metallicity. Lehner et al. (2013) measure the distribution of metallicity of Lyα absorbing clouds around galaxies with redshift up to one. The clouds are observed in absorption against background QSOs. The distribution of metallicity turns out to be bimodal with typical values around 2.5% solar and 50% solar. The high-metallicity branch is expected to represent galaxy outflows, whereas the low-metallicity branch corresponds to inflows. The lowest measured metallicity turns out to be 1% times the solar metallicity. In a completely independent type of work, Filho et al. (2013) studied the H I gas around XMP galaxies. The XMPs happen to be extremely gas-rich, with gas fractions typically in excess of 10. Assuming that all the metals produced by the observed stellar populations have been diluted in their huge H I reservoirs, the metallicity of the H I is again around a few percent of the solar value. The same conclusion has been recently reached by Thuan et al. (2016). Sometimes the metallicity of the H I gas can be measured directly using UV lines in absorption against the stellar light of the galaxy. In the case of IZw 18, Lebouteiller et al. (2013) find that H I region abundances are also around 1% of the solar value.

3.7 Origin of α-enhanced Gas Forming Stars in Local Galaxies

The gas forming stars in some of the local dwarf galaxies has elemental abundances which do not scale with the solar abundances. The gas is α-*enhanced*, using the terminology employed when studying stellar populations in massive galaxies and in the MW halo (e.g., Adibekyan et al. 2011; Vazdekis et al. 2015). The star-forming gas often shows $\log(\mathrm{N/O}) \simeq -1.5$ (e.g., James et al. 2015; Vincenzo et al. 2016) which is much smaller than the solar value (of the order of -0.86; Asplund et al. 2009). Stars and gas with α-enhanced composition are expected to be formed from the ejecta of young stellar populations (e.g., Izotov and Thuan 1999), so that low-mass stars have not had the time to explode as SN, and so, elements like Fe are underrepresented compared to the solar composition. N is one of these elements, even though N is also produced in intermediate-mass stars (e.g., Henry et al. 2000). Figure 11 shows the time evolution of N/O in several different model starbursts with very different SF efficiencies (i.e., with different τ_g in the parlance used in Sect. 1). Vincenzo et al. (2016) compute them to model the relationship between N/O and O/H observed in the galaxies of the local universe. Independently of the SF efficiency, when a starburst is 2 Gyr old it has already reached $\mathrm{N/O} \simeq \mathrm{N/O}_{\odot}$ (see Fig. 11). Something equivalent is shown by Köppen and Hensler (2005) when modeling the evolution of systems that accrete large amounts of metal-free gas. After a transient phase that lasts around 2 Gyr, the system returns to the original $\mathrm{N/O}_{\odot}$. The fact that the gas is α-enhanced is consistent with the thorough study on the heavy element abundances in local star-forming galaxies carried out by Izotov and Thuan (1999). The α-elements Ne, Si, S, and Ar, produced by SN explosions of

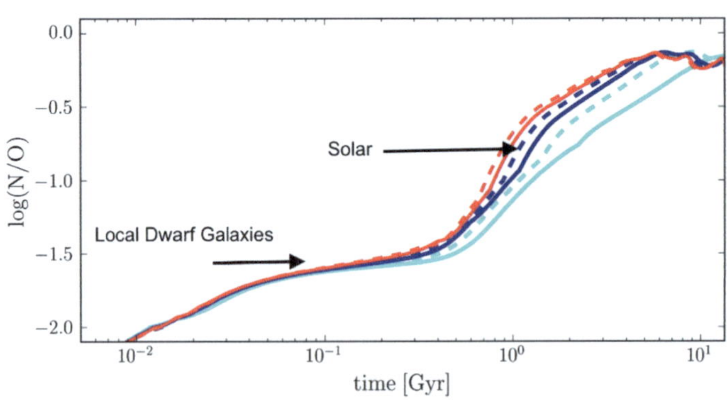

Fig. 11 Time evolution of N/O for chemical evolution models aimed at reproducing the relationship between N/O and O/H exhibited by the galaxies of the local universe. The different curves correspond to different τ_g going from 2 Gyr (*solid magenta line*) to 0.3 Gyr (*red dashed line*). The figure includes the N/O found in the Sun and the characteristic value observed in metal-poor galaxies of the local universe. By the time that the burst is 2 Gyr old, N/O has already reached the solar value in all models. Adapted from Vincenzo et al. (2016)

massive (>10 $M_\odot$) stars that also produce O, show the same α-element/O ratio as the Sun. However, O turns out to be overabundant with respect to N, C, and Fe by factors of around three, which are typical of MW halo stars.

It is well known that all galaxies, including the local dwarf galaxies, have their mass dominated by old stars with ages extending all the way back to the origin of the universe (e.g., Heavens et al. 2004; Sánchez Almeida et al. 2012, see also Fig. 2). The fact that N/O is low in the star-forming gas of some local dwarfs implies that their evolved stellar populations are not the source of the metals. If the metals are not produced by the observed stellar populations, where do they come from? They likely come together with the accreted gas from the IGM. The IGM gas is expected to have α-enhanced chemical composition since the metals it contains were produced by dwarfs galaxies in the early universe when the stellar populations were young (e.g., Adelberger et al. 2005; van de Voort and Schaye 2012; Yates et al. 2013).

The presence of α-enhanced gas forming stars at high redshift seems to be common. Steidel et al. (2016) argue that the unusually high excitation of the gas forming stars in galaxies at redshift around 2 is due to their extreme α-enhanced composition, with O/Fe$\sim$ 5 times the solar ratio. Such extreme conditions yield Fe-poor stars in an ISM of moderate-high metallicity as traced by O. Similar physical conditions in the local universe may be responsible for the presence of high excitation narrow He II lines in the spectra of metal-poor galaxies of the local universe (Shirazi and Brinchmann 2012).

3.8 The Metallicity of the Quiescent BCD Galaxies

The blue compact dwarf (BCD) galaxies are going through an intense starburst phase that cannot be maintained for long. Consequently, there must be many dormant galaxies in a pre- or a post-BCD phase. They are called quiescent BCDs (QBCD). BCDs possess one or a few bright starburst regions on a low-surface host. Masking out the high-surface knots, Amorín et al. (2007, 2009) characterized the photometric properties of the host galaxy, which likely corresponds to the underlying QBCD. Using these properties as reference, Sánchez Almeida et al. (2008) found out that the SDSS catalog contains as many as 30 QBCDs per BCD. BCDs and QBCDs seem to be drawn from a single galaxy population that metamorphose with a cycle where the quiescent phase lasts 30 times longer than the star-forming phase. However, this interpretation presents a difficulty. The gas metallicity of the QBCDs is systematically higher than the metallicity of the BCDs, which cannot happen if the transformation between QBCD and BCD occurs through closed-box evolution, where the precursor always has lower metallicity than the offspring. The problem naturally goes away if the BCD phase is triggered by the accretion of external metal-poor gas that feeds the observed SF episode. The external driving of the BCD phase also explains why the stellar metallicity of BCDs and QBCDs agrees, even though their gas metallicity differs (Sánchez Almeida et al. 2009). The stellar populations of BCDs and QBCDs are statistically the same because only a small fraction of the galaxy stellar mass is built up in each new burst.

3.9 Direct Measurement of Inflows in Star-Forming Galaxies

To the best of our knowledge, there is no direct measurement of metal-poor gas inflows in the CGM of galaxies. Finding whether a particular Doppler shift corresponds to inflows or outflows is tricky because the same signal is provided by inflows or outflows depending on whether the source is in the foreground or the background. Fortunately, the sense of motion can be disambiguated when the gas is absorbing or emitting against the spectrum of the galaxy. In these cases the gas is in the foreground, which breaks the degeneracy.

Thus, gas inflows have been detected in absorption against stellar spectra (e.g., Rubin and MaNGA Team 2016), but they are probably associated with metal-rich gas returning to the galaxy after being ejected by a starburst or an AGN (e.g., Sancisi et al. 2008; Marasco et al. 2012; Fraternali 2014). Neutral gas has been studied in absorption in several metal-poor objects (Aloisi et al. 2003; Cannon et al. 2005; Lebouteiller et al. 2009, 2013). However, the relative velocity between the gas absorption and the systemic velocity of the galaxy is too small to tell whether the gas is falling in or going out. If anything, the gas seems to be flowing out with a mild velocity between 10 and 20 $\mathrm{km\,s^{-1}}$ (Lebouteiller et al. 2013). The origin of this putative metal-poor outflow is puzzling.

Recent observations by Fathivavsari et al. (2016) show signs of infall in DLAs eclipsing QSOs, i.e., dense H I gas so close in redshift to the QSO that it blocks the Lyα emission of the source. Their metallicities are relatively low, and most of them have redshifts larger than that of their background QSO, strengthening the idea that they could be associated with some low metallicity infalling material accreting onto the QSO host galaxies.

Even though the works by Kacprzak et al. (2012, 2015) do not have kinematical information, they are very suggestive of gas inflows in star-forming galaxies. They study the azimuthal distribution of Mg II and O VI absorbers around the galaxy responsible for the absorption. These absorption systems trace gas in the CGM and the IGM, depending on the distance to the galaxy. Kacprzak et al. (2012, 2015) find that the absorption preferentially occurs along the directions pointed out by the minor and the major axes of the central galaxy. The absorbers aligned with the major axis are expected to show gas inflows, with the absorbers in the direction of the minor axis corresponding to gas outflows. In order to secure the whole scenario, one would need to have measurements of the velocity and metallicity of the Mg II systems. Unfortunately, they are not available. However, two independent measurements support the above interpretation. Firstly, outflows prefer the direction perpendicular to the plane of the galaxy because Mg II outflows are faster in face-on galaxies (Bordoloi et al. 2014; Rubin et al. 2014). Secondly, the metallicity of Lyman limit systems (i.e., gas clumps of moderate H I column density) is observed to be bimodal, with one peak at low metallicity and the other at high metallicity (Lehner et al. 2013). This bimodality is to be expected if the observed absorption is produced by metal-poor inflows and metal-rich outflows.

4 Obvious Complications and Future Trends

The evidence discussed in Sect. 3 is all indirect. We have to rely on models to interpret the observables as evidence for gas accretion. The toy model setting up the scene in Sect. 1, and then used in many of the above arguments, oversimplifies many aspects of the accretion process that are important to identify in real galaxies the observational signatures of ongoing cosmological gas accretion. The purpose of this section is to point out some of the obvious complications of a more realistic modeling, and also to point out new observational pathways that may reveal connections between SF and the gas of the IGM.

The CGM and the Galactic Fountain The CGM is a complex region where gas ejected from the galaxy and gas falling into the galaxy coexist. The gas inflow is not only of cosmological origin. Part of the metal-rich materials ejected in previous SF episodes fall back to the disk in what is called galactic fountain (e.g., Fraternali and Binney 2006, 2008; Hobbs et al. 2015). The mixing with metal-rich gas from SF processes speeds up the cooling of the hot metal-poor CGM, and the fountain also returns gas that was never ejected (Melioli et al. 2008; Marinacci et al. 2010; Marasco and Fraternali 2011; Marasco et al. 2012).

The structure of the CGM is extremely complicated according to the current numerical simulations. The morphology of the gas streams becomes increasingly complex at higher resolution, with large coherent flows revealing density and temperature structure at progressively smaller scales, and with no evidence that the substructure is properly captured in the simulations (Nelson et al. 2016). Multiple gas components coexist at the same radius within the halo, making radially averaged analyses misleading. This is particularly true where the hot quasi-static halo interacts with cold rapidly inflowing IGM accretion. Some of the resulting complications are revealed in the study of metal-poor DLAs carried out by Yuan and Cen (2016). The majority of the metal-poor DLAs are far from the central galaxy ($\geq$20 kpc) and result from the cold gas streams from the IGM. In the migration inwards to the galaxy, they mix up with high-metallicity gas from stellar outflows, removing themselves from the metal-poor category. The change from metal-poor to metal-rich complicates the observational identification of the gas coming from the IGM. IGM gas clouds that get mixed with the CGM and become metal-rich are also found in the simulation by Gritton et al. (2014). The difficulties of interpretation are also put forward by Ford et al. (2014). Using cosmological simulations, Ford et al. examine how H I and metal absorption lines trace the dynamical state of the CGM around low-redshift galaxies. Recycled wind material is preferentially found close to galaxies, and is more dominant in low-mass halos. Typical H I absorbers trace unenriched ambient material that is not participating in the baryon cycle, but stronger H I absorbers arise in cool, metal-enriched inflowing gas. Instantaneous radial velocity measurements are generally poor at distinguishing between inflowing and outflowing gas, except in the case of very recent outflows.

Galactic fountains and metal-rich gas produced in previous starbursts are very important because their presence complicates the search for metal-poor gas inflows in the CGM of galaxies. However, their role in sustaining SF should not be overestimated. Most of the gas used to produce stars at any time is pristine. It was never pre-processed by a star. This issue has been recently quantified by Segers et al. (2016) using galaxies from the EAGLE numerical simulation (Schaye et al. 2015). For MW-like galaxies, recycled stellar ejecta account for only 35% of the SFR and 20% of the stellar mass. The contribution was even less important in the past. The toy model in Sect. 1 provides the right order of magnitude for this estimate (see Sánchez Almeida et al. 2014a).

Star Formation Generated by Gas Accretion The current cosmological numerical simulations produce model galaxies that look impressively realistic, and follow most of the well-known scaling relations (e.g., Vogelsberger et al. 2014; Schaye et al. 2015). They provide the theoretical framework to understand the formation of galaxies and the role played by cosmological gas accretion in maintaining SF. However, their limited resolution and the dependence of many predictions on the adopted sub-grid physics make them less reliable to study how individual starbursts grow out of the gas that arrives to the galaxy disk. (Predictions on individual star formation events are discussed in Sec. 3.3.) Improving this aspect is critically important to secure the interpretation of many observables currently used as evidence for cosmological gas accretion.

Imaging the Cosmic Web Much of our knowledge on the CGM and IGM comes from observing absorption lines against background sources that happen to be next to galaxies. However, the observation and analysis of these absorption systems is extremely time-consuming, and even the best cases only provide a very sparse sampling of the CGM and IGM around individual galaxies. A complementary approach is observing the cosmic web gas in emission. The mechanisms to produce such emission are varied. Lyα (as well as Hα) can be produced by electron collisions within a gas stream that releases the gravitational energy gained as gas flows from the IGM into the galaxy halo (Dijkstra and Loeb 2009; Goerdt et al. 2010; Faucher-Giguère et al. 2010). Emission also results from fluorescence induced by an intense UV radiation field such as that produced by a large nearby starburst or a QSO (e.g., Hogan and Weymann 1987; Cantalupo et al. 2012; Ao et al. 2015).

Radio emission is also expected to trace the cosmic web. In this sense, the search for *dark galaxies* (i.e., objects emitting in 21 cm without optical counterpart) is a very revealing and active field of research (e.g., Cannon et al. 2014; Serra et al. 2015; Janesh et al. 2015; Janowiecki et al. 2015). Gas filaments associated with star-forming galaxies are very interesting too (Sancisi et al. 2008; Lelli et al. 2012; Filho et al. 2013). Radio data easily provide kinematical information, which is so important when investigating flows. In this context, a large filament of molecular gas accreting onto a group of massive high redshift galaxies has been recently discovered by Ginolfi et al. (2016, private communication). This observation is intriguing, but it may open up a new way of addressing the search for IGM gas.

Extremely promising is the recent discovery of an extended Lyα blob connected to a QSO (Cantalupo et al. 2014). The emission extends beyond the virial radius of the host galaxy so that it traces gas in the IGM. The existence of extended Lyα emission around QSOs seems to be very common when the observation is deep enough (Borisova et al. 2016). The sensitivity is very much improved using spectrographic observations, which also have the capability of providing the eagerly needed kinematical information (see the chapter by Cantalupo in this book). Extended Lyα halos are common around all kinds of galaxies (e.g., Rauch et al. 2016; Momose et al. 2016).

5 Conclusions

Cosmological numerical simulations predict that the SF in regular disk galaxies is feeding from metal-poor gas accreted from the cosmic web. The observational evidence for a relation between SFR and external gas accretion is both numerous and indirect (Sect. 3). One necessarily has to rely on modeling and analysis to identify the existing hints as actual signs of SF driven by metal-poor gas accretion. Thus, the comparison with numerical simulations and a meticulous interpretation of the observations turn out to be mandatory. There is no reason to believe that the magnitude of the problem will change significantly in the near future. The IGM gas is predicted to be tenuous and ionized, and so extremely elusive observationally. The IGM gas gets mixed with metal-rich outflows, and may loose its distinctive metal-poor character to complicate the study. In addition, metal-rich recycled material is also re-accreted, and often velocities are useless to separate inflows from outflows. Although things have much improved during the last few years, theoretical predictions are still very unspecific as far as the details are concerned. These details are needed to interpret particular observational results as evidence for accretion (Sect. 3.4). Finally, the difficulty to distinguish gas accretion events from gas-rich minor mergers confuses the interpretation even further (Sect. 3.3; Table 1).

Acknowledgements Special thanks are due to Casiana Munõz-Tuñón, Debra Elmegreen, and Bruce Elmegreen, for continuous support and for long thoughtful discussions on almost every topic included in the work. I am also indebted to Mercedes Filho, who pointed out many of the references cited in the work, and shared with me her expertise on H I. The interpretative aspects of the work own much to the collaboration with Claudio Dalla Vecchia and Daniel Ceverino. They were always willing to discuss the physical aspects of the accretion and the star formation process. Thanks are due to R. Amorín for the discussions that led to the writing of Sect. 3.7. Thanks are also due to the editors of the Book for giving me the opportunity to contribute, and for their patience to have my contribution finished. This work has been partly funded by the Spanish Ministry of Economy and Competitiveness, project *Estallidos* AYA2013–47742–C04–02–P and AYA2016- 79724-C4-2-P.

Table 1 List of acronyms and symbols defined and used along the text

Acronym	Expansion	Acronym	Expansion
ADS	NASA Astronomical Data System	$M_\star$	Stellar mass
BCD	Blue compact dwarf	M_g	Gas mass
CGM	Circum-galactic medium	QBCD	Quiescent blue compact dwarf
DLA	Damped Lyman-α absorbers	QSO	Quasar
EAGLE	Evolution and Assembly of GaLaxies and their Environments Schaye et al. (2015)	SDSS	Sloan Digital Sky Survey
		SF	Star formation
		SFR	Star formation rate
FMR	Fundamental metallicity relation	SN, SNe	Supernova, Supernovae
HCM	HII-Chi-Mistry (Pérez-Montero 2014)	SPH	Smoothed particle hydrodynamics
IGM	Inter-galactic medium	UV	Ultraviolet
ISM	Interstellar medium	η	Mass loading factor [Eq. (4)]
IMF	Initial mass function	XMP	Extremely metal poor
MZR	(Stellar) Mass-metallicity relation	Z	Metallicity of the gas [Eq. (9)]
MW	Milky Way		

References

Adelberger, K. L., Shapley, A. E., Steidel, C. C., et al. 2005, ApJ, 629, 636

Adibekyan, V. Z., Santos, N. C., Sousa, S. G., & Israelian, G. 2011, A&A, 535, L11

Aloisi, A., Savaglio, S., Heckman, T. M., et al. 2003, ApJ, 595, 760

Altay, G., Theuns, T., Schaye, J., Booth, C. M., & Dalla Vecchia, C. 2013, MNRAS, 436, 2689

Amorín, R., Aguerri, J. A. L., Muñoz-Tuñón, C., & Cairós, L. M. 2009, A&A, 501, 75

Amorín, R., Sommariva, V., Castellano, M., et al. 2014, A&A, 568, L8

Amorín, R. O., Muñoz-Tuñón, C., Aguerri, J.A.L., Cairós, L.M., & Caon, N. 2007, A&A, 467, 541

Andrews, B. H. & Martini, P. 2013, ApJ, 765, 140

Ao, Y., Matsuda, Y., Beelen, A., et al. 2015, A&A, 581, A132

Ascasibar, Y., Gavilán, M., Pinto, N., et al. 2015, MNRAS, 448, 2126

Asplund, M., Grevesse, N., Sauval, A. J., & Scott, P. 2009, ARA&A, 47, 481

Audouze, J. & Silk, J. 1995, ApJL, 451, L49

Bigiel, F., Leroy, A., Walter, F., et al. 2008, AJ, 136, 2846

Bordoloi, R., Lilly, S. J., Hardmeier, E., et al. 2014, ApJ, 794, 130

Borisova, E., Cantalupo, S., Lilly, S. J., et al. 2016, ApJ, 831, 39

Bothwell, M. S., Maiolino, R., Cicone, C., Peng, Y., & Wagg, J. 2016a, A&A, 595, A48

Bothwell, M. S., Maiolino, R., Kennicutt, R., et al. 2013, MNRAS, 433, 1425

Bothwell, M. S., Maiolino, R., Peng, Y., et al. 2016b, MNRAS, 455, 1156

Bouché, N., Dekel, A., Genzel, R., et al. 2010, ApJ, 718, 1001

Bournaud, F., Perret, V., Renaud, F., et al. 2014, ApJ, 780, 57

Brisbin, D. & Harwit, M. 2012, ApJ, 750, 142

Bromm, V. & Larson, R. B. 2004, ARA&A, 42, 79

Brown, J. S., Martini, P., & Andrews, B. H. 2016, MNRAS, 458, 1529

Cannon, J. M., Johnson, M., McQuinn, K. B. W., et al. 2014, ApJL, 787, L1

Cannon, J. M., Skillman, E. D., Sembach, K. R., & Bomans, D. J. 2005, ApJ, 618, 247

Cantalupo, S., Arrigoni-Battaia, F., Prochaska, J. X., Hennawi, J. F., & Madau, P. 2014, Nature, 506, 63

Cantalupo, S., Lilly, S. J., & Haehnelt, M. G. 2012, MNRAS, 425, 1992

Carigi, L. 1996, RMxAA, 32

Casuso, E. & Beckman, J. E. 2004, A&A, 419, 181

Casuso, E., Beckman, J. E., & Buenrostro, V. 2006, PASP, 118, 833
Ceverino, D., Dekel, A., & Bournaud, F. 2010, MNRAS, 404, 2151
Ceverino, D., Sánchez Almeida, J., Muñoz Tuñón, C., et al. 2016, MNRAS, 457, 2605
Chiappini, C., Matteucci, F., & Romano, D. 2001, ApJ, 554, 1044
Christensen, C. R., Davé, R., Governato, F., et al. 2016, ApJ, 824, 57
Croxall, K. V., van Zee, L., Lee, H., et al. 2009, ApJ, 705, 723
Cullen, F., Cirasuolo, M., McLure, R. J., Dunlop, J. S., & Bowler, R. A. A. 2014, MNRAS, 440, 2300
Dalla Vecchia, C. & Schaye, J. 2008, MNRAS, 387, 1431
Dalla Vecchia, C. & Schaye, J. 2012, MNRAS, 426, 140
Davé, R., Finlator, K., & Oppenheimer, B. D. 2011, MNRAS, 416, 1354
Davé, R., Finlator, K., & Oppenheimer, B. D. 2012, MNRAS, 421, 98
Davé, R. and Katz, N. and Oppenheimer, B. D. and Kollmeier, J. A. and Weinberg, D. H. 2013, MNRAS, 434, 2645
Dayal, P., Ferrara, A., & Dunlop, J. S. 2013, MNRAS, 430, 2891
de Avillez, M. A. & Mac Low, M.-M. 2002, ApJ, 581, 1047
de los Reyes, M. A., Ly, C., Lee, J. C., et al. 2015, AJ, 149, 79
De Rossi, M. E., Theuns, T., Font, A. S., & McCarthy, I. G. 2015, MNRAS, 452, 486
Dekel, A., Birnboim, Y., Engel, G., et al. 2009, Nature, 457, 451
Dekel, A. & Mandelker, N. 2014, MNRAS, 444, 2071
Dekel, A., Zolotov, A., Tweed, D., et al. 2013, MNRAS, 435, 999
Dijkstra, M. & Loeb, A. 2009, MNRAS, 400, 1109
Divoy, C., Contini, T., Pérez-Montero, E., et al. 2014, A&A, 569, A64
Edmunds, M. 2005, Astronomy and Geophysics, 46, 4.12
Edmunds, M. G. 1990, MNRAS, 246, 678
Ellison, S. L., Patton, D. R., Simard, L., & McConnachie, A. W. 2008, ApJL, 672, L107
Elmegreen, D. M., Elmegreen, B. G., Rubin, D. S., & Schaffer, M. A. 2005, ApJ, 631, 85
Elmegreen, D. M., Elmegreen, B. G., Sánchez Almeida, J., et al. 2012, ApJ, 750, 95
Fathivavsari, H., Petitjean, P., Noterdaeme, P., et al. 2016, MNRAS, 461, 1816
Faucher-Giguère, C.-A., Kereš, D., Dijkstra, M., Hernquist, L., & Zaldarriaga, M. 2010, ApJ, 725, 633
Feldmann, R. 2013, MNRAS, 433, 1910
Filho, M. E., Sánchez Almeida, J., Amorín, R., et al. 2016, ApJ, 820, 109
Filho, M. E., Winkel, B., Sánchez Almeida, J., et al. 2013, A&A, 558, A18
Finlator, K. & Davé, R. 2008, MNRAS, 385, 2181
Forbes, D. A. & Kroupa, P. 2011, PASA, 28, 77
Forbes, J. C., Krumholz, M. R., Burkert, A., & Dekel, A. 2014a, MNRAS, 438, 1552
Forbes, J. C., Krumholz, M. R., Burkert, A., & Dekel, A. 2014b, MNRAS, 443, 168
Ford, A. B., Davé, R., Oppenheimer, B. D., et al. 2014, MNRAS, 444, 1260
Fraternali, F. 2014, in IAU Symposium, Vol. 298, Setting the scene for Gaia and LAMOST, ed. S. Feltzing, G. Zhao, N. A. Walton, & P. Whitelock, 228
Fraternali, F. & Binney, J. J. 2006, MNRAS, 366, 449
Fraternali, F. & Binney, J. J. 2008, MNRAS, 386, 935
Fraternali, F. & Tomassetti, M. 2012, MNRAS, 426, 2166
Genzel, R., Tacconi, L. J., Gracia-Carpio, J., et al. 2010, MNRAS, 407, 2091
Gnedin, N. Y., Tasker, E. J., & Fujimoto, Y. 2014, ApJL, 787, L7
Goerdt, T., Dekel, A., Sternberg, A., et al. 2010, MNRAS, 407, 613
Gritton, J. A., Shelton, R. L., & Kwak, K. 2014, ApJ, 795, 99
Grønnow, A. E., Finlator, K., & Christensen, L. 2015, MNRAS, 451, 4005
Guglielmo, V., Poggianti, B. M., Moretti, A., et al. 2015, MNRAS, 450, 2749
Guseva, N. G., Izotov, Y. I., Fricke, K. J., & Henkel, C. 2015, A&A, 579, A11
Guseva, N. G., Papaderos, P., Meyer, H. T., Izotov, Y. I., & Fricke, K. J. 2009, A&A, 505, 63
Harwit, M. & Brisbin, D. 2015, ApJ, 800, 91
Haurberg, N. C., Rosenberg, J., & Salzer, J. J. 2013, ApJ, 765, 66

Heavens, A., Panter, B., Jimenez, R., & Dunlop, J. 2004, Nature, 428, 625
Henry, R. B. C., Edmunds, M. G., & Köppen, J. 2000, ApJ, 541, 660
Hirschauer, A. S., Salzer, J. J., Skillman, E. D., et al. 2016, ApJ, 822, 108
Hobbs, A., Read, J., & Nicola, A. 2015, MNRAS, 452, 3593
Hogan, C. J. & Weymann, R. J. 1987, MNRAS, 225, 1P
Izotov, Y. I., Guseva, N. G., Fricke, K. J., & Henkel, C. 2014, A&A, 561, A33
Izotov, Y. I., Guseva, N. G., Fricke, K. J., & Henkel, C. 2015, MNRAS, 451, 2251
Izotov, Y. I., Guseva, N. G., Fricke, K. J., & Papaderos, P. 2009, A&A, 503, 61
Izotov, Y. I. & Thuan, T. X. 1999, ApJ, 511, 639
Izotov, Y. I., Thuan, T. X., & Guseva, N. G. 2012, A&A, 546, A122
James, B. L., Koposov, S., Stark, D. P., et al. 2015, MNRAS, 448, 2687
Janesh, W., Rhode, K. L., Salzer, J. J., et al. 2015, ApJ, 811, 35
Janowiecki, S., Leisman, L., Józsa, G., et al. 2015, ApJ, 801, 96
Jimmy, Tran, K.-V., Saintonge, A., et al. 2015, ApJ, 812, 98
Kacprzak, G. G., Churchill, C. W., & Nielsen, N. M. 2012, ApJL, 760, L7
Kacprzak, G. G., Muzahid, S., Churchill, C. W., Nielsen, N. M., & Charlton, J. C. 2015, ApJ, 815, 22
Kacprzak, G. G., van de Voort, F., Glazebrook, K., et al. 2016, ApJL, 826, L11
Kashino, D., Renzini, A., Silverman, J. D., & Daddi, E. 2016, ApJL, 823, L24
Kazantzidis, S., Zentner, A. R., Kravtsov, A. V., Bullock, J. S., & Debattista, V. P. 2009, ApJ, 700, 1896
Kennicutt, R. C. & Evans, N. J. 2012, ARA&A, 50, 531
Kennicutt, Jr., R. C. 1983, ApJ, 272, 54
Kennicutt, Jr., R. C. 1998, ApJ, 498, 541
Kobulnicky, H. A. & Skillman, E. D. 1996, ApJ, 471, 211
Köppen, J. & Hensler, G. 2005, A&A, 434, 531
Kröger, D., Hensler, G., & Freyer, T. 2006, A&A, 450, L5
Kunth, D. & Lebouteiller, V. 2011, in EAS Publications Series, Vol. 48, EAS Publications Series, ed. M. Koleva, P. Prugniel, & I. Vauglin, 95–96
Kunth, D. & Sargent, W. L. W. 1986, ApJ, 300, 496
Lara-López, M. A., Cepa, J., Bongiovanni, A., et al. 2010, A&A, 521, L53
Larson, R. B. 1972, Nature Physical Science, 236, 7
Lebouteiller, V., Heap, S., Hubeny, I., & Kunth, D. 2013, A&A, 553, A16
Lebouteiller, V., Kunth, D., Thuan, T. X., & Désert, J. M. 2009, A&A, 494, 915
Legrand, F., Tenorio-Tagle, G., Silich, S., Kunth, D., & Cerviño, M. 2001, ApJ, 560, 630
Lehner, N., Howk, J. C., Tripp, T. M., et al. 2013, ApJ, 770, 138
Leitner, S. N. & Kravtsov, A. V. 2011, ApJ, 734, 48
Lelli, F., Verheijen, M., & Fraternali, F. 2014, MNRAS, 445, 1694
Lelli, F., Verheijen, M., Fraternali, F., & Sancisi, R. 2012, A&A, 537, A72
Levesque, E. M., Berger, E., Soderberg, A. M., & Chornock, R. 2011, ApJ, 739, 23
Lian, J. H., Li, J. R., Yan, W., & Kong, X. 2015, MNRAS, 446, 1449
Lilly, S. J., Carollo, C. M., Pipino, A., Renzini, A., & Peng, Y. 2013, ApJ, 772, 119
López-Sánchez, Á. R. 2010, A&A, 521, A63
Ly, C., Malkan, M. A., Nagao, T., et al. 2014, ApJ, 780, 122
Ly, C., Rigby, J. R., Cooper, M., & Yan, R. 2015, ApJ, 805, 45
Lynden-Bell, D. 1975, Vistas in Astronomy, 19, 299
Magrini, L., Stanghellini, L., Corbelli, E., Galli, D., & Villaver, E. 2010, A&A, 512, A63
Maier, C., Lilly, S. J., Ziegler, B. L., et al. 2014, ApJ, 792, 3
Maier, C., Ziegler, B. L., Lilly, S. J., et al. 2015, A&A, 577, A14
Malinie, G., Hartmann, D. H., Clayton, D. D., & Mathews, G. J. 1993, ApJ, 413, 633
Mandelker, N., Dekel, A., Ceverino, D., et al. 2014, MNRAS, 443, 3675
Mannucci, F., Cresci, G., Maiolino, R., Marconi, A., & Gnerucci, A. 2010, MNRAS, 408, 2115
Marasco, A. & Fraternali, F. 2011, A&A, 525, A134
Marasco, A., Fraternali, F., & Binney, J. J. 2012, MNRAS, 419, 1107

Marinacci, F., Binney, J., Fraternali, F., et al. 2010, MNRAS, 404, 1464
Martin, C. L. 1999, ApJ, 513, 156
Martin, C. L., Shapley, A. E., Coil, A. L., et al. 2012, ApJ, 760, 127
Martinelli, A. & Matteucci, F. 2000, A&A, 353, 269
Melioli, C., Brighenti, F., D'Ercole, A., & de Gouveia Dal Pino, E. M. 2008, MNRAS, 388, 573
Mollá, M., Díaz, Á. I., Gibson, B. K., Cavichia, O., & López-Sánchez, Á.-R. 2016, MNRAS, 462, 1329
Momose, R., Ouchi, M., Nakajima, K., et al. 2016, MNRAS, 457, 2318
Morales-Luis, A. B., Sánchez Almeida, J., Aguerri, J. A. L., & Muñoz-Tuñón, C. 2011, ApJ, 743, 77
Muratov, A. L., Gnedin, O. Y., Gnedin, N. Y., & Zemp, M. 2013, ApJ, 772, 106
Nelson, D., Genel, S., Pillepich, A., et al. 2016, MNRAS, 460, 2881
Newman, S. F., Genzel, R., Förster-Schreiber, N. M., et al. 2012, ApJ, 761, 43
Olmo-García, A., Sánchez Almeida, J., Muñoz-Tuñón, C., et al. 2017, ApJ, 834, 181
Pagel, B. E. J. 2001, in Cosmic evolution, ed. E. Vangioni-Flam, R. Ferlet, & M. Lemoine, 223
Papaderos, P., Guseva, N. G., Izotov, Y. I., & Fricke, K. J. 2008, A&A, 491, 113
Papaderos, P., Izotov, Y. I., Guseva, N. G., Thuan, T. X., & Fricke, K. J. 2006, A&A, 454, 119
Pasquali, A., Gallazzi, A., & van den Bosch, F. C. 2012, MNRAS, 425, 273
Pawlik, M. M., Wild, V., Walcher, C. J., et al. 2016, MNRAS, 456, 3032
Peeples, M. S., Pogge, R. W., & Stanek, K. Z. 2009, ApJ, 695, 259
Peeples, M. S. & Shankar, F. 2011, MNRAS, 417, 2962
Peng, Y.-j. & Maiolino, R. 2014, MNRAS, 443, 3643
Pérez-Montero, E. 2014, MNRAS, 441, 2663
Petit, A. C., Krumholz, M. R., Goldbaum, N. J., & Forbes, J. C. 2015, MNRAS, 449, 2588
Pezzulli, G. & Fraternali, F. 2016, MNRAS, 455, 2308
Pilyugin, L. S., Grebel, E. K., & Zinchenko, I. A. 2015, MNRAS, 450, 3254
Pipino, A., Lilly, S. J., & Carollo, C. M. 2014, MNRAS, 441, 1444
Rahmati, A., Schaye, J., Crain, R. A., et al. 2016, MNRAS, 459, 310
Rauch, M., Becker, G. D., & Haehnelt, M. G. 2016, MNRAS, 455, 3991
Reichard, T. A., Heckman, T. M., Rudnick, G., et al. 2009, ApJ, 691, 1005
Richards, S. N., Schaefer, A. L., López-Sánchez, Á. R., et al. 2014, MNRAS, 445, 1104
Roberts, M. S. 1963, ARA&A, 1, 149
Rocha-Pinto, H. J. & Maciel, W. J. 1996, MNRAS, 279, 447
Rodríguez-Puebla, A., Primack, J. R., Behroozi, P., & Faber, S. M. 2016, MNRAS, 455, 2592
Romeo Velonà, A. D., Sommer-Larsen, J., Napolitano, N. R., et al. 2013, ApJ, 770, 155
Rubin, K. & MaNGA Team 2016, AAS Abstracts, Vol. 227, 312.10
Rubin, K. H. R., Prochaska, J. X., Koo, D. C., et al. 2014, ApJ, 794, 156
Salim, S., Lee, J. C., Ly, C., et al. 2014, ApJ, 797, 126
Sánchez, S. F., Rosales-Ortega, F. F., Jungwiert, B., et al. 2013, A&A, 554, A58
Sánchez Almeida, J., Aguerri, J. A. L., Muñoz-Tuñón, C., & Vazdekis, A. 2009, ApJ, 698, 1497
Sánchez Almeida, J., Elmegreen, B. G., Muñoz-Tuñón, C., & Elmegreen, D. M. 2014a, A&A Rev., 22, 71
Sánchez Almeida, J., Elmegreen, B. G., Muñoz-Tuñón, C., et al. 2015, ApJL, 810, L15
Sánchez Almeida, J., Morales-Luis, A. B., Muñoz-Tuñón, C., et al. 2014b, ApJ, 783, 45
Sánchez Almeida, J., Muñoz-Tuñón, C., Amorín, R., et al. 2008, ApJ, 685, 194
Sánchez Almeida, J., Muñoz-Tuñón, C., Elmegreen, D. M., Elmegreen, B. G., & Méndez-Abreu, J. 2013, ApJ, 767, 74
Sánchez Almeida, J., Pérez-Montero, E., Morales-Luis, A. B., et al. 2016, ApJ, 819, 110
Sánchez Almeida, J., Terlevich, R., Terlevich, E., Cid Fernandes, R., & Morales-Luis, A. B. 2012, ApJ, 756, 163
Sancisi, R., Fraternali, F., Oosterloo, T., & van der Hulst, T. 2008, A&A Rev., 15, 189
Sanders, R. L., Shapley, A. E., Kriek, M., et al. 2015, ApJ, 799, 138
Sargent, W. L. W. & Searle, L. 1970, ApJL, 162, L155
Schaye, J., Crain, R. A., Bower, R. G., et al. 2015, MNRAS, 446, 521

Schaye, J. & Dalla Vecchia, C. 2008, MNRAS, 383, 1210
Schaye, J., Dalla Vecchia, C., Booth, C. M., et al. 2010, MNRAS, 402, 1536
Schmidt, M. 1959, ApJ, 129, 243
Schmidt, M. 1963, ApJ, 137, 758
Segers, M. C., Crain, R. A., Schaye, J., et al. 2016, MNRAS, 456, 1235
Serra, P., Koribalski, B., Kilborn, V., et al. 2015, MNRAS, 452, 2680
Shen, S., Madau, P., Aguirre, A., et al. 2012, ApJ, 760, 50
Shirazi, M. & Brinchmann, J. 2012, MNRAS, 421, 1043
Silk, J. & Mamon, G. A. 2012, Research in Astronomy and Astrophysics, 12, 917
Skillman, E. D., Kennicutt, R. C., & Hodge, P. W. 1989, ApJ, 347, 875
Somerville, R. S. & Davé, R. 2015, ARA&A, 53, 51
Steidel, C. C., Strom, A. L., Pettini, M., et al. 2016, ApJ, 826, 159
Stott, J. P., Sobral, D., Swinbank, A. M., et al. 2014, MNRAS, 443, 2695
Telford, O. G., Dalcanton, J. J., Skillman, E. D., & Conroy, C. 2016, ApJ, 827, 35
Terlevich, R., Melnick, J., Masegosa, J., Moles, M., & Copetti, M. V. F. 1991, AAPs, 91, 285
Thompson, T. A. & Krumholz, M. R. 2016, MNRAS, 455, 334
Thöne, C. C., Christensen, L., Prochaska, J. X., et al. 2014, MNRAS, 441, 2034
Thuan, T. X., Goehring, K. M., Hibbard, J. E., Izotov, Y. I., & Hunt, L. K. 2016, MNRAS
Thuan, T. X. & Izotov, Y. I. 2005, ApJS, 161, 240
Tinsley, B. M. 1980, FCPh, 5, 287
Tremonti, C. A., Heckman, T. M., Kauffmann, G., et al. 2004, ApJ, 613, 898
Troncoso, P., Maiolino, R., Sommariva, V., et al. 2014, A&A, 563, A58
van de Voort, F. & Schaye, J. 2012, MNRAS, 423, 2991
van den Bergh, S. 1962, AJ, 67, 486
Vazdekis, A., Coelho, P., Cassisi, S., et al. 2015, MNRAS, 449, 1177
Veilleux, S., Cecil, G., & Bland-Hawthorn, J. 2005, ARA&A, 43, 769
Verbeke, R., De Rijcke, S., Koleva, M., et al. 2014, MNRAS, 442, 1830
Vincenzo, F., Belfiore, F., Maiolino, R., Matteucci, F., & Ventura, P. 2016, MNRAS, 458, 3466
Vogelsberger, M., Genel, S., Springel, V., et al. 2014, MNRAS, 444, 1518
Werk, J. K., Putman, M. E., Meurer, G. R., et al. 2010, ApJ, 715, 656
Woolf, V. M. & West, A. A. 2012, MNRAS, 422, 1489
Worthey, G., Dorman, B., & Jones, L. A. 1996, AJ, 112, 948
Wu, Y.-Z., Zhang, S.-N., Zhao, Y.-H., & Zhang, W. 2016, MNRAS, 457, 2929
Wuyts, E., Wisnioski, E., Fossati, M., et al. 2016, ApJ, 827, 74
Yabe, K., Ohta, K., Iwamuro, F., et al. 2012, PASJ, 64
Yang, C.-C. & Krumholz, M. 2012, ApJ, 758, 48
Yates, R. M., Henriques, B., Thomas, P. A., et al. 2013, MNRAS, 435, 3500
Yates, R. M., Kauffmann, G., & Guo, Q. 2012, MNRAS, 422, 215
Yuan, S. & Cen, R. 2016, MNRAS, 457, 487
Zahid, H. J., Kashino, D., Silverman, J. D., et al. 2014, ApJ, 792, 75

Gas Accretion Traced in Absorption in Galaxy Spectroscopy

Kate H.R. Rubin

1 Introduction

Spectroscopy of galaxy continua has been used for decades as a powerful probe of the kinematics of gas in the foreground. Absorption transitions sensitive to cool, diffuse material trace its motion with respect to the galaxy's stellar component along the line of sight. The galaxy continuum "beam" is absorbed by gas over a broad range of scales, from the $\sim$100–200 pc scale heights of the dense interstellar medium (Langer et al. 2014) to the $>$100 kpc extent of the diffuse gas reservoir filling the galaxy halo (Prochaska et al. 2011; Tumlinson et al. 2011; Werk et al. 2014) and beyond.

The detection of absorption lines in a galaxy's spectrum (i.e., "down the barrel") which are *redshifted* with respect to its rest frame is unequivocal evidence of gas flow toward the continuum source. And while this signature arises from material at any location along the sightline to the galaxy, such that the flow may not ultimately reach the galaxy itself, redshifted absorption has been broadly interpreted as strong evidence for gas inflow toward or accretion onto the background host. Over the past 10 years, as high signal-to-noise (S/N) galaxy spectroscopy covering the rest-frame optical into the near-ultraviolet has become more routine, this technique has become sensitive to the inflow of material over a broad range of densities and temperatures: redshifted Ca II H & K $\lambda\lambda 3934, 3969$ or Na I D $\lambda\lambda 5891, 5897$ absorption probes cold, mostly neutral gas infall at a temperature $T < 1000$ K; redshifts in low-ionization transitions such as Mg II $\lambda\lambda 2796, 2803$ or Fe II $\lambda\lambda 2586, 2600$ trace the

K.H.R. Rubin (✉)
Department of Astronomy, San Diego State University, San Diego, CA 92182, USA
e-mail: krubin@sdsu.edu

A. Fox, R. Davé (eds.), *Gas Accretion onto Galaxies*, Astrophysics and Space Science Library 430, DOI 10.1007/978-3-319-52512-9_5

inflow of cool, photoionized gas at $T \sim 10^4$ K; and the detection of redshifted C IV $\lambda\lambda1548, 1550$ or Si IV 1394, 1402 absorption would in principle trace yet warmer ($T \sim 10^5$ K) gas accretion.

Furthermore, unlike background QSO absorption line experiments, which typically must adopt the assumption that inflowing gas has a relatively low metallicity ($Z/Z_\odot < 1$; Lehner et al. 2013) in order to disentangle accreting systems from those enriched by galactic outflows, redshifted self-absorption naturally traces metal-rich inflow, and may even trace pristine inflow in rare cases where spectral coverage of Lyα is available (Fathivavsari et al. 2016). Moreover, while studies searching for the signature of gas accretion in emission from the 21 cm transition of neutral hydrogen are currently limited by the faint surface brightness of such features to galaxies within a few hundred Mpc of our own (Martin et al. 2010), the "down the barrel" technique has been used to detect gas accretion onto galaxies as distant as $z \sim 1$ (Coil et al. 2011; Rubin et al. 2012; Martin et al. 2012).

In spite of these clear advantages, however, the first report of redshifted absorption observed down the barrel toward a sample of more than a single object did not occur until 2009 (Sato et al. 2009). Even today, secure detections of this phenomenon have been reported for only ~60–80 systems. Instead, measurements of *blueshifted* absorption tracing cool gas outflow have dominated the literature (Heckman et al. 2000; Martin 2005; Rupke et al. 2005b; Weiner et al. 2009; Steidel et al. 2010; Chen et al. 2010b). Large-scale galactic winds, thought to be driven by processes associated with star formation, are now known to arise ubiquitously in star-forming objects from the local universe to $z > 2$ (Heckman et al. 2000; Ajiki et al. 2002; Martin et al. 2012; Rubin et al. 2014). These winds are likewise traced by all of the metal-line absorption features listed above, and are observed to have velocities ranging from $\sim -50\,\mathrm{km\,s^{-1}}$ to $< -800\,\mathrm{km\,s^{-1}}$ (Steidel et al. 2010; Rubin et al. 2014; Du et al. 2016). As the free-fall velocity of material within the virial radius of a typical massive, star-forming galaxy halo at $z \sim 0$ (with halo mass $M_h \sim 10^{11-12} M_\odot$) is expected to be only ~ 100–$200\,\mathrm{km\,s^{-1}}$ (Goerdt and Ceverino 2015), the preponderance of winds covering the sightlines to galaxies must necessarily obscure the detection of material falling inward at such comparatively modest velocities. Indeed, this issue is compounded by the low spectral resolution of the vast majority of spectroscopic surveys useful for these analyses (Weiner et al. 2009; Steidel et al. 2010; Martin et al. 2012; Rubin et al. 2014).

In this chapter, we review the works in which the few *bona fide* instances of gas inflow were reported, beginning with the first detections via transitions in the rest-frame optical in Sect. 2. Due to technical limitations (described below), the focus of these early studies was on red, early-type galaxies and/or galaxies exhibiting signs of AGN activity. Reports of inflow onto galaxies hosting the most luminous AGN (i.e., bright QSOs) are discussed in Sect. 2.3. In Sect. 3, we describe the first detections of inflow onto actively star-forming systems facilitated by high-S/N spectroscopic galaxy surveys in the rest-frame ultraviolet. The biases inherent in the use of the "down the barrel" technique given the ubiquity of galactic winds are described in Sect. 4.1, and unique constraints on the morphology of gas inflow which will soon be facilitated by ongoing spatially resolved spectroscopic surveys are discussed in Sect. 4.2. Section 5 presents a summary, and offers some recommendations for future experiments which will leverage the full potential of this technique in the detection and characterization of the process of gas accretion onto galaxies.

2 First Detections: Gas Accretion in Late-Stage Galaxy Evolution

For galaxy spectroscopy to successfully constrain the kinematics of absorption lines along the line of sight, it must have sufficient continuum S/N in the spectral range surrounding the transition in question. The limiting S/N depends in detail on a number of factors, including the typical equivalent width of the transition and the complexity of the absorption line analysis employed. Studies testing this limit and adopting a simple, "single-component" analysis have required between $S/N \sim 5\,\text{Å}^{-1}$ (for analysis of Mg II; Rubin et al. 2014) and $S/N \sim 15\,\text{Å}^{-1}$ (for analysis of Na I D; Sato et al. 2009).

Many of the first works to attempt constraints on the kinematics of such absorption transitions *in galaxies not specifically selected to be starbursts*—i.e., in magnitude-limited galaxy samples representative of "typical" star-forming or quiescent objects—used datasets which did not broadly satisfy these S/N requirements (Weiner et al. 2009; Rubin et al. 2010; Chen et al. 2010b; Steidel et al. 2010). Each of these studies were leveraging spectroscopic samples obtained for the primary purpose of conducting a redshift survey, rather than an assessment of cool gas kinematics. As a result, these works often resorted to co-adding subsamples of tens or hundreds of spectra to achieve the S/N required for absorption line analysis. Moreover, this co-added spectroscopy yielded absorption profiles which were universally asymmetric with excess equivalent width blueward of the galaxies' rest frame. The first detections of redshifted absorption profiles were reported only in later studies which obtained high-S/N spectroscopy of *individual* galaxies.

2.1 *First Reports of Inflow Observed Down the Barrel*

The very first detections of redshifted self-absorption made use of some of the highest-S/N spectroscopy obtained during the execution of the DEEP2 redshift survey (Davis et al. 2003). In their discussion of a massive ultraluminous infrared galaxy with an X-ray-bright central AGN at $z = 1.15$ located in the AEGIS survey field (Davis et al. 2007), Le Floc'h et al. (2007) commented on strong Ca II H & K absorption lines which were redshifted by $\sim$150–200 km s^{-1} relative to the velocity of [O II] emission from the host.

Shortly thereafter, Sato et al. (2009) significantly expanded the sample of redshifted self-absorption detections through their analysis of Na I D kinematics in an S/N-limited subsample of the DEEP2 survey of the Extended Groth Strip. The parent DEEP2 sample was magnitude-selected to $R < 24.1$, such that it contained substantial populations of both star-forming and passive galaxies. However, Sato et al. (2009) found that they required at least $\mathrm{S/N} \sim 6.5\ \mathrm{pix}^{-1}$ in the rest-frame continuum around Na I D to constrain its velocity to within a 68% confidence interval spanning less than $\sim$200 km s^{-1}. This limited their analysis sample to 205 objects at $0.11 < z < 0.54$, about 75% of which are "red-sequence" galaxies. Outflows and inflows are reported in this work if the fitted Na I centroid is shifted from systemic velocity by more than $\pm$50 km s^{-1}, and occur in nearly equal numbers: outflows are detected in 32 objects, while inflows are detected with high confidence in 31 objects.

The Na I profiles shown in the left-hand panel of Fig. 1 demonstrate the typical quality of the data used in this analysis. As the two transitions in the Na I D doublet are separated by only $\sim$300 km s^{-1}, the lines are blended in these spectra due to the large intrinsic velocity of both the stellar and gas components giving rise to the observed absorption. The velocity $v(\mathrm{Na\,I\,D}) = 0\,\mathrm{km\,s^{-1}}$ on the x-axis corresponds to the rest wavelength of the 5897 Å doublet transition. The authors show their best-fit absorption line model for each spectrum with colored curves, and mark the corresponding best-fit velocity of the λ5897 Å transition with vertical dotted lines. The top-most spectrum exhibits significantly redshifted absorption with a fitted velocity of $\sim +80$ km s^{-1} (Fig. 1).

Sato et al. (2009) note that while a significant fraction of the galaxies in their sample which exhibit outflows are star-forming and lie in the so-called blue cloud (blue symbols; Fig. 1, right), all but one of the objects exhibiting inflows occupy the red sequence (red symbols; Fig. 1, right). In interpreting this finding, they caution that there may be additional absorption features arising from the stellar populations in these systems which are not yet understood, and which could in principle shift the minimum of the stellar continuum flux near $\lambda_{\mathrm{obs}} \sim$ 5890–5900 Å redward of the rest wavelengths of Na I D. Line profile fitting of such features could in turn be mistakenly interpreted as redshifted interstellar absorption. However, the authors also report that many of the inflow galaxies in their sample exhibit optical emission line ratios ([N II]/Hα) consistent with Seyfert or LINER activity, and draw an intriguing comparison to the inflows observed in neutral hydrogen in the inner regions of radio-bright elliptical galaxies (van Gorkom et al. 1989). They speculate that the observed absorbing gas may in fact feed this central activity.

2.2 *Inflows onto AGN-Host Galaxies*

Since the publication of this novel and important work, there have been several studies corroborating the detection of inflow toward galaxies hosting active AGN. Krug et al. (2010) explored Na I D kinematics in a sample of Seyfert galaxies

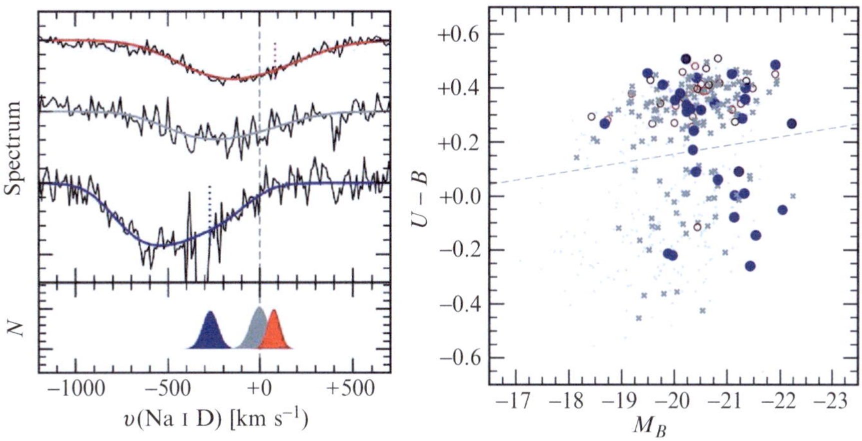

Fig. 1 Spectroscopy and photometry of the sample analyzed in Sato et al. (2009), the first study to report and characterize gas accretion onto distant galaxies. *Left*: Continuum-normalized spectra showing the region around the Na I doublet (*black*). $v = 0\,\mathrm{km\,s^{-1}}$ is set at the systemic velocity of the λ5897 doublet transition. The *colored curves* show the best-fit absorption line models, and the *vertical dotted lines* show the corresponding best-fit velocity of the λ5897 line. The *bottom panel* shows the marginalized probability distributions of the model profile velocity for each spectrum, obtained from a Markov Chain Monte Carlo sampling of the absorption line profile parameters. *Right*: Rest-frame color-magnitude diagram showing the full sample. Objects exhibiting outflows are marked with *blue filled circles*; those exhibiting redshifted absorption are marked with *red open circles*; and *gray crosses mark* objects which exhibit neither significant blueshifts nor redshifts in their Na I profiles. The *light gray dots* indicate objects whose spectra lack the S/N required to constrain their absorption line kinematics. The *dashed line* divides the blue cloud and red sequence populations. Panels are reproductions of Figs. 4 (left) and 10 (right) from the article "AEGIS: The Nature of the Host Galaxies of Low-Ionization Outflows at $z < 0.6$", by Sato et al. (2009, ApJ, 696, 214). ©AAS. Reproduced with permission

selected to be infrared-faint (with $10^{9.9} < L_{\mathrm{IR}}/L_{\odot} < 10^{11}$). Previous studies of Na I D absorption toward nearby galaxy samples had targeted infrared-bright starburst or starburst/AGN composite systems (Rupke et al. 2005a,b; Martin 2005), and had detected outflows in the majority of these objects. A primary goal of Krug et al. (2010) was to develop a comparison sample of objects without ongoing starbursts to determine the relative contributions of star formation vs. AGN activity in driving these winds. The authors obtained spectroscopy of 35 galaxies using the RC Spectrograph on the Kitt Peak 4m telescope. With a median S/N near Na I D of $\sim 85\,\text{Å}^{-1}$ and a spectral resolution of $85\,\mathrm{km\,s^{-1}}$, they performed both single- and double-velocity component absorption line model fits with typical central velocity uncertainties of $<50\,\mathrm{km\,s^{-1}}$.

In contrast to previous studies of starbursting systems, Krug et al. (2010) detected outflows traced by Na I in only 4 galaxies (11% of their sample) and instead detected inflows in over a third (13 galaxies) of their sample. The central velocities of these flows ranged from just over +50 to $+140\,\mathrm{km\,s^{-1}}$; however, no significant correlation between inflow kinematics and host galaxy mass, infrared luminosity, or inclination was observed. From this analysis, the authors concluded that star formation rather than AGN activity makes the dominant contribution to the driving of outflows (although this conclusion does not apply to Type 1 Seyferts; Krug et al. 2010). Furthermore, they speculated that the high observed inflow velocities are suggestive of material located close to the galaxy nuclei rather than in the outskirts of the disks. They also searched for signs of nuclear morphological features which could indicate that the inflowing material is undergoing angular momentum loss, finding that five galaxies in their inflow sample exhibit nuclear dust spirals, bars, or rings. Five additional inflow galaxies have nearby companions with which they may be interacting. Such phenomena are thought to be required in order to facilitate the inflow of gas toward galaxy nuclei and its ultimate accretion onto the central black hole.

The high quality of these data also permitted a rough estimate of the rate of mass inflow onto these systems via constraints on the Na I column density. Assuming a factor of 10 ionization correction (i.e., that $N(\mathrm{Na})/N(\mathrm{Na\,I}) = 10$), a metallicity approximately twice the solar metallicity, and that the absorbing gas is at a distance of $r = 1$ kpc from the nucleus, it is estimated to be flowing inward at $(1-5)\frac{r}{1\,\mathrm{kpc}}$ $M_{\odot}\,\mathrm{yr}^{-1}$. This rate is approximately two orders of magnitude larger than the mass accretion rate required to power a typical Seyfert nucleus (Crenshaw et al. 2003; Krug et al. 2010). These cold gas inflows may therefore serve as an important fueling mechanism for AGN activity in these objects.

Yet further evidence for massive inflows of gas onto AGN-host galaxies was reported by Stone et al. (2016), who studied molecular gas kinematics in a sample of 52 local AGN selected from the *Swift*-Burst Alert Telescope Survey of hard X-ray objects. The Stone et al. (2016) targets thoroughly sample the AGN luminosity function to its brightest end. The authors analyzed spectroscopy of these systems covering the OH 119 μm feature obtained with the *Herschel*/PACS far-IR interferometer, detecting the transition in absorption in 17 sources. As in the Krug et al. (2010) study discussed above, only a handful (four) of these 17 AGN exhibited molecular outflows, while the OH absorption feature was redshifted by $>50\,\mathrm{km\,s}^{-1}$ in seven of their targets (corresponding to a detection rate of $\sim$40%). The authors suggested that the significantly lower detection rates of inflows toward IR-luminous galaxies may be due to the disruption of these accretion flows by the faster winds driven by their central starbursts.

2.3 *Inflows on the Smallest Scales: Feeding Luminous QSOs?*

Spectroscopy of bright quasars also probes gas flows toward the central source (i.e., the AGN), and can be obtained with much greater efficiency than faint galaxy spectroscopy. Indeed, the SDSS-I/II and SDSS-III BOSS redshift surveys have now revealed several instances of redshifted absorption observed toward bright QSOs

(Hall et al. 2002, 2013). These profiles occur in quasars with "broad" absorption lines (BAL), defined to extend over thousands of $\mathrm{km\,s^{-1}}$ in velocity (Allen et al. 2011). In addition, the absorption typically only partially covers the emitting source. The vast majority of BAL QSOs exhibit absorption troughs lying entirely at velocities blueward of systemic, and are thought to arise from AGN-driven feedback/outflow. However, in their search through more than 100,000 SDSS/BOSS QSO spectra, Hall et al. (2013) discovered 19 BAL objects in which the trough in at least one transition extends to velocities $v > 3000\ \mathrm{km\,s^{-1}}$. These authors estimated that such redshifts occur in approximately 1 in every 1000 BAL quasars.

This work discusses several different physical scenarios which may explain the observed extreme absorption. The accretion of material onto the host dark matter halo or host galaxy at kpc-scale distances from the central source may certainly contribute to the redshifted absorption signal; however, the free-fall velocity of such material is expected to be only a few hundred $\mathrm{km\,s^{-1}}$. For gas to achieve an infall velocity $v > 3000\ \mathrm{km\,s^{-1}}$, it must reach scales as small as a few hundred Schwarzschild radii (Hall et al. 2013). It remains to be established whether infalling gas clumps can maintain sufficiently high densities at such small distances to give rise to absorption in the observed transitions (e.g., Si IV, C IV, Al III $\lambda\lambda1854, 1862$, Mg II). Hydrodynamical simulations of accretion flows onto a supermassive black hole may be used to address this question, e.g., Li et al. (2013); however, detailed predictions of absorption profile shapes in multiple transitions are needed for a quantitative comparison to the observations. Alternative scenarios which could give rise to redshifted, broad absorption include gravitational redshifting of a spherically symmetric wind; rotating accretion disk winds observed toward an extended emission source; outflow from a *second* QSO close in the foreground to the first; or even the relativistic Doppler effect of outflowing ions absorbing photons of lower frequency due to time dilation. Hall et al. (2013) concluded that none of these mechanisms can individually explain the observations; however, they suggest that both rotationally dominated accretion disk winds and infalling material are likely contributing to the redshifted broad absorption profiles.

In some instances, these data may also constrain the distance between the absorbing material and the central black hole. Shi et al. (2016) analyzed one of the BAL QSOs discussed in Hall et al. (2013), reporting the detection of not only redshifted Mg II and Fe II line profiles but also redshifted hydrogen Balmer and He I* $\lambda3889$ absorption. The latter transition is highly sensitive to the number of ionizing photons impinging on the gas per ion (U), while Balmer absorption lines generally arise only in very high-density environments ($n(\mathrm{H}) > 10^6\ \mathrm{cm^{-3}}$; Shi et al. 2016). Together with an estimate of the ionizing luminosity of the QSO, constraints on both U and $n(\mathrm{H})$ from analysis of the He I* and Balmer lines imply a distance of $r_{\mathrm{abs}} \sim 4\,\mathrm{pc}$ for this particular absorbing system. Shi et al. (2016) note that this distance is much larger than expected for a rotating disk wind or a gravitationally redshifted AGN outflow, and suggest that it is instead consistent with infall from the inner surface of a dusty torus surrounding the accretion disk (Barvainis 1987). Detailed analysis of a larger sample of similar systems has the potential to more conclusively establish the frequency of such infall events, the mass of material involved, and their overall contribution to the fueling of bright QSOs.

Apart from the BAL phenomenon, QSOs are observed to exhibit numerous other classes of foreground absorbers. One such class—that of "proximate" absorbers—differs from BALs primarily in that they have significantly more narrow velocity widths ($<100\,\mathrm{km\,s^{-1}}$). Proximate absorbers are typically defined to have a central velocity within $<10{,}000\,\mathrm{km\,s^{-1}}$ of the QSO emission line redshift, and are often (though not always) observed to fully cover the emitting source (Ellison et al. 2010). Recently, Fathivavsari et al. (2016) presented a sample of six $z \sim 2$ QSOs in which a proximate damped Lyα (DLA) absorber entirely eclipses the broad Lyα emission from the central AGN. They report that in five of these systems, the absorber has a velocity $\sim$100–1200 $\mathrm{km\,s^{-1}}$ redward of the QSO systemic velocity, speculating that the absorption traces infalling material. However, because this work relies on spectroscopy of broad QSO emission lines to constrain the host galaxy redshift (e.g., C IV, C III] λ1909, He II λ1640), these estimates suffer from significant systematic uncertainties ($\sim$ 200–500 $\mathrm{km\,s^{-1}}$; Shen et al. 2016). Follow-up spectroscopy covering [O II] λ3728 or [O III] λ5007 in the near-infrared will be important for verifying this intriguing finding.

2.4 *Inflows onto Early-Type and Post-Starburst Galaxies*

Concomitant with the assembly of this compelling evidence for gas accretion onto active AGN, two studies offered additional empirical support for inflow onto red-sequence galaxies, or galaxies exhibiting signs of recent quenching. In the first study to obtain rest-frame near-UV galaxy spectroscopy of sufficient depth for analysis of Fe II $\lambda\lambda$2586, 2600 and Mg II $\lambda\lambda$2796, 2803 absorption in individual sightlines, Coil et al. (2011) targeted a sample of 10 X-ray-luminous AGN and 13 post-starburst galaxies at $0.2 < z < 0.8$ selected from the DEEP2 and SDSS redshift surveys. The AGN are moderately luminous, with $\log(L_X/\mathrm{erg\,s^{-1}}) \sim 41$–42, while the post-starbursts were identified via a decomposition of their optical spectroscopy into old and young stellar population components. Those objects for which spectral fitting yielded a relatively luminous young component (contributing at least 25% of the continuum at $\lambda_{\mathrm{rest}} \sim 4500$ Å) and which exhibited little or no Hβ emission were targeted. The Keck/LRIS spectroscopy obtained for this study not only traced blueshifted absorption in 60% of the AGN and 31% of the post-starbursts, but also yielded redshifted absorption profiles in two of the latter. These redshifts were detected in Mg II (and in one case in Mg I λ2852) at speeds of 75–115 $\mathrm{km\,s^{-1}}$, but were not detected securely in Fe II absorption. While it is difficult to draw definitive conclusions from such small sample sizes, the ubiquitous detection of Mg II and Fe II absorption in excess of that predicted for the stellar continuum in these spectra suggests that the mechanism causing the cessation of star formation did not completely deplete the gas supply in these systems.

Most recently, Sarzi et al. (2016) undertook a study of Na I absorption in SDSS spectra of galaxies selected to have high-spatial-resolution 20 cm continuum coverage obtained with the Very Long Baseline Array (Deller et al. 2014). Nearly 60% of this sample are early-type galaxies, and the vast majority of these early types do not exhibit blueshifted Na I. The authors briefly comment on their detection of redshifted absorption in ~10–20 objects, most of which either exhibit very weak optical emission lines or LINER/Seyfert-like emission line ratios. It is suggested that this phenomenon may be the signature of bars or unsettled dust lanes in these systems.

2.5 Summary

In summary, cool gas kinematics have now been assessed in relatively small samples of early-type and/or AGN-host galaxies. Measurements of Na I absorption velocities toward red-sequence systems suggest a ~20% incidence rate of redshifted profiles (Sato et al. 2009). Spectroscopy of Seyfert galaxies and X-ray-selected AGN reveals a yet higher incidence (~40%) of redshifted Na I or OH absorption (Krug et al. 2010; Stone et al. 2016). These rates are approximately the same or higher than the outflow detection rates for the same samples: ~14% for the red-sequence sample of Sato et al. (2009) and ~10–25% for the Seyfert and BAT/AGN samples of Krug et al. (2010) and Stone et al. (2016). While the mass of these flows remains poorly constrained, the similar rates of these phenomena may ultimately point to a "steady state" of gas cycling through such systems (even as they maintain very low rates of star formation).

It should also be noted, however, that the physical scale over which these flows occur may in fact be limited to the regions very close to or within the galaxies' stellar component. The common use of Na I in the aforementioned studies naturally biases these assessments to tracing the motions of cold, dust-enshrouded material, which in red-sequence galaxies may in fact lie in dust lanes or cold molecular disks (Sato et al. 2009; Davis et al. 2011). Spectroscopy of additional transitions tracing gas over a broader range of density and temperature (e.g., Mg II or Fe II) will be important for developing constraints on the physical state and scale of these flows.

3 Tracing Inflows with Rest-Frame Ultraviolet Galaxy Spectroscopy

As discussed above, the initial focus in the literature on Na I absorption kinematics biased these samples toward the highest surface-brightness systems at $\lambda_{\rm rest} \sim 5900$ Å–i.e., infrared-selected starbursts, massive ellipticals, and bright AGN hosts. It was not until large surveys obtaining deep, high-S/N spectroscopy in the rest-frame ultraviolet were performed that the first constraints on cool gas kinematics toward a significant (>50) sample of individual, "normal" *star-forming* galaxies were discussed. Moreover, the Mg II and Fe II absorption transitions covered in these spectra can be significantly stronger than Na I due to a number of factors; e.g., Mg and Fe are more abundant than Na, and their singly ionized transitions trace a much broader range of temperature and density than neutral Na. UV galaxy spectra therefore have the potential to trace more diffuse material, including the halo gas which is known to give rise to Mg II absorption in background QSO sightlines to projected distances of $R_{\perp} \sim 100$ kpc (Bergeron 1986; Steidel et al. 1994; Kacprzak et al. 2007; Chen et al. 2010a). Rest-frame UV transitions are thus significantly more sensitive to both inflow and outflow, and are accessible at observed-frame optical wavelengths for galaxies at $z > 0.3$.

The first unambiguous detection of inflow observed down the barrel toward a sample of star-forming galaxies was reported in Rubin et al. (2012). The galaxies were targeted over the course of a high-S/N Keck/LRIS survey of ~100 galaxies at redshifts $0.3 < z < 1.4$ (Rubin et al. 2014). This sample, selected to a magnitude limit $B_{\rm AB} < 23$, spans the star-forming sequence at $z \sim 0.5$, and is thus representative of the "normal" star-forming galaxy population (Fig. 2, left). In addition, this survey targeted fields with deep *HST*/ACS imaging, facilitating a detailed morphological analysis (Fig. 2, right).

Blueshifted absorption tracing outflows was detected in the majority (~66%) of this sample. Moreover, the detection rate of these winds does not vary significantly with the star formation rate (SFR) or stellar mass of the host, but rather depends primarily on galaxy orientation (Fig. 2, right). This finding is suggestive not only of ubiquitous outflows, but also of an approximately biconical morphology for these flows across the star-forming sequence.

In the same survey, redshifted absorption tracing cool inflow was detected in six of the remaining galaxy spectra (red diamonds in Fig. 2; Fig. 3) with velocities ~80–200 km s^{-1}. The galaxies themselves have SFRs ranging from ~1 to 40 $M_{\odot}$ yr^{-1}, and have stellar masses in the range $9.6 < \log M_*/M_{\odot} < 10.5$. Perhaps most significantly, five of these six galaxies have disk-like morphologies and are viewed in a nearly edge-on orientation (with inclinations >55°; Fig. 2, right). The authors suggest that the preferential detection of inflows toward edge-on galaxies indicates that cool infall is more likely to occur along the plane of galactic disks, rather than along the minor axis. We also note here that higher sensitivity to inflows in more edge-on systems is a natural consequence of biconical winds.

Assuming a metallicity for these flows of $Z = 0.1Z_{\odot}$ and adopting constraints on the inflow Mg II and Fe II column densities from their absorption line analysis, Rubin et al. (2012) estimated lower limits for the mass accretion rates onto this sample of six objects in the range ~0.2–3 $M_{\odot}$ yr^{-1}. These observed flow rates are approximately consistent with the rate of mass flow onto our own Galaxy (Lehner and Howk 2011).

Nearly concurrently, Martin et al. (2012) carried out a similar but completely independent study of cool gas kinematics in Keck/LRIS spectroscopy of ~200 galaxies. The galaxy sample has a somewhat higher median redshift than that of Rubin et al. (2012) ($\langle z \rangle \sim 0.5$ vs. ~ 1) and is selected to a fainter magnitude

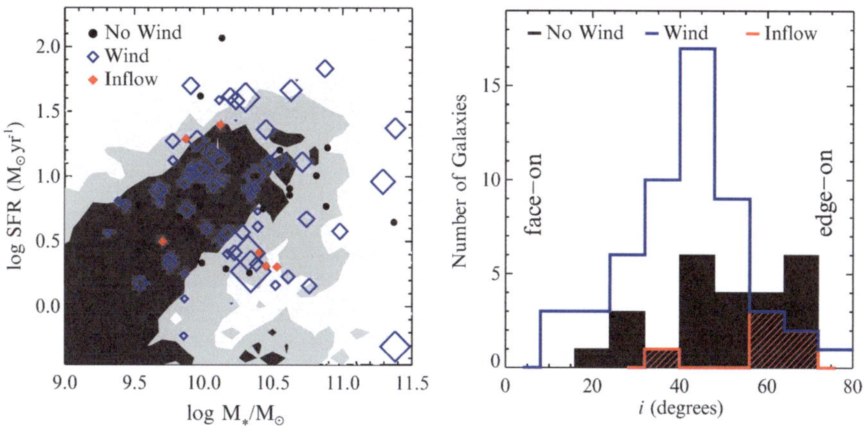

Fig. 2 The first survey to detect cool gas inflow traced by Mg II and Fe II absorption onto a sample of distant star-forming galaxies (Rubin et al. 2012, 2014). *Left*: *Colored* and *black points* show the sample of $\sim$ 100 galaxies targeted in Rubin et al. (2012, 2014), and *gray contours* show the SFR-M_* distribution of the underlying galaxy population (Barro et al. 2011). *Blue open diamonds* indicate galaxies with detected winds, *red filled diamonds* mark objects with inflows, and *black circles* indicate objects with neither winds nor inflow. The size of the *blue diamonds* is scaled with the Mg II equivalent width of the outflow. Winds are detected in $\sim$ 2/3 of the galaxies, while detected inflows occur at a rate $\sim$6%. *Right:* Distribution of inclinations (i) for the disk-like galaxies selected from the same study. The distribution for galaxies with winds is shown in *blue*, and the distributions for galaxies with inflows and without winds/inflows are shown in *red* and *black*, respectively. While face-on galaxies (at low i) are significantly more likely to drive a detected wind than galaxies viewed edge-on, the galaxies with detected inflow are nearly all highly inclined. Panels are adapted from Rubin et al. (2012, 2014)

limit $B_{AB} < 24.0$. However, the vast majority of the targets are star-forming, with SFRs ranging from $\sim$1 to 98 $M_\odot$ yr^{-1} and stellar masses $8.85 < \log M_*/M_\odot < 11.3$. This survey made use of a somewhat lower-resolution spectroscopic setup for $\sim$70% of the sample (having a FWHM resolution element $\sim$435 km s^{-1} vs. $\sim$282 km s^{-1}), which in principle limits sensitivity to lower-velocity flows. In spite of this, redshifted absorption profiles were detected in nine spectra (see Fig. 4), yielding a detection rate ($\sim$4%) consistent with that of Rubin et al. (2012). Martin et al. (2012) also performed a careful analysis of the two-dimensional spectra of these objects, identifying weak nebular line emission offset from the continuum trace at the same velocity as the redshifted absorption in a few (4) cases. They speculated that the inflows in these systems are being fed by relatively dense, star-forming structures (e.g., satellite dwarf galaxies) rather than diffuse accretion streams from the circumgalactic medium.

Martin et al. (2012) found that the galaxies exhibiting inflows span the range in stellar mass and SFR occupied by the parent sample. They also noted that among the four inflow galaxies for which quantitative morphologies are available, only one has a high inclination ($i \sim 61°$); the remaining three galaxies have $i < 55°$ (Kornei et al. 2012). The detection of inflow toward these objects requires that the idea put forth by Rubin et al. (2012) that infall is more likely to be observed along the plane

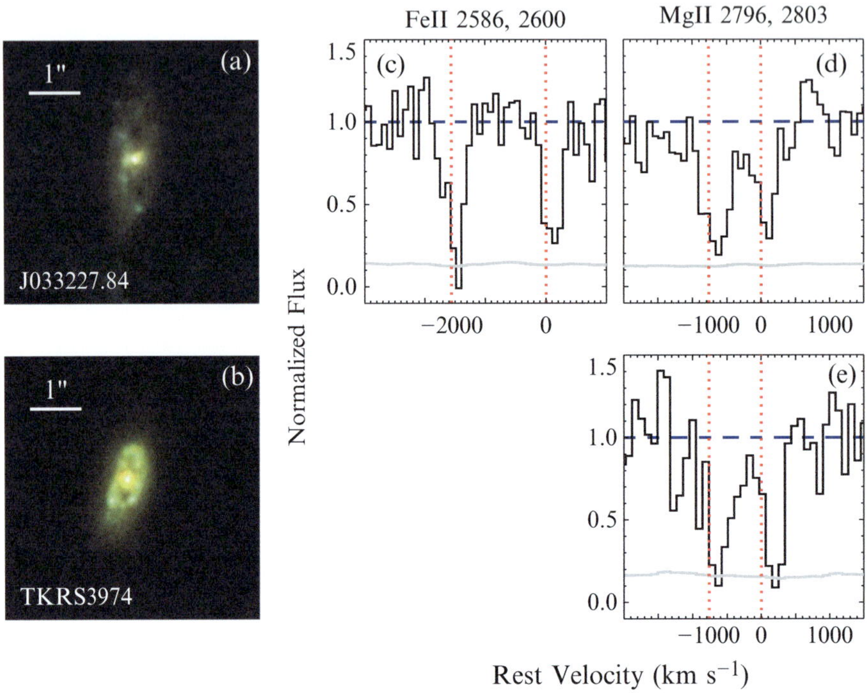

Fig. 3 Imaging and spectroscopy of two galaxies with ongoing gas inflow. The *left column* shows *HST*/ACS images, and the *middle* and *right columns* show Fe II and Mg II transitions in the galaxy spectra. The line profiles are redshifted with respect to systemic velocity (marked with *vertical dotted lines*). Analysis of the imaging indicates that these galaxies are disk-like with edge-on orientations. The figure is a reproduction of a portion of Fig. 1 from the article "The Direct Detection of Cool, Metal-Enriched Gas Accretion onto Galaxies at $z \sim 0.5$", by Rubin et al. (2012, ApJL, 747, 26). ©AAS. Reproduced with permission

of galactic disks be considered more carefully and investigated with a significantly larger spectroscopic sample.

These two studies have provided us with the first, unequivocal evidence for gas accretion onto distant, star-forming galaxies. However, the physical nature of these flows remains an open question. There are numerous potential sources for the enriched material producing the observed absorption, including gas which has been tidally stripped from nearby dwarf galaxies, or wind ejecta from the central galaxy which is being recycled back to the disk. Indeed, such wind recycling is predicted in numerous cosmological galaxy formation simulations (Oppenheimer et al. 2010; Vogelsberger et al. 2013). Current absorption line data cannot distinguish between these alternatives; however, detailed constraints on the covering fraction or cross section of the inflow from upcoming integral field spectroscopic surveys may aid in differentiating between these scenarios. This topic will be discussed further in Sects. 4.2 and 5.

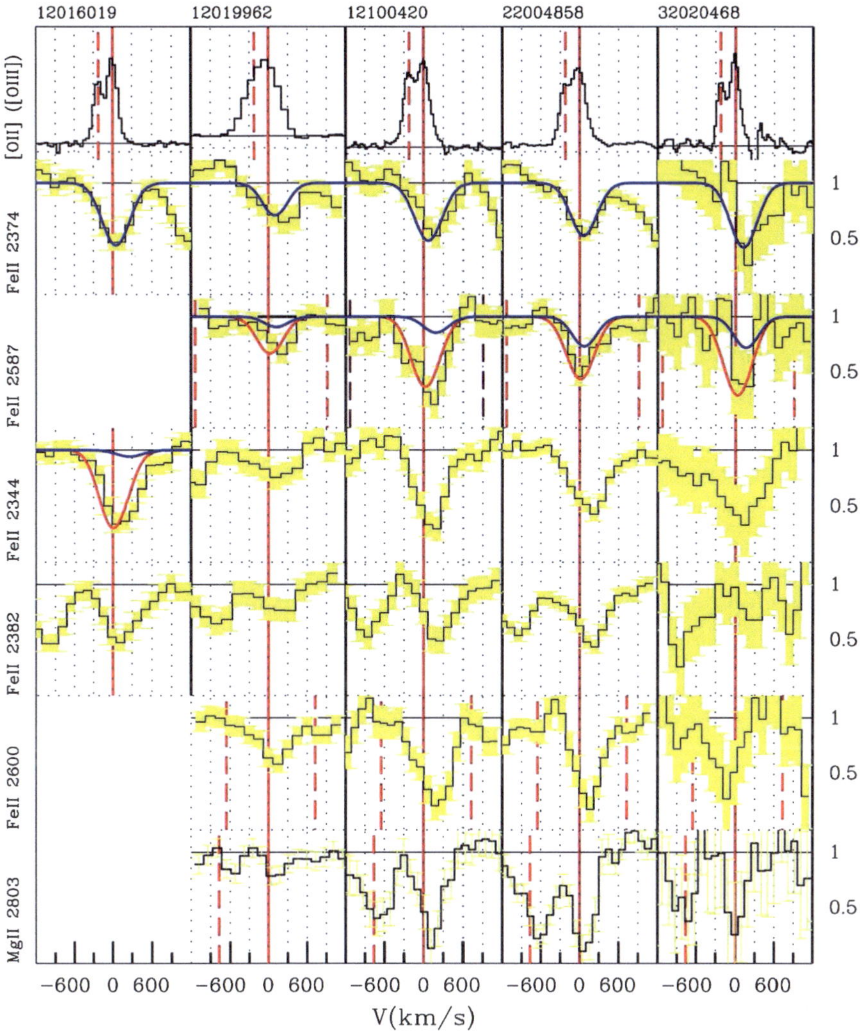

Fig. 4 Resonant absorption lines and [O II] emission profiles for five galaxies with securely detected inflows from Martin et al. (2012). Galaxy ID numbers are indicated at the top of each column. The *solid red vertical lines mark* the rest-frame velocity of each transition, and *dashed vertical lines mark* the wavelengths of nearby resonant transitions. Single-component absorption line fits are shown in the row of Fe II $\lambda 2374$ profiles, and fits including both an absorption component with a velocity fixed at $0\,\mathrm{km\,s^{-1}}$ and a "flow" component are overplotted on the Fe II $\lambda 2586$ profiles. These objects exhibit strong redshifts in most of the absorption transitions shown. The [O II] profiles are used to determine the systemic velocity. This figure is a reproduction of Fig. 16 from the article "Demographics and Physical Properties of Gas Outflows/Inflows at $0.4 < z < 1.4$", by Martin et al. (2012, ApJ, 760, 127). ©AAS. Reproduced with permission

4 Toward Assessment of the Incidence of Inflow

Ultimately, empirical studies characterizing the cool gas flows around galaxies must constrain, e.g., the incidence of accretion in a given galaxy population as a function of the mass inflow rate and age of the universe. Such detailed assessments are required for an incisive test of galaxy formation models (Davé et al. 2012; Nelson et al. 2015). We now consider a few factors which complicate the use of current datasets in developing such constraints. In the following section we discuss future directions which may ameliorate these issues.

4.1 Spectral Confusion

The faintness of the galaxies studied in Rubin et al. (2012) and Martin et al. (2012) forced the selection of a low-resolution spectroscopic setup for these surveys (FWHM ~250–400 km s^{-1}). Thus, the velocity profiles arising from distinct inflowing and outflowing gas structures along the same line of sight are blended in these spectra. Indeed, in none of the Mg II or Fe II line profiles discussed in Rubin et al. (2014) is more than one velocity component resolved. If there is, e.g., a component of gas being accreted at $v \sim +100\,\mathrm{km\,s^{-1}}$, and another wind component foreground to the same galaxy beam moving at $v \sim -200\,\mathrm{km\,s^{-1}}$, a centroid of the resulting line profile (observed at low resolution) will suggest the presence of an outflow with a velocity $v \sim -100\,\mathrm{km\,s^{-1}}$ (while the inflowing component is completely obscured). Or, in the case of two components each at $v \sim +100\,\mathrm{km\,s^{-1}}$ and $-100\,\mathrm{km\,s^{-1}}$, a centroid of the observed line profile will reveal neither outflow nor inflow. Thus, there is a limited set of circumstances in which the centroid of such blended profiles will be significantly redshifted, revealing an inflow.

Indeed, as noted in Rubin et al. (2012), the six objects toward which inflows were detected are unique not for the material moving at positive velocities along the line of sight, but for the *absence of winds*. This is demonstrated in Fig. 5, which compares the equivalent widths measured redward and blueward of systemic velocity in the Fe II λ2600 (left) and Mg II λ2803 (right) transitions in the spectral sample of Rubin et al. (2014). Objects with detected winds are marked with blue crosses and cyan triangles; objects with detected inflows are marked with red crosses; and objects with neither winds nor inflow are marked with black filled circles. A 1:1 relation is indicated with the dashed lines. Inflows were detected in galaxies with larger red EWs than blue EWs; however, there are many other systems which have red EWs larger than the smallest values measured for the inflow sample (i.e., red $EW_{2600} > 0.6$ Å and red $EW_{2803} > 0.8$ Å; see Fig. 5). Based on an accounting of the number of galaxies with red EWs at least as large as those in which inflow was detected, Rubin et al. (2012) estimated that inflow must be occurring in at least 20–40% of the full galaxy sample.

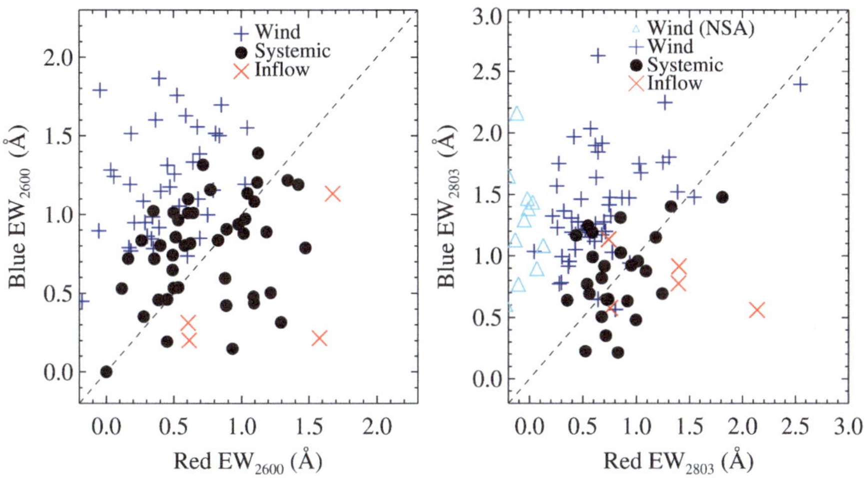

Fig. 5 Assessment of the symmetry of Fe II and Mg II absorption line profiles in low-resolution galaxy spectroscopy. All measurements shown are from analysis of the UV spectroscopy of ~100 star-forming galaxies discussed in Rubin et al. (2012, 2014). *Left*: The equivalent width (EW) of the Fe II $\lambda 2600$ transition, measured at $v < 0\,\mathrm{km\,s^{-1}}$ ("*blue*") vs. the EW of the same transition at $v > 0\,\mathrm{km\,s^{-1}}$ ("*red*") for galaxies with blueshifted Fe II (*blue crosses*), without winds or inflows (*black filled circles*), and with inflows detected in Fe II (*red crosses*). Several galaxies without detected inflows have red EW_{2600} values >0.6 Å, comparable to or larger than the red EW_{2600} for two of the inflow galaxies. *Right*: same as *left-hand panel*, for the Mg II $\lambda 2803$ transition. Spectra which have no detectable absorption at $v > 0\,\mathrm{km\,s^{-1}}$ are marked with *cyan triangles*. As is the case for Fe II $\lambda 2600$, numerous galaxies without detected inflows have red $EW_{2803} > 0.8$ Å, the smallest value measured for the inflow sample. This survey is therefore likely failing to flag ongoing inflow in a significant fraction of those galaxies in the "wind" and "systemic" subsamples. This figure is a reproduction of Fig. 7 from the article "Evidence for Ubiquitous Collimated Galactic-Scale Outflows Along the Star-Forming Sequence at $z \sim 0.5$", Rubin et al. (2014, ApJ, 794, 156). ©AAS. Reproduced with permission

Sensitivity to inflow in the presence of absorption due to winds or interstellar material is certainly improved at higher spectral resolution. The Na I spectroscopy of nearby Seyfert galaxies discussed in Krug et al. (2010) (see Sect. 2.2) has a FWHM velocity resolution of $\sim 85\,\mathrm{km\,s^{-1}}$, and is thus significantly more sensitive to detailed line profile shapes than that of Rubin et al. (2012) or Martin et al. (2012). Such high spectral resolution can be achieved with current technology for distant galaxies which are also strongly gravitationally lensed (e.g., Pettini et al. 2002; Quider et al. 2009, 2010; Dessauges-Zavadsky et al. 2010). If at $z > 2$, high-resolution optical spectroscopy of these targets offers the added advantage of coverage of a rich suite of rest-frame UV transitions, including Lyα, O I $\lambda 1302$, several Si II transitions, Si IV $\lambda\lambda 1393, 1402$, and C IV $\lambda\lambda 1548, 1550$.

Keck/ESI coverage of many of these transitions in a spectrum of the "Cosmic Eye" is shown in Fig. 6, adapted from Quider et al. (2010). The Cosmic Eye is a Lyman break galaxy (LBG) at $z = 3.073$ which is heavily magnified by foreground

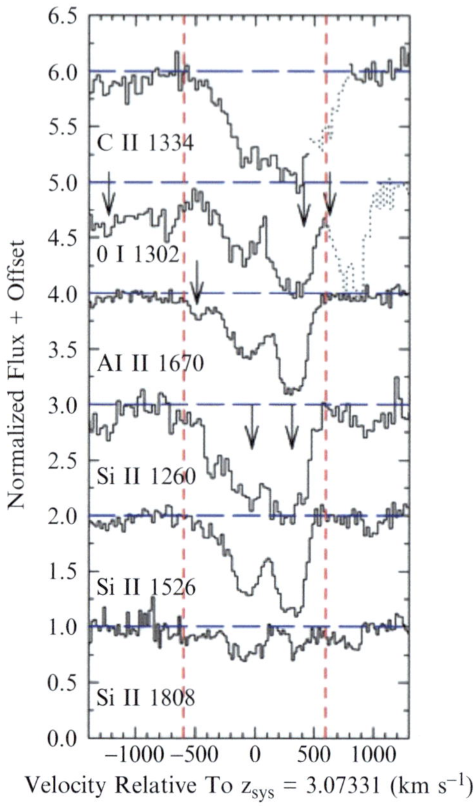

Fig. 6 High resolution Keck/ESI spectrum showing rest-frame UV absorption lines in a lensed Lyman break galaxy at $z_{sys} = 3.073$ published in Quider et al. (2010). *Arrows mark* absorbers unrelated to this system. The values on the x-axis indicate velocity relative to z_{sys}. The absorption transitions shown trace either cool, photoionized material (e.g., C II, Al II, Si II) or neutral gas (O I) along the sightline. The spectrum resolves two absorbing "components," one at $v \sim -70\,\mathrm{km\,s^{-1}}$ and one at $v \sim +350\,\mathrm{km\,s^{-1}}$, suggesting both ongoing outflow and accretion. Spectroscopy of similar quality and fidelity of a significantly larger galaxy sample will reveal the frequency with which these phenomena occur along the same line of sight. This figure reproduces a portion of Fig. 4 from the article "A study of interstellar gas and stars in the gravitationally lensed galaxy 'the Cosmic Eye' from rest-frame ultraviolet spectroscopy" by Quider et al. (2010, MNRAS, 402, 1467). Reproduced with the permission of Oxford University Press

massive structures, producing two overlapping arcs on the sky. The Keck/ESI spectroscopy obtained by Quider et al. (2010) has a FWHM velocity resolution $\sim 75\,\mathrm{km\,s^{-1}}$, and successfully resolves at least two distinct absorption components in several transitions. In fact, the authors note that this spectrum exhibits a strong, redshifted absorption component with $v \sim 350\,\mathrm{km\,s^{-1}}$ in addition to a blueshifted component presumably arising from winds. This may be interpreted as yet another detection of gas accretion, the first at $z \sim 3$. Indeed, the red absorption component in this line profile was unique among the lensed LBGs that had been studied to date, each of which exhibited strongly (and exclusively) blueshifted absorption.

Moreover, these data may in principle be used to constrain the column density, metallicity, and mass of the absorbing gas to a substantially higher level of precision than is possible with spectroscopy covering only the Na I or Mg II and Fe II lines. The absorption in the latter two transitions is nearly always saturated, limiting the line-of-sight column density to be larger than a modest value (e.g., $N_{\rm MgII} \sim 10^{14}\,{\rm cm}^{-2}$ or $N_{\rm H} \sim 10^{18.4}\,{\rm cm}^{-2}$ assuming solar metallicity; Rubin et al. 2012; Martin et al. 2012). Furthermore, rest-frame optical and near-UV spectroscopy cannot be used to measure the column density of hydrogen toward the galaxy, nor does it cover transitions of a given element in multiple levels of ionization for constraints on the ionization state of the material. While the S/N of the spectrum of the Cosmic Eye discussed above is insufficient for such analysis, spectroscopy of other lensed LBGs, e.g., the "Cosmic Horseshoe" (Quider et al. 2009), MS1512-cB58 (Pettini et al. 2002), and the "8 o'clock arc" (Dessauges-Zavadsky et al. 2010) has yielded ionic column density measurements for ISM and outflow absorbing components to a precision of ± 0.1 dex via analysis of weak (and unsaturated) Si II and Fe II transitions. Combined with analysis of the Lyα line profiles in the same spectra, this work has offered some of the only constraints on the metallicity of gas known to be outflowing from distant galaxies. Deeper observations of the Cosmic Eye and expanded samples of high dispersion, rest-frame far-UV spectroscopy of lensed LBGs have the potential to yield important constraints on the mass and metallicity of both outflows and cool galactic gas accretion.

4.2 Spatial Resolution

In addition to increasing spectral resolution to gain sensitivity to gas accretion, spatially resolving the background stellar beams will further improve our constraints on the morphology of inflow. Westmoquette et al. (2013) offered one of the earliest demonstrations of the potential of this technique, targeting the central regions of the M82 starburst with four pointings of the DensePak IFU on the WIYN telescope (Fig. 7). These observations achieved 3-arcsec spatial resolution at a spectral resolution of $\sim 45\,{\rm km\,s}^{-1}$. Combined with CO emission observations from Walter et al. (2002), they facilitated a detailed comparison between the kinematics of Hα emission, CO emission, and Na I absorption.

The authors extracted spectra from a pseudo-slit 7 arcsec in width placed along the major axis of the system, co-adding the data in several spatial bins along the slit. Measurements of the Hα, Na I, and CO velocities in these bins are shown in the right-hand panel of Fig. 7. In general, these velocities are similar at all slit locations; however, there is one component of Na I absorption (green) which has a large velocity offset, appearing to counter-rotate. The authors suggest that this component could be due to infalling tidal debris, which may even have originated in the HI gas filaments populating the circumgalactic medium around this system.

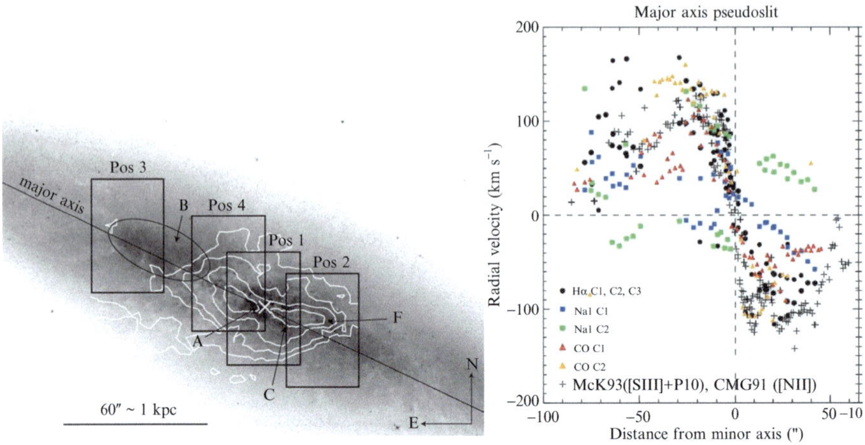

Fig. 7 Spatially resolved spectroscopy of the starburst galaxy M82. *Left*: *HST*/ACS image of M82 with four IFU pointings overlaid (*rectangles*). The *white contours* show the total CO flux from Walter et al. (2002). *Right*: Radial velocity of Na I, Hα, and CO measured in a 7 arcsec-wide pseudo-slit extracted from the IFU data along the major axis. The CO and Hα velocities are similar across the galaxy; however, there is an absorbing component of Na I (*green*) which is offset in velocity by many tens of km s^{-1}. Westmoquette et al. (2013) suggest that this component may be due to accreting tidal debris. Panels are reproductions of Figs. 1 (left) and 7 (right) from the article "Spatially resolved kinematics of the multi-phase interstellar medium in the inner disc of M82" by Westmoquette et al. (2013, MNRAS, 428, 1743). Reproduced with the permission of Oxford University Press

While this interpretation is somewhat speculative, this spatially resolved spectroscopy has nevertheless resolved a heretofore unknown velocity component of cold gas in a system in which gas flows have been studied in great detail for over a decade (Shopbell and Bland-Hawthorn 1998). Moreover, these data provide evidence in support of a picture in which inflows persist even in strongly starbursting systems with powerful ongoing outflows. Upcoming massive IFU surveys such as SDSS-IV/MaNGA (Bundy et al. 2015) will provide qualitatively similar observations for $\sim$10,000 nearby galaxies, resolving Na I kinematics on $\sim$1–2 kpc spatial scales across the face of each object. These data will be extremely sensitive to the morphology and cross section of inflowing streams as they approach the ISM of galaxies.

5 Summary and Future Directions

Deep galaxy spectroscopy has now provided perhaps the only unequivocal evidence for the inflow of gas toward galaxies beyond the local universe. While the earliest spectroscopic datasets probing cool gas kinematics "down the barrel" specifically targeted starbursting galaxies and typically reported only blueshifted (i.e., outflowing) absorption (Heckman et al. 2000; Rupke et al. 2005b; Martin 2005), later surveys achieving the requisite S/N have revealed numerous instances of redshifted metal-line absorption profiles toward a wide variety of host systems. Study of Na I D profiles in spectra of red-sequence galaxies at $z \sim 0.3$ has suggested an incidence rate of inflow of $\sim$20%; i.e., similar to the rate of outflow detections for the same sample (Sato et al. 2009). Spectroscopy of Na I and the OH 119 μm feature in the far-IR in small samples of nearby Seyfert and X-ray-bright AGN hosts points to a yet higher rate of redshifted cool gas absorption ($\sim$40%; Krug et al. 2010; Stone et al. 2016). Such inflows may in fact be necessary to fuel the observed nuclear activity; however, the amount of mass carried in these flows remains poorly constrained, and current observations do not establish the present (or ultimate) location of the gas along the line of sight (although see Shi et al. 2016, for an exception to this generalization).

In the few years following these first detections, deep spectroscopy of Fe II and Mg II transitions in the rest-frame near-UV has finally revealed evidence for inflow onto "normal" star-forming galaxies at $z > 0.3$ (Rubin et al. 2012; Martin et al. 2012). The reported rate of incidence is low ($<$10%), as blueshifted absorption tends to dominate the metal-line profiles in these spectra. Furthermore, Rubin et al. (2012) have presented evidence suggesting that inflow is more likely to be detected in these star-forming systems when they are viewed in an edge-on orientation. However, Martin et al. (2012) reported detections of inflow toward galaxies viewed over a wide range of orientations, and hence do not support this claimed dependence of inflow detection rate on viewing angle.

Such studies have been pivotal for furthering our understanding of the cycling of cool gas through galaxy environments. However, the evidence they offer is anecdotal rather than statistical. A complete, empirical picture of the baryon cycle must establish the incidence, mass, and morphology of gas inflow as a function of host galaxy stellar mass, star formation activity and history, and AGN luminosity. As all of the aforementioned samples are S/N-limited, they cannot assess such quantities regardless of their spectral resolution or the transitions they probe. Surveys of cool gas kinematics in samples selected to be complete to a given stellar mass limit (i.e., unbiased in their distributions of SFR and galaxy orientation) are required if we are to make substantive progress in our development of an empirical model of galactic gas flows.

At the same time, the observational challenges inherent in such an effort are significant. The works discussed in Sect. 3 surveying samples of $\sim$100–200 objects in the rest-frame near-UV together represent an investment of $\sim$17 nights on the Keck 1 Telescope (Rubin et al. 2012; Martin et al. 2012). A stellar mass-complete survey of a significantly larger sample would require a yet more extensive observing campaign. The ongoing SDSS-IV/MaNGA survey (Bundy et al. 2015) will facilitate a major advancement in our constraints on the incidence and morphology of inflow in the near term, as it will obtain high-S/N, spatially resolved spectroscopy of Na I in an unprecedented sample of $\sim$10,000 nearby galaxies. The MaNGA sample selection is carefully designed to be complete to a stellar mass $\log M_*/M_\odot > 9$, and its observing procedure ensures a minimum S/N per fiber of $\sim$6 per pix at 1.5

effective radii. Although its spectral coverage is limited to Ca II and Na I at relatively low velocity resolution ($R \sim 2000$), its ~1–2 kpc spatial resolution will enable detailed mapping of the incidence of cold, dusty gas inflow at speeds >40 km s^{-1} in an extremely large galaxy sample.

To assess the mass and metallicity of these flows, deep, high resolution rest-frame far-UV spectroscopy will be required (Quider et al. 2009, 2010; Dessauges-Zavadsky et al. 2010). Such observations of samples of more than a few objects must await the next generation of wide-field multi-object spectrographs on 30m-class ground-based optical telescopes. As laid out in the Thirty Meter Telescope Detailed Science Case (Skidmore 2015), the prospective instrument WFOS will be capable of obtaining $R \sim 5000$ spectroscopy of more than 100 (unlensed) LBGs at $z \sim 2$–3 with $R_{\rm AB} < 24.5$ simultaneously. S/N >30 will be achieved in just a few hours for these spectra, which will cover all of the ionic transitions discussed in the context of the Cosmic Eye (Sect. 4.1), including Lyα, O I, several Si II transitions, several Fe II transitions, Al III, Si IV, and C IV. Such a dataset will permit detailed constraints on the column densities, ionization states, metallicities, and mass of gas components arising in the ISM, outflows, and in accreting streams at high redshift.

Similarly detailed characterization of gas flows at $z < 1.5$ must await the next UV-sensitive space mission. A prospective high-resolution imaging spectrograph (France et al. 2016) conceived for the Large Ultraviolet/Optical/InfraRed (LUVOIR) surveyor NASA mission concept will not only access the important UV transitions discussed above, but will also do so at a spatial resolution of ~10–100 pc. This instrument will readily differentiate between material flowing inward in accreting streams from ongoing outflow and establish the mass and morphology of this accreting material. These capabilities will ultimately allow us to complete our empirical picture of the cycling of diffuse baryons through galaxy environments.

References

Ajiki, M., Taniguchi, Y., Murayama, T. et al. 2002, ApJ, 576, 25
Allen, J. T., Hewett, P. C., Maddox, N. et al. 2011, MNRAS, 410, 860
Barro, G., Pérez-González, P. G., Gallego, J. et al. 2011, ApJS, 193, 13
Barvainis, R. 1987, ApJ, 320, 537
Bergeron, J. 1986, A&A, 155, L8
Bundy, K., Bershady, M. A., Law, D. R. et al. 2015, ApJ , 798, 7
Chen, H.-W., Helsby, J. E., Gauthier J.-R., et al. 2010, ApJ, 714, 1521
Chen, Y.-M., Tremonti, C. A., Heckman, T. M. et al. 2010, AJ, 140, 445
Coil, A., Weiner, B. J., Holz, D. E., et al. 2011, ApJ, , 743, 23
Crenshaw, D. M., Kraemer, S. B., George, I. M. 2003, ARA&A, 41, 117
Davé, R., Finlator, K., Oppenheimer, B. D. 2012, MNRAS , 421, 98
Davis, M., Faber, S. M., Newman, J. et al. 2003, SPIE , 4834, 161
Davis, M., Guhathakurta, P., Konidaris, N. P., et al. 2007, ApJ, 660, 1
Davis, T. A., Alatalo, K., Sarzi, M., et al. 2011, MNRAS, 417, 882
Deller, A. T., Middelberg, E. 2014, AJ, 147, 14
Dessauges-Zavadsky, M., D'Odorico, S., Schaerer, D. et al. 2010, A&A, 510, 26

Du, X., Shapley, A. E., Martin, C. L., Coil, A. L. 2016, arXiv, 1604.03945
Ellison, S. L., Prochaska, J. X., Hennawi, J.et al. 2010, MNRAS, 406, 1435
Fathivavsari, H., Petitjean, P., Noterdaeme, P. et al. 2016, MNRAS, 461, 1816
France, K., Fleming, B., Hoadley, K. 2016, J. Astron.Telesc. Instrum. Syst., 2(4), 041203
Goerdt, T., Ceverino, D. 2015, MNRAS, 450, 3359
Hall, P. B., Anderson, S. F., Strauss, M. A. et al. 2002, ApJ, 141, 267
Hall, P. B., Brandt, W. N., Petitjean, P. et al. 2013, MNRAS, 434, 222
Heckman, T. M., Lehnert, M. D., Strickland, D. K., Armus, L. 2000, ApJS, 129, 493
Kacprzak, G. G., Churchill, C. W., Steidel, C. C. et al. 2007, ApJ, 662, 909
Kornei, K. A., Shapley, A. E., Martin, C. L. et al. 2012, ApJ, 758, 135
Krug, H. B., Rupke, D. S. N., Veilleux, S. 2010, ApJ , 708, 1145
Langer, W. D., Pineda, J. L., Velusamy, T. 2014, Astronomy & Astrophysics, 564, 101
Le Floc'h, E., Willmer, C. N. A., Noeske, K. et al. 2007, ApJ, 660, 65
Lehner, N., Howk, J. C. 2011, Science, 334, 955
Lehner, N., Howk, J C., Tripp, T. M., et al. 2013, ApJ, 770, 138
Li, J., Ostriker, J., Sunyaev, R. 2013, ApJ, 767, 105
Martin, C. L. 2005, ApJ, 621, 227
Martin, A. M., Papastergis, E., Giovanelli, R. et al. 2010, ApJ, 723, 1359
Martin, C. L., Shapley, A. E., Coil, A. L. et al. 2012, ApJ, 760, 127
Nelson, D., Genel, S., Vogelsberger, M. et al. 2015, MNRAS, 448, 59
Oppenheimer, B. D., Dave, R., Kereš, D. et al. 2010, MNRAS, 406, 2325
Pettini, M., Rix, S. A., Steidel, C. C., et al. 2002, ApJ, 569, 742
Prochaska, J. X., Weiner, B. J., Chen, H.-W., et al. 2011, ApJ, 740, 91
Quider, A. M., Pettini, M., Shapley, A. E., Steidel, C. C. 2009, MNRAS, 398, 1263
Quider, A. M., Shapley, A. E., Pettini, M. et al. 2010, MNRAS, 402, 1467
Rupke, D. S., Veilleux, S., Sanders, D. B. 2005b, ApJS , 160, 115
Rupke, D. S., Veilleux, S., Sanders, D. B. 2005a, ApJ , 632, 751
Rubin, K. H. R., Weiner, B. J., Koo, D. C., et al. 2010, ApJ, 719, 1503
Rubin, K. H. R., Prochaska, J. X., Koo, D. C., Phillips, A. C., 2012, ApJL, 747, 26
Rubin, K. H. R., Prochaska, J. X., Koo, D. C., et al. 2014, ApJ, 794, 156
Sarzi, M., Kaviraj, S., Nedelchev, B., et al. 2016, MNRAS, 456, 25
Sato, T., Martin, C. L., Noeske, K. G., et al. 2009, ApJ, 696, 214
Shen, Y., Brandt, W. N., Denney, K. D., et al. 2016, arXiv, 1602.03894
Shi, X., Jiang, P., Wang, H., et al. 2016, arXiv, 1608.05487
Shopbell, P. L. & Bland-Hawthorn, J. 1998, ApJ, 493, 129
Steidel, C. C., Dickinson, M., Persson, S. E. 1994, ApJ, 437, L75
Steidel, C. C., Erb, D. K., Shapley, A. E., et al. 2010, ApJ, 717, 289
Stone, M., Veilleux, S., Melendez, M. 2016, arXiv, 1605.06512
Thirty Meter Telescope Detailed Science Case 2015, ed. Skidmore, W., arXiv:1505.01195
Tumlinson, J., Thom, C., Werk, J. K. et al. 2011, Science, 334, 948
van Gorkom, J. H., Knapp, G. R., Ekers, R. D., et al. 1989, AJ, 97, 708
Vogelsberger, M., Genel, S., Sijacki, D., et al. 2013, MNRAS, 436, 3031
Walter, F., Weiss, A., Scoville, N. 2002, ApJ , 580, 21
Weiner, B. J., Coil, A. L., Prochaska, J. X., et al. 2009, ApJ, 692, 187
Werk, J. K, Prochaska, J. X., Tumlinson, J., et al. 2014, ApJ, 792, 8
Westmoquette, M. S., Smith, L. J., Gallagher, J. S., Walter, F. 2013, MNRAS, 428, 1743

Gas Accretion via Lyman Limit Systems

Nicolas Lehner

1 Introduction

Among the most pressing problems in galaxy formation is to understand how gas accretion and feedback influence the evolution of galaxies and the intergalactic medium (IGM). Modern cosmological simulations cannot explain the mass-metallicity relationship and color bimodalities observed in galaxies without invoking such large-scale flows (e.g., Tremonti et al. 2004; Kereš et al. 2005). Stars cannot continue to form in galaxies over billions of years without a replenishment of gas in galaxies from the IGM (e.g., Maller and Bullock 2004; Dekel and Birnboim 2006), while feedback from star formation and AGN activity can fuel massive outflows from a gas-rich galaxy that may choke off star formation (e.g., Oppenheimer et al. 2010, 2012; Fumagalli et al. 2011b; Faucher-Giguère et al. 2011; Kacprzak et al. 2008). The competition between these regulating processes is played out at the interface between the galaxies and the IGM, a region now defined as the circumgalactic medium (CGM).

The large-scale outflows, inflows, recycling in the CGM play therefore critical roles in the evolution of galaxies that require empirical characterization. Outflows appear ubiquitous in the universe at any z (e.g., Pettini et al. 2001; Shapley et al. 2003; Steidel et al. 2004, 2010; Weiner et al. 2009; Rubin et al. 2014). The COS-Halos and COS-Dwarfs galaxy-centric experiments have demonstrated that the low-z CGM is a massive reservoir of galactic metals and baryons with at least as much metals and baryons as in the disk of galaxies (Tumlinson et al. 2011; Werk et al. 2014; Peeples et al. 2014; Bordoloi et al. 2014, and see also, e.g., Stocke et al. 2013; Liang and Chen 2014 for other surveys). As discussed at length in this book,

N. Lehner (✉)
Center for Astrophysics, Department of Physics, University of Notre Dame,
225 Nieuwland Science Hall, Notre Dame, IN 46556, USA
e-mail: nlehner@nd.edu

A. Fox, R. Davé (eds.), *Gas Accretion onto Galaxies*, Astrophysics and Space Science Library 430, DOI 10.1007/978-3-319-52512-9_6

gas accretion is crucial for maintaining star formation in the disk of galaxies over billions of years, controlling in part the evolution of the elemental abundances in galaxies, and could be responsible for the mass-metallicity relationship (see, e.g., Kacprzak et al. 2016). However, cold gas accretion that, according to cosmological simulations (e.g., Faucher-Giguère et al. 2011, 2015; Shen et al. 2012; Fumagalli et al. 2011b, 2014b; van de Voort and Schaye 2012; van de Voort et al. 2012; Hafen et al. 2016), fuels the star formation in galaxies have been extremely difficult to find observationally. According to these simulations, cold ($T < 10^{5.5}\,\mathrm{K} \ll T_{\rm vir}$, e.g., van de Voort and Schaye 2012) gas accretion should not be pristine but already enriched by previous episodes of star formation. However, it should not be either too metal-enriched from outflowing or recycled material from galaxies with intense star formation. Cold streams are also thought to contribute to the population of absorbers with $17 \lesssim \log N_{\rm HI} < 19$, a class of absorbers known as Lyman limit systems (LLSs). In this chapter, I will argue that the empirically derived properties of the LLSs and partial LLSs (pLLSs) imply that we have found large reservoirs of metal-poor gas around galaxies, but it is still an open question if this gas is actually cold accretion as seen in simulations or is even accreting.

The pLLSs and LLSs have—by the definition adopted here—H I column densities $16 \lesssim \log N_{\rm HI} < 17.2$ and $17.2 \le \log N_{\rm HI} < 19$, respectively, i.e., on the $N_{\rm HI}$ scale of the universe they are located between the Lyα forest ($\log N_{\rm HI} < 16$) and the damped Lyα absorbers ($\log N_{\rm HI} \ge 20.3$) and sub-DLAs a.k.a. SLLSs ($19 \le \log N_{\rm HI} < 20.3$). The H I column densities of the pLLSs and LLSs and the densities implied for the gas they probe ($-4 \lesssim \log n_{\rm H} \lesssim -1$, e.g., Schaye 2001; Lehner et al. 2013; Fumagalli et al. 2016b) place them at the interface between the IGM (probed by Lyα forest absorbers) and galaxies and their immediate surroundings (probed by SLLSs and DLAs). With an optical depth at the Lyman limit $\tau_{\rm LL} \ge 1$, LLSs are optically thick to Lyman continuum radiation, absorbing the UV radiation at $\lambda \le 912$ Å. Empirically characterizing the UV background has been a scientific driver for studying LLSs as well as their potential connection to galaxies (e.g., Tytler 1982; Sargent et al. 1989; Bahcall et al. 1993; Lanzetta et al. 1995; O'Meara et al. 2007; Prochaska et al. 2010; Ribaudo et al. 2011a). While pLLSs and LLSs can be readily detected in low resolution spectra and their H I column densities can be estimated directly from the optical depth at the Lyman limit (see Fig. 1), these earlier surveys did not have the resolution to detect weak metal lines associated with these absorbers, and hence their properties (metallicity, density, size, mass, etc.) have remained largely unconstrained till more recently.

The ability to determine the properties of the LLSs and pLLSs is indeed critical for using the pLLSs/LLSs as gas tracers of the IGM/galaxy interface. Among all these properties, the metallicity is key since it can be used as a "tracer" of the origins of the gas (Ribaudo et al. 2011b; Fumagalli et al. 2011b; Lehner et al. 2013). The metallicity level of the gas should help differentiating between gas that is accreting from the IGM and the more enriched material produced recently by galaxies at the considered epoch, which could be outflowing or inflowing. If very metal-poor gas is found around a metal-enriched galaxy, it is difficult to find a plausible scenario that would explain how this gas may have come from the galaxy itself. On the other

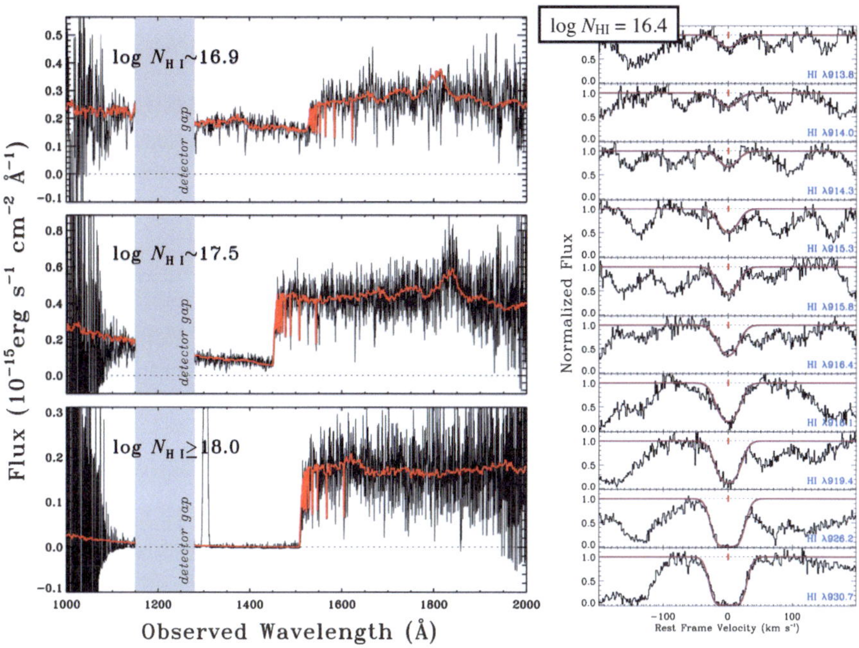

Fig. 1 To measure $N_{\rm HI}$ in pLLSs/LLSs, two methods are available. (1) *left*: we can measure the optical depth at the Lyman limit where $\tau_{\rm LL} \propto N_{\rm HI}$ (these are low resolution COS G140L spectra from Wotta et al. 2016; (2) *right*: we fit a model to the Lyman series to derive $N_{\rm HI}$ (these are normalized Keck HIRES profiles in the redshift frame of the pLLS at $z_{\rm LLS} \sim 3$ from Lehner et al. 2016)

hand, if the circumgalactic gas and the host galaxy have similar metallicities, it is quite natural to invoke large-scale outflows or recycling motions to explain the origin of this gas.

To determine the metallicity without any bias requires selecting the absorbers based on their H I content alone. Contrary to DLAs (and to a lesser extent SLLSs), many (metal-poor) pLLSs and LLSs at $z < 1$ have no detection of Mg II $\lambda\lambda$2797, 2803 absorption down to equivalent widths <10 mÅ; these very weak Mg II absorbers are completely missing in even very deep Mg II surveys (e.g., Churchill et al. 2000). The H I selection is necessary, but not sufficient since, of course, to be able to estimate the metallicity requires being able to accurately estimate $N_{\rm HI}$ and the column densities of metal lines (which can include non-detections of metals if the signal-to-noise—S/N—of the spectra is high enough to place tight constraints on the column densities). It also requires modeling the ionization of the gas since at $\log N_{\rm HI} \lesssim 18.5$ the gas is predominantly ionized and since only $N_{\rm HI}$, not $N_{\rm HII}$, can be measured from the observations.

A first attempt to survey the physical properties of the LLSs was undertaken by Steidel (1990) with 10 strong LLSs ($\log N_{\rm HI} \gtrsim 17$) at $z \sim 3$. He used observations with 35–80 km s^{-1} spectral resolution, and with this low resolution, the metallicities and other physical properties were only crudely estimated with many limits, but he still was able to show that the total H column of these LLSs is often as large as in DLAs, but mostly in form of $N_{\rm HII}$. Prochaska (1999) showed that with high S/N and high resolution ($\sim$9 km s^{-1}) obtained with the 10-m Keck I telescope the properties of the LLSs can be estimated much more accurately. The right-hand side of Fig. 1 shows that with high resolution and S/N spectra, we can also derive $N_{\rm HI}$ from the analysis of the Lyman series transitions; this is also particularly useful for making sure the metals are observed over a similar velocity range (Lehner et al. 2013, 2016).

In this chapter, I will report on recent discoveries within the last $\sim$5–10 years, which in my opinion, have led to several important findings, but have also raised new puzzles. There are two game changers that have occurred in this timeframe, which have enabled transformative progress. As already mentioned high S/N and resolution spectra are important to derive the properties of the pLLSs and LLSs. Equally as important has been the ability to assemble samples of LLSs/pLLSs with more than a handful of absorbers where their properties can be determined. To start to be able to derive some statistically robust trends requires at least a sample size of 25–50. At low redshift, the Cosmic Origins Spectrograph (COS)—installed in 2009 on the *Hubble Space Telescope* (*HST*)—has revolutionized this field, thanks to its great leap in sensitivity with spectral resolution $R \sim 17{,}000$ good enough to accurately estimate $N_{\rm HI}$ and $N_{\rm metals}$. Prior to COS, there was at $z < 1$ only a handful of LLSs with a determination of their metallicities (Lehner et al. 2009 and the references therein). Since the COS installation, the sample has increased to 30–50 (Lehner et al. 2013; Wotta et al. 2016) and will increase to $\gtrsim$100 in the coming year. At high redshift, dedicated surveys (like, e.g., the HD-LLS survey Prochaska et al. 2015) and the availability of large databases of reduced QSO spectra obtained with large telescopes (e.g., KODIAQ—Lehner et al. 2014, 2016; O'Meara et al. 2015; XQ-100—Lopez et al. 2016) have enabled many new discoveries.

This chapter is organized as follows. In Sect. 2, I provide some background on how we determine the metallicities of the pLLSs and LLSs. This process requires making large ionization corrections, but I argue we can do this reasonably well, and well enough to robustly empirically characterize the metallicities of these absorbers. In the next five sections (Sects. 3–7), I review some of the major findings on the metallicities and relative abundances of the pLLSs and LLSs and their evolution over cosmic time. In Sect. 8, I offer a different but complementary perspective when I present some of the recent results on the surveys of O VI in pLLSs and LLSs and their implications for large-scale outflows and inflows through the CGM of galaxies. In Sect. 9, I specifically discuss the findings that may point to cold, metal-poor gas accretion traced by pLLSs and LLSs onto galaxies, and finally in Sect. 10, I provide a succinct summary and some future directions.

2 Metallicity: Methodology and Uncertainties

The metallicity is the key parameter to constrain if one wants to determine the properties of the pLLSs and LLSs, assess their possible origins and their associations with some of the structures of the universe. The most robust approach to estimate the metallicity of the pLLSs and LLSs would be to use the O I/H I ratio since charge exchange reactions with hydrogen ensure that the ionizations of H I and O I are strongly coupled (although in the very diffuse ionized gas probed by these absorbers, some ionization correction may be even necessary when using O I). However, for absorbers with $\log N_{\rm HI} \lesssim 17.5$, O I is rarely detected, and the limit that can be placed on $N_{\rm O\,I}$ is generally not sensitive enough to be interesting. So we are left to compare ionized species with H I, and hence we need to determine $N_{\rm HII}$, which requires undertaking ionization modeling of the gas. LLSs and pLLSs are often multiphase, with absorption seen in different ionization stages. However, the low to intermediate ions (e.g., [Si II], Si III, Si IV, C II, C III) and high ions (O VI) often show distinct kinematics (Crighton et al. 2013; Fox et al. 2013; Lehner et al. 2013, 2014; and see Sect. 8), with the low ions following more closely the velocity structure of the H I (e.g., Lehner et al. 2009, 2013; Crighton et al. 2013, 2015; Fumagalli et al. 2016b). The Doppler broadenings of the H I components associated with the pLLSs and LLSs ($b \simeq 15$–$30\,{\rm km\,s^{-1}}$) also imply typically that the gas temperature is $T \lesssim 1$–4×10^4 K. Hence to estimate the metallicity of the pLLSs and LLSs, one needs to estimate $N_{\rm HII}$ in the $T \lesssim 1$–4×10^4 K photoionized gas.[1]

To model the ionization, it has been customary to use Cloudy (Ferland et al. 2013) where the gas is assumed to have a uniform slab geometry and is photoionized by the Haardt-Madau background radiation field from quasars and galaxies (Haardt and Madau 1996, 2012). For each absorber, the ionization parameter (the ratio of H ionizing photon density to total hydrogen number density—$U = n_\gamma/n_{\rm H}$) and the metallicity[2] are varied to search for models that best fit the observational constraints set by the column densities. On the left-hand side of Fig. 2, I show an example of Cloudy simulation for the pLLS at $z \sim 0.3$ toward PG1630+377 (see Ribaudo et al. 2011b for the entire description) where the Haardt-Madau background radiation field from quasars and galaxies (HM05, as implemented within Cloudy) was used. In that case, the upper/lower limits and measurements of the metals constrain well the metallicity. Solar relative abundances are assumed as a starting point for comparing ions of C, Si, O, Mg, etc., i.e., a priori the effects of dust or nucleosynthesis on the relative abundances are not included. The quoted metallicity in this chapter is, however, always for the α elements (X $=$ Si, O, Mg); I will discuss later that the C/α ratio may not always be solar.

A major drawback to have to correct for ionization is that the characteristics of the photoionized gas depend on the shape and strength of the ionizing spectrum,

[1]Except otherwise stated, the metallicity discussed here is the metallicity of the cool gas.

[2]I use throughout the usual square-bracket notation for the metallicity $[{\rm X/H}] \equiv \log N_{\rm X}/N_{\rm H} - \log({\rm X/H})_\odot$, where X is a given element.

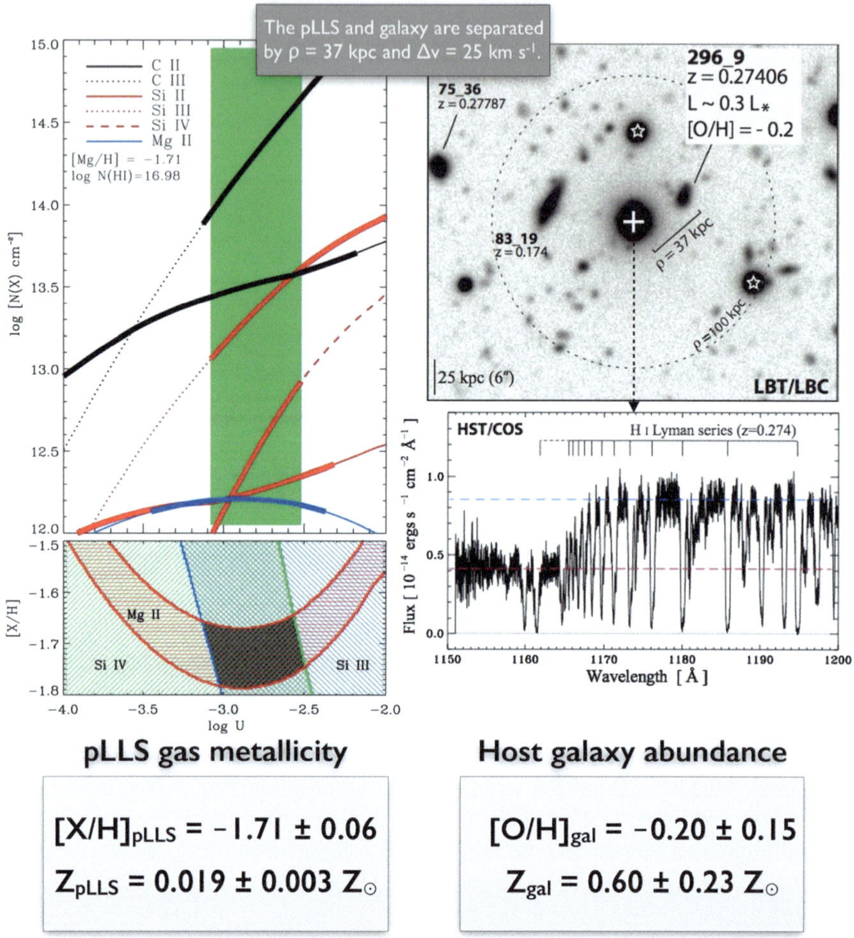

Fig. 2 One of the first examples of a very low-metallicity pLLS found at an impact parameter $\rho = 37$ kpc of a galaxy with the same redshift (results and figures from Ribaudo et al. 2011b). The *top left panel* shows the Cloudy-predicted column densities as a function of ionization parameter for the given metallicity and estimated $N_{\rm HI}$ (from the COS data shown on the *bottom right panel*). The *bold portions* of the curves show where the model column densities and observations are consistent and the *green band* shows the range of U for which the models are consistent for all the different ions. The *lower left panel* shows the allowed [X/H]–U region by the observations. The *top right panel* shows an LBT g-band image of the QSO field where the associated galaxy is labeled 296_9. From a Keck LRIS spectrum, Ribaudo et al. (2011b) were able to estimate an accurate redshift, mass, and abundance of the galaxy. *This low-metallicity gas clearly did not originate from galaxy 296_9, and yet it is well within its virial radius*

and there are some systematic uncertainties arising from the ionization modeling. Several independent groups have assessed the systematic errors connected to these ionization models (Howk et al. 2009; Lehner et al. 2013, 2016; Crighton et al.

2015; Fumagalli et al. 2016b; Wotta et al. 2016). For example, Wotta et al. (2016) used HM12 (Haardt and Madau 2012) to estimate the metallicities of 10 LLSs at $z < 1$ from Lehner et al. (2013) that were initially modeled with HM05. The comparison of the results demonstrates that the harder spectrum of HM12 increases the metallicity on average by $+0.3$ dex those derived using HM05 (with no dependence on the initial metallicity of the gas, i.e., a similar shift in metallicity is observed for both high or low-metallicity absorbers). For the absorber shown in Fig. 2, the metallicity increases by $+0.47$ dex using HM12 instead of HM05, but in the case of HM12, the solution found is not satisfactory because too much [Si II] is predicted by the model ($> +0.2$ dex) while the HM05 background did fit all the observables, including the non-detection of [Si II]. Fumagalli et al. (2016b) provide a thorough analysis of a large sample of LLSs at $z \sim 2$–3 where they use several ionization models (including dust depletion, proximity to local sources, collisions) and Bayesian techniques to derive the physical properties and metallicities of the LLSs. They find as well that the metallicity estimates are typically not too sensitive to the assumptions behind the ionization corrections.

Determining the spectral shape of the ionizing background is an on-going challenge. However, combining all the recent results, there is some consensus that the systematic uncertainties associated with the choice of ionizing background mode are of the order of 0.3–0.5 dex (i.e., a factor 2–3, see references above). The statistical uncertainties for a given model are typically much smaller, especially if the column densities of H I and metals are small. Hence, the systematic uncertainty dominates the error budget. This is not precision cosmology, but these uncertainties are comparable to the systematic uncertainties derived using emission line diagnostics for galaxies (e.g., Berg et al. 2016). It is accurate enough to separate confidently a metal-poor absorber ($[X/H] \lesssim -1$) from an absorber with a near-solar metallicity or to find trends or characterize the metallicity distribution. Furthermore, *the same selection criteria and techniques to analyze the characteristics of the pLLSs/LLSs can be used from $z \lesssim 1$ to $z \sim 4$, allowing us to probe the galaxy/IGM interface over 12 billion years.*

3 The Metallicity of the pLLSs and LLSs at $z \lesssim 1$

Although I cannot yet provide much information about the galaxies associated with the pLLSs/LLSs since this work is still in progress (but see Sect. 9), there are great advantages studying the $z \lesssim 1.5$ universe over the higher redshift universe, especially in terms of determining the properties of the associated galaxies. At high redshift, only the brightest, most massive galaxies can be observed, while at $z \lesssim 1$ luminous and very sub-luminous galaxies can be observed. The $z < 1.5$ epoch is also a time where galaxy transformation is occurring with blue galaxies turning into red galaxies. The morphology, inclination, and geometry of $z < 1$ galaxies can also be robustly determined (e.g., Bordoloi et al. 2011; Kacprzak et al. 2012a; Bouché et al. 2012), which will help us providing additional insights on the actual gas origin (see Sect. 9).

In the example shown in Fig. 2, I present a pLLS with a very low-metallicity $[X/H] \simeq -1.7$ (2% solar metallicity) at $z \simeq 0.3$, which is extremely low metallicity at this redshift. This is quite a bit lower than typical of dwarf galaxies in the low redshift universe (e.g., Skillman et al. 1994; Kunth and Östlin 2000; Kniazev et al. 2003; Hirschauer et al. 2016; Berg et al. 2016). Figure 2 also shows the galaxy environment near the background QSO. A $0.3L*$ galaxy is found at a very similar redshift and with a near-solar abundance, a factor ~30 larger than the metallicity of the pLLSs at an impact parameter of $\rho = 37$ kpc. This pLLS is well within the virial radius of this galaxy and yet this gas does not come from this galaxy (at least not from a recent epoch). What is its origin? What is its fate? Ribaudo et al. (2011b) proposed that this pLLS might probe infalling gas onto galaxies, perhaps related to the cold flow accretion observed in cosmological simulations. The physical and chemical properties of this low-metallicity pLLS are comparable to those predicted by cold flow models, including the temperature, metallicity, and $N_{\rm HI}$ as well as host galaxy properties such as velocity offset and mass. This example shows how the metallicity acts as a key discriminator, and since this discovery, one of the goals has been to increase the sample size to determine the prevalence of metal-poor versus metal-rich pLLSs and LLSs at $z < 1$.

At the time of the Ribaudo et al. (2011b) study, there were only about eight metallicity estimates in LLSs (see, e.g., Lehner et al. 2009; Cooksey et al. 2008; Zonak et al. 2004; Tripp et al. 2011) and about 50% of those had $[X/H] \lesssim -1.5$. This is remarkable because as I will show below the sample of pLLSs and LLSs has increased by factors 3.5 and now 7 and yet the same fraction of very metal-poor gas has been found for the combined sample of pLLSs and LLSs. This is also remarkable because these results show the prevalence of low-metallicity gas at low redshift where for a long time the canonical value of the metallicity for the IGM was set to $[X/H] \simeq -1$. The low metallicity has been essentially missed in studies of the Lyα forest at $z < 1$ because for absorbers with $N_{\rm HI} \lesssim 15$, the S/N typically achieved in *HST* UV spectra is not sufficient to detect metals to determine metallicities with $[X/H] \lesssim -1$.

One of the main goals of the low-z studies has been to build representative samples of the pLLSs and LLSs. The first sizable sample of pLLSs and LLSs was assembled by Lehner et al. (2013). Owing to the original selection of UV-bright QSOs for the high-resolution COS observations, this first sample is heavily biased toward low $N_{\rm HI}$ absorbers with 23 pLLSs and only 5 LLSs. In the second survey, Wotta et al. (2016) used QSOs that were selected to be optically bright, and hence there is no bias against strong pLLSs or LLSs in the UV spectra of these QSOs, providing a more uniform combined sample in $N_{\rm HI}$ over the pLLS and LLS range. Owing to that selection, the QSOs were obtained with the low resolution mode of COS, leading Wotta et al. to develop a new method to derive the metallicity, which used only the comparison of Mg II and H I. They demonstrate that with a prior knowledge of the U distribution (determined from the analysis of Lehner

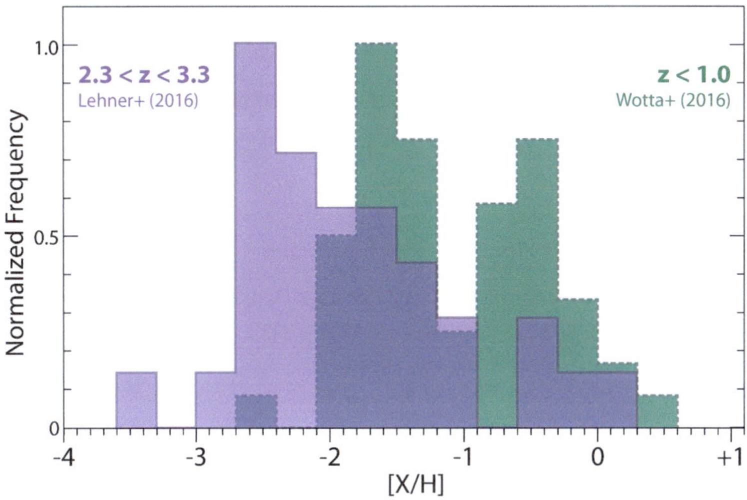

Fig. 3 Evolution of the metallicity distribution of the H I column density of the pLLSs/LLSs with $16.2 \lesssim \log N_{\rm HI} \lesssim 18$ from $z < 1$ (from Wotta et al. 2016) to $2.3 < z < 3.3$ (from Lehner et al. 2016) (figure courtesy of Chris Howk)

et al. 2013), the metallicities of the pLLSs and LLSs at $z \lesssim 1$ can be accurately estimated and the two samples from Lehner et al. (2013) and Wotta et al. (2016) can be combined. I refer the reader to the original papers for more details. Here I use the combined sample from the two surveys (which also include the few LLSs with $\log N_{\rm HI} \geq 17.7$ which were not included in Wotta et al. 2016).

Figures 3 and 4 present two different ways to display the metallicity distribution of the pLLSs and LLSs. The green histogram in Fig. 3 shows the metallicity distribution of the combined sample of pLLSs and LLSs at $z < 1$, with two prominent peaks at $\langle[{\rm X/H}]\rangle \simeq -1.9$ and $\langle[{\rm X/H}]\rangle \simeq -0.3$ (see Wotta et al. 2016 and Lehner et al. 2013 for details; the low-metallicity peak has a mean that appears low relative to its distribution peak because it was estimated using the survival analysis to take into account of the 13 upper limits that are displayed in Fig. 4). A Gaussian mixture modeling (GMM) and a Dip test on the metallicity distribution of the pLLSs and LLSs reject a unimodal distribution at high significance levels (99.6% and 82.4%, respectively). As noted Muratov and Gnedin (2010), the Dip test appears less powerful than GMM based on the significance levels, but it has the benefit of being insensitive to the assumption of Gaussianity and is therefore a true test of modality.

Figure 4 shows the metallicity of the absorbers as a function of $N_{\rm HI}$. This type of figure allows us to make a direct comparison with other $N_{\rm HI}$ absorbers as well as displaying the unbinned metallicity distribution of the absorbers. While the connection between $N_{\rm HI}$ and impact parameter to the galaxies is crude, examining the metallicity for absorbers covering a broad range of H I column densities provides

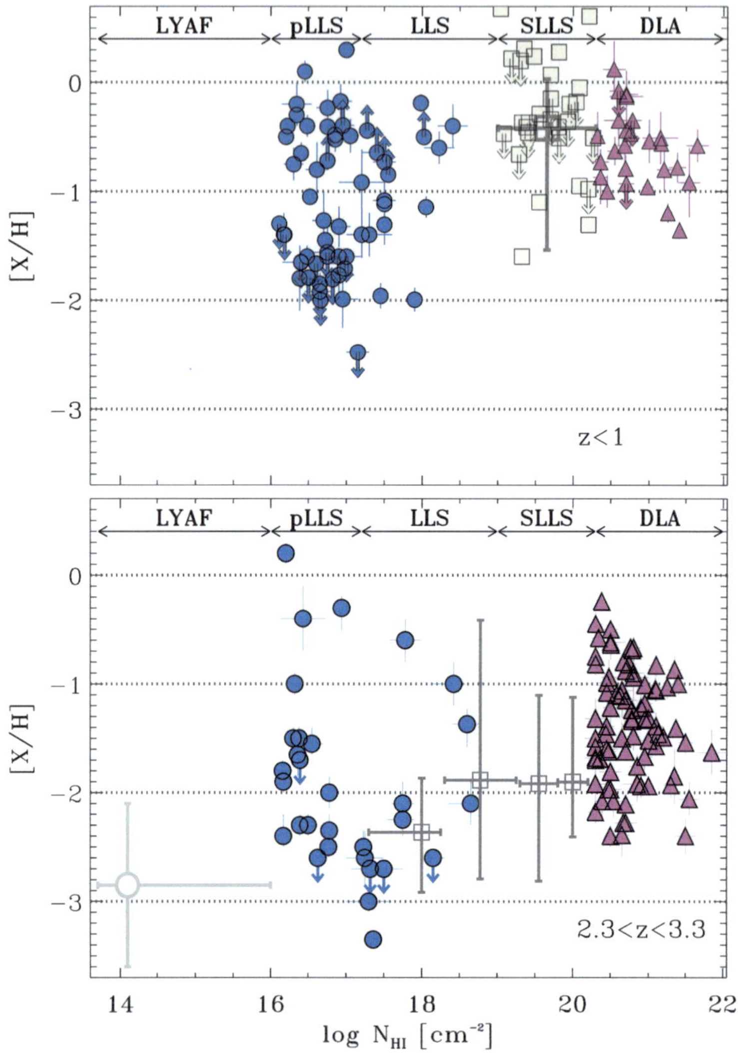

Fig. 4 Metallicity as a function of the H I column density for absorbers at $z < 1$ (*top*, results from Lehner et al. 2013; Wotta et al. 2016 and the references therein) to $2.3 < z < 3.3$ (*bottom*, results from Lehner et al. 2016 and the references therein, and see also the main text). Note that the metallicity-scale in both panels is the same for direct comparison

a measure of how the origins of the gas are changing with environment (impact parameter and density) within galaxy halos. I will discuss this evolution with $N_{\rm HI}$ further in Sect. 5 and focus here solely on the pLLSs and LLSs.

Starting with pLLSs, there is a lack of absorbers in this figure near [X/H] ~ -1 ($-1.2 \lesssim$ [X/H] $\lesssim -0.9$). For the pLLS sample, the Dip test rejects a unimodal distribution at the significance level of 95.5% (>99.9% according to the GMM), i.e., it is more significant than for the combined pLLS and LLS sample.[3] Figure 4 also reveals that while bimodal nature of the metallicity distribution is now pretty well established for the pLLSs,[4] it is less so for the LLSs. In fact there is a strong hint that the metallicity distribution may transition to a broad unimodal distribution in the LLS regime (Wotta et al. 2016): (1) there are more LLSs near the −1 dex dip, and (2) there are significantly less LLSs with [X/H] ≤ -1.4 than pLLSs (5–19% compared to 35–51% for the pLLSs; 68% confidence level). An ongoing survey will double the current sample of the LLSs using the COS archive and show if indeed the metallicity distribution changes in the LLS regime (C. Wotta et al. 2017, in preparation). While there is a lack of LLSs with [X/H]< -1.4, the fraction of LLSs with [X/H]< -1.4 is still quite important (39–61% at the 68% CL).

4 Metallicity Distribution at High z and Redshift Evolution

The great aspect with QSO absorbers is that the same diagnostics and techniques are used at both high and low z, so the evolution of the metallicity and physical conditions of the absorbers can be studied straightforwardly over cosmic time. The cosmic evolution of the DLAs (e.g., Prochaska et al. 2003; Rafelski et al. 2012; Battisti et al. 2012; Jorgenson et al. 2013) and SLLSs (e.g., Som et al. 2013, 2015; Fumagalli et al. 2016b; Quiret et al. 2016) have been studied for several years. Thanks to large archives of good quality (in particular high resolution) spectra of both low and high z QSOs, we can now study the cosmic evolution of the H I-selected pLLSs and LLSs. The same selection criteria and analyses of the pLLSs and LLSs are applied at $z < 1$ and $2.3 < z < 3.3$ by Lehner et al. (2013) (and also Wotta et al. 2016) and Lehner et al. (2016), respectively, which allow for a direct comparison between the two samples as displayed in Figs. 3 and 4. They are H I-selected to have H I column densities between $16 \lesssim \log N_{\rm HI} < 19$. The H I column density for each absorber can be estimated reasonably accurately (within $\sim$0.3 dex, and often better than 0.05–0.10 dex). There is enough information from the metal lines to derive sensitively both high and low metallicities. And finally, the reported metallicities are from α-elements, and i.e., $X = \alpha$ where α is Si and/or O at high z and Si, Mg, S, and/or O at low z.

The reader should, however, keep in mind that the overdensities of the structures of the universe change as a function of z, and therefore absorbers at some given

[3]In the Dip test and GMM, upper and lower limits are treated as values. If one decreases/increases the upper/lower limits by a very large factor $\geq$5 (0.7 dex), the unimodal distribution is still rejected at the $\geq$80% and $\geq$95% significance level for the Dip test and GMM, respectively. Therefore, the metallicity distribution of the pLLSs is very likely at least bimodal (there could be another population of pLLSs at [X/H] $\lesssim -2$ if the actual values of the several of the upper limits are indeed much lower than the current limits set by the observations).

[4]Note as well that the upper and lower limits push the two branches apart, not together.

$N_{\rm HI}$ at high and low redshifts are not necessarily physically analogous (e.g., Schaye 2001; Davé et al. 1999). The change in the overdensity $\delta_{\rm H} \equiv (n_{\rm H} - \bar{n}_{\rm H})/\bar{n}_{\rm H}$ is a factor ~8 between the mean redshifts of the low and high z samples (Lehner et al. 2016). For the Lyα forest absorbers, SLLSs, and DLAs, the redshift evolution of the density does not change the fact that Lyα forest absorbers still trace very diffuse gas ($\delta_{\rm H} \ll 100$) and SLLSs/DLAs trace virialized structures ($\delta_{\rm H} \gg 100$) at any z. However, for the LLSs and especially the pLLSs, while at $z < 1$ they probe gas well within the CGM of galaxies (see Sect. 9), at $z \sim 2.8$, $\delta_{\rm H}$ can be $\lesssim 100$, and hence some pLLSs could probe more diffuse ionized gas at $z > 2$. In terms of high redshift galaxy observations, KBSS shows that only half of the absorbers with $\log N_{\rm HI} > 15.5$ are found in the CGM of (*massive*) galaxies at $z \sim 2$; the other half may probe more diffuse gas or the CGM of dwarf galaxies (Rudie et al. 2012). Observations with MUSE also found no bright, star-forming galaxy in the vicinity of two very metal-poor LLSs (Fumagalli et al. 2016a). However, I emphasize these observations targeted only very metal-poor LLSs, which are just a subset of the LLS population (see below). With the current knowledge, while LLSs and pLLSs are by definition at the interface between the denser and more diffuse gas at any z, high z pLLSs and LLSs may not always trace necessarily the dense CGM of galaxies as their counterparts at $z < 1$.

In Fig. 3, I show the metallicity distribution for the 31 H I-selected pLLSs and LLSs at $2.3 < z < 3.3$ from the KODIAQ Z pilot study (Lehner et al. 2016). Visually and statically, the metallicity distribution at high z is quite different from the low z metallicity distribution: not clearly bimodal and the peak of the distribution has shifted to lower metallicity. Lehner et al. (2016) estimate for the pLLSs *and* LLSs that $\langle[{\rm X/H}]\rangle = -2.00 \pm 0.17$ (where the quoted error is the Kaplan-Meier (KM) error on the mean value obtained from the survival analysis). There is little overlap at $[{\rm X/H}] \le -2$ between the high and low redshift samples, although there are several upper limits around -2, -1.7 dex at $z < 1$ (see Fig. 4). However, at $[{\rm X/H}] \ge -2$, there is a large overlap. Quite remarkably there are also several pLLSs/LLSs (with a fraction ~4–20%, 68% CL) with $-0.5 \lesssim [{\rm X/H}] \lesssim +0.2$ at $z > 2$. There is no evidence of a strong dip anywhere in the metallicity distribution as observed at low redshift (L13, W16) and there is a prominent peak near the mean. The Dip test on the high z metallicity distribution shows that the significance level with which a unimodal distribution can be rejected is only 26%.

In Fig. 3, I also show the result from the HD-LLS survey with the square gray data points from Fumagalli et al. (2016b) over the same redshift interval than probed by the KODIAQ Z sample shown in this figure. The HD-LLS survey targets LLSs and SLLSs with $\log N_{\rm HI} > 17.2$ at $z \sim 2.5$–3.5 (Prochaska et al. 2015), i.e., over a similar redshift interval. There are only 9 LLSs with $\log N_{\rm HI} \sim 17.5$, while all the others have $\log N_{\rm HI} \gtrsim 18$. Yet it is striking how the averaged values for overlapping $N_{\rm HI}$ agree remarkably well (see Fig. 3; I emphasize that point because even though there are similar authors on both the HD-LLS and KODIAQ Z teams, the results were independently derived).

Finally, as I discuss above the pLLSs and LLSs at low z may not sample the same structure as their high redshift analogs. However, the LLSs at $2.3 < z < 3.3$ could evolve into the low z pLLSs. From the results displayed in Fig. 3, it is evident that the metallicity distribution of the LLSs at $2.3 < z < 3.3$ is consistent with a unimodal distribution, significantly different from the bimodal metallicity distribution of the pLLSs at $z \lesssim 1$. Therefore, even considering the redshift evolution of the cosmic structures, there is a significant evolution of the metallicity distribution of the LLSs with z.

5 Metallicities as a Function of $N_{\rm HI}$ over Cosmic Time

As alluded to before, plotting the metallicity as a function of $N_{\rm HI}$ allows for a direct comparison on how the metallicity distributions change with $N_{\rm HI}$ from the most diffuse regions to densest regions of the universe traced by H I, and, with information at different z, we can determine how it evolves over cosmic time. At $z < 1$, the S/N of the COS QSO spectra does not allow us to probe sensitively metallicity below $[{\rm X/H}] \lesssim -1$ for the Lyα forest, i.e., absorbers with $\log N_{\rm HI} \lesssim 15$. However, at $z > 2$ high S/N QSO spectra can be obtained so the metallicities of the Lyα forest can be determined with much less bias (e.g., Ellison et al. 2000; Schaye et al. 2003; Aguirre et al. 2004; Simcoe et al. 2004). In the right-hand side of Fig. 4, I show the mean and standard deviation from Simcoe et al. (2004) who determined in the spectra of 7 QSOs the metallicity using O VI and C IV for absorbers with $13.6 \lesssim \log N_{\rm HI} \lesssim 16$ (most between $13.6 \lesssim \log N_{\rm HI} \lesssim 14.4$ which is highlighted by the asymmetric error on the horizontal axis) at $z \sim 2.5$. In that figure, the gray squares represent the mean metallicity of the strong LLSs and SLLSs from the HD-LLS survey (Fumagalli et al. 2016b), where the horizontal bar indicates the range of the $N_{\rm HI}$ bin and the vertical error bar represents the 25th/75th percentile of the composite posterior PDF. For the DLAs, I use the measurements and compilation from Rafelski et al. (2012). For the SLLSs and DLAs, see also the compilation made by Lehner et al. (2013) and Quiret et al. (2016) and the references therein (for the Quiret et al. sample, please note that there has been no attempt to remove or flag absorbers that were selected with some a priori, e.g., very weak metal features in SDSS spectra).

As displayed in Fig. 4 and reviewed in length by Lehner et al. (2013) and Wotta et al. (2016), it is clear the metallicities vary strongly with $N_{\rm HI}$ at any z. Considering the strong H I absorbers with $\log N_{\rm HI} \geq 16$, it is readily apparent that the prevalence of low-metallicity gas is only observed for the PLLSs and LLSs. We can quantify this further by defining the "very metal-poor" (VMP) absorbers as those with metallicities 2σ below the mean metallicity of the DLAs in any given redshift interval. At $z < 1$, that threshold is $[{\rm X/H}]_{\rm VMP} \leq -1.4$ while at $2.3 < z < 3.3$, it is $[{\rm X/H}]_{\rm VMP} \leq -2.4$. By definition, VMP gas is quasi-absent for DLAs (<3%–10%, 68% CL, which applies for all the percentage given here), but it is also small for the SLLSs (<10% at $z < 1$; ~20% at $2.3 < z < 3.3$). On the other hand, the fraction of VMP pLLSs+LLSs at any z is much larger (20–45%) and with the current statistics relatively invariant over cosmic time from $z < 1$ to $z \sim 2$–3. There are some hints that the pLLSs and LLSs may have different metallicity distributions at both low and high z, but this will require increasing the sample sizes to robustly determine these distributions across the $N_{\rm HI}$ range probed by the pLLSs and LLSs.

Therefore at any z, VMP gas is only observed in pLLSs and LLSs and the Lyα forest (presumably that is the case too at $z < 1$ for the Lyα forest). The VMP gas must have a different origin than the typical DLAs. It demonstrates there are very large reservoirs of VMP gas around $z < 1$ galaxies and in the interface between the IGM and galaxies at $z > 2$. At $3.2 \leq z \leq 4.4$ with a smaller sample probing very strong LLSs ($17.8 \lesssim \log N_{\rm HI} \lesssim 19.5$) and an indirect method, Cooper et al. (2015) found 28–40% of the LLS population could trace VMP gas.[5] The metallicities of the VMP pLLSs/LLSs increase with decreasing redshift, but their fractions appear to remain about the same over 12 billion years.

On the other hand, about half of the sample of pLLSs and LLSs at both high and low z has metallicities that overlap with the DLA and SLLS populations, implying that metals are found in large quantities in large and small overdensities of the universe and near and far from galaxies. Since pLLSs and LLSs are typically probing gas beyond the luminous boundaries of galaxies, the origins of these pLLSs and LLSs must have been originally outflows or ejection of metals via interactions of galaxies. Gas recycling (i.e., infalling gas previously ejected from a galaxy) is also another possibility for these metal rich pLLSs and LLSs.

6 Pristine LLSs

Although cosmological simulations predict that LLSs probing cold flow accretions should have been enriched to some levels (e.g., Fumagalli et al. 2011a; Hafen et al. 2016), with the observations we can estimate the importance of pristine pLLSs and LLSs. My definition of pristine gas is no detection of metals and an upper limit on the metallicity [X/H] < -3.5 dex. At $z < 1$, there is currently no pLLS or LLS that fits this definition. In the Lehner et al. (2013) sample, several pLLSs and LLSs have only upper limits on the metallicity, but the most sensitive one is [X/H] < -2. For this absorber, there is detection of C III and [O III]. In the Wotta et al. (2016) sample, there are also several pLLSs and LLSs with upper limits on [X/H]. However, for that study only Mg II was used to determine the metallicity. Mg II is certainly one of the most critical ions to determine the metallicity at $z < 1$ for the pLLSs and LLSs (Lehner et al. 2013), but its non-detection does not imply that the gas is pristine since in all known cases so far there has been always absorption of C III, Si III, and/or [O III] when these ions are available. And, indeed, in the lowest metallicity pLLS of the Wotta et al. (2016) sample ([X/H] < -2.5), there is likely a detection

[5] Cooper et al. (and see also Glidden et al. 2016) targeted a sample of 17 high $N_{\rm HI}$ LLS (typically $\log N_{\rm HI} \sim 17.5$) at $3.2 \lesssim z \lesssim 4.4$, but selected them on the absence of metal absorption in SDSS spectra, i.e., they single out them to be a priori low-metallicity LLSs.

of C III absorption found in the NUV spectra with follow-up *HST*/COS G185M observations (J.C. Howk et al. 2016, in prep.). Future efforts to improve these limits and probing a larger sample of LLSs at $z < 1$ will be critical to determine the lowest levels of metal enrichment in the low redshift universe.

At $z \sim 3$, there are currently only two examples of LLSs with no evidence of metal absorption down to a limit [X/H] < -4.2 and < -3.8 (Fumagalli et al. 2011a). Lehner et al. (2016) detected several pLLSs and LLSs with no metals that might be reminiscent of these pristine LLSs, but, unfortunately, Si III is contaminated for each of these cases, and no stringent limit on the metallicity can be set. To better understand the level of mixing of metals in the early universe, we will need a larger sample to reliably determine the frequency of pristine gas at $2 < z < 4.5$ in the interface regions between galaxies and the Lyα forest absorbers. From the current sample, the fraction of pLLSs/LLSs with [X/H] < -3.5 with *no* metal detected is $< 7.8\%$ at 90% CL.

7 C/α in pLLSs and LLSs over Cosmic Time

Although we have limited information on the relative abundances, we have some constraints on the C/α ratio, which is a good indicator of the nucleosynthesis history (since in low density, diffuse gas, carbon is not expected to be strongly depleted into dust grains), and hence this ratio provides additional information regarding the origin of the gas. For the pLLSs and LLSs, this ratio was principally derived from the photoionization models, but Lehner et al. (2016) also report direct measurements using the different ionization stages of C and Si [for example, at high z, $\mathrm{C}/\alpha = (N_{\mathrm{CII}} + N_{\mathrm{CIII}} + N_{\mathrm{CIV}})/(N_{\mathrm{SiII}} + N_{\mathrm{SiIII}} + N_{\mathrm{SiIV}})$], and quite remarkably the direct and modeling methods provide consistent results.

For the pLLSs and LLSs, at high z, α is mostly Si, but at low redshift it can also be O, Mg, and/or S depending on the system and redshift. In Fig. 5, I show [C/α] vs. [α/H] for the pLLSs and LLSs from both the high and low redshift samples (from Lehner et al. 2016). For comparison, the results for high redshift DLAs and SLLSs and Milky Way stars are shown. For the DLAs and SLLSs, the results are from Pettini et al. (2008), Penprase et al. (2010), and Cooke et al. (2011) (and the references therein and see also Becker et al. 2012 for $z \gtrsim 5$ measurements). For the Galaxy thin and thick stars, the results from Bensby and Feltzing (2006) are used, and for the halo stars, Fabbian et al. (2010) and Akerman et al. (2004). For the stars, α is O, while for the DLAs and SLLSs, α is O or Si (changing O to Si or vice versa for the DLAs would have little effect on the distribution of these data).

The overall trend observed in Fig. 5 in the stellar and SLLS/DLA samples can be separated in roughly two regions: *region 1* covering $-3 \lesssim [\alpha/\mathrm{H}] \lesssim -1$, where [C/$\alpha$] decreases with increasing metallicity from super-solar to about -0.7 dex; and *region 2* over $-1 \lesssim [\alpha/\mathrm{H}] \lesssim +0.2$, where [C/$\alpha$] increases with increasing metallicity from about -0.6 dex to super-solar. The behavior in region 2 has been well known for some time and is thought to occur as a result of the delayed

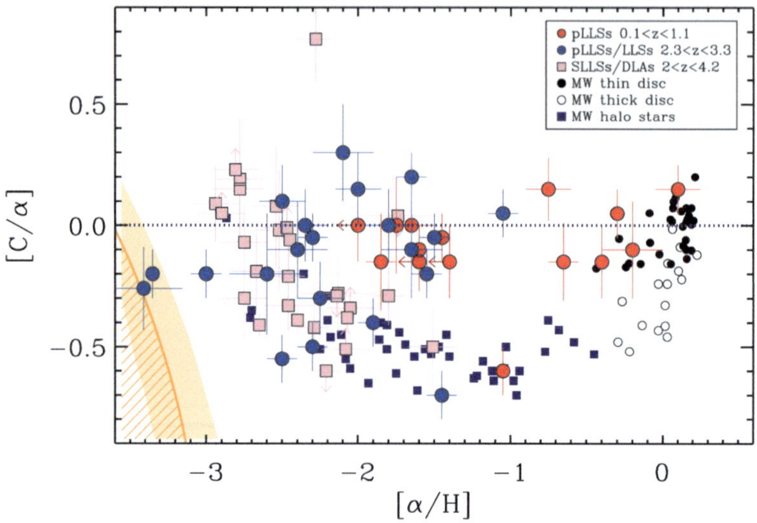

Fig. 5 Evolution of [C/α] as a function of the metallicity [α/H] for various types of absorbers at different z and *stars* indicated in the legend (from Lehner et al. 2016). Any absorbers in the *hatched orange region* may have been polluted by Pop III stars

release of carbon from low- and intermediate-mass stars combined with a strong metallicity dependence of the yields of carbon by massive stars with mass-loss (e.g., Akerman et al. 2004; Fabbian et al. 2010). The increase of C/α to lower metallicity at $[\alpha/\mathrm{H}] \lesssim -1$ has now been confirmed independently in both stellar atmospheres and SLLSs/DLAs. The exact reason for this trend is not yet well understood. One possible interpretation for the high values of C/α could be the leftovers from the enhanced production of C (relative to α-elements, and in particular O) in Population III (Pop III) stars. As shown by Frebel et al. (2007) and Bromm and Loeb (2003), the gas progenitor of Pop III stars must have high C abundance to efficiently cool the gas in order to actually form stars and drive the transition from Pop III to Pop II stars (see also Cooke et al. 2011 for more discussion). This condition is shown in Fig. 5 (hatched orange region), which is defined as the "transition discriminant" criterion. No Pop II stars should be found in that zone, but any gas in this region will likely have been polluted by Pop III stars; as we discuss below and can be seen in Fig. 5, only two LLSs are found in that "forbidden" zone.

For the pLLSs and LLSs, things are quite different. First of all it is worth to note that in the regions of overlapping metallicities, there is no evidence of a difference between the low and high redshift samples, and hence they can be treated as one. About half of the sample of the pLLSs and LLSs follows a similar distribution to that observed for the DLAs and stars over the entire range of metallicities. For these, their chemical enrichment history (at least of C and α-elements) appears to be similar to that of the Galactic stars and the bulk of the SLLSs/DLAs. However, the other half—mostly clustered at $-2.2 \lesssim [\alpha/\mathrm{H}] \lesssim +0.2$—does not follow the trend observed in stars or DLAs. These gas clouds are carbon-enhanced by a factor $\gtrsim$ 2–5 ($\gtrsim$ 0.3–0.7 dex) compared to stars or most DLAs with similar [α/H]. The enhanced C/α ratio in the metallicity range $-2 \lesssim$ [X/H] $\lesssim -0.5$ implies that this gas must have been polluted by preferential ejection of C from low-metallicity galaxies. A recent study in fact shows that at least some local metal-poor dwarf galaxies have also enhanced C/α over similar metallicities (Berg et al. 2016). While their C/α ratios are not as high as observed for the pLLSs and LLSs and their sample is small (12 galaxies), the absence of a clear trend between C/α and α/H is similar to that observed in pLLSs and LLSs.

As noted by Crighton et al. (2016), extremely metal-poor LLSs ([X/H] ~ -3.5 at $z \sim 3$) with detected metal absorption also provide another path to study the Pop III/Pop II metal-enrichment transition. There are now two LLSs at $z \sim 3.4$ with expected [C/α] and [α/H] that are consistent with gas polluted from Pop III stars (Crighton et al. 2016; Lehner et al. 2016; see Fig. 5). The use of both the low metallicity and C/α ratio provides a strong method to find metal-pollution at the transition from Pop III to Pop II star formation (see Cooke et al. 2011).

8 Lyman Limit Systems and O VI

Although I have focused throughout on the metallicity of the cool gas of the pLLSs and LLSs, the surveys described above have also revealed that O VI absorption with overlapping velocities with H I is found with a high frequency at any z. At $z < 1$, among the 23 pLLSs/LLSs with O VI coverage, only 3 have no O VI absorption below the lowest $N_{\rm O\,VI}$ of detected O VI absorption (there are an additional 3, but the limits were not sensitive). Hence the detection rate of O VI absorption associated with pLLSs/LLSs at $z < 1$ is about 70–80% (Fox et al. 2013). At $2.3 < z < 3.6$, the KODIAQ survey found that about 70% of the LLSs also show O VI absorption (Lehner et al. 2014), but this is still based on small-number statistics (there are only 7 LLSs in the KODIAQ I survey, the other absorbers being SLLSs and DLAs, but see also, e.g., Kirkman and Tytler 1997, 1999).

The O VI absorption is particularly important because owing to its high ionization energies (114–138 eV), it is difficult to photoionize by the UV background alone except if the gas is extremely low density ($n_{\rm H} \lesssim 10^{-3.5}$ cm^{-3}, e.g., Simcoe et al. 2002). When O VI is detected, these pLLSs and LLSs have typically multiple gas-phases as evidenced by the presence of low ions (e.g., C II, [Si II], Si III) and O VI (or other high ions) that cannot be explained by a single photoionization model (e.g., Lehner et al. 2009, 2013; Crighton et al. 2013). Furthermore and independent from any ionization modeling, the high-ion and low-ion gas-phases of pLLSs/LLSs are often distinct based on the kinematic comparison of their velocity profiles, which indicate that most of the pLLSs/LLSs have kinematic sub-structure with velocity centroid offsets between the strongest H I (or low ions) and O VI absorption and/or multiple components seen only in the low or high ions (Fox et al. 2013; Crighton et al. 2013; Lehner et al. 2014). The high S/N and resolution of the KODIAQ spectra also demonstrate that the O VI profiles have typically a larger velocity breadth with the bulk of their absorption often offset from the low ions by several tens or hundreds of kilometers per second and are generally smoother (i.e., with many less velocity components) (Lehner et al. 2014). These properties contrast remarkably with those of the Lyα forest ($\log N_{\rm HI} \lesssim 14.5$) where for a majority of systems, O VI and H I coexist in the same gas with similar (and generally much simpler) kinematics (e.g., Tripp et al. 2008; Muzahid et al. 2012; Savage et al. 2014).

Although O VI is traditionally associated with outflows (see below), in cosmological simulations some of the gas accretion is expected to occur in the form of O VI-bearing gas (see, e.g., Shen et al. 2013; Ford et al. 2014, 2016). Specifically, in the Eris2 simulations of a massive star-forming galaxy at $z \sim 3$, outflows and inflows have similar covering factors within $\sim 1R_{\rm vir}$ for O VI with $13.5 \leq \log N_{\rm OVI} < 14.4$, while only outflows have a large covering factor for absorbers with $\log N_{\rm OVI} \gtrsim 14.4$ at any impact parameters within 1–2$R_{\rm vir}$ (Shen et al. 2013). Further insights can be gained by directly comparing different surveys, and an easy and informative way to do this is by considering the total integrated properties of the O VI absorption, i.e., the total column density ($N_{\rm OVI}$) and the total full-width ($\Delta v_{\rm OVI}$). In Fig. 6, I show a summary of the current status of $\log N_{\rm OVI}$ vs. $\Delta v_{\rm OVI}$ for the IGM at both low and high redshift (e.g., Simcoe et al. 2002; Muzahid et al. 2012; Tripp et al. 2008; Savage et al. 2014), the COS-Halo survey (Tumlinson et al. 2011; Werk et al. 2013, 2016), the KODIAQ survey of high-z LLSs (Lehner et al. 2014; Burns 2014; N. Lehner et al. 2017, in prep.), and the $z < 1$ pLLS/LLS survey (Fox et al. 2013; Lehner et al. 2013). The rectangles in this figure show where most of the data lie for a given survey.

For the KODIAQ survey (and the $z < 1$ pLLS/LLS survey), all the absorbers are pre-selected based on their H I content so that they satisfy some $N_{\rm HI}$ threshold (KODIAQ I, $\log N_{\rm HI} \gtrsim 17.5$; KODIAQ II, $\log N_{\rm HI} \gtrsim 16$ with typically most of the absorbers with $16 \lesssim \log N_{\rm HI} \lesssim 17$, i.e., the $N_{\rm HI}$ distribution of the KODIAQ survey II is more similar to that of the $z < 1$ survey).[6] Since the KODIAQ II results are not yet published, I show their actual distribution. When O VI is detected, most of the KODIAQ data are such that $14.2 \lesssim \log N_{\rm OVI} \lesssim 15.5$ and $150 \lesssim \Delta v_{\rm OVI} \lesssim 500\,{\rm km\,s^{-1}}$. More than half of the KODIAQ I+II sample has $\log N_{\rm OVI} \gtrsim 14.4$ and $\Delta v_{\rm OVI} \gtrsim 300\,{\rm km\,s^{-1}}$, i.e., a majority fits in this category of strong and broad O VI absorption that is associated with large-scale outflows in cosmological simulations. As noted by Lehner et al. (2014), the breadth and strength of the O VI absorption in optically thick H I absorbers at $z \sim$ 2–3.5 are striking and quite similar to those observed in starburst galaxies in low redshift universe (see, e.g., Grimes et al. 2009; Tripp et al. 2011; Muzahid et al. 2015), strongly suggesting that indeed the origin of the bulk of the strong O VI absorption associated with pLLSs and LLSs is hot cooling gas of large-scale outflows from high-redshift galaxies at an epoch where galaxies are forming stars at very high rates. Figure 6 also shows that the KODIAQ

[6]Note also that for KODIAQ I, an additional criterion is that the metallicity of the cool gas could be determined, which is not the case for KODIAQ II.

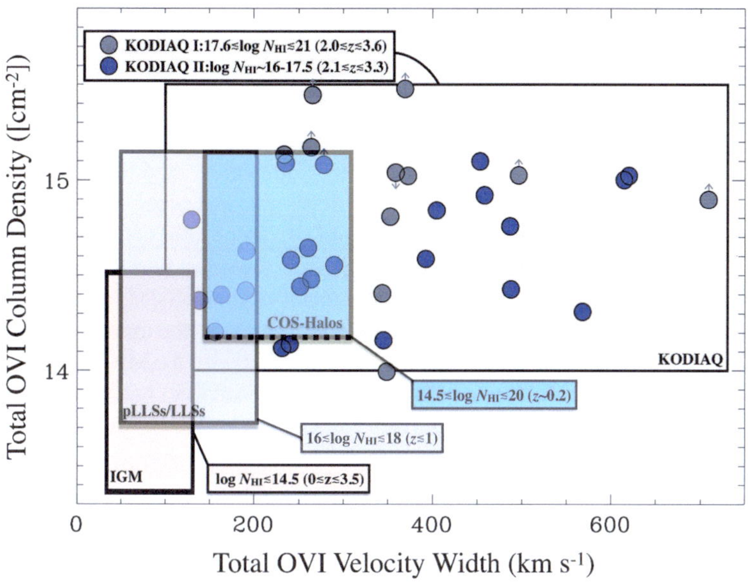

Fig. 6 Total column density of O VI ($\log N_{\rm OVI}$) versus the total width ($\Delta v_{\rm OVI}$) of the O VI absorption found in different surveys. The *rectangles* show where most of the data lie for each survey. For the KODIAQ and the $z < 1$ pLLS/LLS surveys (sample sizes $m \simeq 36, 23$, respectively), the absorbers were H I-selected to satisfy the range of $N_{\rm HI}$ values listed in the figure. For COS-Halos ($m \simeq 40$), absorbers were associated with pre-selected $\sim L*$ galaxies so that they are at impact parameters $\rho \lesssim 160$ kpc (note that *dotted lines* show about the sensitivity level of the COS-Halos survey in $N_{\rm OVI}$; for H I-selected surveys, the sensitivity is typically much better than the lowest values reported in this figure, around $\log N_{\rm OVI} \lesssim 13.2$–$13.5$). For the IGM Ly$\alpha$ forest surveys ($m \gtrsim 50$–100), absorbers are typically O VI-selected found to be associated with low $N_{\rm HI}$ absorbers

and COS-Halos[7] surveys overlap, but are remarkably outside the boundaries set by the Lyα forest.[8] This is another indication that the KODIAQ absorbers most likely trace the CGM of actively star-forming galaxies.

In the previous sections (and see Figs. 3 and 4), we saw that there is a major evolution in the metallicity distribution of the pLLSs and LLSs. There is also a major evolution in the properties of the O VI associated with pLLSs/LLSs. At $z < 1$, O VI absorption in the pLLS sample has typically $50 \lesssim \Delta v_{\rm OVI} \lesssim 150\,{\rm km\,s^{-1}}$

[7]As a reminder, COS-Halos is a dedicated absorption-line survey of $\sim L*$ galaxies at $z \sim 0.2$ where galaxies have various of SFRs and morphologies. The background quasars probed their CGM at impact parameters $10 \lesssim \rho \lesssim 160$ kpc (Tumlinson et al. 2011, 2013; Werk et al. 2012, 2013, 2014, 2016). The O VI absorption is only observed in the CGM of star-forming galaxies with a sensitivity limit $\log N_{\rm OVI} \gtrsim 14.2$.

[8]The O VI absorption in the IGM has been searched from blind O VI surveys. The properties of the O VI absorption are typically such that $13.2 \lesssim \log N_{\rm OVI} \lesssim 14.4$ and $20 \lesssim \Delta v_{\rm OVI} \lesssim 100\,{\rm km\,s^{-1}}$ at any z (Simcoe et al. 2002; Muzahid et al. 2012; Tripp et al. 2008).

and $13.8 \lesssim \log N_{\rm O\,VI} \lesssim 15$ (Fox et al. 2013). There is overlap between the low and high z surveys, but broad and strong O VI absorption associated with LLSs and pLLSs at $z < 1$ is the exception. Indeed, only two absorbers not captured with these boundary limits shown in Fig. 6 are associated each with a massive, large-scale outflow from a massive star-forming galaxy (Tripp et al. 2011; Muzahid et al. 2015). These two absorbers at $z < 1$ have $\Delta v \gtrsim 300\,{\rm km\,s^{-1}}$ and the metallicities of the highly ionized gas and cool ionized gas are similar and super-solar. In the $z < 1$ sample, Fox et al. (2013) also find that strong O VI absorption is not detected when the metallicity of cool gas is low; three of the four broadest and strongest systems at $z < 1$ happen to be three of the four most metal-rich pLLSs/LLSs in the sample, with similar properties (although not as extreme) to those of the absorbers associated with starburst galaxies (Tripp et al. 2011; Muzahid et al. 2015). My interpretation of the major difference in the frequency of strong and broad O VI between the low and high z pLLS/LLS surveys is that typically low-z galaxies are much more quiescent than their high redshift counterparts and, indeed, only in the comparatively rare low-z starburst galaxies can we find the extreme O VI absorption seen commonly at high z. This also suggests at least for the pLLSs/LLSs with strong O VI absorption that they probe the CGM of some actively star-forming galaxies, not the more diffuse IGM.

There is therefore little doubt that the strong and broad O VI absorption associated with pLLSs and LLSs typically traces large-scale outflows from very actively star-forming galaxies. The weaker O VI absorbers have likely a wider range of origins, including outflows, inflows, ambient CGM (e.g., Ford et al. 2014). It is worth noting that high-z weak O VI absorbers and non-detected O VI absorption are only found in LLSs and SLLSs with $[{\rm X/H}] \lesssim -1.5$, i.e., at relatively low metallicity. At $z < 1$, Fox et al. (2013) find that the pLLS bimodal metallicity distribution is reflected to some level in the O VI properties with a lack of low-metallicity pLLSs with strong O VI absorption. The properties of these pLLSs with no detection or of O VI absorption make them good candidates for cold stream inflows according to recent simulations (e.g., Shen et al. 2013). However, a more quantitative and systematic comparison at all z between simulations and observations should be undertaken to gain further insights on these weak O VI absorbers associated with strong H I absorbers.

9 Gas Accretion via pLLSs and LLSs?

I have shown that great progress has been made on our knowledge of the class of absorbers known as the pLLSs and LLSs. This success lies in our ability to study sizable and representative samples with good quality (S/N, resolution) spectra at both low and high z. With this new knowledge, it is clear that very metal-poor gas is not rare at any z at large overdensities, but it is uniquely probed by the pLLSs and LLSs. This gas is typically not pristine, i.e., it contains some metals, but in some cases down to very low level $[{\rm X/H}] \simeq -3.5$. These results provide new stringent empirical results for cosmological hydrodynamical simulations. In particular, there is a strong evolution of the metallicity of the pLLSs/LLSs with z, but also a remarkably constant fraction of VMP pLLSs/LLSs over cosmic times. There is also a major shift in the properties of the O VI absorption associated with pLLSs and LLSs, with extremely strong and broad O VI being quite common at high z but rare at low z. Since these strong and broad O VI absorbers most likely trace large-scale outflows from actively star-forming galaxies, the dramatic shift of the properties of the O VI absorption associated with pLLSs/LLSs with z certainly captures in part the equivalently dramatic change in the star-formation activity level in galaxies with z.

So are the metal-poor pLLSs/LLSs tracing gas accretion? If we use cosmological simulations to guide us, the answer is very likely yes. Simulations have shown that pLLSs/LLSs may be used to trace cold flows (Faucher-Giguère et al. 2011, 2015; Fumagalli et al. 2011b, 2014b; van de Voort and Schaye 2012; van de Voort et al. 2012; Hafen et al. 2016). Simulated pLLSs and LLSs at $z \sim 2$–3 or $z < 1$ often appear, however, to have too many metals compared to the observations. Only in simulations with very mild stellar feedback (Fumagalli et al. 2011a), there is some agreement between the observed and simulated metallicity distributions; in this simulation, cold streams are traced mostly by LLSs within 1 or 2 virial radii of galaxies where the gas has only been enriched to $[X/H] \simeq -1.8$ with similar scatter to that observed at high or low z. However, while mild feedback produces better agreement with the observed metallicity distribution at $z \sim 2$–3, the agreement with the baryon fraction in stars worsens (Fumagalli et al. 2011a). The FIRE zoom-in simulations at $z < 1$ have also recently studied the physical nature of the pLLSs and LLSs (Hafen et al. 2016). These simulations confirm the general interpretation of the bimodal metallicity distribution at $z < 1$: very low-metallicity LLSs are predominantly associated with inflows at $z < 1$, but higher metallicity LLSs trace gas with roughly equal probability of having recycled outflows (inflows) or outflows. However, the model vs. observation comparison also leads to major differences, including that the simulated metallicity distribution is not bimodal and has a metallicity plateau between about -1.3 and -0.5 dex at $z < 1$. Furthermore, while very low-metallicity pLLSs and LLSs are prevalent in the observations, they are not in the FIRE simulations. Possibly recycling and mixing of the gas are too efficient in these simulations to allow for the presence of widespread low-metallicity gas. However, according to all these simulations despite some key disagreements with the observations, there is a consensus that a large fraction of the metal-poor LLSs and pLLSs should probe cold flow accretions onto galaxies, and in fact the covering fraction of the cold flow accretion could peak in the LLS regime. The observations summarized here show that there is a prevalence of metal-poor (and even very metal-poor) pLLSs/LLSs at any z, strongly suggesting that they may be tracers of metal-poor gas accretion.

While there are many positive aspects in using the pLLSs and LLSs to study the gas around galaxies, there is a major limitation: it is complicated and often impossible (especially at high redshift) to determine if the gas is actually inflowing into or outflowing from the host galaxy. All the properties so far derived for the pLLSs and LLSs fit with the gas properties found in the simulated cold flow accretion: the $N_{\rm HI}$ range, the low metallicities, the temperatures, densities, length scales, and even some of the galaxy properties (e.g., Ribaudo et al. 2011b; Kacprzak et al. 2012b; Lehner et al. 2013). However, it is unknown from these observations if the gas is actually accreting onto the galaxies.

The only experiment where we can directly determine the direction of motion along the line-of-sight is the down-the-barrel experiment whereby the galaxy itself is used as the background target: observations of the blueshifted or redshifted absorption relative to the systemic velocity of the galaxy directly inform us if the gas is outflowing or inflowing (e.g., Heckman et al. 2000; Weiner et al. 2009; Martin et al. 2012; Rubin et al. 2012, and see also the chapter by Kate Rubin in this book). In this method, strong metal transitions (like Fe II, Mg II) are, however, used, and, hence, it remains ambiguous to know if the gas that is accreting onto a galaxy as recycled gas or from the IGM since the metallicity of the gas is generally unknown.

Another method for indirectly determining the origin of the gas has been to characterize geometry of the galaxy relative to the QSO line of sight (see for a full description of this method the chapter in this book by Glenn Kacprzak). Several Mg II surveys have demonstrated that a class of strong Mg II absorbers at small impact parameters ($\rho \lesssim 40$ kpc) are primarily found along the galaxies' minor axes, consistent with an origin from large-scale outflows (e.g., Bordoloi et al. 2011; Kacprzak et al. 2012a; Bouché et al. 2012). However, Kacprzak et al. (2012b) have also shown one example of a very metal-poor LLS observed projected along the minor axis of the associated galaxy albeit at a projected distance of 104 kpc. Evidently all absorption detected in the spectra of QSOs probing gas along the projected minor axis of galaxies may not be produced by winds, especially at these large impact parameters. Metal-poor gas may also not necessarily align with the projected major axis of the galaxy as expected from some simulations (e.g., Stewart et al. 2011), i.e., other mechanisms than cold flow accretion might be at play, as, e.g., the precipitation model proposed by Voit et al. (2015).

One can also undertake a combination of both methods just described above whereby absorption is observed toward a galaxy (down-the-barrel) and toward a QSO transversely projected at some distance (e.g., Kacprzak et al. 2014; Bouché et al. 2016). I suspect the number of galaxy-QSO pairs will increase dramatically in the future with IFU observations and larger aperture telescope.

Cosmological simulations show a preference for inflowing streams to align with the major axis of galaxies, metal-enriched outflows along the minor axis, and tidal debris more uniformly distributed (e.g., Stewart et al. 2011; van de Voort and Schaye 2012; Shen et al. 2013). This combined with the empirical results from Mg II absorber-galaxy survey strongly motivated the Bimodal Absorption System Imaging Campaign (BASIC; PI: Lehner), which aimed to jointly determined the correlation (if any) between the geometry of the galaxies with the metallicities of the pLLSs/LLSs. This program combines newly acquired and archival ACS images as well as ground-based galaxy imaging and spectroscopy (LBT/MODS-LBC; MUSE; Keck LRIS) of all the fields from the Lehner et al. (2013) sample. With the ongoing BASIC survey, we will measure the geometry, morphology, environment of galaxies associated with $\sim$30 pLLSs/LLSs and in particular test if the bimodal metallicity

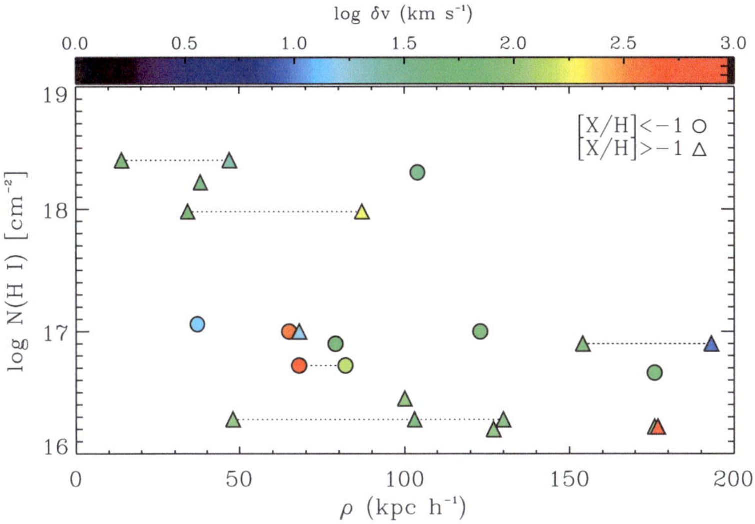

Fig. 7 The H I column density as a function of the impact parameter for the pLLSs and LLSs at $z < 1$ from the Lehner et al. (2013) sample. The data are *color-coded* in velocity separation between the redshift of the absorbers and associated galaxy $\delta v = c|z_{\rm abs} - z_{\rm gal}|/(1 + z_{\rm abs})$. Data connected with a *dotted line* indicate that more than one candidate galaxy was found (note that the *triangles* nearly overlapping at $\log N_{\rm HI} \simeq 16.2$ and $\rho \simeq 176$ kpc are also a pair of candidate galaxies for the same absorber)

distribution is connected to the geometric orientation of galaxies as predicted by simulations.

While it is too early to report on any specific results from BASIC, in Fig. 7, I present an updated figure from Lehner et al. (2013) showing $N_{\rm HI}$ as a function of the impact parameter ρ of the possible associated galaxies at $z < 1$. This figure differs from Fig. 9 in Lehner et al. (2013) in several important aspects. First, in Lehner et al. (2013) only the "closest" galaxy in redshift and impact parameter to a given absorber was considered, while here I show all the potential candidate galaxies for a given absorber (not just one). Second, I differentiate the low and high metallicity absorbers. Third, the velocity separation $\delta v = c|z_{\rm abs} - z_{\rm gal}|/(1 + z_{\rm abs})$ between the redshift of the absorbers and associated galaxy is indicated. I stress that this is still very preliminary since not the entire sample has been obtained and the galaxy completeness has not been systematically assessed yet. However, as Lehner et al. (2013) indicated at $z < 1$ for 11 LLSs, now for 16 pLLSs/LLSs (7 metal-poor absorbers, 9 metal-rich absorbers) where their fields have been observed sufficiently deeply ($\gtrsim 0.1L*$), at least one galaxy is found within $\rho \lesssim 200$ kpc and typically $\delta v \lesssim 100\,{\rm km\,s^{-1}}$ (except for a few cases, see Fig. 7). From Fig. 7, there is no hint of a difference between low and high metallicity absorbers with impact parameters. With the current sample, there is, however, a strong hint that [X/H] ≥ -1 pLLSs/LLSs are found associated with more than one galaxy, while [X/H] < -1 pLLSs/LLSs are all but one associated with a single galaxy. Further studies will be needed to interpret the galaxy/LLS connection, but it is evident that we will learn soon more about the environments and hence the nature of these systems. With the current data, it is reasonable to conclude that the VMP pLLSs/LLSs at $z < 1$ are indeed probed of the extended of the CGM of galaxies (not necessarily of a single galaxy at least for metal-rich absorbers).

At $z > 2$, the geometry of the galaxies will be difficult to decipher, but observations with the VLT/MUSE or Keck/KCWI of the fields of view of the pLLSs and LLSs will shed light on the galaxy environments of the pLLSs and LLSs. These observational efforts are in their infancy, but recent MUSE observations show that the connection between very metal-poor LLSs and star-forming galaxies fed by cold streams is quite plausible at high z (Fumagalli et al. 2016a) (at least for these VMP LLSs).

10 Conclusions and Future Directions

As a massive reservoir of galactic baryons and metals, the mediator of galactic feedback, and the fuel for star formation, the CGM plays a major role in galaxy evolution, yet the full nature of that role and how it evolves with cosmic time remains to be determined. With the pLLS and LLS surveys at $z < 1$ and $z > 2$, we have made critical advances on our understanding of the CGM of galaxies over the last few years. It is obvious from these studies that pLLSs and LLSs trace different processes and have several origins. It is also quite obvious now that, at the high overdensities probed by pLLSs and higher column density systems, the gas is very rarely pristine. There is a large fraction of pLLSs and LLSs that have metallicities similar to those observed in SLLSs and DLAs. However, there is also a substantial fraction (20–45%) of very metal-poor gas at all studied z intervals. At $z < 1$, there is also little doubt that the pLLSs and LLSs are tracing the dense regions of the CGM of galaxies, while at higher z the question is still open (although pLLSs or LLSs associated with strong O VI absorption are quite likely to probe the CGM of some very actively star-forming galaxies). At $z < 1$, these findings indicate that galaxies have large reservoir of metal-poor gas at projected physical distances $\lesssim$ 100–200 kpc, which may eventually feed the galaxies, but the exact mechanisms on how this occurs (and if this really occurs) are still an open question.

Hence as other techniques to study the gas around galaxies, the evidence for large-scale outflows is quite clear and unambiguous (pLLSs and LLSs with strong and broad O VI absorption trace large-scale outflows from starburst galaxies), but the evidence for accretion is still not as equivocal. A good way to conclude this chapter on LLSs and accretion is by quoting Michele Fumagalli from his talk at the IGM@50 *Is the Intergalactic Medium Driving Star Formation* Spineto conference last year: "So far, there is no [major] empirical evidence against the association between [pLLS]/LLSs and cold accretion as put forward by theory." I would add, however, that we should not dismiss yet other accreting gas models that might be at play (e.g., Marasco et al. 2013; Voit et al. 2015).

I predict the next 5 years are going to be equally as rich in new discoveries as the previous 5 years if not more:

- At all z, samples where we will be able to robustly study the metallicity distribution will increase by a factor 2–4 at $z < 1$ ($\gtrsim$100 absorbers) and a factor >5–10 at $z > 1.5$ (>150–300 absorbers), thanks to the growing archives at MAST, KOA, and ESO. Not only the samples will increase, but there will also be better sampling in redshift space (e.g., finer redshift sampling will be possible at $2 < z < 4$ as well as in the redshift interval $1 < z < 2$ where currently there is no information) and in $N_{\rm HI}$ (e.g., the LLS sample will increase at $z < 1$ as well as the weak pLLS sample at the transition of the Lyα forest at both $z < 1$ and $z > 2$).
- HST images and integral field spectroscopy with moderate to high spectral resolution will transform our knowledge of the LLS and pLLS galaxy environments. We are now in position to study the galaxies associated with absorbers where some of their key properties are determined (the metallicity of course, but also their various gas-phases—including the highly ionized probed by O VI, kinematics, etc.). The recent and ongoing advent of IFU spectroscopy (and future AO) systems on ground-based 8–10 m class telescopes (MUSE, KCWI) will speed up the acquisition of the galaxies, but more critically galaxies at $z \lesssim 1$ can be spatially and kinematically resolved providing a new wealth of information extremely difficult to acquire in the past (see, e.g., Fumagalli et al. 2014a, 2016a; Bouché et al. 2016; Peroux et al. 2017). As these efforts will be undertaken at both low and high z, these galaxy surveys will enable the study of their evolution over much of the age of the universe. On a longer time frame, extremely large telescopes, with their huge light gathering power coupled with diffraction-limited angular resolution, will even further revolutionize our understanding of the CGM of galaxies at both high and low z.
- Finally, observers and simulators are working more and more together and I expect from these collaborations more insights will be gained on the nature of the gas probed by the pLLSs and LLSs. In particular, new tools are being developed so that results from the observations and simulations can be more directly comparable and more easily interpretable (e.g., by treating metallicities of individual absorbers as posterior probability distribution functions, see Fumagalli et al. 2016b).

Acknowledgements It is a pleasure to thank all my collaborators, current and past graduate students who all have been key on shedding light on these (now much less mysterious) absorbers over the last few years, and, in particular, Chris Howk, John O'Meara, and Xavier Prochaska who have been critical players on many aspects to push forward these projects at both low and high redshifts. I also thank Chris Howk for reading and providing useful comments on this manuscript and Lorrie Straka for reducing the MUSE data and providing the galaxy redshifts that help to make Fig. 7. The writing and some of the research presented in this work has been supported by NASA through the Astrophysics Data Analysis Program (ADAP) grant NNX16AF52G, HST-AR-12854, and HST-GO-14269 from the Space Telescope Science Institute, which is operated by the Association of Universities for Research in Astronomy, Incorporated, under NASA contract NAS5-26555.

References

Aguirre, A., Schaye, J., Kim, T.-S., et al. 2004, ApJ, 602, 38
Akerman, C. J., Carigi, L., Nissen, P. E., Pettini, M., & Asplund, M. 2004, A&A, 414, 931
Bahcall, J. N., Bergeron, J., Boksenberg, A., et al. 1993, ApJS, 87, 1
Battisti, A. J., Meiring, J. D., Tripp, T. M., et al. 2012, ApJ, 744, 93
Becker, G. D., Sargent, W. L. W., Rauch, M., & Carswell, R. F. 2012, ApJ, 744, 91
Bensby, T., & Feltzing, S. 2006, MNRAS, 367, 1181
Berg, D. A., Skillman, E. D., Henry, R. B. C., Erb, D. K., & Carigi, L. 2016, ApJ, 827, 126
Bordoloi, R., Lilly, S. J., Knobel, C., et al. 2011, ApJ, 743, 10
Bordoloi, R., Tumlinson, J., Werk, J. K., et al. 2014, ApJ, 796, 136
Bouché, N., Hohensee, W., Vargas, R., et al. 2012, MNRAS, 426, 801
Bouché, N., Finley, H., Schroetter, I., et al. 2016, ApJ, 820, 121
Bromm, V., & Loeb, A. 2003, Nature, 425, 812
Burns, V. 2014, *A High Resolution Study of Circumgalactic O VI Absorbers at* $2 < z < 4$, University of Notre Dame Senior Honors Thesis
Churchill, C. W., Mellon, R. R., Charlton, J. C., et al. 2000, ApJS, 130, 91
Cooke, R., Pettini, M., Steidel, C. C., Rudie, G. C., & Nissen, P. E. 2011, MNRAS, 417, 1534
Cooksey, K. L., Prochaska, J. X., Chen, H.-W., Mulchaey, J. S., & Weiner, B. J. 2008, ApJ, 676, 262
Cooper, T. J., Simcoe, R. A., Cooksey, K. L., O'Meara, J. M., & Torrey, P. 2015, ApJ, 812, 58
Crighton, N. H. M., Hennawi, J. F., & Prochaska, J. X. 2013, ApJL, 776, L18
Crighton, N. H. M., Hennawi, J. F., Simcoe, R. A., et al. 2015, MNRAS, 446, 18
Crighton, N. H. M., O'Meara, J. M., & Murphy, M. T. 2016, MNRAS, 457, L44
Davé, R., Hernquist, L., Katz, N., & Weinberg, D. H. 1999, ApJ, 511, 521
Dekel, A., & Birnboim, Y. 2006, MNRAS, 368, 2
Ellison, S. L., Songaila, A., Schaye, J., & Pettini, M. 2000, AJ, 120, 1175
Fabbian, D., Khomenko, E., Moreno-Insertis, F., & Nordlund, Å. 2010, ApJ, 724, 1536
Faucher-Giguère, C.-A., Hopkins, P. F., Kereš, D., et al. 2015, MNRAS, 449, 987
Faucher-Giguère, C.-A., Kereš, D., & Ma, C.-P. 2011, MNRAS, 417, 2982
Ferland, G. J., Porter, R. L., van Hoof, P. A. M., et al. 2013, RMxAA, 49, 137
Ford, A. B., Davé, R., Oppenheimer, B. D., et al. 2014, MNRAS, 444, 1260
Ford, A. B., Werk, J. K., Davé, R., et al. 2016, MNRAS, 459, 1745
Fox, A. J., Lehner, N., Tumlinson, J., et al. 2013, ApJ, 778, 187
Frebel, A., Johnson, J. L., & Bromm, V. 2007, MNRAS, 380, L40
Fumagalli, M., Cantalupo, S., Dekel, A., et al. 2016a, MNRAS, 462, 1978
Fumagalli, M., Fossati, M., Hau, G. K. T., et al. 2014a, MNRAS, 445, 4335
Fumagalli, M., Hennawi, J. F., Prochaska, J. X., et al. 2014b, ApJ, 780, 74
Fumagalli, M., O'Meara, J. M., & Prochaska, J. X. 2011a, Science, 334, 1245
—. 2016b, MNRAS, 455, 4100
Fumagalli, M., Prochaska, J. X., Kasen, D., et al. 2011b, MNRAS, 418, 1796
Glidden A., Cooper T. J., Cooksey K. L., Simcoe R. A., O'Meara J. M., 2016, ApJ, 833, 270
Lehner N., O'Meara J. M., Howk J. C., Prochaska J. X., Fumagalli M., 2016, ApJ, 833, 283
Werk J. K., et al., 2016, ApJ, 833, 54
Grimes, J. P., Heckman, T., Aloisi, A., et al. 2009, ApJS, 181, 272
Haardt, F., & Madau, P. 1996, ApJ, 461, 20
—. 2012, ApJ, 746, 125
Hafen, Z., Faucher-Giguere, C.-A., Angles-Alcazar, D., et al. 2016, ApJ, submitted, arXiv:1608.05712
Heckman, T. M., Lehnert, M. D., Strickland, D. K., & Armus, L. 2000, ApJS, 129, 493
Hirschauer, A. S., Salzer, J. J., Skillman, E. D., et al. 2016, ApJ, 822, 108
Howk, J. C., Ribaudo, J. S., Lehner, N., Prochaska, J. X., & Chen, H.-W. 2009, MNRAS, 396, 1875

Jorgenson, R. A., Murphy, M. T., & Thompson, R. 2013, MNRAS, 435, 482
Kacprzak, G. G., Churchill, C. W., & Nielsen, N. M. 2012a, ApJL, 760, L7
Kacprzak, G. G., Churchill, C. W., Steidel, C. C., & Murphy, M. T. 2008, AJ, 135, 922
Kacprzak, G. G., Churchill, C. W., Steidel, C. C., Spitler, L. R., & Holtzman, J. A. 2012b, MNRAS, 427, 3029
Kacprzak, G. G., Martin, C. L., Bouché, N., et al. 2014, ApJL, 792, L12
Kacprzak, G. G., van de Voort, F., Glazebrook, K., et al. 2016, ApJL, 826, L11
Kereš, D., Katz, N., Weinberg, D. H., & Davé, R. 2005, MNRAS, 363, 2
Kirkman, D., & Tytler, D. 1997, ApJL, 489, L123
—. 1999, ApJL, 512, L5
Kniazev, A. Y., Grebel, E. K., Hao, L., et al. 2003, ApJL, 593, L73
Kunth, D., & Östlin, G. 2000, A&A Rev., 10, 1
Lanzetta, K. M., Bowen, D. V., Tytler, D., & Webb, J. K. 1995, ApJ, 442, 538
Lehner, N., O'Meara, J. M., Fox, A. J., et al. 2014, ApJ, 788, 119
Lehner, N., O'Meara, J. M., Howk, J. C., Prochaska, J. X., & Fumagalli, M. 2016, ApJ, in press, arXiv:1608.02588
Lehner, N., Prochaska, J. X., Kobulnicky, H. A., et al. 2009, ApJ, 694, 734
Lehner, N., Howk, J. C., Tripp, T. M., et al. 2013, ApJ, 770, 138
Liang, C. J., & Chen, H.-W. 2014, MNRAS, 445, 2061
Lopez, S., D'Odorico, V., Ellison, S. L., et al. 2016, A&A, arXiv:1607.08776
Maller, A. H., & Bullock, J. S. 2004, MNRAS, 355, 694
Marasco, A., Marinacci, F., & Fraternali, F. 2013, MNRAS, 433, 1634
Martin, C. L., Shapley, A. E., Coil, A. L., et al. 2012, ApJ, 760, 127
Muratov, A. L., & Gnedin, O. Y. 2010, ApJ, 718, 1266
Muzahid, S., Kacprzak, G. G., Churchill, C. W., et al. 2015, ApJ, 811, 132
Muzahid, S., Srianand, R., Bergeron, J., & Petitjean, P. 2012, MNRAS, 421, 446
O'Meara, J. M., Prochaska, J. X., Burles, S., et al. 2007, ApJ, 656, 666
O'Meara, J. M., Lehner, N., Howk, J. C., et al. 2015, AJ, 150, 111
Oppenheimer, B. D., Davé, R., Katz, N., Kollmeier, J. A., & Weinberg, D. H. 2012, MNRAS, 420, 829
Oppenheimer, B. D., Davé, R., Kereš, D., et al. 2010, MNRAS, 406, 2325
Peeples, M. S., Werk, J. K., Tumlinson, J., et al. 2014, ApJ, 786, 54
Penprase, B. E., Prochaska, J. X., Sargent, W. L. W., Toro-Martinez, I., & Beeler, D. J. 2010, ApJ, 721, 1
Peroux, C., Rahmani, H., Quiret, S., et al. 2017, MNRAS, 464, 2053
Pettini, M., Shapley, A. E., Steidel, C. C., et al. 2001, ApJ, 554, 981
Pettini, M., Zych, B. J., Steidel, C. C., & Chaffee, F. H. 2008, MNRAS, 385, 2011
Prochaska, J. X. 1999, ApJL, 511, L71
Prochaska, J. X., Gawiser, E., Wolfe, A. M., Castro, S., & Djorgovski, S. G. 2003, ApJL, 595, L9
Prochaska, J. X., O'Meara, J. M., Fumagalli, M., Bernstein, R. A., & Burles, S. M. 2015, ApJS, 221, 2
Prochaska, J. X., O'Meara, J. M., & Worseck, G. 2010, ApJ, 718, 392
Quiret, S., Péroux, C., Zafar, T., et al. 2016, MNRAS, 458, 4074
Rafelski, M., Wolfe, A. M., Prochaska, J. X., Neeleman, M., & Mendez, A. J. 2012, ApJ, 755, 89
Ribaudo, J., Lehner, N., & Howk, J. C. 2011a, ApJ, 736, 42
Ribaudo, J., Lehner, N., Howk, J. C., et al. 2011b, ApJ, 743, 207
Rubin, K. H. R., Prochaska, J. X., Koo, D. C., & Phillips, A. C. 2012, ApJL, 747, L26
Rubin, K. H. R., Prochaska, J. X., Koo, D. C., et al. 2014, ApJ, 794, 156
Rudie, G. C., Steidel, C. C., & Pettini, M. 2012, ApJL, 757, L30
Sargent, W. L. W., Steidel, C. C., & Boksenberg, A. 1989, ApJS, 69, 703
Savage, B. D., Kim, T.-S., Wakker, B. P., et al. 2014, ApJS, 212, 8
Schaye, J. 2001, ApJL, 562, L95
Schaye, J., Aguirre, A., Kim, T.-S., et al. 2003, ApJ, 596, 768
Shapley, A. E., Steidel, C. C., Pettini, M., & Adelberger, K. L. 2003, ApJ, 588, 65

Shen, S., Madau, P., Aguirre, A., et al. 2012, ApJ, 760, 50
Shen, S., Madau, P., Guedes, J., et al. 2013, ApJ, 765, 89
Simcoe, R. A., Sargent, W. L. W., & Rauch, M. 2002, ApJ, 578, 737
—. 2004, ApJ, 606, 92
Skillman, E. D., Televich, R. J., Kennicutt, Jr., R. C., Garnett, D. R., & Terlevich, E. 1994, ApJ, 431, 172
Som, D., Kulkarni, V. P., Meiring, J., et al. 2013, MNRAS, 435, 1469
—. 2015, ApJ, 806, 25
Steidel, C. C. 1990, ApJS, 74, 37
Steidel, C. C., Erb, D. K., Shapley, A. E., et al. 2010, ApJ, 717, 289
Steidel, C. C., Shapley, A. E., Pettini, M., et al. 2004, ApJ, 604, 534
Stewart, K. R., Kaufmann, T., Bullock, J. S., et al. 2011, ApJ, 738, 39
Stocke, J. T., Keeney, B. A., Danforth, C. W., et al. 2013, ApJ, 763, 148
Tremonti, C. A., Heckman, T. M., Kauffmann, G., et al. 2004, ApJ, 613, 898
Tripp, T. M., Sembach, K. R., Bowen, D. V., et al. 2008, ApJS, 177, 39
Tripp, T. M., Meiring, J. D., Prochaska, J. X., et al. 2011, Science, 334, 952
Tumlinson, J., Thom, C., Werk, J. K., et al. 2011, Science, 334, 948
—. 2013, ApJ, 777, 59
Tytler, D. 1982, Nature, 298, 427
van de Voort, F., & Schaye, J. 2012, MNRAS, 423, 2991
van de Voort, F., Schaye, J., Altay, G., & Theuns, T. 2012, MNRAS, 421, 2809
Voit, G. M., Donahue, M., Bryan, G. L., & McDonald, M. 2015, Nature, 519, 203
Weiner, B. J., Coil, A. L., Prochaska, J. X., et al. 2009, ApJ, 692, 187
Werk, J. K., Prochaska, J. X., Thom, C., et al. 2012, ApJS, 198, 3
—. 2013, ApJS, 204, 17
Werk, J. K., Prochaska, J. X., Tumlinson, J., et al. 2014, ApJ, 792, 8
Werk, J. K., Prochaska, J. X., Cantalupo, S., et al. 2016, ApJ, in press, arXiv:1609.00012
Wotta, C. B., Lehner, N., Howk, J. C., O'Meara, J. M., & Prochaska, J. X. 2016, ApJ, 831, 95
Zonak, S. G., Charlton, J. C., Ding, J., & Churchill, C. W. 2004, ApJ, 606, 196

Gas Accretion in Star-Forming Galaxies

Glenn G. Kacprzak

1 Introduction

Cosmological simulations unequivocally predict that cold accretion is the primary growth mechanism of galaxies. The accretion of cold gas occurs via cosmic filaments that transports metal-poor gas onto galaxies, providing their fuel to form stars. Observationally, it is quite clear that galaxy gas-consumption timescales are short compared to the age of the Universe, therefore galaxies must acquire gas from the surroundings to continue to form stars. However, observational data have yet to conclusively show that gas accretion flows are ubiquitously occurring in-and-around galaxies at any epoch. Since the first discoveries of circumgalactic gas around star-forming galaxies (Boksenberg and Sargent 1978; Kunth and Bergeron 1984; Bergeron 1986), we have wondered where does this gas come from and if/how it drives galaxy growth and evolution.

This Chapter reviews the current observational evidence and signatures of cold accretion onto star-forming galaxies. In the following sections, we review the data that suggest accretion is occurring and cover the main topics of circumgalactic gas spatial distribution, kinematics, and metallicity. In a few cases, there exists the combination of all of the above which provides the most tantalizing evidence of cold accretion to date. This chapter primarily focuses on observations using background quasars or galaxies as probes of the circumgalactic medium around intervening foreground galaxies.

G.G. Kacprzak (✉)
Center for Astrophysics & Supercomputing, Swinburne University of Technology, Hawthorn, VIC 3122, Australia
e-mail: gkacprzak@swin.edu.au

A. Fox, R. Davé (eds.), *Gas Accretion onto Galaxies*, Astrophysics and Space Science Library 430, DOI 10.1007/978-3-319-52512-9_7

2 The Spatial Distribution of the Circumgalactic Medium

In order for galaxies to continuously form stars throughout the age of the Universe, they must acquire a sufficient amount of gas from their surroundings. In fact, roughly 50% of a galaxy's gas mass is in the circumgalactic medium (Zheng et al. 2015; Wolfe et al. 2005). Using background quasars as probes of gas surrounding foreground galaxies, we have discovered that galaxies have an abundance of multi-phased circumgalactic gas.

2.1 *Circumgalactic Gas Radial Distribution*

The quantity and extend of the circumgalactic medium has been traced using a range of absorption features such as Lyα (Tripp et al. 1998; Chen et al. 2001b; Wakker and Savage 2009; Chen and Mulchaey 2009; Steidel et al. 2010; Stocke et al. 2013; Richter et al. 2016), Mg II (Steidel et al. 1994; Steidel 1995; Guillemin and Bergeron 1997; Zibetti et al. 2007; Kacprzak et al. 2008; Chen et al. 2010; Nielsen et al. 2013a), C IV (Chen et al. 2001a; Adelberger et al. 2005; Steidel et al. 2010; Bordoloi et al. 2014a; Liang and Chen 2014; Burchett et al. 2015; Richter et al. 2016), and O VI (Savage et al. 2003; Sembach et al. 2004; Stocke et al. 2006; Danforth and Shull 2008; Tripp et al. 2008; Wakker and Savage 2009; Prochaska et al. 2011; Tumlinson et al. 2011; Johnson et al. 2013, 2015; Stocke et al. 2013; Kacprzak et al. 2015). These studies all have shown that regardless of redshift (at least between $z =$ 0–3), galaxies typically have hydrogen gas detected out to $\sim$500 kpc with "metal-enriched" gas within 100–200 kpc. Furthermore, the data show an anti-correlation with equivalent width and impact parameter, with the covering fraction being unity close to the galaxy and declining with increasing distance. This is demonstrated in Fig. 1, which shows the $\sim$200 kpc extent of Mg II absorbing gas around typical galaxies from the MAGIICAT catalog (Nielsen et al. 2013b) along with well fit anti-correlation between equivalent width and impact parameter ($\log W_r(2796) = 0.015 \times D + 0.27$). Note also that the gas covering fraction, for an equivalent width limit of 0.3 Å, is roughly unity near the galaxy and decreases to about 20% beyond 100 kpc.

Interestingly, Richter (2012) demonstrated, using the distribution of high-velocity clouds around the Milky Way and M31, that high-velocity clouds could give rise to the majority of the absorption systems seen around other galaxies. The accretion rate of high-velocity gas at $z = 0$ is almost equivalent to the star formation density of the local Universe and thus, at least at low redshifts, high-velocity clouds could provide a significant fraction of the gas mass accreted onto galaxies (see chapter by Philipp Richter for further discussion of gas accretion onto the Milky Way).

Simulations predict that gas accretion should occur via a "hot" or "cold" mode, which is dependent on a galaxy being above or below a critical halo mass ranging

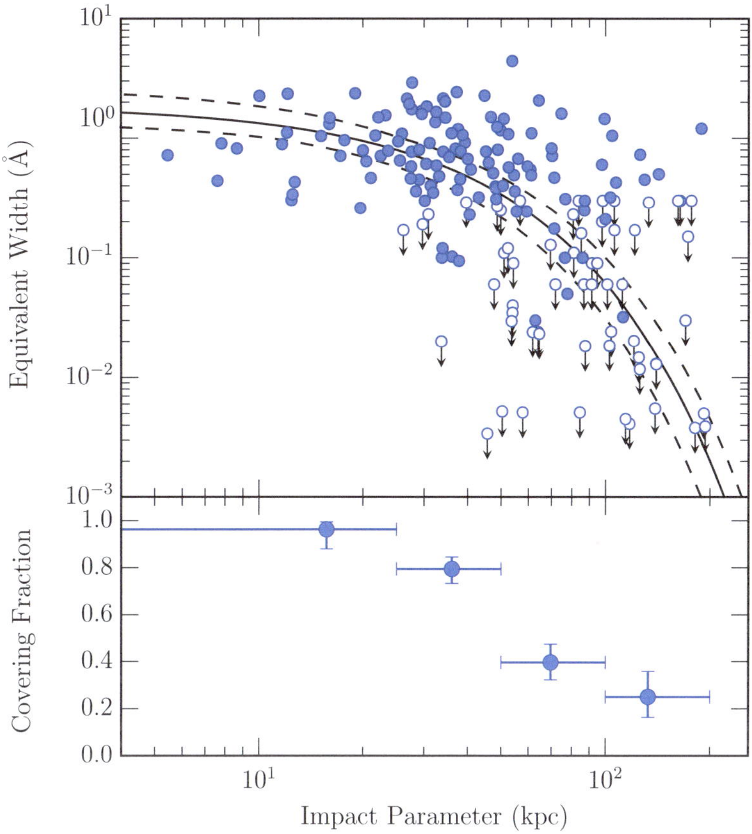

Fig. 1 *Top*: the rest-frame Mg II equivalent width as a function of impact parameter for the MAGIICAT catalog for "isolated" galaxies (Nielsen et al. 2013a,b). The *closed symbols* are detections while *open symbols* are 3σ are upper limits. The fit and 1σ confidence levels are shown ($\log W_r(2796) = [0.015 \pm 0.002] \times D + [0.27 \pm 0.11]$). Note the large extent of gas surrounding galaxies, which begs the question of what is the origin of this gas and is it some combination of gas outflows and accretion. *Bottom*: the radial decline of the gas covering fraction as a function of impact parameter for an equivalent detection limit of 0.3 Å. Image courtesy of Nikole Nielsen

between $\log(M_h) = 11$–12 (Birnboim and Dekel 2003; Kereš et al. 2005; Dekel and Birnboim 2006; Ocvirk et al. 2008; Brooks et al. 2009; Dekel et al. 2009; Kereš et al. 2009; Stewart et al. 2011; van de Voort et al. 2011). A repercussion of these models is that the covering fraction of cool accreting gas should drop significantly to almost zero for massive galaxies (Stewart et al. 2011). However, observational evidence shows that the covering fraction of cold gas is constant over a larger range of halo masses $10.7 \leq \log(M_h) \leq 13.9$ within a given impact parameter (or impact parameter normalized by the virial radius) and that gaseous galaxy halos are self-similar (Churchill et al. 2013a,b). This is suggestive that either outflows and/or other substructures contribute to absorption in high-mass halos such that low- and high-mass gas halos are observationally indistinguishable or the data indicate that predictions of a mass-dependent shutdown of cold-mode accretion may require revision. This area needs to be examined further in order to address the discrepancy between observations and simulations.

Although the average radial gas profiles around star-forming galaxies are well quantified, it does not provide much insight into the nature of the circumgalactic gas. Thus, it is critical to determine the geometric distribution of gas relative to its host galaxy to help improved our understanding of its origins.

2.2 Circumgalactic Gas Spatial Distribution

In the mid-1990s, the exploration of circumgalactic medium geometry started with Steidel (1995) who acquired a large sample of 51 galaxy-quasar pairs. The data were suggestive that, independent of galaxy spectroscopic/morphological type, Mg II gas resided within a spherical halo with unity gas covering fraction. The data fit well to a Holmberg-like luminosity scaling between a characteristic halo radius and galaxy K-band luminosity. However, even Steidel noted that spherical halos were likely a tremendous over-simplification of the true situation, however, the data did not disprove it. Using simple geometric models, Charlton and Churchill (1996) determined that both spherical halos and extended monolithic thick-disk models could be made consistent with the current data. They suggested that the kinematic structure of the absorption profiles could be used to further constrain the gas geometry, which we further discuss in Sect. 3.

Cosmological simulations commonly show that gas accretion should occur along filaments that are co-planar to the galaxy disk, whereas gas outflows are expected to be expelled along the galaxy projected minor axis (e.g., Shen et al. 2012; Stewart et al. 2013). Reminiscent of Charlton and Churchill (1996), Kacprzak et al. (2011b) reported that the Mg II equivalent width measured from high-resolution quasar spectra was dependent on galaxy inclination, suggesting that the circumgalactic medium has a co-planer geometry that is coupled to the galaxy inclination. It was noted, however, that the absorbing gas could arise from tidal streams, satellites, filaments, etc., which could also have somewhat co-planer distributions. By stacking over 5000 background galaxies to probe over 4000 foreground galaxies, Bordoloi et al. (2011) found a strong azimuthal dependence of the Mg II absorption within 50 kpc of inclined disk-dominated galaxies (also see Lan et al. 2014). They found elevated equivalent width along the galaxy minor axis and lower equivalent width along the major axis. Their data are indicative of bipolar outflows with possible flows along the major axis. Later, Bordoloi et al. (2014b) presented models of the circumgalactic medium azimuthal angle distribution by using joint constraints from: the integrated Mg II absorption from stacked background galaxy spectra (Bordoloi et al. 2011) and Mg II absorption from individual galaxies as seen from background quasar spectra (Kacprzak et al. 2011b). They determined that either

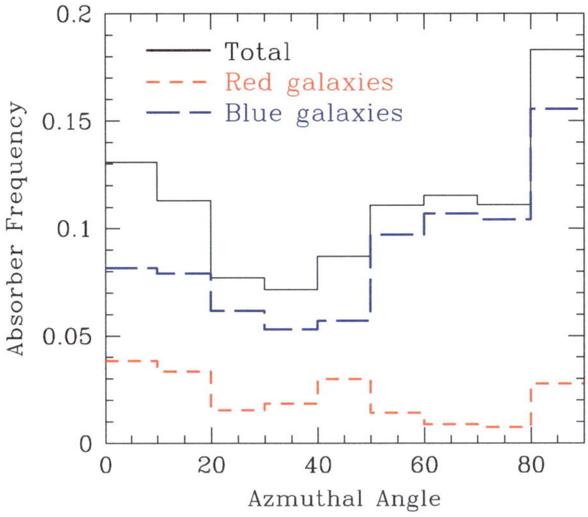

Fig. 2 Binned azimuthal angle mean probability distribution function (PDF) for Mg II absorbing galaxies (*solid line*). The binned PDFs are normalized such that the total area is equal to unity, yielding an observed frequency in each azimuthal bin. Absorption is detected with increased frequency towards the major ($\Phi = 0°$) and minor ($\Phi = 90°$) axes. Also shown is the galaxy color dependence of the distribution split by a $B - R \leq 1.1$ representing late-type galaxies (*dashed blue line*) and $B-R > 1.1$ representing early type galaxies (*dotted red line*). The data are consistent with star-forming galaxies accreting gas and producing large-scale outflows while quiescent galaxies have much less gas activity

composite models consisting of a bipolar outflow component plus a spherical or disk component, or a single highly softened bipolar distribution, could well represent data within 40 kpc.

Bouché et al. (2012), using ten galaxies, first showed that the azimuthal angle distribution of absorbing gas traced by Mg II appeared to be bimodal with half of the Mg II sight-lines showing a co-planar geometry. Kacprzak et al. (2012a) further confirmed the bimodality in the azimuthal angle distribution of gas around galaxies, where cool dense circumgalactic gas prefers to exist along the projected galaxy major and minor axes where the gas covering fraction is enhanced by 20–30% as shown in Fig. 2. Also shown in Fig. 2 is that blue star-forming galaxies drive the bimodality while red passive galaxies may contain gas along their projected major axis. The lower equivalent width detected along the projected major axis is suggestive that accretion would likely contain metal-poor gas with moderate velocity width profiles.

The aforementioned results provide a geometric picture that is consistent with galaxy evolution scenarios where star-forming galaxies accrete co-planer gas within narrow streams with opening angles of about 40°, providing fuel for new stars that produce metal-enriched galactic scale outflows with wide opening angles of 100°, while red galaxies exist passively due to reduced gas reservoirs. These conclusions are based on Mg II observations, however, both infalling gas and outflowing are expected to contain multi-phased gas.

Mathes et al. (2014) first attempted to address the azimuthal angle dependence for highly ionized gas traced by O VI and found it to have a spatially uniform distribution out to 300 kpc. Using a larger sample of O VI absorption selected galaxies, Kacprzak et al. (2015) reported a bimodality in the azimuthal angle distribution of gas around galaxies within 200 kpc. Similar to Mg II, they found that O VI is commonly detected within opening angles of 20°–40° of the galaxy projected major axis and within opening angles of at least 60° along the projected minor axis. Again similar to Mg II, weaker equivalent width systems tend to reside on along the project major axis. This would be expected for either lower column density, lower kinematic dispersion, or low metallicity (or a combination thereof) gas accreting towards the galaxy major axis. Different from the Mg II results, non-detections of O VI exist almost exclusively between 20° and 60°, suggesting that O VI is not mixed throughout the circumgalactic medium and remains confined within the accretion filaments and the gas outflows.

Further supporting this bimodality accretion/outflow picture is the recent work using H I 21-cm absorption to probe the circumgalactic medium within impact parameters of <35 kpc around $z < 0.4$ galaxies. (Dutta et al. 2016) found that the majority of their absorbers (nine) exist along the projected major axis and a few (three) exists along the projected minor axis. The data are supportive of high column density co-planer H I thick-disk around these galaxies. In addition, the three minor axis absorption systems all reside within 15 kpc, therefore they conclude that these low impact parameter minor axis systems could originate from warps in these thick and extended H I disks.

Although gas geometry is highly suggestive of (and consistent with) our exception of gas accretion onto star-forming galaxies, it alone is not sufficient enough to determine if gas is actually fact accreting onto galaxies. Gas and relative gas-galaxy kinematics can provide additional data that can be used to address whether we are detecting gas accretion or not.

3 Circumgalactic Gas Kinematics

Absorption systems produced by the circumgalactic medium hold key kinematic signatures into unlocking the behavior of gas around galaxies. High-resolution spectroscopy of the background quasars is critical to resolving the velocity substructures within these complex absorption systems. These data can be used to differentiate between scenarios of gas accretion, disk rotation, and outflows.

3.1 Internal or Intrinsic Gas Kinematics

For the most part, metal-line absorption systems are not composed of a uniform velocity distribution of "clouds," but tend to exist in groupings closer together in velocity with occasional higher velocity clouds offset from the groupings. For a handful high-resolution Mg II absorption profiles, Lanzetta and Bowen (1992) inferred that their velocity structure was dominated by coherent motions as oppose to random. They further showed that the absorption profiles are consistent with a rotating ensembles of clouds similar to a co-rotating disk. With a larger sample of high-resolution absorption profiles, Charlton and Churchill (1998) applied statistical tests for a variety of kinematic models and concluded that pure disk rotation and pure accretion models are likely ruled out. However, models with contributions from both a rotating disk and infall/halo can reproduce velocities that are nearly consistent with the observed kinematics.

Similar work focusing on the low-ion transitions (such as Si II) associated with damped Lyα systems, Prochaska and Wolfe (1997) examined if a range of models such as rotating cold disks, slowly rotating hot disks, massive isothermal halos, and a hydrodynamic spherical accretion models could explain the observed absorption kinematics. They determined that thick rapidly rotating disks are the only model consistent with the data at high confidence levels. Their tests suggest that disk rotation speeds of around 225 km s^{-1} are preferred, which is typical for Milky Way-like galaxy rotation speeds. Furthermore, the gas is likely to be cold since ratio of the gas velocity dispersion over the disk rotation speed must be less than 0.1. These data are suggestive of thick, cold, and possibly accreting disks surrounding galaxies.

All these studies are based on absorption-line data alone and therefore it is important to quantify how the absorption kinematics changes with galaxy properties.

Using primarily Lyman limits systems, Borthakur et al. (2015) found Lyα absorption from the circumgalactic medium has similar velocity spreads to that of their host galaxy's interstellar medium as observed via H I emission. The combination of the correlation between the galaxy gas fraction and the impact-parameter-corrected Lyα equivalent width is consistent with idea that the H I disk is fed by circumgalactic gas accretion (Borthakur et al. 2015). Furthermore, they find a correlation between impact-parameter-corrected Lyα equivalent width and the galaxy specific star formation rate suggesting a link between gas accretion driving star formation (Borthakur et al. 2016).

Nielsen et al. (2015) have quantified the Mg II absorber velocity profiles using pixel-velocity two-point correlation functions and determined how absorption kinematics vary as a function of the galaxy orientation and other physical properties as shown in Fig. 3. While they find that absorption profiles with the largest velocity dispersion are associated with blue, face-on galaxies probed along the projected minor axis, which is suggestive out outflows, they find something different for edge-on galaxies. For edge-on galaxies probed along the major axis, they find large Mg II absorber velocity dispersions and large column density clouds at low velocity regardless of galaxy color, which is used as a tracer of star formation

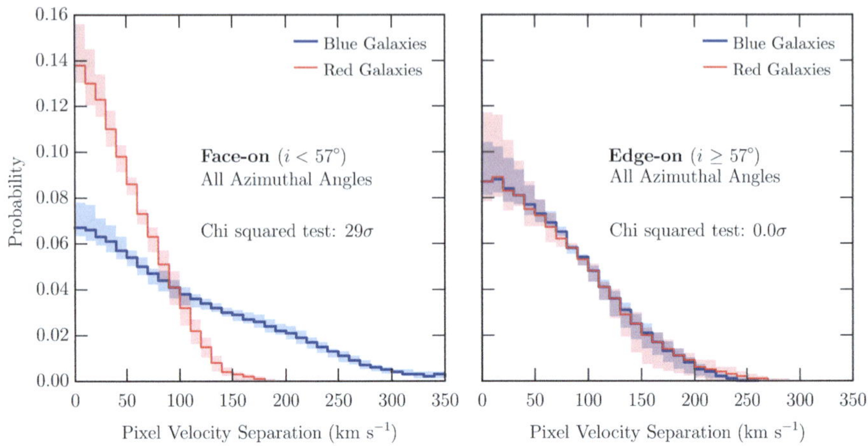

Fig. 3 Shown are pixel-velocity two-point correlation functions (TPCFs) examining how the spread in the pixel velocities of absorption profiles differ for different galaxy properties. The TPCFs presented give the probability of having pixel velocities at a given velocity offset from the absorption redshift defined by the optical depth weighed mean. The *shaded regions* are the 1σ bootstrap uncertainties. For all azimuthal angles, the TPCFs are shown for blue and red face-on galaxies (*left*) and edge-on galaxies (*right*). Note the dramatic differences between the absorber velocity dispersions for blue and red galaxies for face-on orientations (*left*). The larger velocity spread is suggestive of high-velocity outflows being ejected from the galaxy, while red galaxies are less active. On the other hand, there is no difference between blue and red for edge-on orientations (*right*). The velocity spread is comparable to the rotation speed of galaxies, which is consistent with gas accretion and/or gaseous disk co-rotation around the galaxies

rate. It is suggested that the large absorber velocity dispersions seen for edge-on galaxies (Fig. 3—right) may be caused by gas rotation/accretion where the line-of-sight velocity is maximized for edge-on galaxy inclination. Furthermore, the large cloud column densities may indicate that co-rotating or accreting gas is fairly coherent along the line-of-sight. The only way to test this scenario is to compare the absorption velocities relative to the rotation velocity of the host galaxies.

In addition, some evolution in the circumgalactic medium has also been observed. Blue galaxies do not show an evolution in the velocity dispersions and cloud column densities with redshift (between $0.3 \leq z \leq 1$), while red galaxies have a circumgalactic medium that becomes more kinematically quiescent with time (Nielsen et al. 2016). This is suggestive that the gas cycle in blue star-forming galaxies is active, be it via accretion or outflows, while red galaxies exhibit little-to-no gas activity. This is consistent with the little-to-no O VI found around $z \sim 0.25$ quiescent galaxies from the COS–Halos survey (Tumlinson et al. 2011).

3.2 Relative Gas-Galaxy Kinematics

It is interesting that the internal velocity structure of absorption systems are reflective of their host galaxy type, orientation, and redshift, however, the question then arises of how/if the circumgalactic medium is kinematically connected to their host galaxies. The most direct measure of gas accretion is observing it down-the-barrel by using the host galaxy as the background source. This method is ideal since there are no degeneracies in the line-of-sight direction/velocity. Although this method does have its difficulties too since metal-enriched outflows and the interstellar medium will tend to dominate the observed absorption over the metal-poor accreting gas. These down-the-barrel gas accretion events have been observed in a few cases with absorption velocity shifts relative to the galaxy of 80–200 km s^{-1} (Martin et al. 2012; Rubin et al. 2012). These down-the-barrel gas accretion observations are discussed in detail in the chapter by Kate Rubin. It is still unknown if these observations are signatures of cold accretion or recycled material falling back onto galaxies, yet they are the most direct measure of accreting gas to date.

Using quasar sight-lines, we find that the distribution of velocity separations between Mg II absorption and their host galaxies tends to be Gaussian with a mean offset of 16 km s^{-1} and dispersion of about 140 km s^{-1} (Chen et al. 2010). Although there are much higher velocity extremes, typically expected for outflows, it is interesting that the velocity range is more typical of galaxy rotation speeds having masses close-to or less-than that of the Milky Way.

When galaxy masses are known, one can compare the relative galaxy and absorption velocities to those of the escape velocities of their halos. Figure 4 shows escape velocities, computed for spherically symmetric Navarro–Frenk–White dark matter halo profile (Navarro et al. 1996), as a function of halo mass. It can be seen that very few absorption velocity centroids exceed the estimated galaxy halo escape velocities. This is surprisingly true for a full range of galaxy masses from dwarf galaxies probed by C IV absorption (Bordoloi et al. 2014a) to more massive galaxies probed by O VI (Tumlinson et al. 2011). Consistently, it has been found that the vast majority of absorption systems that reside within one galaxy virial radius are bound to the dark matter halos of their hosts (Stocke et al. 2013; Mathes et al. 2014; Tumlinson et al. 2011; Bordoloi et al. 2014a; Ho et al. 2016). It is worth noting that some of the velocity ranges covered by the absorption profiles are comparable to the escape velocities but are typically for the far wings of the profiles where the gas column densities are the lowest. Therefore two possible scenarios, or a combination thereof, can be drawn: (1) gas that is traced by absorption can be driven into the halo by star formation driven outflows and eventually fall back onto the galaxy (known as recycled winds) and/or (2) the gas is new material accreting from the intergalactic medium. With these data alone, we likely cannot distinguish between these scenarios.

To test how the circumgalactic gas is kinematically coupled to their galaxy hosts, Steidel et al. (2002) presented the first rotation curves of five intermediate-redshift Mg II selected absorbing galaxies. Interestingly, they found that for four

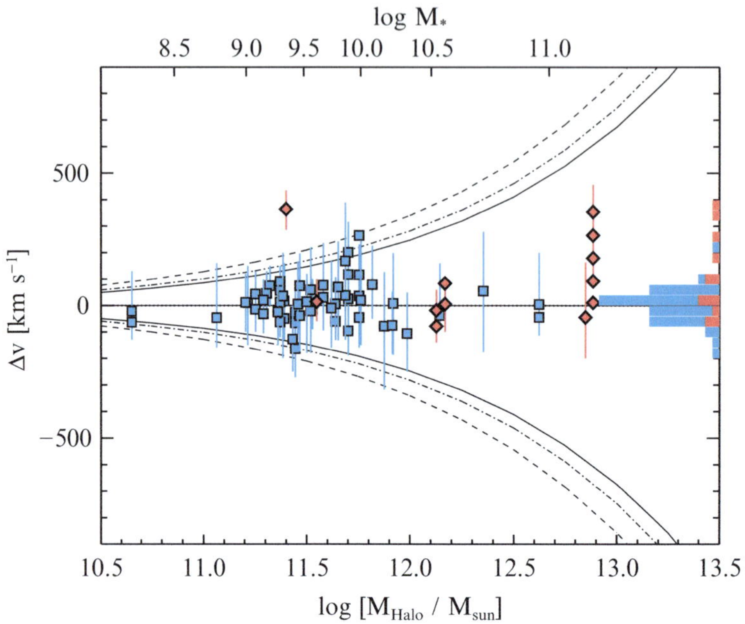

Fig. 4 Combined C IV (Bordoloi et al. 2014a) and O VI (Tumlinson et al. 2011) absorption velocity centroids with respect to the systemic redshift of their host galaxies as a function of the inferred dark matter halo mass for star-forming (*blue squares*) and passive (*red diamond*) galaxies. The range bars indicate the maximum projected kinematic extent of each absorption system. The histogram represents the distribution of individual component velocities. The *dashed lines* show the mass-dependent escape velocities at $R = 50$, 100, and 150 kpc, respectively. Note that all absorption-line systems appear to be bound to their halos and have velocities (and velocity ranges) comparable to galaxy circular velocities. This means that both outflowing and accreting gas could give rise to the observed kinematics. Image courtesy of Jason Tumlinson, Rongmon Bordoloi and the COS−Halos team

of the five cases, the absorption velocities lie entirely to one side of the galaxy systemic redshift and consistent with the side expected for rotation. Using simple thick disk-halo models, they concluded that the bulk of Mg II gas velocities could be explained by an extension of disk rotation with some velocity lag (Steidel et al. 2002). This was further confirmed by Kacprzak et al. (2010a) who also showed that infalling gas or lagging rotation is required to explain the gas kinematics. Using cosmological hydrodynamical galaxy simulations to replicate their data allowed them to concluding that coherently rotating accreting gas is likely responsible for the observed kinematic offset.

There have now been over 50 galaxies/absorbers pairs that have been compared this way and the vast majority exhibit disk-like and/or accretion kinematics (Steidel et al. 2002; Chen et al. 2005; Kacprzak et al. 2010a, 2011a; Bouché et al. 2013, 2016; Burchett et al. 2013; Keeney et al. 2013; Jorgenson and Wolfe 2014; Diamond-Stanic et al. 2016; Ho et al. 2016) and some show outflowing wind signatures (Ellison et al. 2003; Kacprzak et al. 2010a; Bouché et al. 2012; Schroetter et al. 2015, 2016; Muzahid et al. 2016) while some exhibiting group dynamics (Lehner et al. 2009; Kacprzak et al. 2010b; Bielby et al. 2016; Péroux et al. 2016). Even for systems with multiple quasar sight-lines (Bowen et al. 2016) or for multiply lensed quasars near known foreground galaxies (Chen et al. 2014) provide the same kinematic evidence that a co-rotating disk with either some lagging rotation or accretion is required to reproduce the observed absorption kinematics. One caveat is that above works have a range of galaxy inclinations and quasar sight-line azimuthal angles, which could complicate the conclusions drawn. It is likely best to select galaxies where the quasar is located along the projected major axis where accreted gas is expected to be located.

Ho et al. (2016) designed an experiment where they selected Mg II absorbers associated with highly inclined ($i > 43°$) star-forming galaxies with quasars sight-lines passing within 30° of the projected major axis. Presented in Fig. 5 are the rotation curves and the velocity spread of the absorption shown as a function of distance within the galaxy viral radius. It is clear that there is a strong correlation between the Mg II absorption velocities and the galaxy rotation velocities. The majority of the Mg II equivalent widths are detected at velocities less than the actual rotation speed of the dark matter halo (blue squares), while the Keplerian fall-off from the measured rotation curve provides lower limits on the rotation speed of the circumgalactic medium (dashed, black line). The cyan curves illustrate constant $Rv_{\rm rot}(R)$ and show that the infalling gas would have specific angular momentum at least as large as that in the galactic disk, for which some of the gas has comparable specific angular momentum. The Mg II absorption-line velocity widths cannot be generated with circular disk-like orbit and a simple disk model with a radial inflowing accretion reproduced the data quite well (Ho et al. 2016).

We have shown that lagging or infalling gas appears to be a common kinematic signature of the circumgalactic gas near star-forming galaxies from $0.1 \leq z \leq 2.5$ and is consistently seen for a range of galaxy inclination and position angles. These observations are consistent with current simulations that show that large co-rotating gaseous structures in the halo of the galaxy that are fueled, aligned, and kinematically connected to filamentary gas infall along the cosmic web (Stewart et al. 2011, 2013, 2016; Danovich et al. 2012, 2015). The predictions and results from simulations are discussed further in the chapters by Kyle Stewart and Claude-André Faucher-Giguère. Stewart et al. (2013) demonstrated that there is a qualitative agreement among the majority of cosmological simulations and that the buildup of high-angular momentum halo gas and the formation of cold-flow disks are likely a robust prediction of ΛCDM. These simulations naturally predict that accreted gas can be observationally distinguishable from outflowing gas from its kinematic signature of large one-sided velocity offsets. Thus, it is plausible we have already observed gas accretion through these kinematic velocity offsets mentioned above.

It is also important to note that the vast majority of systems discussed above have column densities typical of Lyman Limit Systems (N(H I)$> 10^{17.2}$ cm^{-2}), which are exclusively associated with galaxies. At the lowest column densities of N(H I)$<$ 10^{14} cm^{-2}, Lyα absorption was found to not mimic the rotation of, and/or accretion onto, galaxies derived from H I observations (Côté et al. 2005). Maybe this is not

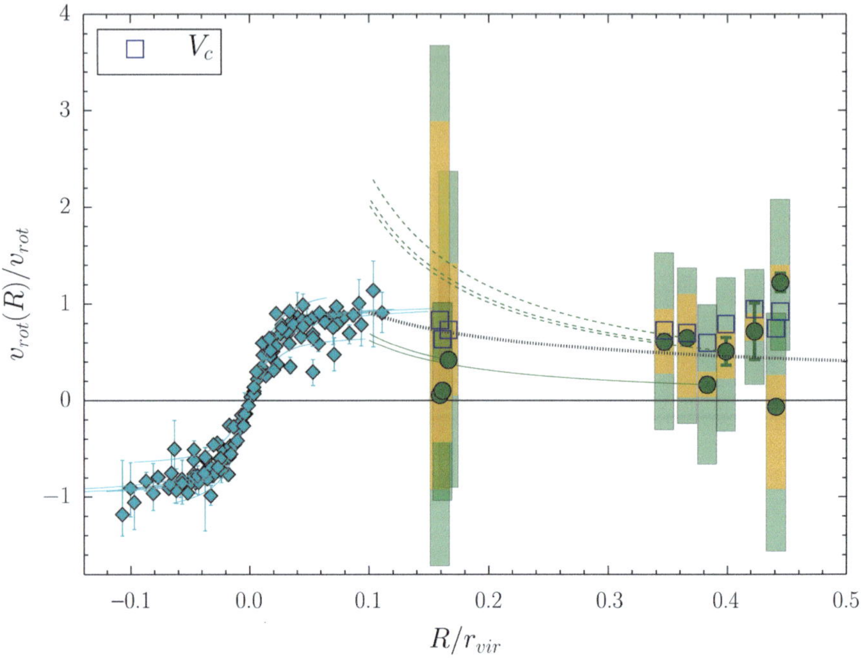

Fig. 5 Galaxy rotation curves of ten galaxies, normalized by the rotation speed as a fraction of the halo virial radius, are compared to the kinematics of their circumgalactic gas. The Mg II absorption velocities are deprojected such that the velocity shown represents the tangential motion in the disk plane that would give to the observed sight-line velocities. The measured and intrinsic velocity range of each Mg II absorption system after deprojection is indicated by the *green* and *orange bars* with the Mg II absorption velocity along the quasar sight-line (*green circles*). Note that the absorption systems align with the expected side for extended disk rotation. Also shown are dark matter halo rotation speed models (*blue squares*) and the Keplerian fall-off from the galaxy rotation curves, which sets a lower limits on the rotation speed in the circumgalactic medium (*dashed*, *black line*). The *cyan curves* illustrate constant $Rv_{\rm rot}(R)$ and indicates that the infalling gas would have specific angular momentum at least as large as that in the galactic disk

unexpected since this low column density gas is not likely to be associated with galaxies and likely associated with the Lyα forest (gas associated with galaxies have N(H I)> $10^{14.5}$ cm^{-2}– Rudie et al. 2012).

4 Circumgalactic and Galaxy Gas-Phase Metallicities

It is possible that metallicity can be used to determine the origins of absorbing gas observed around galaxies since outflows are expected to be metal-enriched while accreted gas should have lower metallicity (e.g., Shen et al. 2012). Gas accretion is expected to be metal poor but not purely pristine given that the first generations of Population III stars have likely enriched the gas to 10^{-4} $Z_{\odot}$ by a redshift of 15–20 (e.g., Yoshida et al. 2004). In fact, there is a metallicity floor whereby it is rare to find absorption systems with metallicities much lower than 10^{-3} $Z_{\odot}$ even out to $z \sim 5$ (Prochaska et al. 2003; Penprase et al. 2010; Cooke et al. 2011, 2015; Battisti et al. 2012; Rafelski et al. 2012; Jorgenson et al. 2013; Cooper et al. 2015; Fumagalli et al. 2016a; Lehner et al. 2016; Quiret et al. 2016). Although a few systems do have metallicities $<10^{-3}$ $Z_{\odot}$ (Fumagalli et al. 2011b; Cooke et al. 2011, 2015; Lehner et al. 2016) and possibly have Population III abundance patterns (Crighton et al. 2016). Cosmological simulations predict gas accretion metallicities should be between $10^{-3} and 10^{-0.5}$ $Z_{\odot}$, which is dependent on redshift and halo mass (Fumagalli et al. 2011a; Oppenheimer et al. 2012; van de Voort and Schaye 2012; Shen et al. 2013; Kacprzak et al. 2016), however, this metallicity range does have some overlap with the metallicities of recycled outflowing gas.

There has been an abundance of studies that have identified galaxies with circumgalactic gas metallicity measurements in an effort to determine the source of the absorption. The general census shows that absorption systems near galaxies are either metal-poor with metallicities between $-2 <$[X/H]< -1 (Tripp et al. 2005; Cooksey et al. 2008; Kacprzak et al. 2010b, 2014; Ribaudo et al. 2011; Thom et al. 2011; Churchill et al. 2012; Bouché et al. 2013, 2016; Crighton et al. 2013, 2015; Stocke et al. 2013; Muzahid et al. 2015; Fumagalli et al. 2016b; Rahmani et al. 2016) or metal-enriched with metallicities of [X/H]> -0.7 (Chen et al. 2005; Lehner et al. 2009; Péroux et al. 2011, 2016; Bregman et al. 2013; Krogager et al. 2013; Meiring et al. 2013; Stocke et al. 2013; Crighton et al. 2015; Muzahid et al. 2015, 2016; Rahmani et al. 2016). Determining the fraction of metal-rich and metal-poor systems is complicated since all the aforementioned studies have a range of observational biases due to the way the targets were selected—some selected from the presence of metal-lines. A clear complication of tracing the circumgalactic gas using metal-lines as tracers may bias you towards metal enriched systems. Ideally, the best way to avoid such a biases is to select absorption systems by hydrogen only.

Selecting only by hydrogen (but not necessarily with known galaxy hosts) has shown that the metallicity distribution of all Lyman limit systems below $z < 1$ appear to have a bimodal distribution (Lehner et al. 2013; Wotta et al. 2016). The shape of Lyman limit systems (17.2 $<$log N(H I) $<$ 17.7 in their study) metallicity bimodality distribution could be explained by outflows producing the high metallicity peak ([X/H]~ -0.3) while accreting/recycled gas could produce the low metallicity peak ([X/H] ~ -1.9) (Wotta et al. 2016). Interestingly, between $2 < z < 3.5$, the metallicity distribution of Lyman limit systems uni-modal peaking at [X/H]$= -2$, in contrast to the bimodal distribution seen at $z < 1$ (Fumagalli et al. 2011b; Lehner et al. 2016). Therefore it is likely that there exists a vast reservoir of metal-poor cool gas that can accrete onto galaxies at high redshift and outflows build up the circumgalactic medium at a later time. These results are discussed in detail in the chapter by Nicolas Lehner.

The overall knowledge of the metallicity distribution of the circumgalactic medium provides critical clues to the physics of gas cycles of galaxies. The

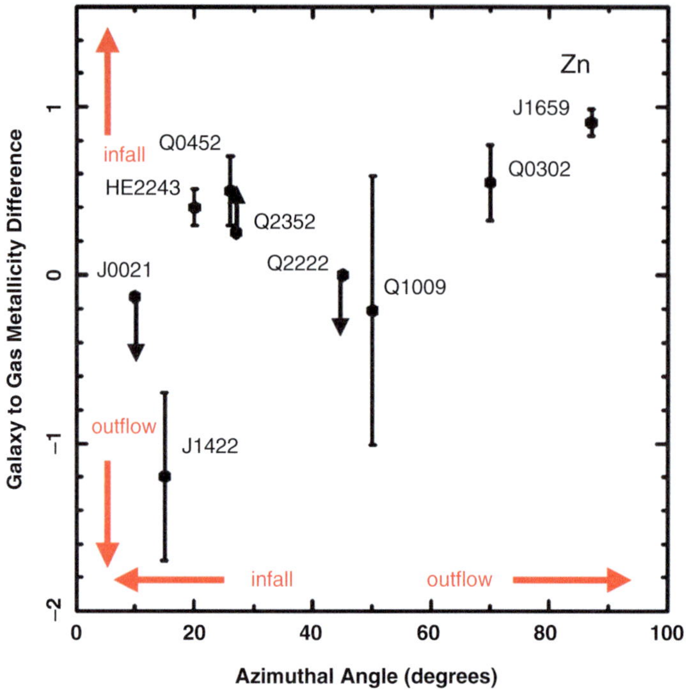

Fig. 6 Metallicity difference between the host galaxy and absorber as a function of azimuthal angle. In this plot, accreted gas is expected to reside in the upper-left corner (for low/co-planer azimuthal angle and high metallicity difference), while outflowing gas should reside in the lower-right corner (high/minor axis azimuthal angle and metallicity similar to the host galaxy) as indicated by the *red arrows*. Note outflowing gas appears to be metal-poor, while accreting gas exhibits a range in metallicity. Additional observations are necessary to better relate metallicity and geometry in gas flows as only a minor anti-correlation is currently measured. Image from Péroux et al. (2016)

bimodal metallicity distribution is suggestive that we are observing both outflows and accretion, but our assumptions rely strongly on predictions from simulations, which still have some issues with modeling the circumgalactic medium since it is extremely feedback dependent. One way to determine if we are observing outflows and gas accretion is to combine our expectation of the gas geometry of accretion being along the minor axis and metal poor, while outflows are metal-enrich and ejected along the minor axis.

Preliminary work by Péroux et al. (2016), using nine galaxies, indicate that there is a very weak anti-correlation with metallicity and azimuthal angle. Figure 6 shows the relative metallicity of the absorbing gas with respect to the host galaxy metallicity as a function of azimuthal angle, which are two independent indicators of gas flow origins. Note that in the figure a positive difference in metallicity indicates a circumgalactic medium metallicity lower than that of the host galaxy's HII regions (expected for metal-poor accreting gas) while a negative value indicates a higher circumgalactic medium metallicity than the host galaxy (expected for metal-enriched outflows). For the few objects shown in Fig. 6, there is not clear correlation as expected under simple geometric and metallicity assumptions. Note that, different from expectations, there does not appear to be any high metallicity gas at high azimuthal angles. At low azimuthal angles, there are a range of metallicity differences including negative values, which is unexpected for accreting gas. Given the few number of systems and only a weak anti-correlation, more systems are required to understand if there is a relation between the spatial location of the circumgalactic medium and metallicity. This is an active area of research and may be the most promising avenue to peruse in the future.

5 Putting It All Together

The previous sections describe the individual geometric, kinematic, and metallicity indicators as evidence for cold-mode accretion. On their own, they are quite suggestive that we have detected signatures of gas accretion, however, combining all of these accretion indicators together can provide quite compelling evidence.

There are a few of such examples that exist where some point to their circumgalactic medium originating from metal enriched outflows along the galaxy minor axis (Kacprzak et al. 2014; Muzahid et al. 2015) or the circumgalactic medium is kinematically consistent with gas arising from tidal/streams or interacting galaxy groups (Kacprzak et al. 2011b; Muzahid et al. 2016; Péroux et al. 2016) or even enriched gas that is being recycled along the galaxy major axis (Bouché et al. 2016).

Bouché et al. (2013) presented a nice example of a moderately inclined $z = 2.3$ star-forming galaxy where the quasar sight-line is within 20° of the galaxy's projected major axis. They derived the galaxy rotation field using IFU observations and found that the kinematics of circumgalactic medium at 26 kpc away could be reproduced by a combination of an extended rotating disk and radial gas accretion. The metallicity of the circumgalactic medium is -0.72, which is typically for gas accretion metallicities from cosmological simulations at these redshifts (van de Voort and Schaye 2012; Kacprzak et al. 2016). The mass inflow rate was estimated to be between 30 and 60 $M_{\odot}yr^{-1}$, which is similar to the galaxy star-formation rate of 33 $M_{\odot}yr^{-1}$ and suggestive that there is a balance between gas accretion and star-formation activity. This particular system is described in detail in the chapter by Nicolas Bouché.

In a slightly different case, a star-forming galaxy at $z = 0.66$ was examined where the quasar sightline is within 3° of the minor axis at a distance of 104 kpc (Kacprzak et al. 2012b). Contrary to expectations, they identify a cool gas phase with metallicity ~ -1.7 that has kinematics consistent with a accretion or lagging halo model. Furthermore, they also identify a warm collisionally ionized phase that also has low metallicity (~ -2.2). The warm gas phase is kinematically consistent with both radial outflows or radial accretion. Given the metallicities and kinematics, they conclude that the gas is accreting onto the galaxy, however, this is contrary to the previously discussed interpretations that absorption found along the projected minor axis is typically associated outflows. This could be a case where there is a miss-alignment with the accreting filaments and the disk or is an example of the three filaments typically predicted by simulations (e.g., Dekel et al. 2009; Danovich et al. 2012), thus making it unlikely that all accreting gas is co-planer.

Detailed studies like these are one of the best ways of constraining the origins of the circumgalactic medium and to help us understand how gas accretion works. Larger samples are required to build up a statistical sample of systems to negate cosmic variance. Large IFU instruments like the Multi Unit Spectroscopic Explorer (MUSE) will provide ample full field imaging/spectra datasets and help us build up these larger samples with less observational time. Gas accretion studies using IFUs are further discussed in the chapter by Nicolas Bouché.

6 Direct Imaging of Gas Accretion

All of the aforementioned efforts are possibly direct observations of gas accretion, yet the scientific community is not satisfied since it is difficult to prove conclusively that we have detected cold gas accretion onto galaxies. The only way to conclusively do so is by direct spectral imaging of cosmic flows onto galaxies. Obviously this is quite difficult due to the faintness and low gas column densities of these cold-flow filaments. However, we may be on the verge of being able to detect these cold gas flows using the new generation of ultra sensitive instruments such as the Keck Comic Web Imager.

It is also possible that we have already observed comic accretion onto two separate quasar hosts (Martin et al. 2015, 2016) with one example shown in Fig. 7. The figure shows a narrow-band image obtained with the Palomar Comic Web Imager. The extended nebula and filaments are likely illuminated by the nearby quasar. The streams, as indicted in the figure, extend out to $\sim$160–230 kpc. It was determined that this nebula/disk and filaments are well fit to a rotating disk model with are dark matter halo mass of $\sim 10^{12.5}\,M_\odot$ with a circular velocity of $\sim$350 km s^{-1} at the viral radius of 125 kpc (Martin et al. 2016). They further found that by adding gas accretion with velocities of 80–100 km s^{-1} improved their kinematic model fit. Furthermore, they estimate the baryonic spin parameter is three times higher than that of the dark matter halo and has an orbital period of 1.9 Gyr. The high-angular momentum and the well-fitted inflowing stable disk are consistent with the predictions of cold accretion from cosmological simulations (Stewart et al. 2011, 2013, 2016; Danovich et al. 2012, 2015).

It is possible that we have direct evidence of cold accretion already, but it has yet to be observed and quantified for typical star-forming galaxies (which is much harder). These types of ultra-sensitive instruments will potentially allow us to directly image cosmic accretion in the near future.

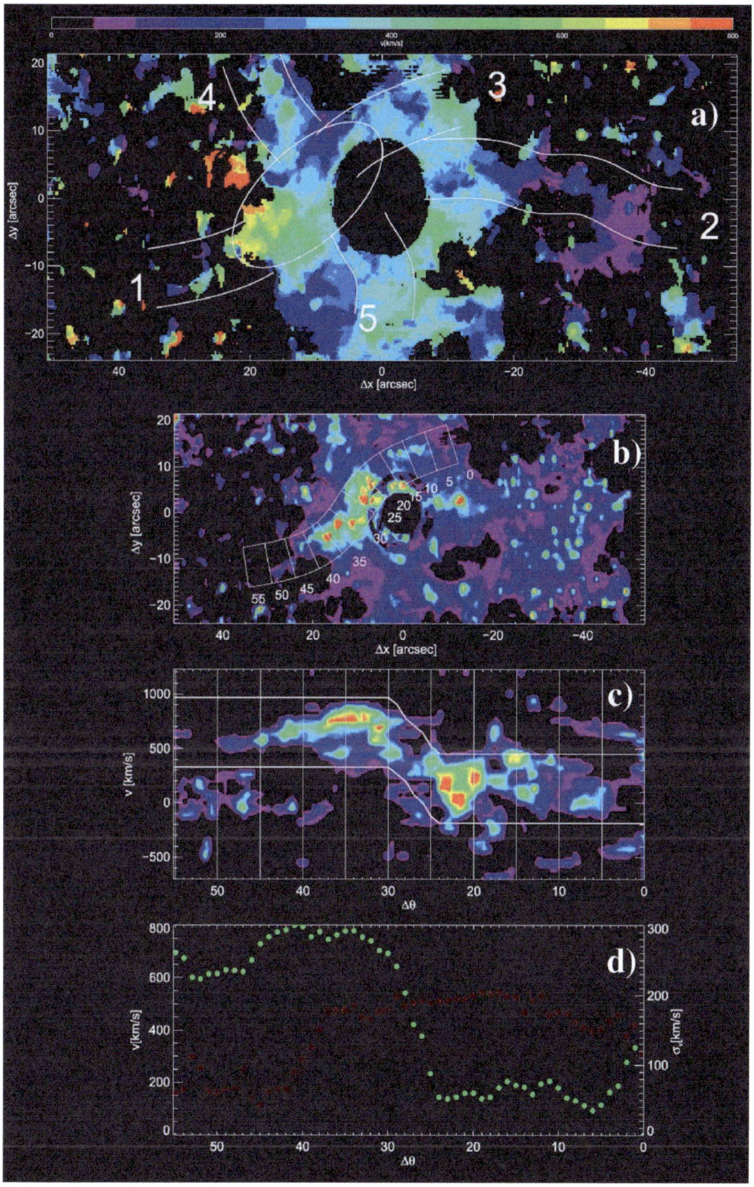

Fig. 7 (**a**) Mean velocity obtained with an intensity-weighted velocity moment within a narrow 7 Å band image shown in panel (**b**). The data were obtained using Palomar cosmic web imager. The disk and filamentary candidates are indicated. (**b**) A pseudo slit is placed over the narrow-band image covering the disk and filaments 1 and 3 indicated by the *curved white lines with tick marks* indicating the distance in arcseconds along the slit. (**c**) The narrow-band image produced using the sheared velocity window slit indicating the distance along the slit as in panel (**b**)–(**d**) The mean velocity (*green*) and velocity dispersion (*red*) from the above panels. Note observed rotation with high-velocity dispersions seen along the filament. This is also seen for the remaining filaments (not shown here). This is potentially one of the first direct images of cold gas accretion. Image courtesy of Chris Martin and modified from Martin et al. (2016)

7 Summary

Although we only have candidate detections of gas accretion, the community has provided a body of evidence indicating that cosmic accretion exists and is occurring over a large range of redshifts. Individual studies on their own are only suggestive that this accretion is occurring, but taking all the aforementioned aspects together does paint a nice picture of low metallicity gas accreting along co-planar co-rotating disk/filaments. The picture seems nice and simple but understanding how previously ejected gas is also recycled back onto the galaxy is also important since it can possibly mimic gas accretion signatures (Oppenheimer et al. 2012; Ford et al. 2014).

Simulations predict the cross-section of cold flows to be as low as 5% (Faucher-Giguère and Kereš 2011; Fumagalli et al. 2011b; Kimm et al. 2011; Goerdt et al. 2012) and that the metal-poor cold-flow streams signatures should be overwhelmed by the metal-rich outflows signatures detected in absorption spectra. However, signatures of intergalactic cold gas accretion seem to be quite frequent. Thus, either the simulations are under-producing gas accretion cross-sections due to various reasons such as dust, resolution effects, self-shielding, and/or magnetic fields, or the observations are selecting more than just outflowing and accreting gas. Trying to address how recycled winds fit into the picture, and understanding how we can differentiate them from gas accretion, will possibly aid in resolving this issue. It is great to see, however, how the simulations and observations have been working together in our community to try and understand the gas cycles of galaxies.

Aside from observing accretion directly, the way forward now is to combine geometry, kinematics, and metallicity at a range of epochs to try to understand how gas accretion occurs. Hopefully in the near future, we will be able to directly image gas accretion, putting to rest one of the most debated issues in the circumgalactic field.

Acknowledgements Glenn G. Kacprzak acknowledges the support of the Australian Research Council through the award of a Future Fellowship (FT140100933). Thanks to Nikole Nielsen for providing comments, and to Nikole Nielsen, Stephanie Ho, and Rongmon Bordoloi for providing additional new plots for this chapter.

References

Adelberger, K. L., Shapley, A. E., Steidel, C. C., et al. 2005, ApJ, 629, 636
Battisti, A. J., Meiring, J. D., Tripp, T. M., et al. 2012, ApJ, 744, 93
Bergeron, J. 1986, A&A, 155, L8
Bielby, R., Crighton, N. H. M., Fumagalli, M., et al. 2016, arXiv:1607.03386
Birnboim, Y., & Dekel, A. 2003, MNRAS, 345, 349
Boksenberg, A., & Sargent, W. L. W. 1978, ApJ, 220, 42
Bordoloi, R., Lilly, S. J., Knobel, C., et al. 2011, ApJ, 743, 10
Bordoloi, R., Lilly, S. J., Kacprzak, G. G., & Churchill, C. W. 2014, ApJ, 784, 108
Bordoloi, R., Tumlinson, J., Werk, J. K., et al. 2014, ApJ, 796, 136

Borthakur, S., Heckman, T., Tumlinson, J., et al. 2015, ApJ, 813, 46
Borthakur, S., Heckman, T., Tumlinson, J., et al. 2016, arXiv:1609.06308
Bouché, N., Finley, H., Schroetter, I., et al. 2016, ApJ, 820, 121
Bouché, N., Hohensee, W., Vargas, R., et al. 2012, MNRAS, 426, 801
Bouché, N., Murphy, M. T., Kacprzak, G. G., et al. 2013, Science, 341, 50
Bowen, D. V., Chelouche, D., Jenkins, E. B., et al. 2016, ApJ, 826, 50
Bregman, J. N., Miller, E. D., Seitzer, P., Cowley, C. R., & Miller, M. J. 2013, ApJ, 766, 57
Brooks, A. M., Governato, F., Quinn, T., Brook, C. B., & Wadsley, J. 2009, ApJ, 694, 396
Burchett, J. N., Tripp, T. M., Prochaska, J. X., et al. 2015, ApJ, 815, 91
Burchett, J. N., Tripp, T. M., Werk, J. K., et al. 2013, ApJ, 779, L17
Charlton, J. C., & Churchill, C. W. 1996, ApJ, 465, 631
Charlton, J. C., & Churchill, C. W. 1998, ApJ, 499, 181
Chen, H.-W., Kennicutt, R. C., Jr., & Rauch, M. 2005, ApJ, 620, 703
Churchill, C. W., Kacprzak, G. G., Steidel, C. C., et al. 2012, ApJ, 760, 68
Churchill, C. W., Nielsen, N. M., Kacprzak, G. G., & Trujillo-Gomez, S. 2013, ApJ, 763, L42
Churchill, C. W., Trujillo-Gomez, S., Nielsen, N. M., & Kacprzak, G. G. 2013, ApJ, 779, 87
Chen, H.-W., Helsby, J. E., Gauthier, J.-R., et al. 2010, ApJ, 714, 1521
Chen, H.-W., Gauthier, J.-R., Sharon, K., et al. 2014, MNRAS, 438, 1435
Chen, H.-W., Lanzetta, K. M., & Webb, J. K. 2001, ApJ, 556, 158
Chen, H.-W., Lanzetta, K. M., Webb, J. K., & Barcons, X. 2001, ApJ, 559, 654
Chen, H.-W., & Mulchaey, J. S. 2009, ApJ, 701, 1219
Côté, S., Wyse, R. F. G., Carignan, C., Freeman, K. C., & Broadhurst, T. 2005, ApJ, 618, 178
Cooke, R. J., Pettini, M., & Jorgenson, R. A. 2015, ApJ, 800, 12
Cooke, R., Pettini, M., Steidel, C. C., Rudie, G. C., & Nissen, P. E. 2011, MNRAS, 417, 1534
Cooksey, K. L., Prochaska, J. X., Chen, H.-W., Mulchaey, J. S., & Weiner, B. J. 2008, ApJ, 676, 262
Cooper, T. J., Simcoe, R. A., Cooksey, K. L., O'Meara, J. M., & Torrey, P. 2015, ApJ, 812, 58
Crighton, N. H. M., Hennawi, J. F., & Prochaska, J. X. 2013, ApJ, 776, L18
Crighton, N. H. M., Hennawi, J. F., Simcoe, R. A., et al. 2015, MNRAS, 446, 18
Crighton, N. H. M., O'Meara, J. M., & Murphy, M. T. 2016, MNRAS, 457, L44
Danforth, C. W., & Shull, J. M. 2008, ApJ, 679, 194
Danovich, M., Dekel, A., Hahn, O., & Teyssier, R. 2012, MNRAS, 422, 1732
Danovich, M., Dekel, A., Hahn, O., Ceverino, D., & Primack, J. 2015, MNRAS, 449, 2087
Dekel, A., & Birnboim, Y. 2006, MNRAS, 368, 2
Dekel, A., Birnboim, Y., Engel, G., et al. 2009, Nature, 457, 451
Diamond-Stanic, A. M., Coil, A. L., Moustakas, J., et al. 2016, ApJ, 824, 24
Dutta, R., Srianand, R., Gupta, N., et al. 2016, arXiv:1610.05316
Ellison, S. L., Mallén-Ornelas, G., & Sawicki, M. 2003, ApJ, 589, 709
Faucher-Giguère, C.-A., & Kereš, D. 2011, MNRAS, 412, L118
Ford, A. B., Davé, R., Oppenheimer, B. D., et al. 2014, MNRAS, 444, 1260
Fumagalli, M., Cantalupo, S., Dekel, A., et al. 2016, MNRAS, 462, 1978
Fumagalli, M., Prochaska, J. X., Kasen, D., et al. 2011, MNRAS, 418, 1796
Fumagalli, M., O'Meara, J. M., & Prochaska, J. X. 2011, Science, 334, 1245
Fumagalli, M., O'Meara, J. M., & Prochaska, J. X. 2016, MNRAS, 455, 4100
Goerdt, T., Dekel, A., Sternberg, A., Gnat, O., & Ceverino, D. 2012, MNRAS, 424, 2292
Guillemin, P., & Bergeron, J. 1997, A&A, 328, 499
Ho, S. H., Martin, C. L., Kacprzak, G. G., & Churchill, C. W. 2016, arXiv:1611.04579. http://adsabs.harvard.edu/abs/2016arXiv161104579H
Johnson, S. D., Chen, H.-W., & Mulchaey, J. S. 2015, MNRAS, 449, 3263
Johnson, S. D., Chen, H.-W., & Mulchaey, J. S. 2013, MNRAS, 434, 1765
Jorgenson, R. A., Murphy, M. T., & Thompson, R. 2013, MNRAS, 435, 482
Jorgenson, R. A., & Wolfe, A. M. 2014, ApJ, 785, 16
Kacprzak, G. G., Churchill, C. W., Barton, E. J., & Cooke, J. 2011, ApJ, 733, 105
Kacprzak, G. G., Churchill, C. W., Ceverino, D., et al. 2010, ApJ, 711, 533

Kacprzak, G. G., Churchill, C. W., Evans, J. L., Murphy, M. T., & Steidel, C. C. 2011b, MNRAS, 416, 3118
Kacprzak, G. G., Churchill, C. W., & Nielsen, N. M. 2012, ApJ, 760, L7
Kacprzak, G. G., Churchill, C. W., Steidel, C. C., Spitler, L. R., & Holtzman, J. A. 2012, MNRAS, 427, 3029
Kacprzak, G. G., Churchill, C. W., Steidel, C. C., & Murphy, M. T. 2008, AJ, 135, 922–927
Kacprzak, G. G., Martin, C. L., Bouché, N., et al. 2014, ApJ, 792, L12
Kacprzak, G. G., Murphy, M. T., & Churchill, C. W. 2010, MNRAS, 406, 445
Kacprzak, G. G., Muzahid, S., Churchill, C. W., Nielsen, N. M., & Charlton, J. C. 2015, ApJ, 815, 22
Kacprzak, G. G., van de Voort, F., Glazebrook, K., et al. 2016, ApJ, 826, L11
Keeney, B. A., Stocke, J. T., Rosenberg, J. L., et al. 2013, ApJ, 765, 27
Kereš, D., Katz, N., Weinberg, D. H., & Davé, R. 2005, MNRAS, 363, 2
Kereš, D., Katz, N., Fardal, M., Davé, R., & Weinberg, D. H. 2009, MNRAS, 395, 160
Kimm, T., Slyz, A., Devriendt, J., & Pichon, C. 2011, MNRAS, 413, L51
Krogager, J.-K., Fynbo, J. P. U., Ledoux, C., et al. 2013, MNRAS, 433, 3091
Kunth, D., & Bergeron, J. 1984, MNRAS, 210, 873
Lan, T.-W., Ménard, B., & Zhu, G. 2014, ApJ, 795, 31
Lanzetta, K. M., & Bowen, D. V. 1992, ApJ, 391, 48
Lehner, N., Howk, J. C., Tripp, T. M., et al. 2013, ApJ, 770, 138
Lehner, N., O'Meara, J. M., Howk, J. C., Prochaska, J. X., & Fumagalli, M. 2016, ApJ, 833, 283 http://adsabs.harvard.edu/abs/2016ApJ...833..283L
Lehner, N., Prochaska, J. X., Kobulnicky, H. A., et al. 2009, ApJ, 694, 734
Liang, C. J., & Chen, H.-W. 2014, MNRAS, 445, 2061
Martin, C. L., Shapley, A. E., Coil, A. L., et al. 2012, ApJ, 760, 127
Martin, D. C., Matuszewski, M., Morrissey, P., et al. 2015, Nature, 524, 192
Martin, D. C., Matuszewski, M., Morrissey, P., et al. 2016, ApJ, 824, L5
Mathes, N. L., Churchill, C. W., Kacprzak, G. G., et al. 2014, ApJ, 792, 128
Meiring, J. D., Tripp, T. M., Werk, J. K., et al. 2013, ApJ, 767, 49
Muzahid, S., Kacprzak, G. G., Charlton, J. C., & Churchill, C. W. 2016, ApJ, 823, 66
Muzahid, S., Kacprzak, G. G., Churchill, C. W., et al. 2015, ApJ, 811, 132
Navarro, J. F., Frenk, C. S., & White, S. D. M. 1996, ApJ, 462, 563
Nielsen, N. M., Churchill, C. W., Kacprzak, G. G., & Murphy, M. T. 2013, ApJ, 776, 114
Nielsen, N. M., Churchill, C. W., & Kacprzak, G. G. 2013, ApJ, 776, 115
Nielsen, N. M., Churchill, C. W., Kacprzak, G. G., Murphy, M. T., & Evans, J. L. 2015, ApJ, 812, 83
Nielsen, N. M., Churchill, C. W., Kacprzak, G. G., Murphy, M. T., & Evans, J. L. 2016, ApJ, 818, 171
Ocvirk, P., Pichon, C., & Teyssier, R. 2008, MNRAS, 390, 1326
Oppenheimer, B. D., Davé, R., Katz, N., Kollmeier, J. A., & Weinberg, D. H. 2012, MNRAS, 420, 829
Penprase, B. E., Prochaska, J. X., Sargent, W. L. W., Toro-Martinez, I., & Beeler, D. J. 2010, ApJ, 721, 1
Péroux, C., Bouché, N., Kulkarni, V. P., York, D. G., & Vladilo, G. 2011, MNRAS, 410, 2237
Péroux, C., Rahmani, H., Quiret, S., et al. 2016, arXiv:1609.07389
Prochaska, J. X., Gawiser, E., Wolfe, A. M., Cooke, J., & Gelino, D. 2003, ApJS, 147, 227
Prochaska, J. X., Weiner, B., Chen, H.-W., Mulchaey, J., & Cooksey, K. 2011, ApJ, 740, 91
Prochaska, J. X., & Wolfe, A. M. 1997, ApJ, 487, 73
Quiret, S., Péroux, C., Zafar, T., et al. 2016, MNRAS, 458, 4074
Rafelski, M., Wolfe, A. M., Prochaska, J. X., Neeleman, M., & Mendez, A. J. 2012, ApJ, 755, 89
Rahmani, H., Péroux, C., Turnshek, D. A., et al. 2016, MNRAS, 463, 980
Ribaudo, J., Lehner, N., Howk, J. C., et al. 2011, ApJ, 743, 207
Richter, P. 2012, ApJ, 750, 165
Richter, P., Wakker, B. P., Fechner, C., et al. 2016, A&A, 590, A68

Rubin, K. H. R., Prochaska, J. X., Koo, D. C., & Phillips, A. C. 2012, ApJ, 747, L26
Rudie, G. C., Steidel, C. C., Trainor, R. F., et al. 2012, ApJ, 750, 67
Savage, B. D., Sembach, K. R., Wakker, B. P., et al. 2003, ApJS, 146, 125
Schroetter, I., Bouché, N., Péroux, C., et al. 2015, ApJ, 804, 83
Schroetter, I., Bouché, N., Wendt, M., et al. 2016, arXiv:1605.03412
Sembach, K. R., Tripp, T. M., Savage, B. D., & Richter, P. 2004, ApJS, 155, 351
Shen, S., Madau, P., Aguirre, A., et al. 2012, ApJ, 760, 50
Shen, S., Madau, P., Guedes, J., et al. 2013, ApJ, 765, 89
Steidel, C. C. 1995, QSO Absorption Lines, 139
Steidel, C. C., Dickinson, M., & Persson, S. E. 1994, ApJ, 437, L75
Steidel, C. C., Erb, D. K., Shapley, A. E., et al. 2010, ApJ, 717, 289
Steidel, C. C., Kollmeier, J. A., Shapley, A. E., et al. 2002, ApJ, 570, 526
Stewart, K. R., Brooks, A. M., Bullock, J. S., et al. 2013, ApJ, 769, 74
Stewart, K. R., Kaufmann, T., Bullock, J. S., et al. 2011, ApJ, 738, 39
Stewart, K., Maller, A., Oñorbe, J., et al. 2016, arXiv:1606.08542
Stocke, J. T., Keeney, B. A., Danforth, C. W., et al. 2013, ApJ, 763, 148
Stocke, J. T., Penton, S. V., Danforth, C. W., et al. 2006, ApJ, 641, 217
Thom, C., Werk, J. K., Tumlinson, J., et al. 2011, ApJ, 736, 1
Tripp, T. M., Jenkins, E. B., Bowen, D. V., et al. 2005, ApJ, 619, 714
Tripp, T. M., Lu, L., & Savage, B. D. 1998, ApJ, 508, 200
Tripp, T. M., Sembach, K. R., Bowen, D. V., et al. 2008, ApJS, 177, 39
Tumlinson, J., Thom, C., Werk, J. K., et al. 2011, Science, 334, 948
van de Voort, F., & Schaye, J. 2012, MNRAS, 423, 2991
van de Voort, F., Schaye, J., Booth, C. M., Haas, M. R., & Dalla Vecchia, C. 2011, MNRAS, 414, 2458
Wakker, B. P., & Savage, B. D. 2009, ApJS, 182, 378
Wolfe, A. M., Gawiser, E., & Prochaska, J. X. 2005, ARA&A, 43, 861
Wotta, C. B., Lehner, N., Howk, J. C., O'Meara, J. M., & Prochaska, J. X. 2016, arXiv:1608.02584
Yoshida, N., Bromm, V., & Hernquist, L. 2004, ApJ, 605, 579
Zheng, Y., Putman, M. E., Peek, J. E. G., & Joung, M. R. 2015, ApJ, 807, 103
Zibetti, S., Ménard, B., Nestor, D. B., et al. 2007, ApJ, 658, 161

The Circumgalactic Medium in Massive Halos

Hsiao-Wen Chen

1 Introduction

In the standard picture of galaxy formation and evolution, primordial gas first cools and condenses within dark matter halos to form stars (e.g., White and Rees 1978; Blumenthal et al. 1984). Gas in high-mass halos with dark matter halo mass exceeding $M_h \approx 10^{12}\,M_\odot$ is expected to be shock-heated to high temperatures (e.g., Birnboim and Dekel 2003; Kereš et al. 2005, 2009; Dekel and Bernboim 2006). Within the hot gas halos, thermal instabilities can induce the formation of pressure-supported cool clouds (e.g., Mo and Miralda-Escudé 1996; Maller and Bullock 2004; Sharma et al. 2012; Voit et al. 2015). In lower mass halos, cool filaments from the intergalactic medium can reach deep into the center of the halo without being shock-heated. Both condensed cool clouds of hot halos and cool filaments can, in principle, supply the fuels necessary to sustain star formation in galaxies. As new stars form and evolve, the surrounding interstellar and circumstellar gas is expected to be heated and enriched by heavy elements ejected from massive stars, regulating subsequent star formation.

Several semi-analytic studies have been carried out in searching for a general prescription that connects galaxies found in observations to dark matter halos formed in theoretical frameworks. These studies seek to establish a mean relation between galaxy stellar mass ($M_{\rm star}$) and host halo mass (M_h) by matching the observed space density of galaxies with the expected abundance of dark matter halos as a function of mass (e.g., Vale and Ostriker 2004; Shankar et al. 2006). The mean galaxy mass and halo mass ratio is found to peak at $M_h \sim 10^{12}\,M_\odot$ with $M_{\rm star}/M_h \approx 0.04$ and declines rapidly both toward higher and lower masses

H.-W. Chen (✉)
Department of Astronomy & Astrophysics and Kavli Institute for Cosmological Physics,
The University of Chicago, Chicago, IL 60637, USA
e-mail: hchen@oddjob.uchicago.edu

A. Fox, R. Davé (eds.), *Gas Accretion onto Galaxies*, Astrophysics and Space
Science Library 430, DOI 10.1007/978-3-319-52512-9_8

(e.g., Moster et al. 2010; Guo et al. 2010; Behroozi et al. 2010). The declining $M_{\rm star}/M_h$ indicates a reduced star formation efficiency in both low- and high-mass halos. Different feedback mechanisms are invoked in theoretical models in order to match the observed low star formation efficiency in low- and high-mass halos. As supernova-driven winds are thought to suppress star formation efficiency in low-mass dwarf galaxies (e.g., Larson 1974; Dekel and Silk 1986), feedback due to active galactic nuclei (AGN) powered by supermassive black holes is invoked to quench star formation in high-mass halos, resulting in massive quiescent galaxies (e.g., Bower et al. 2006; Croton et al. 2006; Dubois et al. 2013). While blueshifted broad absorption and emission lines are commonly seen in luminous quasars, indicating the presence of high-speed outflows, direct observational evidence of AGN feedback on large scales ($\sim$10–100 kpc) remains scarce (e.g., Alexander et al. 2010; Greene et al. 2012; Maiolino et al. 2012) and see also Fabian (2012) for a review.

Observations of the circumgalactic medium (CGM) in massive halos offer complementary and critical constraints for the extent of feedback and gas accretion (e.g., Somerville and Davé 2015). In particular, the circumgalactic space within the halo radius, $R_{\rm vir}$, lies between galaxies, where star formation takes place, and the intergalactic medium (IGM), where 90% of all baryonic matter in the universe resides (Rauch et al. 1997). As a result, CGM properties are shaped by the complex interactions between IGM accretion and outflows driven by energetic feedback processes in the galaxies.

Imaging observations of the cool ($T \sim 10^4$ K) CGM are only feasible in 21 cm surveys at $z \lesssim 0.2$, because, with few exceptions (e.g., Cantalupo et al. 2014; Fernández et al. 2016; Hayes et al. 2016), the gas density is typically too low to be detected in emission. Absorption-line spectroscopy of background quasars provides a powerful, alternative tool for studying this tenuous gas in the distant universe based on the absorption features imprinted in the quasar spectra (e.g., Lanzetta et al. 1995; Bowen et al. 1995; Steidel et al. 2002). But, because both quasars and massive galaxies are rare, close pairs of massive galaxies and background quasars by chance projection are even rarer. Studying the CGM around massive galaxies using quasar absorption spectroscopy therefore requires a large spectroscopic sample of galaxies and quasars over a substantial volume in order to assemble a statistical sample of massive galaxy and quasar pairs. The Sloan Digital Sky Survey (SDSS; York et al. 2000) has produced a large spectroscopic archive of distant galaxies and quasars, facilitating the assembly of a statistically significant sample of close massive galaxy and quasar pairs, as well as a large sample of projected quasar pairs. These pair samples have enabled systematic studies of halo gas beyond the nearby universe using absorption spectroscopy.

This chapter presents a review of the current state of knowledge on the cool CGM properties in massive halos of $M_h \approx 10^{12-14}\,{\rm M}_\odot$ at $z \approx 0.2$–2. Specifically, the review will focus on massive quiescent galaxies with $M_{\rm star} \gtrsim 10^{11}\,{\rm M}_\odot$ at $z \lesssim 1$, with additional coverage on quasar host halos. Empirical studies of cool/warm gas in galaxy groups and clusters of $M_h \gtrsim 10^{14}\,{\rm M}_\odot$ have been carried out for a small sample (e.g., Lopez et al. 2018; Yoon et al. 2012; Andrews et al. 2013; Stocke et al.

2014), but a detailed understanding of the gas phase in galaxy cluster and group environments relies primarily on X-ray studies of the hot plasma. Extensive reviews on the X-ray properties of intragroup and intracluster gas can be found in Mulchaey (2000), Mathews and Brighenti (2003), and Kravtsov and Borgani (2012).

The emphasis on quiescent galaxies in high-mass halos is motivated by the empirical finding that more than 90% of massive galaxies with $M_{\rm star} \gtrsim 10^{11}\,{\rm M}_\odot$ in the local universe contain primarily evolved stellar populations with little on-going star formation (e.g., Peng et al. 2010; Tinker et al 2013). It is therefore expected that a general understanding of the CGM properties in massive halos can be established based on observations of massive quiescent galaxies. An added bonus in studying massive quiescent galaxies is the unique opportunity to explore other feedback mechanisms for quenching star formation in massive halos, in the absence of complicated starburst driven winds.

The emphasis on quasar host halos is motivated by two factors. First, while determining quasar host mass is difficult due to uncertain host galaxy properties, the large observed clustering amplitude indicates that the mean halo mass of quasar hosts is high, $M_h \gtrsim 10^{12.5}\,{\rm M}_\odot$ (Porciani et al. 2004; White et al. 2012; Shen et al. 2013b). In addition, observations of the CGM properties in quasar host halos directly address the issues concerning the extent of AGN feedback (e.g., Shen et al. 2013a; Fumagalli et al. 2014; Rahmati et al. 2015; Faucher-Giguère et al. 2015, 2016). Most nearby elliptical galaxies are found to host a supermassive black hole at the center (Kormendy and Richstone 1995; Ho 2008), and therefore high-redshift quasar hosts are likely the progenitors of these nearby massive quiescent galaxies. Studies of the CGM in quasar host halos may provide important clues for how the CGM properties are shaped while the galaxy undergoes an active quasar phase.

2 Incidence/Covering Fraction of Cool Gas in Quiescent Halos

Luminous red galaxies (LRGs) uncovered in the SDSS exhibit little on-going star formation and display photometric and spectral properties resembling nearby elliptical galaxies (Eisenstein et al. 2001). They are characterized by a mean luminosity of $\approx 5L_*$ and a mean stellar mass of $M_{\rm star} \approx 3\times10^{11}\,{\rm M}_\odot$ at $z\approx 0.5$ (e.g., Tojeiro et al. 2011), with corresponding dark matter halo mass of $M_h \approx 3\times10^{13}\,{\rm M}_\odot$ (e.g., Zheng et al. 2007; Blake et al. 2008; Padmanabhan et al. 2009), and therefore offer an ideal laboratory for studying the cool gas content in massive quiescent halos.

Roughly 10% of LRGs exhibit [O II] emission features (Eisenstein et al. 2003; Roseboom et al. 2006), suggesting the possible presence of on-going star formation in some of these massive galaxies. However, a more detailed examination of their spectral properties reveals unusually high [N II]/Hα together with low [O III]/[O II] ratios, both of which indicate the presence of low-ionization nuclear emission line

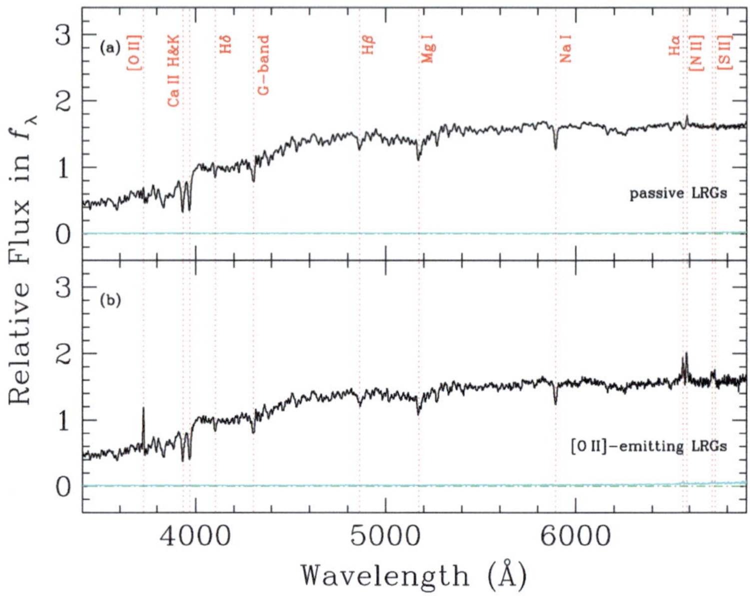

Fig. 1 The quiescent nature of $z \approx 0.5$ luminous red galaxies (LRGs) shown in stacked rest-frame optical spectra. Panel (**a**) shows the median stack of ∼1780 passive LRGs and panel (**b**) shows the median stack of ∼280 LRGs with [O II] emission detected (see Huang et al. (2016) for a detailed description of the sample definition and selection). Prominent absorption and emission lines are labeled in *red*. The corresponding $1 - \sigma$ dispersion in each stack is shown in *cyan* at the bottom of each panel. In both stacked spectra, the spectral properties are characterized by prominent absorption features due to Ca II H and K, G-band, Mg I and Na I that indicate a predominantly old stellar population, as well as a relatively weak Balmer absorption series. The observed high [N II]/Hα emission ratios seen in both passive and [O II]-emitting LRGs, together with the low [O III]/[O II] ratios, suggest the presence of LINER in these LRGs, which could be attributed to the presence of underlying AGNs or evolved AGB stars in these galaxies

regions (LINER) in these galaxies (e.g., Huang et al. 2016). Figure 1 shows stacked spectra of SDSS LRGs grouped into "passive" and [O II]-emitting subsamples. In both panels of Fig. 1, the stacked spectra are characterized by prominent absorption features due to Ca II H and K, G-band, Mg I and Na I, along with LINER-like emission, namely strong [N II]/Hα ratios. The prominent absorption features indicate a predominantly old stellar population, whereas the LINER-like emission indicates the presence of additional ionizing sources in these galaxies. Although the origin of LINER-like emission is not fully understood, the observed slow decline in the spatial profile of LINER-like emission requires spatially distributed ionizing sources, rather than centrally concentrated active galactic nuclei (AGN), in these quiescent galaxies (e.g., Sarzi et al. 2006, 2010; Yan and Blanton 2012; Singh et al. 2013; Belfiore et al. 2016). Likely sources include photo-ionization due to hot, post-asymptotic giant branch (post-AGB) stars or winds from Type Ia supernovae (e.g., Conroy et al. 2015). In the absence of on-going star formation, these LRGs also provide a unique sample for testing additional feedback mechanisms for quenching star formation in massive halos, including gravitational heating (e.g., Johansson et al. 2009) and winds from evolved stars and Type Ia supernovae (e.g., Conroy et al. 2015).

A first step toward understanding the quiescent state of LRGs is to characterize the incidence and covering fraction of cool gas in their host halos. Over the past several years, absorption-line observations of background QSOs have revealed extended cool gas at projected distances $d \approx 10$–300 kpc from LRGs at $z \sim 0.5$. Specifically, the LRG–Mg II cross-correlation function displays an amplitude that is comparable to the LRG auto-correlation on small scales ($\lesssim 450$ comoving kpc), well within the virial radii of the halos. The large clustering amplitude on small scales suggests the presence of Mg II gas inside these massive quiescent halos (e.g., Bouché et al. 2006; Tinker and Chen 2008; Gauthier et al. 2009; Lundgren et al. 2009). In addition, surveys of Mg II $\lambda\lambda$ 2796, 2803 absorption doublets in LRG halos have covered a non-negligible fraction of LRG halos with associated Mg II absorbers (e.g., Bowen and Chelouche 2011; Gauthier and Chen 2011; Gauthier et al. 2014; Zhu et al. 2014; Huang et al. 2016).

These Mg II absorption transitions are commonly seen at projected distances of $d \lesssim 100$ kpc from star-forming galaxies (e.g., Lanzetta and Bowen 1990; Steidel et al. 1994; Nestor et al. 2007; Kacprzak et al. 2008; Chen and Tinker 2008; Chen et al. 2010a; Werk et al. 2013) and are understood to arise primarily in photo-ionized gas of temperature $T \sim 10^4$ K (e.g., Bergeron and Stasińska 1986; Hamann 1997) and neutral hydrogen column density $N(HI) \approx 10^{18} - 10^{22}\,\mathrm{cm}^{-2}$ (e.g., Churchill et al. 2000; Rao et al. 2006). The large rest wavelengths make the Mg II doublet a convenient probe of chemically enriched gas at $z \approx 0.3$–2.3 in optical QSO spectra. In addition, the rest-frame absorption equivalent width $W_r(2796)$ is found to increase proportionally with the number of individual components and the velocity spread of the components (e.g., Petitjean and Bergeron 1990; Churchill et al. 2000, 2003). Measurements of $W_r(2796)$ therefore trace the underlying gas kinematics corresponding to $\sim 100\,\mathrm{km\,s^{-1}}$ per Angstrom. Several authors have proposed that strong Mg II absorbers of $W_r(2796) > 1$ Å originate in starburst driven outflows (e.g., Zibetti et al. 2007; Ménard et al. 2011) and that the large $W_r(2796)$ is driven by non-gravitational motion in the outflowing media (e.g., Bouché et al. 2006). The presence of strong Mg II absorbers near quiescent galaxies is therefore particularly surprising.

Figure 2 compares the observed mean covering fraction of Mg II absorbing gas, $\langle\kappa\rangle_{\mathrm{Mg\,II}}$, in halos around typical L_*- and sub-L_*-type, star-forming galaxies (blue triangles) with what is seen in LRG halos (green and red symbols) at $z \approx 0.3$–0.5. For a representative comparison across a broad luminosity (or mass) range, the gas covering fraction is determined within a fiducial halo gas radius, R_{gas}, that scales with galaxy luminosity according to $R_{\mathrm{gas}} \approx 107\,(L_B\,/\,L_B^*)^{0.35}$ kpc (Kacprzak et al. 2008; Chen et al. 2010a). In addition, κ is estimated for halo clouds with absorption strength exceeding $W_0 = 0.3$ Å, a detection threshold that is allowed by the quality of the absorption spectra. Previous studies have shown that typical L_* galaxies have $R_{\mathrm{gas}} \approx 130$ kpc and R_{gas} scales with luminosity according to $R_{\mathrm{gas}} \propto L^{0.35}$ (see Chen and Tinker 2008; Chen et al. 2010a). Following this scaling relation, LRGs with a mean luminosity of $\approx 3.6\,L_*$ are expected to have $R_{\mathrm{gas}} \approx 206$ kpc.

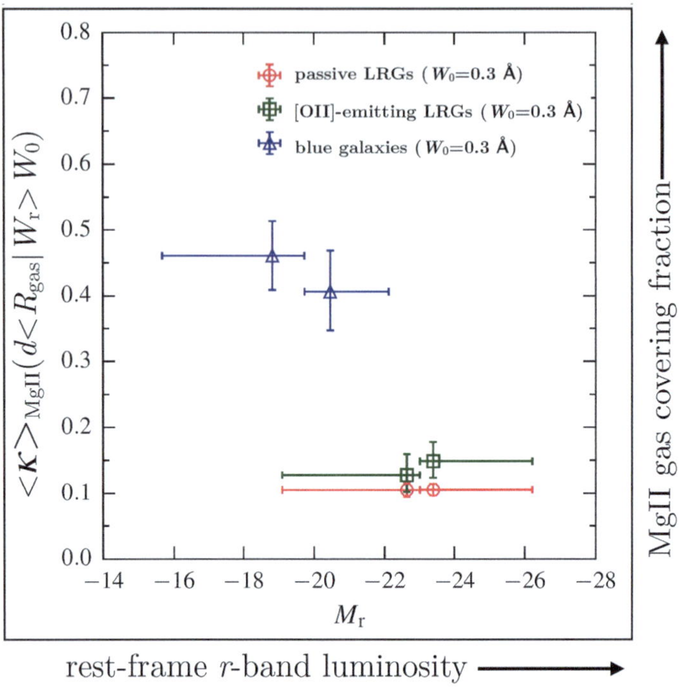

Fig. 2 Mass dependence of mean covering fraction of chemically enriched cool gas in galaxy halos (Chen et al. 2010a; Gauthier and Chen 2011; Huang et al. 2016). The mean covering fraction, $\langle\kappa\rangle$, is calculated based on observations of Mg II absorbing gas within a fiducial halo gas radius, $R_{\rm gas}$. Rest-frame r-band magnitude is adopted here as a proxy for the underlying total stellar mass (see Liang and Chen 2014). Star-forming galaxies at $z \approx 0.3$ are shown in *blue triangles*, while luminous red galaxies (LRGs) at $z \approx 0.5$ are shown in *squares/circles*. The covering fraction is determined for gas with absorption strength exceeding $W_0 = 0.3$ Å, a detection threshold that is allowed by the quality of the absorption spectra. The *horizontal bars* represent the full range of M_r for galaxies included in each bin, and the *vertical error bars* represent the 68% confidence interval. In comparison to star-forming galaxies (Chen et al. 2010a), LRGs, being old and massive, exhibit a much reduced covering fraction of chemically enriched cool gas as probed by the Mg II absorption features (Gauthier and Chen 2011; Huang et al. 2016). However, observations also show a definitive presence of cool gas in these LRG halos, demonstrating that they are not completely devoid of star formation fuels (Huang et al. 2016). Furthermore, roughly 10% of LRGs exhibit [O II] emission features and LINER-like spectra. These [O II]-emitting LRGs display a slightly elevated cool gas covering fraction from passive LRGs but the mean value remains significantly lower than what is seen in lower-mass, star-forming galaxies (Huang et al. 2016). Similar trends in the mean gas covering fraction have also been reported for H I Lyα and high-ionization species probed by the O VI $\lambda\lambda$ 1031,1037 doublet transitions. Massive quiescent galaxies also consistently display non-zero covering fractions of Lyα and O VI absorbing gas (Tumlinson et al. 2011, 2013; Thom et al. 2012; Werk et al. 2013)

Interpreting the rest-frame r-band absolute magnitude as a proxy for the underlying stellar mass (e.g., Liang and Chen 2014), Fig. 2 displays a strong mass dependence in the observed incidence and covering fraction of Mg II absorbing gas.

Specifically, the observed Mg II covering fraction in LRG halos is more than four times lower than what is found in halos around lower-mass, star-forming galaxies. However, in spite of this substantially reduced cool gas content in massive quiescent halos, the distinctly non-zero Mg II gas covering fraction around LRGs demonstrates the definitive presence of chemically enriched cool gas around evolved galaxies. Finally, it is interesting to note that while [O II]-emitting LRGs display a slightly elevated overall cool gas content than passive LRGs, the mean gas covering fraction remains under 20% (Huang et al. 2016).

Figure 2 is based on two separate studies designed to characterize the CGM properties of L_*/sub-L_* galaxies using ~260 galaxies at $z \approx 0.25$ (Chen et al. 2010a) and those of massive quiescent galaxies using ~38,000 LRGs (Huang et al. 2016), respectively. The COS-Halos survey was designed for a systematic study of gaseous halos over a broad range of mass and star formation history, and the sample contains 28 blue sub-L_* galaxies with a median mass of $\langle M_{\rm star}({\rm blue}) \rangle \approx 2 \times 10^{10}\,{\rm M}_\odot$ and 16 massive red galaxies with a median mass of $\langle M_{\rm star}({\rm red}) \rangle \approx 10^{11}\,{\rm M}_\odot$ (Tumlinson et al. 2013; Werk et al. 2013). While the uncertainties are large due to a smaller sample size, the COS-Halos galaxies confirm the observed mass dependence of Mg II gas covering fraction in Fig. 2 (Werk et al. 2013).

A declining gas covering fraction from low-mass, star-forming galaxies to massive quiescent galaxies is also observed in H I and highly ionized species probed by the O VI $\lambda\lambda$ 1031,1037 doublet transitions. Specifically, the covering fraction of moderately strong Lyα absorbers of $W_r(1215) > 0.1$ Å declines from $\gtrsim$90% around blue, star-forming galaxies (Tumlinson et al. 2013) to ≈40–50% around red passive galaxies (Thom et al. 2012), and the covering fraction of O VI absorbing gas decreases from ≈90% to ≈30% (Tumlinson et al. 2011; Werk et al. 2013). In both cases, massive quiescent galaxies consistently display non-zero covering fractions of Lyα and O VI absorbing gas.

Deep 21 cm and CO imaging surveys have also revealed that roughly 40% of elliptical galaxies in the nearby universe contain cool neutral gas (e.g., Oosterloo et al. 2010; Young et al. 2014). While on-going star formation is observed at low levels, $\lesssim 1\,{\rm M}_\odot\,{\rm yr}^{-1}$, in many early type galaxies of $M_{\rm star} \approx 5 \times 10^{10}\,{\rm M}_\odot$, it is rarely seen in more massive systems of $M_{\rm star} \gtrsim 10^{11}\,{\rm M}_\odot$ (e.g., Salim and Rich 2010). Combining H I/CO imaging surveys of local elliptical galaxies and absorption-line observations of distant early type galaxies at $z \approx 0.2$–0.5 demonstrates that cool gas is indeed present in some, although not all, massive quiescent halos.

A declining cool gas fraction with increasing halo mass is expected in theoretical models that attribute the observed Mg II absorbers to infalling gas from either thermally unstable hot halos or the intergalactic medium (e.g., Maller and Bullock 2004; Kereš et al. 2009). In the presence of cool gas, it is expected that these halo clouds could provide the fuels necessary for sustaining star formation in the LRGs. The lack of on-going star formation in these massive galaxies over such an extended cosmological time period ($\gtrsim$2 Gyr; Gauthier and Chen 2011) therefore suggests a prolonged duty cycle for the underlying heat sources.

3 Radial Profiles of Absorbing Gas

A key advantage of 21 cm and CO imaging surveys is their ability to resolve the spatial distribution of the gas around individual galaxies. The morphologies of the detected H I and CO gas span a broad range, from regular disk- or ring-like structures to irregular distributions of clumps and/or streams (e.g., Oosterloo et al. 2007; Serra et al. 2012) with roughly 1/4 displaying centralized disk or ring-like structures (Serra et al. 2012). Such a wide range of morphologies suggests different origins of the gas in different galaxies, including leftover materials from previous mergers and newly accreted gas from the CGM/IGM. On the other hand, QSO absorption spectroscopy, while offering extremely sensitive probes of tenuous gas beyond the limits of 21 cm/CO surveys, reveals gas properties along a single sightline per halo and does not provide constraints for spatial variations in individual halos. Nevertheless, considering an ensemble of close QSO and galaxy pairs over a wide range of projected separations provides a spherical average of halo absorption properties for the entire galaxy sample.

Figure 3 shows the spatial distribution of the observed Mg II absorption strength [$W_r(2796)$] and annular average of Mg II gas covering fraction ($\langle\kappa\rangle_{\rm MgII}$) as a function of projected distance (d) for a sample of 13330 LRGs at $z \approx 0.5$ from Huang et al. (2016). These LRGs are selected to have a background QSO within $d \approx 500$ kpc, the virial radius of a typical LRG halo, with sufficiently high signal-to-noise (S/N) absorption spectra available in the SDSS archive for detecting Mg II absorbers of $W_r(2796) \gtrsim 0.3$ Å. The LRGs are grouped into passive and [O II]-emitting subsamples based on the observed strength of [O II] emission lines (Fig. 1). A number of unique features are apparent in the observed spatial distribution of chemically enriched cool gas in LRG halos:

- a relatively flat trend is seen in the observed $W_r(2796)$ versus d distribution for individual absorbers uncovered over the entire projected distance range $d \lesssim 500$ kpc of both passive and [O II]-emitting LRG halos, although the scatter is large (*left* panel);
- a constant gas covering fraction, $\langle\kappa\rangle_{\rm Mg\,II} \approx 15\%$, is found at $d < 120$ kpc around passive LRGs, while $\langle\kappa\rangle_{\rm Mg\,II}$ increases from $\langle\kappa\rangle_{\rm Mg\,II} \approx 15\%$ at $d \approx 120$ kpc to $\langle\kappa\rangle_{\rm Mg\,II} \approx 40\%$ (though with a large error bar) at $d \lesssim 40$ kpc from [O II]-emitting LRGs;
- the gas covering fraction declines rapidly at $d > 120$ kpc, reaching the expected rate of incidence due to random background absorbers at $d \gtrsim 300$ kpc (orange dashed line in the *right* panel);
- strong Mg II absorbers of $W_r(2796) > 1$ Å appear to be common throughout the LRG halos at $d \lesssim 500$ kpc.

The observed flat $W_r(2796)$ versus d trend for Mg II absorbers and the large fraction of strong Mg II absorbers with $W_r(2796) > 1$ Å in LRG halos (see also Bowen and Chelouche 2011) are both in stark contrast to what is known for star-forming, L_* and sub-L_* galaxies. Specifically, CGM observations of star-forming

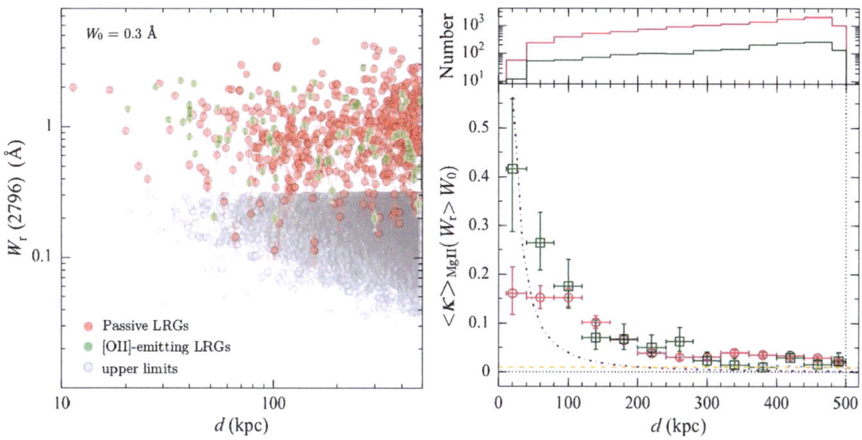

Fig. 3 Spatial distribution of the observed Mg II absorption strength ($W_r(2796)$; *left panel*) and covering fraction ($\langle\kappa\rangle_{\rm MgII}$; *right panel*) as a function of projected distance d from LRGs at $z \approx 0.5$. The figures are adapted from Huang et al. (2016). The LRGs are selected to have a background QSO at $d < 500$ kpc (the virial radius of a typical LRG halo) with sufficiently high signal-to-noise (S/N) absorption spectra available for constraining the presence/absence of Mg II absorbers with absorption strength exceeding $W_0 = 0.3$ Å. A total of 13330 LRGs in the SDSS data archive satisfy this criterion, 1575 of these exhibit [O II]-emission features and the rest are labeled as passive galaxies. For non-detections, a $2 - \sigma$ upper limit of $W_r(2796)$ is presented for the LRG. For a few sightlines, the QSO spectra are of sufficiently high S/N for detecting Mg II absorption lines as weak as $W_r(2796) \approx 0.1$ Å (*red* and *green data points* with $W_r(2796) < 0.3$ Å in the *left panel*). These are by definition included in the covering fraction calculations in the *right panel*, although LRGs with $W_r(2796) < 0.3$ Å are considered non-absorbing galaxies at the $W_0 = 0.3$-Å threshold. The *upper-right panel* displays the numbers of LRGs, passive (in *red*) and [O II]-emitting (in *green*), contributing to the $\langle\kappa\rangle_{\rm MgII}$ calculation in each 40-kpc bin (*horizontal bars* in the *right panel*). Uncertainties in $\langle\kappa\rangle_{\rm MgII}$ are shown in *vertical error bars*, which represent the 68% confidence intervals. Possible contributions due to random background absorbers and gas-rich satellite galaxies are also shown in the *orange dashed line* and the *purple dash-dotted curve*, respectively

galaxies at both low and high redshifts have shown steadily declining absorption strength, both in hydrogen and in heavy ions, with increasing projected distance (e.g., Lanzetta and Bowen 1990; Bowen et al. 1995; Chen et al. 1998, 2001a; Steidel et al. 2010; Liang and Chen 2014). Strong Mg II absorbers with $W_r(2796) > 1$ Å are only found at $d \lesssim 60$ kpc from isolated L_* galaxies (e.g., Chen and Tinker 2008; Chen et al. 2010a). In addition, the extent of halo gas at a fixed absorption equivalent width threshold, $R_{\rm gas}$, is found to scale with galaxy luminosity (e.g., Chen et al. 1998, 2010a, 2001b; Kacprzak et al. 2008; Chen and Tinker 2008). Therefore, the relation between $W_r(2796)$ versus $d/R_{\rm gas}$ exhibits a significantly smaller scatter than the relation between $W_r(2796)$ and d. This luminosity–$R_{\rm gas}$ scaling relation is understood as such that more luminous (and presumably more massive) halos possess more extended halo gas. For the LRGs, however, the scatter in the observed $W_r(2796)$ versus luminosity-normalized d relation is the same as in the observed

$W_r(2796)$ versus d distribution. Figure 3 shows that, aside from a declining $\langle \kappa_{\rm Mg\,II} \rangle$ at $d > 120$ kpc, the cool clouds that give rise to the observed Mg II absorption features do not appear to depend on either the luminosity (or equivalently mass) or projected distance of the LRGs.[1]

The lack of dependence of the observed Mg II absorbing clouds on the LRG properties appears to rule out the possibility of these absorbers being connected to outgoing materials from the LRGs. However, the observed flat mean covering fraction of Mg II absorbers at $d \lesssim 120$ kpc for passive LRGs is also inconsistent with the expectation for accreted materials from the IGM, which are expected to show an increasing covering fraction with decreasing projected distance (e.g., Faucher-Giguère and Kereš 2011; Fumagalli et al. 2011; Shen et al. 2013a; Ford et al. 2014). Such a discrepancy indicates that IGM accretion alone cannot explain the spatial distribution of chemically enriched cool gas in these massive quiescent halos, at least not in the inner 120 kpc.

Contributions from the interstellar and circumgalactic gas of satellite galaxies to the observed Mg II absorbers in LRG halos have also been considered (e.g., Gauthier et al. 2010; Huang et al. 2016). In particular, observations have shown that LRGs reside in overdense environments with neighboring satellites (e.g., Tal et al. 2012). Roughly 20% of satellite galaxies are blue and presumably gas-rich (e.g., Hansen et al. 2009; Prescott et al. 2011). Under the assumption that these blue satellites can retain their gaseous halos, the expected contributions to the observed covering fraction of Mg II gas can be estimated based on the known luminosity scaling relation (e.g., Chen and Tinker 2008). The expectation is shown in the purple curve in the right panel of Fig. 3. It is clear that blue satellites alone cannot account for the observed 15% covering fraction of Mg IIabsorbing gas at $d \sim 100$ kpc in LRG halos.

A promising explanation for the observed Mg II absorbers in LRG halos are cool clumps condensing out of thermally unstable hot halos (e.g., Mo and Miralda-Escudé 1996; Maller and Bullock 2004; Voit et al. 2015). In particular, hot halos are a common feature among the LRG population, because efforts in search of a Sunyaev-Zel'dovich signal (Sunyaev and Zeldovich 1970) around high-mass LRGs ($M_h \gtrsim 8 \times 10^{13}\,{\rm M}_\odot$) have yielded detections in all mass bins studied (Hand et al. 2011). In addition, Mg II absorbers are frequently resolved into multiple components with the number of components being proportional to both the rest-frame absorption equivalent width $W_r(2796)$ (e.g., Petitjean and Bergeron 1990; Churchill et al. 2003) and the minimum line-of-sight velocity spread (e.g., Churchill

[1]Using stacked QSO spectra at the rest frames of all LRGs, it has been shown that the mean absorption equivalent width $\langle W_0 \rangle$ of Mg II gas declines steadily with increasing projected distance in LRG halos (Zhu et al. 2014). The steady decline of $\langle W_0 \rangle$ does not necessarily contradict the relatively flat $W_r(2796)$ versus d distribution for individual Mg II absorbers in Fig. 3, because $\langle W_0 \rangle$ represents a weighted average of absorber strength and gas covering fraction. A direct comparison between $\langle W_0 \rangle$ and Fig. 3 requires knowledge of the frequency distribution function of Mg II absorbers in LRG halos, $f(W_r)$, and the underlying number density ratio between absorbing and non-absorbing LRGs.

et al. 2000). The Mg II absorption equivalent width is therefore driven by the line-of-sight cloud motion rather than by the underlying total gas column density. Recall also that the Mg II absorbers found in the vicinities of passive and [O II]-emitting LRGs display a constant $W_r(2796)$ across the entire projected distance range shown in Fig. 3 with a mean and dispersion of $\langle \log W_r(2796)/\text{Å} \rangle = -0.050 \pm 0.25$.

Interpreting the scatter as due to Poisson noise in the number of individual clumps intercepted along a line of sight leads to an estimate of the mean number of clumps per galactic halo per sightline, $n^{\rm clump}$ (e.g., Lanzetta and Bowen 1990; Chen et al. 2010a). In this simple toy model, each absorber is characterized as $W_r = n^{\rm clump} \times \omega_0$, where ω_0 is the mean absorption equivalent width per component. Following Poisson counting statistics, the intrinsic scatter $\delta(\log W_r) \equiv \delta(W_r)/(W_r \ln 10)$ is related to the mean number of clumps $\langle n^{\rm clump} \rangle$ according to

$$\delta(\log W_r) = \frac{1}{\ln 10} \frac{\sqrt{\langle n^{\rm clump} \rangle + 1}}{\langle n^{\rm clump} \rangle}, \quad (1)$$

which leads to $\langle n^{\rm clump} \rangle \sim 3.8$ for $\langle W_r(2796) \rangle \approx 1$ Å in the LRG halos with a corresponding mean absorption equivalent width per component of $\omega_0 = 0.24$ Å. The inferred $\langle n^{\rm clump} \rangle$ per sightline for LRG halos is more than four times smaller than what was inferred for lower-mass, star-forming galaxies at $d \approx 40$ kpc (Chen et al. 2010a). With a halo size twice as big as L_* galaxies, the significantly lower $\langle n^{\rm clump} \rangle$ therefore suggests that the volume filling factor of Mg II absorbing gas is $\sim$10 times lower in LRG halos than in L_* halos.

4 Kinematics

While obtaining spatially resolved velocity maps is challenging for both stellar and gaseous components of individual high-redshift galaxies, important insights into the physical nature of cool halo clouds can be gained from comparing the distribution of relative velocities between absorbers and their associated galaxies. One important observable of cool halo gas is the line-of-sight velocity dispersion within the host halo. Using a sample of galaxy and absorber pairs, it is possible to constrain the ensemble average of the velocity distribution of absorbing gas relative to their host galaxies.

The distribution of relative velocity between Mg II absorbers and their host galaxies is shown in Fig. 4 for both passive and [O II]-emitting LRGs. For the two LRG subsamples, the relative motion of Mg II absorbing gas is well centered around the systemic redshifts of the galaxies. While the velocity distribution around [O II]-emitting galaxies is well characterized by a single Gaussian of dispersion $\sigma_v = 167\,\text{km s}^{-1}$, a double Gaussian profile is required to characterize the relative motion of Mg II gas around passive LRGs with a narrow component of $\sigma_v^n = 163\,\text{km s}^{-1}$ for the majority of the sample and a broad component of $\sigma_v^b = 415\,\text{km s}^{-1}$.

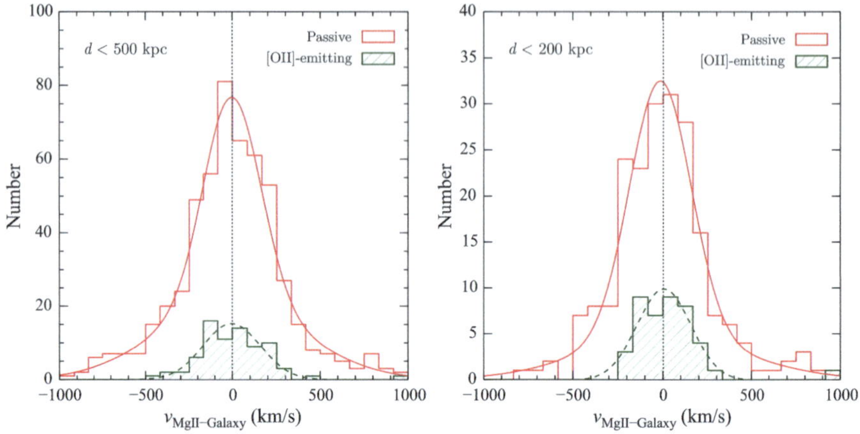

Fig. 4 Velocity distributions of Mg II-absorbing clumps relative to the systemic redshifts of LRGs based on an ensemble of Mg II absorber and LRG pairs (adapted from Huang et al. 2016). The relative velocity distribution of Mg II-absorbing gas relative to passive LRGs is shown in the *red open histogram*, while the velocity distribution of Mg II absorbing gas relative to [O II]-emitting LRGs is shown in the *green hatched histogram*. Including all Mg II absorbers at $d < 500$ kpc of an LRG (*left panel*), the velocity distribution of the gas is well characterized by a single Gaussian distribution centered at $\langle v_{\rm Mg\,II\text{-}Galaxy}\rangle = -5\,{\rm km\,s^{-1}}$ with a dispersion of $\sigma_v = 167\,{\rm km\,s^{-1}}$ (*green, dashed curve*). At the same time, a double Gaussian profile is required to better characterize the velocity distribution of Mg II absorbing gas around passive LRGs (*red solid curve*) with a narrow component centered at $\langle v_{\rm Mg\,II\text{-}Galaxy}\rangle = -3\,{\rm km\,s^{-1}}$ and $\sigma_v^n = 163\,{\rm km\,s^{-1}}$ and a broad component centered at $\langle v_{\rm Mg\,II\text{-}Galaxy}\rangle = -17\,{\rm km\,s^{-1}}$ and $\sigma_v^b = 415\,{\rm km\,s^{-1}}$. Roughly 12% of passive LRGs have a Mg II absorber found at velocities exceeding $500\,{\rm km\,s^{-1}}$ from the systemic redshifts of the galaxies. Consistent velocity distributions are found for Mg II absorbers detected at $d < 200$ kpc from passive and [O II]-emitting galaxies (*right panel*), showing little variation in the overall gas motion between inner and outer halos

Roughly 12% of passive LRGs are found in the broad component with velocities exceeding $500\,{\rm km\,s^{-1}}$. These properties apply to both inner ($d < 200$ kpc) and outer ($200 < d < 500$ kpc) halos (see Huang et al. 2016, for a detailed discussion).

The implications of Fig. 4 are twofold. First of all, the similarity in the relative velocity distribution of Mg II absorbing gas around passive and [O II]-emitting LRGs reaffirms the findings from the previous section (Sect. 3) that the observed Mg II absorbers are unlikely to originate in outgoing materials from the LRGs. Secondly, the mean halo mass of LRGs is $M_h \approx 3 \times 10^{13}\,M_\odot$ (e.g., Mandelbaum et al. 2008; Gauthier et al. 2009). The expected line-of-sight velocity dispersion for virialized gas in these massive halos is $\sigma_h \approx 265\,{\rm km\,s^{-1}}$. The observed velocity dispersion of Mg II absorbing gas in ≈90% of the LRG sample (the narrow component in Fig. 4) is merely 60% of what is expected from virial motion (see also Zhu et al. 2014), namely

$$\sigma_v \approx 0.6\,\sigma_h. \tag{2}$$

Such suppression in gas motion not only indicates that the gas is gravitationally bound in the LRG halo but also that the kinetic energy of the gas is being dissipated. Similar suppression in the line-of-sight velocity dispersion is also found for Lyα and O VI absorbing gas around COS-Halos red galaxies with $M_{\rm star} \gtrsim 10^{10.7}\,{\rm M}_\odot$ (Tumlinson et al. 2011, 2013).

In the presence of a hot halo (e.g., Hand et al. 2011), cool clumps should experience a ram-pressure drag force and slow down. For ram-pressure to drive decelerations in the clump motion, these clumps cannot be too massive. The observed suppression in the velocity dispersion of the absorbing gas therefore places a mass limit on these clumps. Following the formalism of Maller and Bullock (2004), a maximum clump mass can be estimated for the LRG halos according to $m_{\rm cl} \approx 5.1 \times 10^4\,M_\odot\,T_6^{-3/8}\,(\Lambda_z\,t_8)^{1/2}$, where T_6 is the halo gas temperature in units of 10^6 K, Λ_z is the cooling parameter that varies with the gas metallicity Z_g, and $t_8 = t_f(c_h)/8$ Gyr is the halo formation time that depends on the halo concentration c_h. For a halo temperature of $T \sim 6 \times 10^6$ K for the LRGs and $t_8 \sim 8.9$ Gyr using $c_h \sim 10$ from the halo mass–concentration relation (e.g., Mandelbaum et al. 2008), the estimated mass of individual Mg II absorbing clumps is $m_{\rm cl} \approx 5 \times 10^4\,M_\odot$ for solar metallicity and lower for lower-metallicity (see Huang et al. 2016, for more details).

Beyond an ensemble average over a large sample of LRG-Mg II pairs, comparisons of gas and satellite kinematics in individual halos are possible when deep galaxy survey data are available (e.g., Steidel et al. 1997; Whiting et al. 2006; Nestor et al. 2007; Chen and Mulchaey 2009). Previous studies that combined absorber and galaxy data have revealed a number of metal-line absorbers associated with overdensities of galaxies over a wide range of mass scales (e.g., Mulchaey and Chen 2009; Kacprzak et al. 2010; Nestor et al. 2011; Gauthier 2013; Johnson et al. 2013). Comparing the observed line-of-sight velocity dispersion of the absorbing gas with the velocity distribution of galaxy group members offers additional clues for the origin of chemically enriched gas in overdense environments. In particular, imaging and spectroscopic surveys of galaxies in the vicinities of ultra-strong Mg II absorbers, $W_r(2796) \gtrsim 3$ Å, have revealed multiple galaxies at small projected distances and velocity separations from these absorbers (e.g., Nestor et al. 2007, 2011; Gauthier 2013). With the expected velocity spread of $\sim$100 km s^{-1} per Angstrom for Mg II absorbers, the large $W_r(2796)$ implies a line-of-sight velocity spread of the components $\Delta\,v > 300$ km s^{-1}. Attributing the observed gas motion to the underlying gravitational potential of the host dark matter halo would lead to a halo mass exceeding $M_h \approx 3 \times 10^{12}\,{\rm M}_\odot$.

Figure 5 showcases an example of an ultra-strong Mg II absorber of $W_r(2796) = 4.2$ Å identified at $z = 0.5624$ along the sightline toward a background QSO at $z_{\rm QSO} = 1.30$ (Gauthier 2013). Four galaxies are spectroscopically identified at close projected distances and velocity separations from the absorber, including a 3.5-L_* LRG at $d = 246$ kpc and $\Delta\,v = -385$ km s^{-1}, a 1.8-L_* passive galaxy (A) at $d = 55$ kpc and $\Delta\,v \approx 0$ km s^{-1}, a 0.3-L_* galaxy (B) at $d = 38$ kpc and $\Delta\,v \approx -100$ km s^{-1}, and a 0.5-L_* galaxy (G) at $d = 209$ kpc and $\Delta\,v \approx 0$ km s^{-1}.

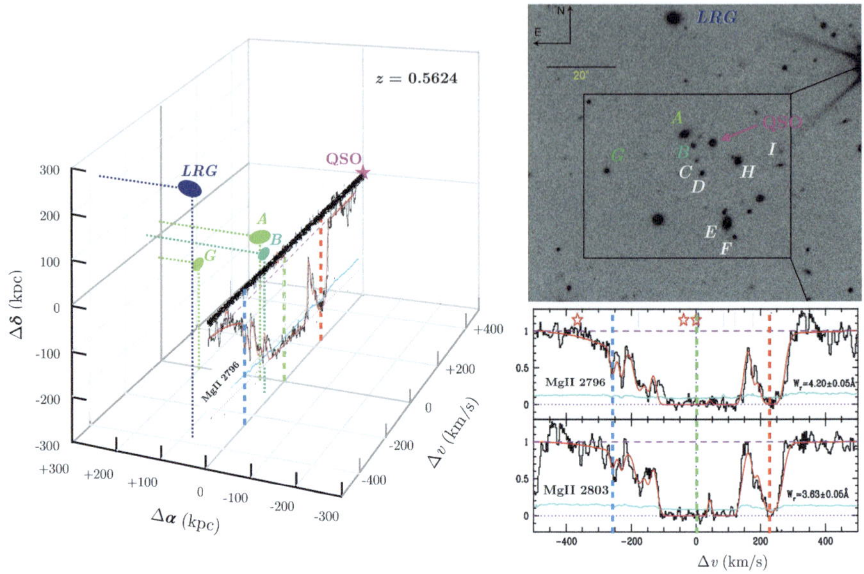

Fig. 5 Comparison of galaxy and Mg II absorption gas kinematics observed in a group of galaxies at $z = 0.5624$ (Gauthier 2013). The imaging panel and absorption-line profiles on the *right* are adapted from Gauthier (2013). An ultra-strong Mg II absorber of $W_r(2796) = 4.2$ Å (*lower-right panel*) is identified at $z = 0.5624$ along the sightline toward a background QSO at $z_{\rm QSO} = 1.30$. Follow-up spectroscopy of galaxies in this field (*upper-right panel*) revealed an LRG at $d = 246$ kpc and $\Delta v = -385\,{\rm km\,s^{-1}}$, along with three additional galaxies, *A*, *B*, and *G* at closer distances and relative velocities from the Mg II absorber. Galaxies $C - F$, *H*, and *I* are found to be at lower redshifts from the galaxy group at $z = 0.56$. The *left panel* displays the spatial and relative velocity distribution of the LRG and three neighboring galaxies relative to the Mg II absorber. The luminosities of the LRG and galaxies *A*, *B*, and *G* are 3.5, 1.8, 0.3, and $0.5\,L_*$, respectively. The galaxies are *color-coded* according to the relative velocities from the Mg II absorber at $\Delta v = 0\,{\rm km\,s^{-1}}$. Galaxies *A* and *G*, marked in *pale green*, occur at $\Delta v \approx 0\,{\rm km\,s^{-1}}$, while the LRG, marked in *blue*, displays the largest blueshifted relative velocity with $\Delta v = -385\,{\rm km\,s^{-1}}$. The most blueshifted and redshifted Mg II components at $\Delta v \approx -250$ and $230\,{\rm km\,s^{-1}}$ are also marked in *pale blue* and *red*, respectively, to guide the visual comparison

Only galaxies *B* and *G* exhibit traces of on-going star formation with an inferred star formation rate of $\approx 3\,{\rm M_\odot\,yr^{-1}}$. The total stellar masses of the galaxies range from $M_{\rm star} \gtrsim 4 \times 10^{10}\,{\rm M_\odot}$ for galaxy *A* and the LRG to $M_{\rm star} \approx 3 \times 10^9\,{\rm M_\odot}$ for galaxies *B* and *G*. The close proximity and the observed massive stellar content qualify these galaxies as parts of a galaxy group with a light-weighted center at $d_{\rm group} \approx 156$ kpc and $\Delta v_{\rm group} = -200\,{\rm km\,s^{-1}}$. The total dark matter halo mass of the group estimated based on the stellar mass of the LRG is $M_h \approx 10^{13}\,{\rm M_\odot}$ (e.g., Behroozi et al. 2010).

The star formation rate intensity observed in galaxies *B* and *G* is small (Gauthier 2013), which indicates that super galactic winds are unlikely to form and drive outflows in these galaxies (e.g., Heckman 2002; Heckman et al. 2015). On the other hand, the large line-of-sight velocity spread of the absorbing gas, from $\Delta v \approx -250$

to $\Delta v \approx 230\,\mathrm{km\,s^{-1}}$ (lower-right panel), is comparable to the projected velocity dispersion expected from virial motion, $\sigma_{\rm vir} \approx 190\,\mathrm{km\,s^{-1}}$, in a halo of $M_h \approx 10^{13}\,\mathrm{M_\odot}$ at $z = 0.5$. This agreement suggests a physical connection between group dynamics and ultra-strong Mg II absorbers. A comparable velocity dispersion between absorbing gas and group galaxies has also been found for a Mg II absorber of $W_r(2796) = 1.8\,\text{Å}$ at $z = 0.3$, for which a group of five sub-L_* galaxies are found at $d < 250\,\mathrm{kpc}$ and $|\Delta v| < 250\ \mathrm{km\,s^{-1}}$ (Kacprzak et al. 2010). The dynamical mass of the host dark matter halo, based on the velocity dispersion of the group members $\sigma_v = 115\,\mathrm{km\,s^{-1}}$, is $M_h \approx 3 \times 10^{12}\,\mathrm{M_\odot}$ for the galaxy group at $z = 0.3$. In both cases, where a strong Mg II is found in a galaxy group, dissipation is not observed in the relative motion between the absorbing gas and associated galaxies. The large velocity dispersion, similar to the broad velocity component seen in Fig. 4, can be understood if the ram-pressure drag force is subdominant in comparison to the gravitational force on the gas clumps. Stripped gas due to tidal interactions between neighboring galaxies is a likely explanation for the dynamically complex Mg II absorbers found in group or cluster environments (e.g., Chynoweth et al. 2008; Marasco et al. 2016), which also results in H I depleted galaxies in these massive cluster and group halos (e.g., Verdes-Montenegro et al. 2001; Chung et al. 2009).

Galaxy surveys have been carried out for two more ultra-strong Mg II absorbers at $z \approx 0.7$. Both are found in a group environment with $M_h = 10^{12\text{–}13}\,\mathrm{M_\odot}$ (Nestor et al. 2011). However, high-resolution spectra are not available for these two absorbers and consequently comparisons of gas and galaxy kinematics are not possible.

5 Chemical Enrichment

Figure 3 clearly indicates that despite being in a quiescent state, halo gas around massive LRGs has been enriched with heavy elements out to the virial radius. The chemical enrichment history in these massive halos, which can be broadly characterized by (1) gas metallicity and (2) relative abundance pattern, offers independent clues for the origin of the observed cool halo gas. Specifically, gas metallicity quantifies the overall heavy element production level, and nearby galaxies are observed to follow a mass–metallicity relation with more massive galaxies displaying on average higher metallicities both in stars (Gallazzi et al. 2005; Kirby et al. 2013) and in the ISM (Tremonti et al. 2004; Lee et al. 2006). Therefore, a natural expectation is that gas ejected from massive galaxies should be highly metal-enriched, while gas ejected from low-mass satellites should exhibit a lower-metallicity. Gas accreted from the IGM should contain still lower metallicities (e.g., Lehner et al. 2013; Kacprzak et al. 2014). However, uncertainties arise as a result of poorly understood processes involving mixing and transport of heavy elements (e.g., Tumlinson 2006). If mixing is effective as metal-enriched materials propagate into the low-density, and presumably lower-metallicity, halo gas, then the overall metallicity is expected to be reduced. If mixing is ineffective, then large variations in chemical abundances are expected in single objects (e.g., Scale and Elmegreen 2004). Consequently, metallicity provides only a weak indicator for possible origins of chemically enriched cool gas in LRG halos.

On the other hand, the relative abundance pattern presents a fossil record for the primary source of heavy element production. Specifically, α-elements (such as O, Mg, Si, and S) are primarily produced in massive stars and core-collapse supernovae (SNe), while a significant fraction of heavier elements such as Fe and Ni are produced in Type Ia SNe over longer timescales (e.g., Tsujimoto et al. 1995; Nomoto et al. 2006). The observed [α/Fe] relative abundance ratio therefore constrains the relative contributions of core-collapse SNe and SNe Ia to the chemical enrichment history (Tinsley 1979; Gilmore and Wyse 1991; Tolstoy et al. 2003). Both distant galaxies (Wolfe et al. 2005) and nearby evolved stars (Matteucci 2014) display an α-element enhanced abundance pattern, indicating a chemical enrichment process driven by core-collapse SNe in the early universe. At the same time, the ISM of nearby elliptical galaxies (e.g., Mathews and Brighenti 2003; Humphrey and Buote 2006) and intracluster medium (e.g., de Plaa et al. 2007) display super-solar Fe/α abundance ratios that reveal increased contributions from SNe Ia in the circumgalactic gas of massive, evolved galaxies. Relative abundance measurements between Fe and α-elements therefore provide a powerful constraint for the origin of chemically enriched gas around galaxies.

At $z \gtrsim 0.3$, a series of Fe II absorption transitions are observable on the ground, which enable an accurate measurement of Fe II column density N(Fe II) on a component by component basis (e.g., Rigby et al. 2002; Narayanan et al. 2008; Zahedy et al. 2016). Similarly, the Mg II doublet transitions enable accurate Mg II column density measurements N(Mg II) for all but a few saturated components. In principle, the total elemental abundances can be determined from the observed N(Fe II) and N(Mg II), if the ionization fractions of these ions are known. Gas metallicity can also be derived if the total hydrogen content can be inferred from measurements of the neutral hydrogen column density, N(H I). In practice, observations of N(H I) at low redshifts require UV spectroscopy carried out in space, and have only been done for a relatively small sample of galaxies (e.g., Chen and Prochaska 2000; Werk et al. 2014). Furthermore, uncertainties in the derived ionization fractions of individual ions can be large and difficult to estimate under simple ionization models.

Observations of the relative abundance ratio between Mg^+ and Fe^+ ions are particularly useful, because Mg^+ and Fe^+ share similar ionization potentials (15 and 16.2 eV, respectively) and are the dominant ionization states of the respective elements in both neutral and cool photo-ionized gas. While the elemental abundances of iron and magnesium may be uncertain, the relative abundance ratio (Fe/Mg) can be inferred with high confidence from the observed $N(\mathrm{Fe\,II})/N(\mathrm{Mg\,II}$ following $\log\,(\mathrm{Fe/Mg}) = \log\,N(\mathrm{Fe\,II})/N(\mathrm{Mg\,II}) - \log\,(f_{\mathrm{Fe}^+}/f_{\mathrm{Mg}^+})$, where f_{Fe^+} is the fraction of Fe in the singly ionized state and f_{Mg^+} is the fraction of Mg in the singly ionized state. The ratio of ionization fractions, $f_{\mathrm{Fe}^+}/f_{\mathrm{Mg}^+}$, estimated based on a suite of photo-ionization calculations is shown in the left panel of Fig. 6.

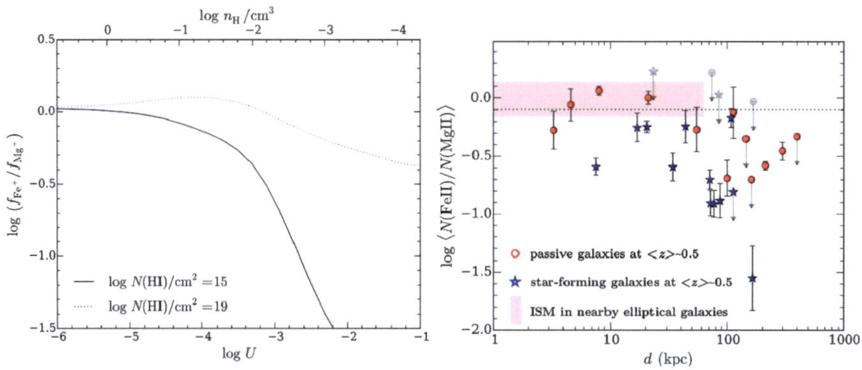

Fig. 6 Enhanced Fe abundances in halos around massive quiescent galaxies in comparison to those around star-forming galaxies (Zahedy et al. 2016, 2017). *Left*: Expected ionization fractions of Mg^+ and Fe^+ ions from Cloudy photo-ionization calculations (Ferland et al. 2013) for photo-ionized gas of temperature $T = 10^4$ K and 0.1 solar metallicity in optically thin (with neutral hydrogen column density $N(\text{H I}) = 10^{15}\,\text{cm}^{-2}$) and optically thick (with $N(\text{H I}) = 10^{19}\,\text{cm}^{-2}$) regimes (Zahedy et al. 2016). Changing metallicity does not alter the results. An updated version of the Haardt and Madau (2001) ionizing radiation field at $z = 0.5$ is adopted for computing the ionization parameter U at different gas density n_H displayed at the top of the panel. These calculations demonstrate that, independent of gas metallicity, the observed $N(\text{Fe II})/N(\text{Mg II})$ in the *right panel* represents a lower limit to the underlying (Fe/Mg) relative abundances. *Right*: Observed mean Fe II to Mg II column density ratio versus projected distance based on absorption spectroscopy of halo gas around quiescent (*red circles*) and star-forming (*blue stars*) galaxies at $\langle z \rangle \approx 0.5$. For comparison, ISM measurements based on X-ray observations of 19 nearby elliptical galaxies (Humphrey and Buote 2006) are also included in the *left panel*, and the *dotted line* shows the solar value. The fractional contribution of Type Ia supernovae to the chemical enrichment history in the inner 60 kpc of massive quiescent halos is found to be similar to what is observed in the solar neighborhood

The photo-ionization models are computed using the Cloudy software (Ferland et al. 2013). These models assume a photo-ionized gas of temperature $T = 10^4$ K and 0.1 solar metallicity in optically thin (with neutral hydrogen column density $N(\text{H I}) = 10^{15}\,\text{cm}^{-2}$) and optically thick (with $N(\text{H I}) = 10^{19}\,\text{cm}^{-2}$) regimes (Zahedy et al. 2016). An updated version of the Haardt and Madau (2001) ionizing radiation field at $z = 0.5$ is adopted for computing the ionization parameter U (defined as the number of ionizing photons per atom) at different gas density n_H displayed at the top of the panel. The ionization fraction of Fe^+ remains roughly equal to that of Mg^+ in the optically thick regime and lower in optically thin gas for the full range of U explored. Consequently, $\log(f_{Fe^+}/f_{Mg^+}) \lesssim 0$ and the observed $N(\text{Fe II})/N(\text{Mg II})$ marks a lower limit to the underlying Fe to α-element abundance ratio:

$$\log\left(\frac{\text{Fe}}{\text{Mg}}\right) > \log\frac{N(\text{Fe II})}{N(\text{Mg II})}. \tag{3}$$

Experimenting with different gas metallicity does not change the predicted ionization fraction, and accounting for differential dust depletion of iron and magnesium would further increase the inferred [Fe/Mg] (Savage and Sembach 1996). This exercise demonstrates that useful empirical constraints for the relative Fe and Mg abundances can be obtained even in the absence of accurate measurements of $N(\mathrm{H\,I})$ and gas metallicity.

A recent study utilizing multiply lensed QSOs for probing gas at small projected distances, $d \lesssim 20\,\mathrm{kpc}$ or $\approx$1–2 effective radii r_e, from the lensing galaxies has uncovered important new clues for the origin of chemically enriched cool gas in massive halos (Zahedy et al. 2016). These lensing galaxies at $z = 0.4$–0.7 share similar properties concerning both the quiescent state and halo mass scales ($M_h \gtrsim 10^{13}\,\mathrm{M}_\odot$), but display distinct absorption-line profiles between different lensing galaxies and between different sightlines of individual lenses. The apparent large scatter in the observed absorption profiles is consistent with the large scatter displayed in Fig. 3 and discussed in Sect. 3. Most interestingly, all Mg II absorbers detected near these lensing galaxies are strong and resolved into 8–15 individual components over a line-of-sight velocity range of $\Delta\,v \approx 300$–$600\,\mathrm{km\,s^{-1}}$. The Mg II absorption is accompanied with even stronger Fe II absorption with matching kinematic profiles. Comparing the relative absorption strengths between individual components also yields uniformly large $N(\mathrm{Fe\,II})/N(\mathrm{Mg\,II})$ ratios over the full range of velocity spread, $\Delta\,v$, with a median of $\langle\,\log N(\mathrm{Fe\,II})/N(\mathrm{Mg\,II})\,\rangle \approx 0$ and a scatter of < 0.1 dex.

Following Eq. (3), the observed $\langle\,\log N(\mathrm{Fe\,II})/N(\mathrm{Mg\,II})\,\rangle \approx 0$ naturally leads to a super-solar Fe/Mg abundance ratio near these massive lensing galaxies. This is in stark contrast to an α-element enhanced chemical composition found in young, star-forming galaxies (e.g., Dessauges-Zavadsky et al. 2004; Crighton et al. 2013; Fox et al. 2014, and see Zahedy et al. 2016 for a detailed comparison), and clearly indicates different origins between gas associated with star-forming galaxies and with massive quiescent galaxies.

Previous studies have shown that the spatial distribution of SNe Ia in nearby early type galaxies follows the stellar light out to $r \sim 4\,r_e$ (e.g., Förster and Schawinski 2008) and that the ISM of these massive quiescent galaxies exhibits an iron-enhanced abundance pattern (e.g., Mathews and Brighenti 2003; Humphrey and Buote 2006). Therefore, it seems likely that the observed Fe-rich gas at $d \approx r_e$ from the lenses originates in the ISM of these massive galaxies, where SNe Ia play a dominant role in driving the observed large velocity width and Fe-rich abundance pattern (Zahedy et al. 2016). If the ISM of massive galaxies at intermediate redshift is locally enriched by SNe Ia, then the observed Fe/Mg is expected to decline with increasing projected distance.

To test this hypothesis, a sample of 13 massive quiescent galaxies (including two lensing galaxies) and 14 star-forming galaxies at intermediate redshifts ($\langle\,z\,\rangle \approx 0.5$) has been assembled (Zahedy et al. 2017). These galaxies were selected to have absorption spectra of background QSOs at $d \approx 10$–$400\,\mathrm{kpc}$ available for constraining the radial profile of Fe/Mg. The right panel of Fig. 6 displays the mean $\langle\,N(\mathrm{Fe\,II})/N(\mathrm{MgII})\rangle$ averaged over all individual components per halo versus

d for both star-forming galaxies (blue stars) and the quiescent galaxy population (red circles). Although the sample is still small, the distinction between quiescent and star-forming halos is already apparent in Fig. 6. While the mean Fe/Mg ratio is consistent with an α-element enhanced pattern in the outer halo at $d > 100$ kpc for both quiescent and star-forming galaxies, halo gas at $d < 100$ kpc from quiescent galaxies exhibits an elevated iron abundance in comparison to star-forming galaxies.

The observed iron enrichment level in the inner halo of $z \approx 0.5$ quiescent galaxies is consistent with the solar value, similar to what is observed in the ISM of nearby elliptical galaxies (e.g., Mathews and Brighenti 2003; Humphrey and Buote 2006) and in the intracluster medium (e.g., de Plaa et al. 2007). The radial decline of the Fe/Mg relative abundances supports the hypothesis that the gas is locally enriched by SNe Ia. The minimum fractional contribution of SNe Ia to the chemical enrichment in the inner halos of massive quiescent galaxies is found to be $f_{\rm Ia} \approx 15$–20% based on the expected yields for Type Ia and core-collapse SNe (Iwamoto et al. 1999) and the observed Fe/Mg ratio at $d < 100$ kpc.

6 Quasar Host Halos

In comparison to LRGs which are massive and quiescent, quasars powered by supermassive black holes also reside in massive halos of $M_h \gtrsim 3 \times 10^{12}\,{\rm M}_\odot$ (Porciani et al. 2004; White et al. 2012; Shen et al. 2013b) but in an active phase that lasts $\approx$10–100 Myr (Martini 2004). While the physical processes that drive the fueling and feedback of the central supermassive black holes are not well understood (e.g., Hopkins and Hernquist 2009; Heckman and Best 2014), quasar feedback is a critical ingredient in shutting down star formation in high-mass halos and producing red-and-dead elliptical galaxies in all theoretical frameworks (e.g., Benson et al. 2003; Croton et al. 2006). Indeed, most nearby early type galaxies are found to harbor a supermassive black hole at the center (Kormendy and Richstone 1995; Ho 2008). Therefore, studying quasar hosts is an integral part of the effort to understand the growth of massive galaxies and observations of quasar host halos provide important clues for how the halo gas properties are shaped while the galaxies undergo an active quasar phase.

It has long been recognized that the intense radiation field from luminous quasars is sufficient to fully ionize their surrounding medium and to keep the IGM ionized over most of the cosmic history of the universe. Near the quasar redshifts, the incidence of Lyα absorption lines in the quasar spectra is reduced due to an enhanced radiation field from the quasars in the proximity zone (e.g., Bajtlik et al. 1988). The line-of-sight quasar proximity effect, which is also observed in different ionic transitions such as Mg II and C IV, offers a unique means of constraining both the metagalactic ionizing radiation and the extent of the ionizing bubble local to the quasar (Dall'Aglio et al. 2008; Wild et al. 2008). However, a reduced incidence of absorption features is not observed along the *transverse* direction from quasars (e.g., Fernandez-Soto et al. 1995; Schirber et al. 2004). In addition, there also appears

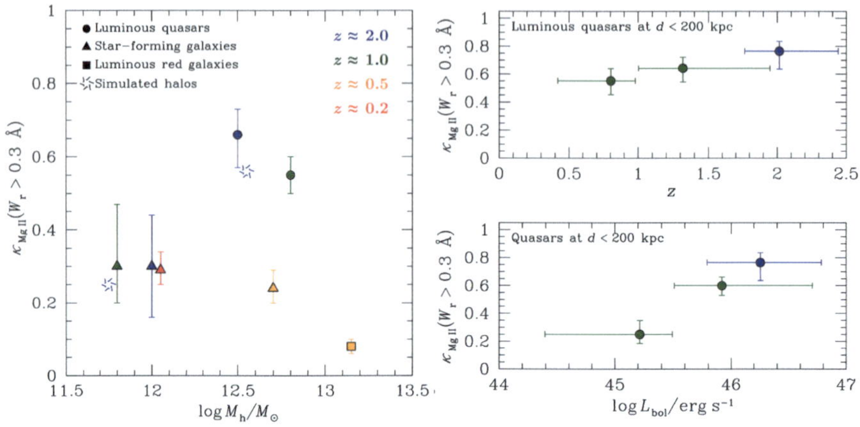

Fig. 7 Elevated cool gas covering fraction in quasar host halos. *Left*: Mean gas covering fraction κ as a function of halo mass for galaxies and quasar hosts at different redshifts. The measurements for galaxies and quasars at $z < 2$ are based on surveys of Mg II absorbers with W_r(Mg II) $>$ 0.3 Å and at $d < R_{\rm vir}$, where $R_{\rm vir}$ represents the virial radius of the host dark matter halos. Measurements for $z \approx 2$ star-forming galaxies (Rudie et al. 2012) and quasars (Prochaska et al. 2014) are inferred from the observed incidence of optically thick H I absorbers and C II absorption observations, respectively. *Closed triangles* represent star-forming galaxies (Chen et al. 2010a; Lovegrove and Simcoe 2011; Rudie et al. 2012; Werk et al. 2014), *squares* represent quiescent LRGs (Huang et al. 2016), and *circles* represent quasar hosts (Prochaska et al. 2014; Johnson et al. 2015b). *Star symbols* are simulation predictions for $z \approx 2$ objects (Faucher-Giguère et al. 2016). Different colors indicate different redshift ranges of each dataset with *red*, *orange*, *green*, and *blue* corresponding to $z \approx 0.2$, 0.5, 1.0, and 2.0, respectively. The mean gas covering fraction in quasar host halos is observed to be more than doubled from what is seen in star-forming galaxies at both low and high redshifts. Numerical simulations have attributed the elevated cool gas content in quasar hosts to either AGN feedback (Rahmati et al. 2015) or star formation feedback from neighboring satellite galaxies (Faucher-Giguère et al. 2016). *Right*: Mean gas covering fraction at $d < 200$ kpc for quasar hosts at different redshifts (*top*) and with different bolometric luminosity (*bottom*). Colors denote different redshift ranges as defined in the *left panel*. While the mean gas covering fraction in quasar hosts is found to evolve only weakly with redshift, a steep correlation is observed between cool gas covering fraction and quasar luminosity. The elevated cool halo gas covering fraction observed in quasar hosts appears to be driven primarily by the most luminous quasar population (Figure credit: Sean Johnson)

to be an excess of cool clouds in quasar halos (Bowen et al. 2006; Hennawi and Prochaska 2007; Farina et al. 2013, 2014; Prochaska et al. 2014).

The left panel of Fig. 7 shows the mean covering fraction of cool halo gas κ as a function of halo mass for galaxies (closed triangles and squares) and quasars (closed circles) at different redshifts. For comparison, expectations from numerical simulations are also included (open star symbols). Observations of halos around galaxies and quasars at $z < 2$ are based on surveys of Mg II absorbers with W_r(Mg II) > 0.3 Å at projected distances less than the virial radius $R_{\rm vir}$ of the host dark matter halos. Observations of $z \approx 2$ star-forming galaxies (Rudie et al. 2012) and quasars (Prochaska et al. 2014) are inferred from the observed incidence of optically thick H I absorbers and C II absorption features, respectively. While halos around normal galaxies display a roughly constant cool gas covering fraction of $\kappa \approx 30\%$ for the star-forming population at all redshifts studied (e.g., Chen et al. 2010a; Lovegrove and Simcoe 2011; Rudie et al. 2012) and a suppressed cool gas content with $\kappa \lesssim 10\%$ for the quiescent population at $z \approx 0.5$ (e.g., Gauthier and Chen 2011; Huang et al. 2016), there appears to be a surge of cool gas with $\kappa \approx 60\%$ in active quasar halos (Prochaska et al. 2014; Farina et al. 2013, 2014; Johnson et al. 2015a) at both low and high redshifts.

While state-of-the-art numerical simulations can reproduce the observed incidence of cool halo gas around star-forming galaxies and quasars at $z \approx 2$, the underlying physical mechanisms are fundamentally different. The elevated cool gas content in quasar hosts are attributed to either AGN feedback (Rahmati et al. 2015) or star formation feedback from neighboring satellite galaxies (Faucher-Giguère et al. 2016).

The observed line-of-sight proximity effect, together with an absence of transverse proximity effect, indicates that the quasar radiation field is not isotropic and that cool clouds outside of the radiation zone can still survive in the host halos. Understanding whether and how the incidence of cool gas varies with quasar properties provides further insights into the origin of the observed excess of cool halo gas around quasars. Using a sample of 195 projected quasar pairs separated by less than 300 kpc in projected distance, it has been shown that the mean gas covering fraction (measured based on observations of Mg II absorption transitions) correlates strongly with the bolometric luminosity $L_{\rm bol}$ of the foreground quasar (Johnson et al. 2015a). Specifically, quasars that are ten times more luminous display on average three to four times higher cool gas covering fraction than low-luminosity quasars at $d < 200$ kpc in the halos (lower-right panel of Fig. 7). The elevated cool halo gas covering fraction observed in quasar hosts appears to be driven primarily by the most luminous quasars in the sample.

The observed strong correlation between κ and $L_{\rm bol}$ is unlikely to be driven by an underlying mass dependence (c.f. Chen et al. 2010b), because the clustering amplitude of quasars remains roughly constant in different luminosity intervals (Shen et al. 2013b; Eftekharzadeh et al. 2015). The observed $L_{\rm bol}$ dependence in cool halo gas covering fraction therefore has profound implications for the connection between halo gas on the 100 kpc scale and quasar activities in the central parsec scale.

Possible explanations for the observed excess of cool gas along transverse sightlines include: (1) overdensity of galaxies in the quasar environment which is expected from the large clustering amplitude observed for luminous quasars, (2) outflows from the quasar/nuclear starburst, and (3) debris from galaxy interactions or mergers that trigger the luminous quasar phase. Of these scenarios, quasar outflows offer the most promising explanation but caveats remain. Significant contributions from correlated galaxies sharing the same large-scale overdensity peak can be ruled out based on the diminishing κ observed at $d \gtrsim 200$ kpc from quasars (Johnson et al. 2015a). Merger remnants are unlikely to explain the observed cool gas at $d \sim 100$ kpc, given that the quasar lifetime is shorter than the dynamical time.

Spatially extended outflows have been detected in [O III] emission out to 10–30 kpc around luminous quasars at $z \sim 0.5$ with outflow velocities as high as $v_{\rm out} \approx 1000\,{\rm km\,s^{-1}}$ (e.g., Greene et al. 2012; Liu et al. 2013, 2014). The observed spherical morphologies in these [O III] emitting nebulae suggest that outflows in these quasar hosts are not well-collimated. It is therefore possible that the absorbers detected at $d \sim 100$ kpc along transverse sightlines in quasar halos originate in these extended outflows with densities too low to be detected in emission.

An outflow origin also provides a natural explanation for the observed strong correlation between κ and $L_{\rm bol}$ with more luminous quasars driving higher mass outflow rates (e.g., Carniani et al. 2015). This is at least qualitatively consistent with the extreme kinematics displayed in some ($\approx$10%) quasar absorbers that spread over a velocity interval of $|\Delta\, v| \gtrsim 1000\,{\rm km\,s^{-1}}$ (e.g., Johnson et al. 2015a). However, free-expanding outflows, traveling at $v_{\rm out} = 1000\,{\rm km\,s^{-1}}$ at 15 kpc from the quasar, are expected to reach 100 kpc in $\approx$100 Myr and longer if the outflows decelerate due to the gravitational potential of the host halo. With a typical quasar lifetime of 10–100 Myr (e.g., Martini 2004), the required outflow speed would need to exceed $1000\,{\rm km\,s^{-1}}$ in order for outflows near the peak of an active quasar phase to explain the observed excess of cool gas at $d \sim 100$ kpc. With the energetics associated with fast-moving outflows, the gas is expected to be heated and highly ionized. If the cool clumps form within the hot outflows due to efficient cooling (e.g., Costa et al. 2015), then a strong correlation is expected between the incidence of cool gas and the presence of highly ionized gas as probed by either [O III] emission (e.g., Greene et al. 2012) or O VI/C IV absorption (e.g., Grimes et al. 2009). Observations of highly ionized gas associated with the low-ionization gas detected in quasar host halos will provide new insights into the origin of the κ vs. $L_{\rm bol}$ correlation.

Alternatively, the cool gas could originate in outflows from star-forming satellites as suggested by a recent simulation study (Faucher-Giguère et al. 2016). In this scenario, the distance required for the outflows to travel within the quasar lifetime would be $\sim$10 kpc, rather than $\sim$100 kpc, and the required outflow energetics would be less extreme. However, the implication would be a more quiescent satellite environment around low-luminosity quasars in order to explain a reduced κ. Comparing the star formation histories of satellites around luminous and low-luminosity quasars will provide a necessary test for this scenario.

7 Summary and Future Prospects

Significant progress has been made over the last decade in characterizing the cool circumgalactic gas in massive halos of $M_h > 10^{12}\,{\rm M}_\odot$ at $z \approx 0.2$–2 using absorption spectroscopy. This progress is facilitated by the unprecedentedly large galaxy and quasar samples available in the SDSS spectroscopic archive. Both massive galaxies and luminous quasars are rare. As a result, finding a background quasar in close projected distances for absorption-line studies of these rare objects requires a large survey volume. The large galaxy and quasar spectroscopic archive help the assembly of statistically significant samples of close quasar and quiescent galaxy pairs and projected quasar pairs. These pair samples have enabled systematic studies of low-density gas beyond the nearby universe. Key findings from various studies can be summarized as follows:

1. Chemically enriched cool gas of $T \sim 10^4$ K is present in massive quiescent halos at $z \sim 0.5$, with a declining annular average of covering fraction from $\langle \kappa \rangle \gtrsim 15\%$ at $d < 120$ kpc to $\langle \kappa \rangle \approx 5\%$ out to the virial radius $R_{\rm vir}$. The improved statistics help rule out definitively an absence of cool gas in massive quiescent halos at $z \sim 0.5$, extending observations of cold gas in local early type galaxies (e.g., Young et al. 2014) to those at intermediate redshifts.
2. Strong Mg II absorbers of $W_r(2796) > 1$ Å produced by photo-ionized cool gas are *not* uncommon throughout quiescent halos from $d < 100$ kpc to the virial radius $d \approx R_{\rm vir}$ and the observed Mg II absorbing strength in these halos does not depend on either galaxy luminosity or mass. The lack of correlation between $W_r(2796)$ and galaxy properties in quiescent halos suggests that the observed cool gas is likely to originate in infalling materials from the IGM, rather than outflowing gas from these early type galaxies.
3. The velocity dispersion of Mg II absorbing gas around the majority ($\approx$90%) of massive quiescent galaxies is suppressed, at $\approx$60% of what is expected from the virial motion. Dissipation is expected if these Mg II absorbers originate in cool clumps condensed out of the hot halo through thermal instabilities and the clumps decelerate due to ram-pressure while moving through the hot halo. In this simple cloud model, the volume filling factor of the clumps is small in these massive halos with a mean number of $n_{\rm clump} \sim 4$ per sightline in order to explain the large scatter found in the $W_r(2796)$ versus d distribution, and a mean clump mass of $m_{\rm cl} \approx 5 \times 10^4\,{\rm M}_\odot$ in order to explain the suppression in velocity distribution.
4. While gas metallicity alone is insufficient for distinguishing between infalling and outflowing gas due to an unknown degree of chemical mixing in the CGM, the observed chemical composition of the gas offers important clues for the chemical enrichment history. The chemical composition of cool halo gas at $d \lesssim 100$ kpc from massive quiescent galaxies displays an elevated iron abundance level that differs from an α-element enhancement typically found in star-forming galaxies and in the IGM. The observed Fe/Mg ratio implies a fractional contribution of SNe Ia to the total (Type Ia and core-collapse combined) of $f_{\rm Ia} \approx 15$–20% in these inner massive halos. Beyond $d \approx 100$ kpc, the observed Fe/Mg ratio recovers the typical α-element enhanced level.
5. There exists a strong correlation between the cool halo gas covering fraction κ in quasar host halos and quasar bolometric luminosity $L_{\rm bol}$, leading to a surge of cool gas in halos about luminous quasars at both low and high redshifts. The strong κ–$L_{\rm bol}$ correlation suggests a physical link between cool gas content on scales of 100 kpc and quasar activities on sub-parsec scales, but interpreting this strong correlation remains challenging. The primary difficulty lies in the relatively short quasar lifetime of $\approx$10–100 Myr in comparison to the long dynamical time necessary to move gaseous clouds over a large distance of

$\approx$200 kpc. Direct imaging of quasar-driven outflows and observations of highly ionized gas associated with cool gas at $d \approx 200$ kpc are necessary to establish direct connections between outflows and cold gas detected at large distances.

Together, these findings suggest that infalling clouds from external sources are likely a dominant source of cool gas detected at $d \gtrsim 100$ kpc from massive quiescent galaxies. The origin of the gas closer in is currently less certain, but SNe Ia driven winds appear to contribute significantly to cool gas found at $d < 100$ kpc. In contrast, cool gas observed at $d \lesssim 200$ kpc from luminous quasars appears to be intimately connected to the on-going quasar activities. The observed strong correlation between cool gas covering fraction in quasar host halos and quasar bolometric luminosity remains a puzzle.

With new instruments and new survey data becoming available, continuing progress is expected in a number of areas over the next few years for a better understanding of the CGM in massive halos. In particular, spatially resolved observations of quasar outflows in the inner 10–30 kpc region, combined with absorption-line kinematics at $\sim$100 kpc from the quasar, will provide key insights into the strong correlation between κ and $L_{\rm bol}$ in quasar host halos. Integral field unit (IFU) spectrographs available on large ground-based telescopes provide a powerful tool for imaging quasar outflows based on observations of high-ionization lines.

In addition, while measurements of chemical compositions provide a unique constraint for the physical origin of chemically enriched gas in massive quiescent halos, measurements of N(HI) are necessary for direct comparisons between observations and state-of-the-art cosmological simulations. The Cosmic Origins Spectrograph (COS; Green et al. 2012) on board the *Hubble Space Telescope* provides the spectral coverage necessary for N(HI) measurements at $z \lesssim 1$. With an increasing number of $z \approx 0.5$ LRGs found near the sightline of a UV bright QSO, there will soon be a statistical sample of massive quiescent halos with known N(HI) at different projected distances for testing simulation predictions.

Furthermore, understanding the roles of satellites and satellite interactions in producing chemically enriched cool clumps in low-density halos requires deep galaxy survey data in quasar fields. With several wide-field integral field spectrographs being installed on ground-based telescopes, deep galaxy survey data in a large number of quasar fields will soon be available for systematic studies of the galaxy environments of kinematically complex absorbers.

Finally, little is known regarding the CGM properties and galactic environments of massive starburst galaxies with $M_{\rm star} \gtrsim 10^{11}\,{\rm M}_\odot$ (e.g., Borthakur et al. 2013). Although these galaxies are very rare, contributing to roughly 10% of the massive galaxy population (with the rest being quiescent LRGs), they are intrinsically UV luminous and massive stars in these galaxies serve as the backlight for probing the internal star-forming ISM in front of the massive young stars. With numerous large-scale galaxy surveys expected in the coming years (e.g., Zhu et al. 2015), combining intrinsic absorption-line observations with absorption spectroscopy along transverse sightlines (e.g., Rubin et al. 2010; Kacprzak et al. 2014) will be feasible for a statistically significant sample of massive starburst galaxies. These new data will offer an important empirical understanding of the impact of starbursts on the CGM in massive halos.

Acknowledgements The author wishes to thank Sean Johnson, Rebecca Pierce, Michael Rauch, and Fakhri Zahedy for providing helpful input and comments. In preparing this review, the author has made use of NASA's Astrophysics Data System Bibliographic Services.

References

Alexander D. M., Swinbank A. M., Smail I., et al. 2010, MNRAS, 402, 2211
Andrews H., Barrientos L. F., López S., et al. 2013, ApJ, 774, 40
Bajtlik S., Duncan R. C., Ostriker J. P., 1988, ApJ, 327, 570
Behroozi P. S., Conroy C., Wechsler R. H., 2010, ApJ, 717, 379
Belfiore F., Maiolino R., Maraston C., et al. 2016, MNRAS, 461, 3111
Benson A. J., Bower R. G., Frenk C. S., et al. 2003, ApJ, 599, 38
Bergeron J., Stasińska G., 1986, A&A, 169, 1
Birnboim Y., Dekel A., 2003, MNRAS, 345, 349
Blake C., Collister A., Lahav O., 2008, MNRAS, 385, 1257
Blumenthal G. R., Faber S. M., Primack J. R., Rees M. J., 1984, Nature, 311, 517
Borthakur S., Heckman T., Strickland D., Wild V., Schiminovich D., 2013, ApJ, 768, 18
Bouché N., Murphy M. T., Péroux C., Csabai I., Wild V., 2006, MNRAS, 371, 495
Bowen D. V., Blades J. C., Pettini M., 1995, ApJ, 448, 634
Bowen D. V., Chelouche D., 2011, ApJ, 727, 47
Bowen D. V., Hennawi J. F., Ménard B., et al. 2006, ApJ, 645, L105
Bower R. G., Benson A. J., Malbon R., et al. 2006, MNRAS, 370, 645
Cantalupo S., Arrigoni-Battaia F., Prochaska J. X., et al. 2014, Nature, 506, 63
Carniani S., Marconi A., Maiolino R., et al. 2015, A&A, 580, A102
Chen H.-W., Helsby J. E., Gauthier J.-R., et al. 2010a, ApJ, 714, 1521
Chen H.-W., Lanzetta K. M., Webb J. K., 2001a, ApJ, 556, 158
Chen H.-W., Lanzetta K. M., Webb J. K., Barcons X., 1998, ApJ, 498, 77
—, 2001b, ApJ, 559, 654
Chen H.-W., Mulchaey J. S., 2009, ApJ, 701, 1219
Chen H.-W., Prochaska J. X., 2000, ApJ, 543, L9
Chen H.-W., Tinker J. L., 2008, ApJ, 687, 745
Chen H.-W., Wild V., Tinker J. L., Gauthier J.-R., et al. 2010b, ApJ, 724, L176
Chung A., van Gorkom J. H., Kenney J. D. P., Crowl H., Vollmer B., 2009, AJ, 138, 1741
Churchill C. W., Mellon R. R., Charlton J. C., et al. 2000, ApJ, 543, 577
Churchill C. W., Vogt S. S., Charlton J. C., 2003, AJ, 125, 98
Chynoweth K. M., Langston G. I., Yun M. S., et al. 2008, AJ, 135, 1983
Conroy C., van Dokkum P. G., Kravtsov A., 2015, ApJ, 803, 77
Costa T., Sijacki D., Haehnelt M. G., 2015, MNRAS, 448, L30
Crighton N. H. M., Hennawi J. F., Prochaska J. X., 2013, ApJ, 776, L18
Croton D. J., Springel V., White S. D. M., et al. 2006, MNRAS, 365, 11
Dall'Aglio A., Wisotzki L., Worseck G., 2008, A&A, 491, 465
de Plaa J., Werner N., Bleeker J. A. M., et al. 2007, A&A, 465, 345
Dekel A., Birnboim Y., 2006, MNRAS, 368, 2
Dekel A., Silk J., 1986, ApJ, 303, 39
Dessauges-Zavadsky M., Calura F., Prochaska J. X., et al. 2004, A&A, 416, 79
Dubois Y., Gavazzi R., Peirani S., Silk J., 2013, MNRAS, 433, 3297
Eftekharzadeh S., Myers A. D., White M., et al. 2015, MNRAS, 453, 2779
Eisenstein D. J., Annis J., Gunn J. E., et al. 2001, AJ, 122, 2267
Eisenstein D. J., Hogg D. W., Fukugita M., et al. 2003, ApJ, 585, 694
Fabian A. C., 2012, ARA&A, 50, 455
Farina E. P., Falomo R., Decarli R., et al. 2013, MNRAS, 429, 1267

Farina E. P., Falomo R., Scarpa R., et al. 2014, MNRAS, 441, 886
Faucher-Giguère C.-A., Feldmann R., Quataert E., et al. 2016, MNRAS, 461, L32
Faucher-Giguère C.-A., Hopkins P. F., Kereš D., et al. 2015, MNRAS, 449, 987
Faucher-Giguère C.-A., Kereš D., 2011, MNRAS, 412, L118
Ferland G. J., Porter R. L., van Hoof P. A. M., et al. 2013, RMxAA, 49, 137
Fernández X., Gim H. B., van Gorkom J. H., et al. 2016, ApJ, 824, L1
Fernandez-Soto A., Barcons X., Carballo R., Webb J. K., 1995, MNRAS, 277, 235
Ford A. B., Davé R., Oppenheimer B. D., et al. 2014, MNRAS, 444, 1260
Förster F., Schawinski K., 2008, MNRAS, 388, L74
Fox A., Richter P., Fechner C., 2014, A&A, 572, A102
Fumagalli M., Hennawi J. F., Prochaska J. X., et al. 2014, ApJ, 780, 74
Fumagalli M., Prochaska J. X., Kasen D., et al. 2011, MNRAS, 418, 1796
Gallazzi A., Charlot S., Brinchmann J., et al. 2005, MNRAS, 362, 41
Gauthier J.-R., 2013, MNRAS, 432, 1444
Gauthier J.-R., Chen H.-W., 2011, MNRAS, 418, 2730
Gauthier J.-R., Chen H.-W., Cooksey K. L., et al. 2014, MNRAS, 439, 342
Gauthier J.-R., Chen H.-W., Tinker J. L., 2009, ApJ, 702, 50
—, 2010, ApJ, 716, 1263
Gilmore G., Wyse R. F. G., 1991, ApJ, 367, L55
Green J. C., Froning C. S., Osterman S., et al. 2012, ApJ, 744, 60
Greene J. E., Zakamska N. L., Smith P. S., 2012, ApJ, 746, 86
Grimes J. P., Heckman T., Aloisi A., et al. 2009, ApJS, 181, 272
Guo Q., White S., Li C., Boylan-Kolchin M., 2010, MNRAS, 404, 1111
Haardt F., Madau P., 2001, in *Clusters of Galaxies and the High Redshift Universe Observed in X-rays*, Neumann D. M., Tran J. T. V., eds.
Hamann F., 1997, ApJS, 109, 279
Hand N., Appel J. W., Battaglia N., et al. 2011, ApJ, 736, 39
Hansen S. M., Sheldon E. S., Wechsler R. H., Koester B. P., 2009, ApJ, 699, 1333
Hayes M., Melinder J., Östlin G., et al. 2016, ApJ, 828, 49
Heckman T. M., 2002, in Astronomical Society of the Pacific Conference Series, Vol. 254, Extragalactic Gas at Low Redshift, Mulchaey J. S., Stocke J. T., eds., p. 292
Heckman T. M., Alexandroff R. M., Borthakur S., et al. 2015, ApJ, 809, 147
Heckman T. M., Best P. N., 2014, ARA&A, 52, 589
Hennawi J. F., Prochaska J. X., 2007, ApJ, 655, 735
Ho L. C., 2008, ARA&A, 46, 475
Hopkins P. F., Hernquist L., 2009, ApJ, 698, 1550
Huang Y.-H., Chen H.-W., Johnson S. D., Weiner B. J., 2016, MNRAS, 455, 1713
Humphrey P. J., Buote D. A., 2006, ApJ, 639, 136
Iwamoto K., Brachwitz F., Nomoto K., et al. 1999, ApJS, 125, 439
Johansson P. H., Naab T., Ostriker J. P., 2009, ApJ, 697, L38
Johnson S. D., Chen H.-W., Mulchaey J. S., 2013, MNRAS, 434, 1765
—, 2015a, MNRAS, 452, 2553
—, 2015b, MNRAS, 449, 3263
Kacprzak G. G., Churchill C. W., Steidel C. C., Murphy M. T., 2008, AJ, 135, 922
Kacprzak G. G., Martin C. L., Bouché N., et al. 2014, ApJ, 792, L12
Kacprzak G. G., Murphy M. T., Churchill C. W., 2010, MNRAS, 406, 445
Kereš D., Katz N., Fardal M., et al. 2009, MNRAS, 395, 160
Kereš D., Katz N., Weinberg D. H., Davé R., 2005, MNRAS, 363, 2
Kirby E. N., Cohen J. G., Guhathakurta P., et al. 2013, ApJ, 779, 102
Kormendy J., Richstone D., 1995, ARA&A, 33, 581
Kravtsov A. V., Borgani S., 2012, ARA&A, 50, 353
Lanzetta K. M., Bowen D., 1990, ApJ, 357, 321
Lanzetta K. M., Bowen D. V., Tytler D., Webb J. K., 1995, ApJ, 442, 538
Larson R. B., 1974, MNRAS, 169, 229

Lee H., Skillman E. D., Cannon J. M., et al. 2006, ApJ, 647, 970
Lehner N., Howk J. C., Tripp T. M., et al. 2013, ApJ, 770, 138
Liang C. J., Chen H.-W., 2014, MNRAS, 445, 2061
Liu G., Zakamska N. L., Greene J. E., 2014, MNRAS, 442, 1303
Liu G., Zakamska N. L., Greene J. E., et al. 2013, MNRAS, 430, 2327
Lopez S., Barrientos L. F., Lira P., et al. 2008, ApJ, 679, 1144
Lovegrove E., Simcoe R. A., 2011, ApJ, 740, 30
Lundgren B. F., Brunner R. J., York D. G., et al. 2009, ApJ, 698, 819
Maiolino R., Gallerani S., Neri R., et al. 2012, MNRAS, 425, L66
Maller A. H., Bullock J. S., 2004, MNRAS, 355, 694
Mandelbaum R., Seljak U., Hirata C. M., 2008, JCAP, 8, 6
Marasco A., Crain R. A., Schaye J., et al. 2016, MNRAS, 461, 2630
Martini P., 2004, Coevolution of Black Holes and Galaxies, 169
Mathews W. G., Brighenti F., 2003, ARA&A, 41, 191
Matteucci F., 2014, *The Origin of the Galaxy and Local Group*, Saas-Fee Advanced Course, Springer-Verlag Berlin Heidelberg, 37, 145
Ménard B., Wild V., Nestor D., et al. 2011, MNRAS, 417, 801
Mo H. J., Miralda-Escudé J., 1996, ApJ, 469, 589
Moster B. P., Somerville R. S., Maulbetsch C., et al. 2010, ApJ, 710, 903
Mulchaey J. S., 2000, ARA&A, 38, 289
Mulchaey J. S., Chen H.-W., 2009, ApJ, 698, L46
Narayanan A., Charlton J. C., Misawa T., et al. 2008, ApJ, 689, 782
Nestor D. B., Johnson B. D., Wild V., et al. 2011, MNRAS, 412, 1559
Nestor D. B., Turnshek D. A., Rao S. M., Quider A. M., 2007, ApJ, 658, 185
Nomoto K., Tominaga N., Umeda H., et al. 2006, Nuclear Physics A, 777, 424
Oosterloo T., Morganti R., Crocker A., et al. 2010, MNRAS, 409, 500
Oosterloo T. A., Morganti R., Sadler E. M., et al. 2007, A&A, 465, 787
Padmanabhan N., White M., Norberg P., Porciani C., 2009, MNRAS, 397, 1862
Peng Y.-j., Lilly S. J., Kovač K., et al. 2010, ApJ, 721, 193
Petitjean P., Bergeron J., 1990, A&A, 231, 309
Porciani C., Magliocchetti M., Norberg P., 2004, MNRAS, 355, 1010
Prescott M., Baldry I. K., James P. A., et al. 2011, MNRAS, 417, 1374
Prochaska J. X., Lau M. W., Hennawi J. F., 2014, ApJ, 796, 140
Rahmati A., Schaye J., Bower R. G., et al. 2015, MNRAS, 452, 2034
Rao S. M., Turnshek D. A., Nestor D. B., 2006, ApJ, 636, 610
Rauch M., Miralda-Escudé J., Sargent W. L. W., et al. 1997, ApJ, 489, 7
Rigby J. R., Charlton J. C., Churchill C. W., 2002, ApJ, 565, 743
Roseboom I. G., Pimbblet K. A., Drinkwater M. J., et al. 2006, MNRAS, 373, 349
Rubin K. H. R., Prochaska J. X., Koo D. C., Phillips A. C., Weiner B. J., 2010, ApJ, 712, 574
Rudie G. C., Steidel C. C., Trainor R. F., et al. 2012, ApJ, 750, 67
Salim S., Rich R. M., 2010, ApJ, 714, L290
Sarzi M., Falcón-Barroso J., Davies R. L., et al. 2006, MNRAS, 366, 1151
Sarzi M., Shields J. C., Schawinski K., et al. 2010, MNRAS, 402, 2187
Savage B. D., Sembach K. R., 1996, ARA&A, 34, 279
Scalo J., Elmegreen B. G., 2004, ARA&A, 42, 275
Schirber M., Miralda-Escudé J., McDonald P., 2004, ApJ, 610, 105
Serra P., Oosterloo T., Morganti R., et al. 2012, MNRAS, 422, 1835
Shankar F., Lapi A., Salucci P., De Zotti G., Danese L., 2006, ApJ, 643, 14
Sharma P., McCourt M., Quataert E., Parrish I. J., 2012, MNRAS, 420, 3174
Shen S., Madau P., Guedes J., Mayer L., Prochaska J. X., Wadsley J., 2013a, ApJ, 765, 89
Shen Y., McBride C. K., White M., et al. 2013b, ApJ, 778, 98
Singh R., van de Ven G., Jahnke K., et al. 2013, A&A, 558, A43
Somerville R. S., Davé R., 2015, ARA&A, 53, 51
Steidel C. C., Dickinson M., Meyer D. M., et al. 1997, ApJ, 480, 568

Steidel C. C., Dickinson M., Persson S. E., 1994, ApJ, 437, L75
Steidel C. C., Erb D. K., Shapley A. E., et al. 2010, ApJ, 717, 289
Steidel C. C., Kollmeier J. A., Shapley A. E., et al. 2002, ApJ, 570, 526
Stocke J. T., Keeney B. A., Danforth C. W., et al. 2014, ApJ, 791, 128
Sunyaev R. A., Zeldovich Y. B., 1970, Ap&SS, 7, 3
Tal T., Wake D. A., van Dokkum P. G., 2012, ApJ, 751, L5
Thom C., Tumlinson J., Werk J. K., et al. 2012, ApJ, 758, L41
Tinker J. L., Chen H.-W., 2008, ApJ, 679, 1218
Tinker J. L., Leauthaud A., Bundy K., et al. 2013, ApJ, 778, 93
Tinsley B. M., 1979, ApJ, 229, 1046
Tojeiro R., Percival W. J., Heavens A. F., Jimenez R., 2011, MNRAS, 413, 434
Tolstoy E., Venn K. A., Shetrone M., et al. 2003, AJ, 125, 707
Tremonti C. A., Heckman T. M., Kauffmann G., et al. 2004, ApJ, 613, 898
Tsujimoto T., Nomoto K., Yoshii Y., et al. 1995, MNRAS, 277, 945
Tumlinson J., 2006, ApJ, 641, 1
Tumlinson J., Thom C., Werk J. K., et al. 2013, ApJ, 777, 59
Tumlinson J., Thom C., Werk J. K., et al. 2011, Science, 334, 948
Vale A., Ostriker J. P., 2004, MNRAS, 353, 189
Verdes-Montenegro L., Yun M. S., Williams B. A., et al. 2001, A&A, 377, 812
Voit G. M., Donahue M., Bryan G. L., McDonald M., 2015, Nature, 519, 203
Werk J. K., Prochaska J. X., Thom C., et al. 2013, ApJS, 204, 17
Werk J. K., Prochaska J. X., Tumlinson J., et al. 2014, ApJ, 792, 8
White M., Myers A. D., Ross N. P., et al. 2012, MNRAS, 424, 933
White S. D. M., Rees M. J., 1978, MNRAS, 183, 341
Whiting M. T., Webster R. L., Francis P. J., 2006, MNRAS, 368, 341
Wild V., Kauffmann G., White S., et al. 2008, MNRAS, 388, 227
Wolfe A. M., Gawiser E., Prochaska J. X., 2005, ARA&A, 43, 861
Yan R., Blanton M. R., 2012, ApJ, 747, 61
Yoon J. H., Putman M. E., Thom C., Chen H.-W., Bryan G. L., 2012, ApJ, 754, 84
York D. G., Adelman J., Anderson Jr. J. E., et al. 2000, AJ, 120, 1579
Young L. M., Scott N., Serra P., et al. 2014, MNRAS, 444, 3408
Zahedy F. S., Chen H.-W., Rauch M., Wilson M. L., Zabludoff A., 2016, MNRAS, 458, 2423
Zahedy F. S., Chen H.-W., Gauthier J.-R., Rauch M., 2017, MNRAS, 466, 1071
Zheng Z., Coil A. L., Zehavi I., 2007, ApJ, 667, 760
Zhu G., Ménard B., Bizyaev D., et al. 2014, MNRAS, 439, 3139
Zhu G. B., Comparat J., Kneib J.-P., et al. 2015, ApJ, 815, 48
Zibetti S., Ménard B., Nestor D. B., et al. 2007, ApJ, 658, 161

Gas Accretion and Giant Lyα Nebulae

Sebastiano Cantalupo

1 Introduction

This chapter reviews the properties of extended Lyα emission in the high-redshift universe ($z > 2$) and their connection with gas flows in and around galaxies and their halos. In particular, the attention will be focused on giant Lyα nebulae, i.e., emission with sizes exceeding 100 kpc (unless otherwise noticed, I will always use "physical" units in this chapter) and in general on Lyα emission extending on scales that are significantly larger than individual galaxies. In terms of energetics, these systems have integrated Lyα luminosities ($L_{\mathrm{Ly}\alpha}$) larger than 10^{43} erg s^{-1}. For space reason, smaller and lower luminosities Lyα halos found around individual Lyman α emitters (LAE), e.g., Wisotski et al. (2016), or in stacking analysis, e.g., Steidel et al. (2011) will not be covered in this chapter although their origin in some cases may be connected with larger nebulae (see, e.g., Borisova et al. 2016, for discussion). Similarly, I will not review here the detection and study of giant nebulae in metal emission line in the low redshift universe (e.g., Bergeron et al. 1983; Veilleux et al. 2003).

The chapter is organized as follows. In Sect. 2, I provide an observationally oriented and historical overview describing the discovery and characterization of high-redshift Lyα Nebulae. In Sect. 3, a detailed discussion about the origin of the Lyα emission is provided in terms of the atomic processes responsible for the production of Lyα photons. In Sect. 4, I review the kinematics of the nebulae and the connection with the physical origin and fate of the gas including accretion onto galaxies and their halos. I summarize the chapter in Sect. 5. Throughout this

S. Cantalupo (✉)
Institute for Astronomy, ETH Zurich, Wolfgang-Pauli-Strasse 27, CH-8093 Zurich, Switzerland
e-mail: cantalupo@phys.ethz.ch

A. Fox, R. Davé (eds.), *Gas Accretion onto Galaxies*, Astrophysics and Space Science Library 430, DOI 10.1007/978-3-319-52512-9_9

chapter, a ΛCDM cosmology with $\Omega_m = 0.3$, $\Omega_\Lambda = 0.7$, and $H = 70\,\mathrm{km\,s^{-1}}$ is assumed. For reference, 1 arcsec at $z = 3$ corresponds to about 7.6 kpc with these cosmological parameters.

2 Observations of Giant Lyα Nebulae

2.1 Quasar Lyα Nebulae

Being the first and more luminous high-redshift object discovered above $z > 2$, quasars have been obvious signposts since the mid-1980s to look for putative galactic and gaseous "companions" in Lyα emission. One of the first attempts reported in the literature is the pioneering narrow-band observation of Djorgovski et al. (1985) on the Lick Observatory 3 meter telescope centered on the radio-loud QSO PKS1614+051 at $z \sim 3.2$ that resulted in the discovery of a companion Lyα emitting galaxy (a narrow-line AGN) at about 5″ from the quasar. Later observations by Hu and Cowie (1987) using the same technique on the 3.6 m Canada France Hawaii Telescope (CFHT) showed the presence of a "bridge" of Lyα emission between the quasar and the companion galaxy. In the same year, Schneider et al. (1987) reported the discovery of "companion" Lyα emission to the triply lensed radio-loud quasar Q2016+112.

These initial discoveries prompted an intense effort to search for Lyα emission around quasars, mostly of which radio-quiet, in subsequent years but despite the large number of quasars observed (about 50), no detectable Lyα candidates were found (these results were mostly unpublished, see discussion in Hu et al. 1991). The situation changed in the early 1990s when observational surveys focused on the much smaller sub-sample of radio-loud quasars (Hu et al. 1991; Heckman et al. 1991a) reported a detection rate of compact or extended companion Lyα emission close to 100%. In particular, Heckman et al. (1991a) reported the discovery of Lyα Nebulae with sizes of about 100 kpc for 15 of the 18 radio-loud quasars observed. The contemporarily discovery of large Lyα Nebulae around non-QSO radio-sources (e.g., McCarthy et al. 1987, at $z \sim 1.8$) as I will discuss in Sect. 2.2, suggested to these authors a possible link between the radio and the Lyα emission. At the same time, searches around radio-loud quasars were also motivated by the possibility to test the hypothesis that radio-loud quasars and radio-galaxies were the same class of objects viewed along different angles with respect to the radio axis.

One of the first detection of "companion" Lyα emission to radio-quiet quasars was due to a serendipitous observation by Steidel et al. (1991) at the Palomar 5 meter telescope: originally searching for the continuum counterpart of a $z \sim 0.8$ MgII absorber in the spectrum of the radio-quiet QSO Q1548+0917, they found instead a narrow and extended Lyα emission line at the same redshift of the quasar in the spectrum of a faint continuum source ∼5″ away from the QSO. Also, additional narrow-band imaging suggested the presence of extended Lyα emission around the quasar. In the following year, Bremer et al. (1992) reported the discovery of extended ($\sim$5″) emission in long-slit spectra of two radio-quiet quasars at $z \sim 3.6$. These results showed that “companion” Lyα emission was not restricted to radio-loud quasars only.

Giant Lyα Nebulae with sizes larger than 10″ (or larger than about 100 kpc) around radio-quiet quasars remained, however, elusive for more than two decades after the survey of Heckman et al. (1991a) around radio-loud quasars. The only exception was the serendipitous discovery of Bergeron et al. (1999) around the $z \sim 2.2$ radio-quiet quasar J2233-606 located in one of the parallel fields of the Hubble Deep Field South. The field was observed during science verification of the VLT-UT1 Test Camera using broadband filters and a narrow-band filter centered on the quasar Lyα emission. Despite some large-scale residual of the flat-fielding procedure, the narrow-band image seemed to show extended emission with a maximum projected size of about 12″ (about 100 kpc with current cosmological parameters) around the quasar.[1] On the other hand, during the same years a few individual detections of small nebulae extending up to a few arcsec around radio-quiet quasars were reported (Fried 1998; Møller et al. 2000; Bunker et al. 2003; Weidinger et al. 2004) and in some cases associated with intergalactic gas (e.g., Weidinger et al. 2004, 2005). By the beginning of the 2010a, thanks to long-slit spectroscopic surveys and small Integral-Field-Unit (IFU) observations, a common picture emerged that associated only relatively small nebulae (i.e., <60–70 kpc) to about 50% of radio-quiet quasars between $2 < z < 5$ (Christensen et al. 2006; Courbin et al. 2008; North et al. 2012; Hennawi and Prochaska 2013; but see Herenz et al. 2015). However, as I discuss below, these results may have been limited by the small sizes of the IFU Field-of-View (FOV) and by the use of long-slit spectroscopy that cannot capture the full extent of asymmetric nebulae.

2.1.1 Rise of the Giants

The last few years witnessed a complete revolution in our knowledge of giant Lyα nebulae around radio-quiet quasars, thanks to serendipitous discoveries by means of NB imaging with custom-built filters (Cantalupo et al. 2014; Martin et al. 2014; Hennawi et al. 2015) and dedicated surveys with VLT/MUSE (e.g., Borisova et al. 2016). Two decades after Hu et al. (1991) and Heckman et al. (1991a), a new narrow-band campaign on quasar fields was initiated in order to search for “dark galaxies” and fluorescently illuminated intergalactic gas following the prediction, e.g., of Cantalupo et al. (2005). Two pilot programs using a spectroscopic “multi-slit plus filter” technique (Cantalupo et al. 2007) and deep NB imaging ($\sim$20 h) on VLT/FORS using a custom-built filter for a quasar at $z \sim 2.4$ (Cantalupo et al. 2012) revealed a dozen of compact Lyα sources with no detectable continuum and

[1] A very recent and deeper observation with GMOS seems to confirm at least part of this nebula up to a size of about 70 kpc (Arrigoni-Battaia et al, in prep.).

Equivalent Widths (EW) larger than 240 Å, the best candidates for "dark galaxies" illuminated by the quasar. Circumgalactic gas was also detected in emission extending by several tens of kpc around a few bright galaxies but the quasar did not show evidence for extended nebulae, in agreement with previous findings.

Stimulated by these initial results, Cantalupo et al. (2014) initiated a campaign using Keck/LRIS and custom-built filters to search for "dark galaxies" around about ten radio-quiet quasars at $z \sim 2$. Surprisingly, the first quasar observed at Keck/LRIS, i.e., UM287, showed clear evidences of extended emission over scales larger than $10''$ after the first 20 min exposure was obtained. At the end of the total integration of 10 h and detailed data reduction, a giant Lyα nebula extending to about $55''$ ($\sim$460 kpc) was found around this quasar (Cantalupo et al. 2014), see Fig. 1, and named "Slug Nebula" (given its morphology and in honor of the mascot of the University of California, Santa Cruz). This discovery was surprising for several reasons: (1) despite its association with a radio-quiet quasar, it is at least twice as large as any previously detected Lyα Nebulae including the much more common

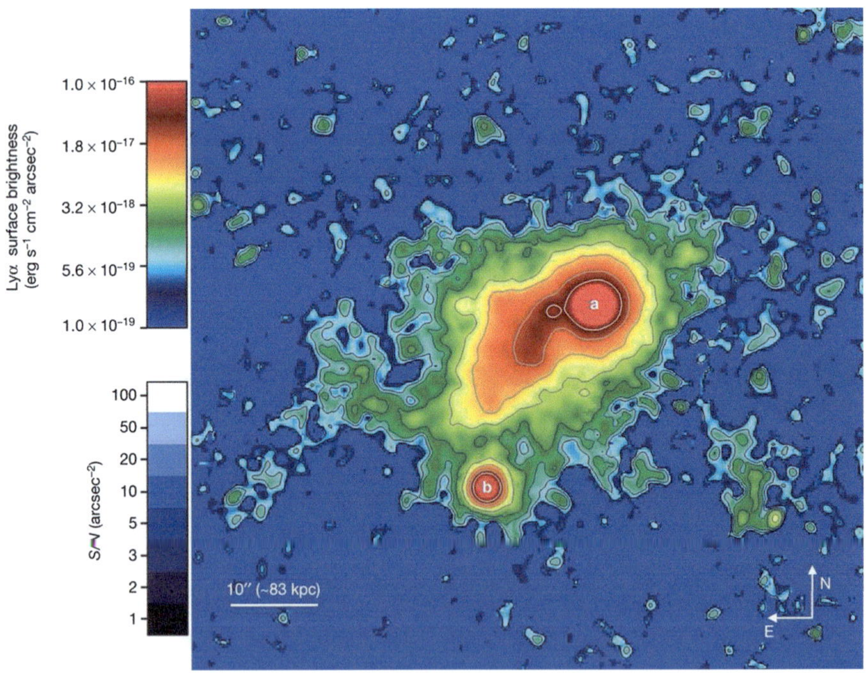

Fig. 1 Continuum-subtracted narrow-band image of the Slug Nebula (Cantalupo et al. 2014) discovered around the bright, radio-quiet quasar UM287 (labeled "*a*" in the image) in a 10-h deep observation made with a custom filter installed on Keck/LRIS. The nebula shows filamentary emission extending on a total projected length of about 55 arcsec ($\sim$460 kpc) and is currently one of the largest and more luminous Lyα nebulae discovered to date. Its surprising properties in terms of extension and high values of Surface Brightness are discussed in details in Sect. 2.1.1. (Figure reproduced with permission from Cantalupo et al. 2014)

radio-galaxy halos and Lyα blobs as discussed below; (2) given its size, it extends well beyond the viral halo of the quasar into the intergalactic medium; (3) it shows a very high Surface Brightness over very large scales that cannot be easily explained unless a large gas clumping factor within intergalactic gas is invoked (see Sect. 3.1). During the same night, a second quasar field was observed using a different custom-built filter, selected among the quasar-pair sample of Hennawi and Prochaska (2013) and showing some hints of extended emission, and another giant Lyα nebula with a size of about 300 kpc was discovered (Hennawi et al. 2015). The particularity of this discovery included also the presence of a physically associated quasar quartet and a large overdensity of Lyα emitting galaxies (differently than the Slug Nebula). Named "Jackpot Nebula" given the rarity of such systems, it traces likely a very peculiar region of the Universe and possibly a proto-cluster. In the same year of these discoveries, Martin et al. (2014) presented the detection of another giant nebula originally found around one of the six quasars observed in narrow-band imaging as a part of the Keck Baryonic Structure Survey (e.g., Trainor and Steidel 2012). Subsequent observations of other quasars as a part of the same survey and including results obtained on GMOS/Gemini (Arrigoni Battaia et al. 2016) showed once again, however, that such detection of giant Lyα nebulae are apparently rare, i.e., with a frequency less than 10%.

The installation of the MUSE Integral-Field-Spectrograph on the Very Large Telescope in 2014 provided new opportunities for the detection and study of Giant Lyα nebulae around quasars, thanks to its a large FOV of $1' \times 1'$ (about $450 \times 450\,\text{kpc}^2$; individual spatial elements have a size of $0.2'' \times 0.2''$) and because, by design, it does not suffer from either narrow-band filter losses or spectroscopic slit losses. Also, the large number of spatial and spectral elements allows for better quasar PSF estimation and removal with respect to NB surveys. Because accurate systemic redshifts are not needed for MUSE observations (as for any other spectroscopic survey) any quasar with Lyα redshifted between the blue and red edges of the MUSE wavelength range ($2.9 < z < 6.5$) can be observed. In one of the first exploratory study as a part of the MUSE Guaranteed Time Observations (GTO), Borisova et al. (2016) observed with short total exposure times (1 h) 17 of the brightest radio-quiet quasars in the Universe at $3.1 < z < 3.7$ and complemented them with two radio-loud quasar at the same redshifts. The picture emerging from these MUSE observations is very different than that based on previous surveys, in that giant nebulae with sizes larger than 100 pkpc are found around essentially every quasar above a surface brightness level of about $10^{-18}\,\text{erg}\,\text{s}^{-1}\,\text{cm}^{-2}\,\text{arcsec}^{-2}$. The nebulae detected with MUSE present a large range in sizes and morphologies, ranging from circular nebulae with a projected diameter of about 110 pkpc to filamentary structures with a projected linear size of 320 pkpc (see Fig. 2). Despite these differences, the circularly averaged SB profiles show a strong similarity between all the giant Quasar Nebulae (including the Slug Nebula at $z \sim 2$ once corrected for redshift-dimming) with very few exceptions, both in terms of slope and normalization, suggesting a similar origin for these systems.

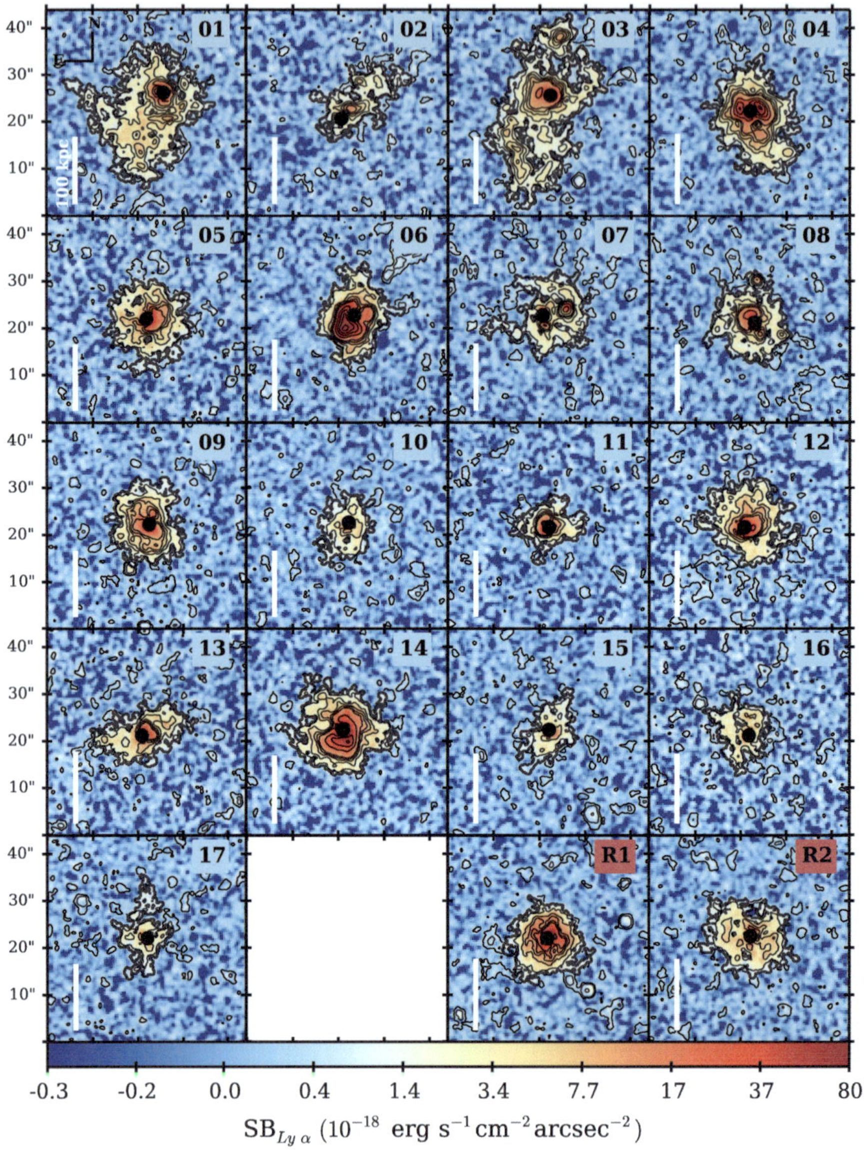

Fig. 2 "Optimally extracted" narrow-band images of the sample of luminous quasars at $3 < z < 4$ observed with MUSE by Borisova et al. (2016) as a part of a "snapshot" survey using short integration times, i.e., 1 h per field (see Borisova et al. 2016 for details on the detection and extraction of these images from the datacubes). The quasars are all radio-quiet with the exception of the two fields labeled "R1" and "R2". All nebulae are larger than 100 kpc and extending in some cases up to at least 320 kpc (e.g., the nebula number 3 or "MQN03") with various morphologies including filamentary structure. This survey showed that giant Lyα nebulae are ubiquitous around bright quasars, including radio-quiet ones, in contrast to previous observations at $z \sim 2$ as discussed in detail in Sect. 2.1 (Figure reproduced with permission from Borisova et al. 2016)

This 100% detection rate of giant nebulae around radio-quiet quasars obtained with MUSE (see also, Fumagalli et al. 2016) is in stark contrast with previous results in the literature as I have reviewed in this section. While the asymmetric morphology of the MUSE nebulae may explain the discrepancy with spectroscopic surveys using a single slit position, the difference with the detection rate of NB surveys at $z \sim 2$ cannot be completely and easily explained by observational limitations such as NB filter losses, uncertainties in quasar systemic redshifts, and Quasar Point-Spread-Function (PSF) removal errors. Redshift and quasar luminosities may therefore play a role in the appearance and properties of Lyα nebulae around quasars and future IFU studies extending to lower redshifts (e.g., with the Keck Cosmic Web Imager) and to lower quasar luminosities are necessary to properly address these open questions.

2.2 Radio-Galaxy Lyα Halos

As mentioned in the previous section, follow-up observations of radio-loud sources provided already in the second half of the 1980s the first evidences for giant Lyα nebulae. In this section, I will focus on non-QSO radio sources at $z > 1$, i.e., High-z Radio Galaxies (HzRG).

Although many different classification exists (see, e.g., McCarthy 1993 and Antonucci 2012 for a review), the most obvious distinction between the QSO and non-QSO class of radio source traces its roots from the appearance of the optical morphology, i.e., from the presence of a quasi-stellar point source versus a spatially resolved galaxy. Because this distinction may be difficult to be applied at high redshift a more physical definition can be made instead from the properties of the emission-line spectra. Indeed, classical radio-galaxies have spectra with relatively narrow (FWHM<2000 km/s) permitted lines with respect to the broader lines showed by quasars (FWHM>5000 km/s). These are sometimes called "Narrow Line Radio Galaxies" or "type 2". Moreover, high-redshift quasars have a much brighter, "thermal" continuum than "type 2" radio-galaxies with a clear non-stellar origin. It is important to notice, however, that a small fraction of "spatially resolved" radio-galaxies show broad lines as well as bright thermal continuum. These are sometimes called "Broad Line Radio Galaxies." To avoid confusion, in this section and in the reminder of this chapter we will always refer to radio-galaxies as "type 2". Like their lower redshift counterparts, HzRG often show extended radio lobes that have been associated with bi-polar jets (e.g., McCarthy et al. 1987) and show a high degree of polarization. The observation of relativistic beaming and smaller, one-sided radio-jets in radio-loud quasars have already suggested in the 1980s that radio-galaxies and radio-loud quasars may be part of the same population of AGN but seen at different orientations (e.g., Barthel 1989; see Antonucci 1993 for a review and Antonucci 2012 for more recent discussion).

Narrow-band imaging and spectroscopy of HzRGs, soon after their discovery in the 1980s and until very recently, has produced a large literature of detections and studies of large Lyα nebulae, similarly to radio-loud quasars. The first observations by McCarthy et al. (1987) of the radio-galaxy 3C 326.1 at $z \sim 1.8$ at the Lick Observatory revealed Lyα emission surrounding the radio lobes and extending by about 70 kpc (with current cosmological parameters). Few years later, McCarthy et al. (1990) reported the discovery of the first giant Lyα nebula extending over 120 kpc around the $z \sim 1.8$ radio-galaxy 3C 294. This nebula is highly elongated and well aligned with the inner radio source axis. Extended C IV and He II emission was also detected. One of the first detection of extended Lyα emission around a radio-galaxy above redshift of three was reported by Eales et al. (1993) studying the $z = 3.22$ radio-galaxy 6C 1232+39, a system with a "classical" double radio structure oriented along the same direction of the Lyα emission but showing some differences in the morphology and symmetry of the optical emission with respect to the radio.

Larger sample of HzRG observations were started to be obtained in the 1990s as a part of several campaigns that resulted in the discovery of several tens of HzRG at $z > 2$ (e.g., Röttgering et al. 1994). The most notably discoveries included a giant Lyα halo extending over 140 kpc around a HzRG at $z \sim 3.6$ (van Ojik et al. 1996) showing a complex kinematical structure spatially correlated with the radio jet: large velocity widths (FWHM$\sim$15000 km/s) in proximity of the radio jets and a aligned, low SB component with narrow velocities (FWHM$\sim$250 km/s) extending 40 kpc beyond both sides of the radio source. I will discuss and compare in detail in Sect. 4, the kinematical properties of radio-loud and radio-quiet nebulae, including this system.

Subsequent observational campaigns found almost ubiquitous detection of extended Lyα emission around HzRG extending in several cases above 100 kpc in size (e.g., Pentericci et al. 1997; Kurk et al. 2000; Reuland et al. 2003; Villar-Martín et al. 2003, 2007; Venemans et al. 2007; Humphrey et al. 2008; Sánchez and Humphrey 2009; Roche et al. 2014) and showing a wealth of morphological structures, including filaments, clumpy regions, and cone-shaped structures. The common features of all these detections include: (1) apparent alignment between Lyα emission and the radio jet axis (e.g., Villar-Martín et al. 2007), (2) broad kinematics (with some exceptions at the far edge of the nebulae), (3) associated extended C IV and He II emission, (4) invariable association with large overdensity as traced by companion Lyα galaxies (e.g., Venemans et al. 2007) or multi-wavelength observations from X-ray to the infrared (Pentericci et al. 1997). In terms of luminosities and surface brightness the HzRG halos show similar values with respect to radio-loud and radio-quiet quasar nebulae, while UV line ratios and kinematics seem to be quite distinct from the radio-quiet quasar Nebulae (see, e.g., Borisova et al. 2016) as I will discuss in Sect. 4.

2.3 Lyα Blobs

The historically distinct category of giant Lyα Nebulae called "Lyα blobs" made their appearance with the deep narrow-band observations of Steidel et al. (2000) of a region containing a high overdensity of galaxies both in projected and velocity space in the so-called Small Selected Area 22^h (SSA22) field (Lilly et al. 1991; Steidel et al. 1998). While searching for candidates Lyα emitting galaxies in this field at $z = 3.09$, Steidel et al. (2000) found to their surprise "two very luminous, very extended regions of line emission which we descriptively call 'blobs'." Extending by about 15 to 17 arcsec (115 to 132 kpc with current cosmology), they were among the largest Lyα Nebulae found at that epoch, with similar extent and luminosities to radio-loud quasar and radio-galaxy Nebulae. However, these "blobs" were apparently lacking any associated, bright continuum or radio source and therefore considered as a possibly different category of objects. In particular, Steidel et al. (2000) considered the possibility of a "cooling flow" origin for the emission, by analogy with Hα emission from observed cooling-flow clusters or photo-ionization by heavily obscured, highly star forming galaxies. However, no firm conclusion was possible with the available data for that time. The "mysterious" nature of these Lyα blobs attracted the attention of several theoretical and numerical studies, right after their discoveries until very recently, that tried to explain the peculiarity of these systems with a variety of physical explanations that did not require a bright photo-ionizing source such as an AGN. These models ranged from galactic superwinds (including e.g., Taniguchi and Shioya 2000) to cooling radiation from the so-called cold-mode accretion (including, e.g., Haiman and Rees 2001; Fardal et al. 2001; Dijkstra and Loeb 2009) as I discuss more in detail in Sect. 3.

From an observational and historical point of view, a Lyα "blob" (LAB) could then be defined as an extended Lyα emission over scales significantly larger than a single galaxy that does not seem to contain an AGN (at the time of their discovery). Extended Lyα nebulae around galaxies that did not contain optically bright or radio-loud AGN in overdense fields were already known before the observations of Steidel et al. (2000), e.g., Francis et al. (1996) and later re-observations published in Francis et al. (2001), and Keel et al. (1999), although in most cases hints of obscured AGN were present in these studies. Steidel "blobs," however, were the first one in this category to exceed the "giant Lyα Nebula" size of 100 kpc. Later Subaru observations of the same SSA22 field made by Matsuda et al. (2004) exploring a much larger area than the original survey by Steidel et al. (2000) found several extended Lyα nebulae around galaxies, i.e., 33 nebulae with area exceeding 15 arcsec^2 in addition to the Steidel "blobs" (also called LAB1 and LAB2). However, none of those new detections were close to LAB1 and LAB2 in terms of overall sizes. For instance, the largest new detection by Matsuda et al. (2004) had an area that was only half the one of LAB2 and a maximum projected size of about 75 kpc. Nevertheless, this overdense field showed clearly an excess of extended sources with respect to "blank" observation suggesting a possible relation between galaxy (or AGN) overdensities and extended Lyα emission. Similarly, deep narrow-band imaging obtained by Palunas et al. (2004) of the overdense field at $z \sim 2.38$ found by Francis et al. (2001), revealed a large Lyα blob with sizes compatible with Steidel's LAB1 and several smaller nebulae around galaxies.

LAB1 was the first LAB to be studied in detail in multi-wavelength observations and revealed the presence of a strong submillimeter source with a bolometric luminosity in excess of $10^{13} L_{\odot}$ (e.g., Chapman et al. 2001; Geach et al. 2005) but no evidence from deep Chandra X-ray observations of a clear X-ray counterpart (Chapman et al. 2004). However, these limits did not exclude the possibility of a luminous AGN but heavily obscured along our line of sight. LAB2 instead showed, in addition to a (fainter) submillimeter source (Chapman et al. 2001), clear evidences for hard X-ray emission (e.g., Basu-Zych and Scharf 2004) and therefore for the presence of a partially unobscured AGN. This link between LABs and luminous star forming galaxies stimulated new observational campaigns to detect LABs by searching, e.g., around luminous infrared sources. During one of such campaigns, Dey et al. (2005) discovered a 160 kpc Lyα emitting nebula around a luminous mid-infrared source first detected with the Spitzer Space Telescope. This nebula shared many similarities with the previously detected LAB1 and LAB2. Although X-ray information was not available at that time, the presence of narrow and centrally concentrated C IV and He II emission within the Nebula suggested again an association with a Type-2 AGN. Differently from radio-galaxy halos, this nebula showed, however, relatively narrow Lyα emission in velocity space and possibly a ordered velocity shear. As we will discuss in Sect. 4, this is compatible with other nebulae associated with radio-quiet AGN.

Giant LABs with sizes above 100 kpc are mainly discovered by targeting known overdense regions or bright infrared and submillimeter galaxies. Blank-field surveys using narrow-band imaging confirmed that giant LABs were indeed extremely rare (e.g., Saito et al. 2006; Yang et al. 2009), i.e., presenting a comoving number density less than 10^{-6} Mpc^3 (e.g., Yang et al. 2009). Broadband surveys plus spectroscopic follow-ups were slightly more successfully, detecting one LAB with a size of about 100 kpc and three smaller nebulae in a volume of about 10^8 Mpc^3 (Prescott et al. 2013). It is instructive to compare these numbers with the comoving number densities at $2 < z < 3$ of bright X-ray selected AGN, i.e., $n_X \sim 10^{-6}$ Mpc^{-3} for $L_X > 10^{44.5}$ erg s^{-1} (Ueda et al. 2003), optically bright quasars at $z \sim 3$, i.e., $n_{QSO} \sim 10^{-7}$ Mpc^{-3} for $M_i(z = 2) < -26.7$ (Shen et al. 2007), and HzRG, i.e., 4×10^{-8} Mpc^{-3} for $L_{2.7GHz} > 10^{33}$ erg s^{-1} Hz^{-1} sr^{-1} (see Venemans et al. 2007).

Detection of smaller nebulae with sizes up to 40–70 kpc are less rare in blind narrow-band surveys and about ten of such discoveries were reported in the past, including Nilsson et al. (2006), Yang et al. (2009), Smith et al. (2009), and Prescott et al. (2009). Deeper and larger narrow-band observations around the SSA22 fields and including some blank fields over a total volume of about 10^6 Mpc^3 as a part of the "Subaru Lyα blob survey" resulted in the discovery of about seven new giant LAB with sizes around 100 kpc (Matsuda et al. 2006). An estimation of the number densities of these systems in this survey, however, could be affected by the presence of the very overdense region in SSA22. In almost all cases, evidences for associated submillimeter or AGN sources were found at the time of the discovery or with subsequent multi-wavelength observations (e.g., Geach et al. 2009; Overzier et al. 2013; Hine et al. 2016; see also Scarlata et al. 2009 and Ao et al. 2015).

Despite the different techniques and volume probed, clustering analysis suggests that the sizes of detected LABs could be positively correlated with the environment overdensity, although the statistic is small (e.g., Yang et al. 2009; Matsuda et al. 2006). The very recent detection reported by Cai et al. (2016) of an LAB with a projected size of about $\sim$440 kpc at the center of one of the largest overdensity known at $z \sim 2.3$ seem to provide additional support for this suggestion. However, also in this case, there are evidences for at least one associated AGN (see also Valentino et al. 2016, for another example).

As I have review in this section, several decades of observations and discoveries have produced a large literature of extended and giant Lyα nebulae that have been classified in various ways and with different nomenclatures depending on the technique and target of the original surveys, i.e., quasars, radio-galaxies, overdense regions, or "apparently blank" fields. Their comparable volume densities and the almost invariable association with AGN or massively star forming galaxies seem to suggest, however, that most of these distinction may be artificial. There are, however, some indication that the kinematical properties and line ratios (e.g., considering He II, C IV, and Lyα) are distinct among radio-quiet and radio-loud nebulae, as well as for some LAB, as I will discuss in Sect. 4.

3 Origin of the Emission

In this section, I will review the three physical processes that are able to produce extended and bright Lyα emission: (1) recombination radiation following hydrogen photoionization (by a quasar or star forming galaxy), (2) "continuum-pumping" or Lyα scattering of the photons produced in the quasar broad-line-region or within the Interstellar Medium (ISM) of a star forming galaxy, and (3) Lyα collisional excitation and recombination radiation following collisional ionization (the so-called Lyα cooling radiation). All these mechanisms require the presence of "cool" gas (i.e., with temperatures well below 10^5 K). In addition, recombination radiation and "continuum-pumping" also require that the gas is "illuminated" by the quasar or by some other bright source of UV photons. In this case, the resulting Lyα emission is sometimes called "fluorescent" in the literature (e.g., Cantalupo et al. 2005, 2012, 2014; Kollmeier et al. 2010; Borisova et al. 2016).

3.1 Recombination Radiation

In this case, Lyα photons are produced in a (photo-)ionized medium as a consequence of radiative recombination cascades and the Lyα emissivity will be simply proportional to the hydrogen recombination rate times the ionized hydrogen density squared. Because the hydrogen recombination rate has a relatively mild dependence from temperature, i.e., about linear (e.g., Osterbrock 1989) around 10^4 K, this is the easiest case to model. The actual number of Lyα photons produced by each recombination event (and therefore by each ionizing photon, in photoionization equilibrium) will depend on the details of the recombination cascades that populate the 2 P2 state that will decay to the 1 S2 producing a Lyα photon. A good approximation, even in the low-density Intergalactic Medium, is to assume that every Lyman-line photon from $n > 2$ levels is converted into lower series photons plus either Lyα or two-photon continuum (Case B approximation, Baker and Menzel 1938). In this case, the fraction of Lyα photons for each recombination event ranges between 0.68 and 0.61 for $10^4 < (T/K) < 10^{4.7}$ (Cantalupo et al. 2008) showing again only a mild dependence on temperature. For reference, in the opposite case (Case A) when all Lyman-series photons leave the cloud, the fraction of Lyα photons for each recombination will be reduced by about a half. Combining these information with the Case B hydrogen recombination coefficient, one can obtain the effective Lyα recombination coefficient $\alpha^{\rm eff}_{\rm Ly\alpha}$ for Case B, that can be approximated with the expression (Cantalupo et al. 2008):

$$\alpha^{\rm eff}_{\rm Ly\alpha} \sim 2 \times 10^{-13} T_4^{-1.26}\ {\rm cm^3 s^{-1}}, \tag{1}$$

where, $T_4 = 10^4$ K, for $10^3 < (T/K) < 10^4$. For a fully ionized cloud of hydrogen and helium, the Lyα integrated volume emissivity is therefore:

$$4\pi j_{\rm Ly\alpha} \sim 3.7 \times 10^{-24} n^2 T_4^{-1.26}\ {\rm erg\,s^{-1}\,cm^{-3}}, \tag{2}$$

where n is the gas density in units of atoms per cm^{-3}. For a uniform gas slab with depth L in units of kpc at redshift z, the observed surface emissivity—equivalent to the surface brightness ignoring radiative transfer effects—will be given therefore by:

$$\mathrm{SB}^{\rm em}_{\rm Ly\alpha} \sim 8 \times 10^{-17} n^2 T_4^{-1.26} L \cdot [(1+z)/4]^{-4}\ {\rm erg\,s^{-1}\,cm^{-2}\,arcsec^{-2}}. \tag{3}$$

The observed values of the SB of giant Lyα nebulae range from a few times 10^{-18} erg s^{-1} cm^{-2} arcsec^{-2} in the external parts (at distances of 50–100 kpc from the central sources or the central peak of the Lyα emission) to 10^{-16} erg s^{-1} cm^{-2} arcsec^{-2} in the central regions. From this simple calculation, it is easy to understand that the observed SB would inevitably require large gas densities ($n > 0.1$) in order to be explained by pure recombination emission. These high density estimates were obtained and claimed since the discoveries of the first nebulae around radio-loud quasars and galaxies (e.g., Heckman et al. 1991a).

Are these densities compatible with our picture of cold gas within massive halos? It is instructive to rewrite the previous expression rewriting the density n in terms of the baryonic overdensity with respect to the canonical value of 200 for collapsed objects (δ_{200}). Making use of the definition of clumping factor, i.e., $C_L = \langle n^2 \rangle_L / \langle n \rangle_L^2$ where the average is made over a cylinder with depth L and projected area of 1 arcsec^2 on the sky, we obtain:

$$\mathrm{SB}^{\mathrm{em}}_{\mathrm{Ly}\alpha} \sim 5\times10^{-20} C_L \langle \delta_{200} \rangle^2_{L_{100}} L_{100} T_4^{-1.26} \cdot [(1+z)/4]^2 \ \mathrm{erg\,s^{-1}\,cm^{-2}\,arcsec^{-2}}, \quad (4)$$

where $L_{100} = L/100\,\mathrm{kpc}$. In this formula, no assumptions have been made about the density distribution and density inhomogeneities are encoded in the clumping factor term C_L. Assuming that 100 kpc is the typical projected length of the cold gas distribution in the halo, it is then straightforward to see that, even if all the halo gas is in the form of a fully ionized and uniform medium at $T = 10^4$ K, the predicted SB will be smaller than the observed values by several orders of magnitude. In this case, the observed SB constrains the clumping factor on scales corresponding to 1 arcsec^2 on the sky plane, i.e., on a square with side of about 7 kpc at $z \sim 3$ and on 100 kpc along the line of sight to be at least $C_L \sim 20$–2000. These simple estimates are confirmed by detailed radiative transfer and cosmological simulations, e.g., Cantalupo et al. (2014). For a series of small clouds with uniform density in a empty medium, the clumping factor is equal to the inverse of the volume filling factor f_v, and therefore the implied values of f_v are smaller than 10^{-3} and likely much lower (i.e., 10^{-4}–10^{-5}), considering that, if a hot medium is present as it should be in these massive halos, most of the mass and density will be in the hot component rather than the cold one used for the estimate above. These small filling factors would imply again that the individual densities of the clouds should be higher than $n > 0.1$ and likely closer to $n \sim 1$–$10\,\mathrm{cm^{-3}}$ considering the presence of a hot medium.

A similar analysis can be made considering the total Lyα luminosity, see, e.g., the classical approach used by McCarthy et al. (1990). In this case, integrating the Lyα emissivity given above one would simply obtain:

$$L_{\mathrm{Ly}\alpha} \sim 3.7 \times 10^{-24} n^2 T_4^{-1.26} f_v V \ \mathrm{erg\,s^{-1}}, \quad (5)$$

where V is the total volume in cm^3, or more conveniently expressed in terms of the halo virial masses:

$$L_{\mathrm{Ly}\alpha} \sim 2.6 \times 10^{47} n^2 T_4^{-1.26} f_v M_{12} \cdot [(1+z)/4]^{-3} \ \mathrm{erg\,s^{-1}}, \quad (6)$$

where M_{12} is the virial mass in units of $10^{12} M_{\odot}$. In this case, the value of f_v in the literature is typically fixed to the one derived from extended line emission in local AGN (10^{-6} to 10^{-4}, e.g., Heckman et al. (1989). McCarthy et al. (1990) adopted f_v to 10^{-5} "as a guess," as reported in their paper, and this value was then adopted later in the literature as a sort of "standard" value. As discussed above, our SB consideration lead to similar values. Once f_v is fixed, the individual densities of the clouds are therefore constrained to be of the order of 1–100 cm^3 given the Lyα luminosities obtained by the various studies in the literature ($10^{43} <$

$L_{\mathrm{Ly}\alpha}/(\mathrm{erg\,s^{-1}}) < 10^{45}$) and assuming that the nebulae are associated with massive halos ($1 < M_{12} < 10$). Interestingly enough, similar densities are required in the pure recombination case to explain the reported limits on He II emission compared to Lyα for some cases, e.g., for the Slug Nebula (Arrigoni Battaia et al. 2015) and for the MUSE radio-quiet quasar nebulae (Borisova et al. 2016).

The analysis above shows that recombination radiation is a viable scenario to explain the observed Lyα luminosities and SB if the nebulae are composed by dense ($n > 1\,\mathrm{cm^{-3}}$), highly ionized, and cold ($T \sim 10^4$) clumps with volume filling factors smaller than 10^{-3} or, analogously, if the gas clumping factor is larger than about a thousand on projected scales with size of about 5–8 kpc. Assuming that such a population of clouds—not resolved by current cosmological simulation—do exist within massive halos, we should examine if they can be indeed ionized by the observed sources within or around the Lyα Nebulae. The required ionizing luminosities Q_{ion} are simply derived using photoionization equilibrium and considering that about 68% of recombinations produce Lyα photons (Case B) at $T \sim 10^4$ as discussed above:

$$Q_{\mathrm{ion}} \sim f_c^{-1} L_{\mathrm{Ly}\alpha} \times 10^{11}\ \mathrm{photons\,s^{-1}}, \tag{7}$$

where f_c is the covering factor of the gas as seen by the ionizing source. Assuming $f_c \sim 1$ this would imply that the required ionization luminosities to explain the Lyα emission of the giant Lyα nebulae ($10^{43} < L_{\mathrm{Ly}\alpha}/(\mathrm{erg\,s^{-1}}) < 10^{45}$) are in the range 10^{54}–$10^{56}\,\mathrm{s^{-1}}$. These ionization luminosity could be explained easily by AGN, even with modest luminosities, or by starbursts with star formation rates of at least a few hundred to few thousand solar masses per year (depending on the ionizing photon escape fraction from the galaxy ISM and on the stellar Initial Mass Function, e.g., Leitherer et al. 1999). As a reference, the quasar associated with the Slug Nebula, the largest and more luminous Lyα nebula discovered to date (Cantalupo et al. 2014) has an expected $Q_{\mathrm{ion}} \sim 10^{57.5}$ photons $\mathrm{s^{-1}}$ using the measured rest-frame UV luminosity and assuming a standard for this type of quasars (e.g., Lusso et al. 2015). Given the ubiquity of AGN and massively star forming galaxies (traced, e.g., in submm) within or around giant Lyα nebulae, there are therefore enough ionizing photons for the recombination scenario. In Sect. 4, I will review the possible origin of the illuminated gas in terms of kinematical signatures of inflows or outflows.

3.2 Continuum Pumping (Scattering)

Due to the resonant nature of the Lyα emission, the detected photons from extended nebulae may also "originate" from the ISM of embedded galaxies or from the broad line region of the associated AGN, rather than being produced "in situ" by recombination processes. A "scattering" scenario for some nebulae and in particular for LABs have been proposed in the past as a possible origin of the level of polarization detected in some sources, e.g., in SSA22-LAB1 by Hayes et al. (2011) and Beck et al. (2016), and in a radio-galaxy nebula by Humphrey et al. (2013). In other cases, radio-quiet nebulae imaged in polarization failed to show any detectable level of polarization, e.g., Prescott et al. (2011). Theoretical prediction for the level of polarization produced by scattering by a central sources have been presented, e.g., by Dijkstra and Loeb (2008) using idealized geometries and velocity fields and extended recently to radiative hydrodynamics simulations by Trebitsch et al. (2016). These models suggest that scattering from a central source is able to produce a larger and steeper polarization profile with respect to scattering by photons produced by "in situ" processes, such as collisional excitation. Comparison with the data cannot clearly exclude one of these two scenarios but current data, at least for SSA22-LAB1, seems more compatible with a "in situ" Lyα photon production rather than scattering from a "central" source.

In case a significant fraction of Lyα photons detected in the extended nebulae are produced by scattering from the "central" source, then all the estimates of masses and densities based on the assumption of pure recombination emission discussed above should be revised. The "scattering" contribution will be of course related to the density of neutral hydrogen atoms (n_{H^0}) rather than the ionized density squared as in the recombination case. The Lyα optical depth at line center is given by:

$$\tau_0 \sim 3.3 \times 10^{-14} T_4^{-1/2} N(HI), \tag{8}$$

where $N(HI)$ is the neutral hydrogen column density in units of cm^{-2}. As an example, using the estimated hydrogen ionization rates of the Slug Nebula quasar (UM287), i.e., $\Gamma \sim 7.7 \times 10^{-9}$ at distances of 100 kpc, we can derive $N(HI)$ in the simple case of uniform small clouds with individual densities of n and filling factor f_v exposed to ionizing radiation as discussed in the previous section. In photoionization equilibrium, the averaged neutral fraction of hydrogen along a path of length L from the quasar will be given by:

$$\langle x^{\rm eq} \rangle_L = L^{-1} \int_0^L x^{\rm eq} dL' \sim L^{-1} \int_0^L n\alpha(T)\Gamma^{-1} dL' \sim 10^{-9} L^2 n, \tag{9}$$

where the last term is valid at $T \sim 10^4$ K, $\alpha(T)$ is the hydrogen recombination coefficient ($\sim 3\times10^{-13}$ at $T_4 \sim 1$), L is the length in units of kpc. Because the $N(HI)$ column density up to a length L is given in this case by $N(HI, L) = \langle x^{\rm eq} \rangle_L \cdot n \cdot f_v \cdot L$, then the Ly$\alpha$ optical depth at line center is simply:

$$\tau_0 \sim 10^{-6} (L/1\,\mathrm{kpc})^3 (n/1\,\mathrm{cm}^{-2})^2 (f_v/10^{-5}). \tag{10}$$

Assuming that most of the scatterings are produced at line center, then a continuum photon produced by the AGN is only scattered into Lyα at distances larger than 100 kpc from the source, even for large clump densities. These distances are only reduced by about a factor of two for sources that are ten times fainter than

UM287. Because most of the emission is concentrated in the inner parts, scattering contribution in the case the gas is illuminated by the ionizing radiation of a source should be therefore negligible and recombination radiation should be the dominant mechanism.

The only case in which scattering may be the dominant production mechanism of the large Lyα emission of the giant nebulae is when the central source is not highly ionizing the gas but still producing copious amounts of Lyα and continuum photons slightly blueward of Lyα. In this case, studied numerically by, e.g., Cantalupo et al. (2014), the gas will be mostly neutral and the optical depths will become very large. The scattering will be extremely efficient and the main problem with this scenario would be to transport out to hundred kpc scales, as observed for the Lyα Nebula, the photons that are resonantly trapped in the inner regions. During scattering, the Lyα photons will perform a random walk both in space and in frequency. Unless the parameters are fine tuned, most of the scatterings in space will be very close to each other and the photons will escape, with a single fly-out, only when scattered in the wings of the line spectral distribution. In order to transport out these photons to another scattering location and increase the size of the Nebula, the column densities need to be large enough to produce scattering in the wings of the line, i.e., $N(HI) > 10^{21}\ \mathrm{cm}^{-2}$, resulting in much larger column densities and masses of neutral gas with respect to what observed in absorption around quasars (e.g., Prochaska et al. 2013; see also Cantalupo et al. 2014 for discussion). Although scattering and a mostly "neutral" scenario for the gas seems less plausible than recombination emission to explain the origin of giant Lyα nebulae, observations of non-resonant lines such He II 1640 and hydrogen Hα from giant Lyα Nebulae may be used in the next future to better constrain the scattering contribution to the emission.

3.3 Collisional Excitation (Cooling)

When both neutral hydrogen and free electrons are present and the electron temperature is around a few times 10^4 K, the collisional excitation rate of Lyα ($q_{\mathrm{Ly}\alpha}$) may be several orders of magnitude larger than the Lyα effective recombination rate (see, e.g., Cantalupo et al. 2008) and therefore produce in principle very strong Lyα emission from lower density gas. Differently than the recombination case, however, the collisional excitation rates are a very strong function of temperature, with an exponential decline for temperatures lower than a few times 10^4 K dropping strongly below the recombination rate just below $T \sim 10^4$ K. On the other hand, when the electron temperature approaches 10^5 K collisional ionization dominates over collisional excitations. As a result, collisional excitation is a very efficient process to produce Lyα photons in a partially ionized medium only for electron temperatures around 2–5 $\times$ 10^4 K. In absence of photoionization and heating sources, e.g., in collisional ionization equilibrium, radiative losses (cooling) due to collisional excited Lyα would quickly reduce the electron temperature effectively suppressing further Lyα emission. A source of heating is therefore required in order to balance the Lyα cooling effects. On the other hand, if heating brings electron temperatures close to 10^5 K, Lyα collisional excitation would be suppressed again, therefore the heating source should be fine tuned in order to produce a stable supply of Lyα emission with collisional excitation.

Since the discovery of the Steidel's LABs in SSA22, a numerous series of theoretical and numerical papers have addressed the possibility that these nebulae—apparently lacking a source of ionization—could be powered instead by Lyα cooling (including, e.g., Haiman and Rees 2001, Fardal et al. 2001; Furlanetto et al. 2005; Dijkstra and Loeb 2009; Goerdt et al. 2010; Faucher-Giguère et al. 2010; Rosdahl and Blaizot 2012). The suggested source of heat in all these studies is given by the conversion of the gravitational potential energy due to cosmological accretion of gas into massive dark matter halos. Alternative sources of energy such as galactic superwinds have also been proposed by Taniguchi and Shioya (2000) with a similar energy budget with respect to the gravitational accretion case but also sharing the same limitations as discussed below.

A simple estimate of the energy associated with cosmological gas accretion for a Navarro–Frenk–White profile with concentration parameter $c = 5$ gives (Faucher-Giguère et al. 2010):

$$\langle \dot{E}_{\rm grav} \rangle \sim 3.8 \times 10^{43} M_{12}^{1.8} [(1+z)/4]^{3.5} \ {\rm erg\,s^{-1}}. \tag{11}$$

For Lyα nebulae associated with very massive halos with $M \sim 10^{13} M_{\odot}$, this order of magnitude estimate for the potential energy is interestingly close to the observed Lyα emission. In our standard picture for the formation of these massive halos, we think, however, that a significant fraction of this potential energy should result in the formation of a hot gaseous halo with temperatures of the order of $T \sim 10^7$ K as a result of the virial shocks. At these temperatures, the gas would more slowly lose energy by other radiation channels rather than Lyα, e.g., free-free continuum radiation in the Extreme UV and X-ray, and, once the temperature decreases, further energy may be radiated away via line emission from He II or from heavier elements, if present. Analytical and numerical models in the past 10 years have nonetheless suggested that some fraction of the densest accreting material could not form a stable shock and therefore could cool and produce Lyα (e.g., Dekel and Birnboim 2006). Cooling Lyα emission models have therefore introduced an efficiency factor that takes into account how much of the potential energy is dissipated by this cold gas phase and required that this efficiency factor is at least few tens of percent in order to obtain enough Lyα photons. Unfortunately, it is very difficult to predict the value of this efficiency factor as it would depend on the details of accretion and cosmological structure formation and on a very accurate estimate of the ionization and temperature state of the gas. Depending on these details, many questions remain open about how and where within the halo the potential energy (or the energy produced by galactic superwinds) could be dissipated in form of Lyα.

The majority of the models discussed so far ignore, however, that sources of ionization do exist within all nebulae observed to date. In most cases, as I have reviewed in Sect. 2, these sources are AGN or massively star forming galaxies. The estimated ionization rates show that there are enough ionizing photons to power the nebulae and therefore it seems current clearer that a pure Lyα cooling origin of the nebulae is not necessarily required by the majority of the observations. This may be confirmed by duty cycle arguments and comparing the number densities of quasars, radio-galaxies, and overdense regions where LABs are found as discussed in previous section (see also Borisova et al. 2016 and Overzier et al. 2013 for further discussion). Questions, however, remain about the possible contribution of collisional Lyα emission to recombination radiation as this would have important consequences on the estimates of densities and masses of the nebulae. Addressing these questions would likely require a new generation of theoretical and numerical models able to resolve the physics and the small scales associated with dense regions within massive halos and, at the same time, the interplay between the gas and the local radiation sources (e.g., in terms of photoionization, photo-heating, and feedback).

4 Origin of the Emitting Gas, Kinematics, and Gas Flows

In the previous sections, we have seen that extended Lyα emission is a common phenomenon around bright high-redshift quasars (with possible differences below $z \sim 3$ for radio-quiet systems), radio-galaxies, and in overdense regions of the Universe. The ubiquity of AGN and massively star forming galaxies associated with these nebulae and simple analytical considerations discussed in Sect. 3 suggest that most of the emission may be due to recombination radiation from dense and cold clouds within the nebulae. Because the emission is sensitive to gas density squared, the emitting gas could be a small fraction of the total gas in and around the massive halos associated with these systems, both in terms of volume and mass. In this section, we will look at the kinematical signatures derived from Lyα spectra of these sources in order to address the questions about the possible origin and fate of this cold gas component.

The proposed origins for the emitting gas and its relation to gas flows in the literature include: (1) cosmological accretion from the IGM, (2) outflowing material from galaxy and AGN feedback, (3) in-situ formation from hot gas condensation. The suggested options about the fate of the gas include: (1) accretion or "recycling" into galaxies (inflow) or into gaseous disks (rotation), (2) expulsion from the galaxy halos towards the IGM, (3) disruption and thermalization into the hot halo gas. Each of these hypothesis could potentially leave an imprint into the observed gas kinematics in terms of Lyα spectral profile shapes, velocity shears, and velocity dispersion. However, it is important to stress that Lyα line is a resonant line and therefore any spectral information may also reflect complex radiative transfer effects rather than kinematics, especially if the gas is not highly ionized as it may be the case for a Lyα cooling origin of the emission.

The first spectroscopic measurements on giant Lyα nebulae were obtained on radio-galaxies and radio-loud quasar nebulae by McCarthy et al. (1987, 1990, 1996) and Heckman et al. (1991b). The properties of these systems from a kinematical point of view appeared remarkably similar: the Full Width Half Maximum (FWHM) of the Lyα emission presented large values typically around 1000–1500 km s^{-1} for all radio-loud nebulae. No hints of velocity shear in excess of 500 km s^{-1} were found, although radio-galaxy nebulae seemed to show a more complex kinematical structure. Given the alignment effect between the radio-jets and Lyα emission discussed in Sect. 2.2, these large FWHM were interpreted preferentially as being associated with powerful outflows from these AGN as a result of the jet-gas interaction (e.g., van Ojik et al. 1997; Humphrey et al. 2006). However, it was early recognized that such large velocity widths could have also been caused by gravitational motion in very massive halos (e.g., Heckman et al. 1991b). More recent spectroscopic observations of radio-loud systems confirmed these large FWHM (e.g., Villar-Martín et al. 2003, 2007; Humphrey et al. 2006; Roche et al. 2014) but found that some radio-galaxies show, in addition, extended and lower-SB halos that appear more kinematically quiet, e.g., with FWHM$\sim$500 km s^{-1}. Moreover, some of these extended "quiescent" regions show indication for velocity shifts of few hundred km s^{-1} that in some cases have been interpreted as a possible sign of rotating disks (e.g., Villar-Martín et al. 2007) or cosmological gas infall (e.g., Humphrey et al. 2007). These models are based on the correlation seen in several radio-galaxy nebulae, i.e., that the more redshifted side of the nebula is the brighter in both Lyα and radio flux, and on the interpretation that this side of the radio-lobes and associated nebula is closer to the observer. This is based on the fact that the radio jet directed towards the observer will be Doppler-boosted and on the assumption that the Lyα emitting gas closer to the observer will be less "absorbed," i.e., scattered to a larger projected area, by neutral gas in the halo. The relative redshift of this near-side relative to the observer would then imply that the emitting nebula has a significant component of infall towards the galaxy.

Kinematical signatures in radio-galaxy and radio-loud quasars are, however, difficult to interpret because of the complex interaction between the possibly accreting gas and the massive energy input of the radio jet. The analysis of extended, rest-frame optical emission lines from several HzRG, e.g., Nesvadba et al. (2006), suggests indeed that kinematics in the brightest parts of the radio-loud nebulae should be dominated by powerful outflows. These observations, together with the measured line ratios using, He II, C IV, O III, and S II with respect to Hα, Hβ, and Lyα, all suggest that this outflowing material is metal rich (solar or super-solar), dusty (A(Hβ) $\sim$1–4 mag), and very dense ($n_e \sim$ 500 cm^{-3}) (see, e.g., Villar-Martín et al. 2003; Nesvadba et al. 2008).

Long-slit and integral field spectroscopy of radio-quiet nebulae show instead that the large majority of these systems have much smaller FWHM, i.e., FWHM$\sim$300–500 km s^{-1} than radio-loud systems (e.g., Weidinger et al. 2005; Christensen et al. 2006; Arrigoni Battaia et al. 2015; Borisova et al. 2016), unless they are associated with very overdense environment like in the case of SSA22-LAB1 and SSA22-LAB2 (e.g., Matsuda et al. 2006) or in the case of the "Jackpot" Nebula (Hennawi et al. 2015). These smaller velocity widths are indicative of better environments where signs of infall or rotation could be studied more in detail without the need to disentangle them from the broad component associated with the radio-jets and powerful outflows.

In particular, some radio-quiet giant Lyα Nebulae, such as the Slug Nebula (Cantalupo et al. 2014), are extended by several hundred of kpc and therefore the kinematics in their external parts are less likely to be "contaminated" by AGN outflows, if present. Long-slit spectroscopy (Arrigoni Battaia et al. 2015) and integral field observations (Martin et al. 2015) have revealed relatively narrow Lyα emission (FWHM$<$ 350 km s^{-1}) at distances larger than 100 kpc from the quasar and apparent velocity shears of about 800 km s^{-1} that seem coherent over the brightest part of the Nebula (i.e., over 200 kpc in projected length). These velocity shears have been interpreted as evidence for rotation and would imply therefore that a massive, gaseous disk-lie structure with size of about 150 kpc could be present in the brightest part of the Nebula (Martin et al. 2015) while the more tenuous, extended part have been interpreted as signatures of gas infall from the Intergalactic Medium. In particular, the velocity profile seems consistent with the rotation curve of a disk within an NFW profile of a massive halo but only if the "center" of the disk-like structure is located 25$''$ away from the main quasar, in a region of low SB. Considering only the high SB regions, the spectra could also be interpreted, however, as arising from projection effects of two different structures located at about 500–800 km s^{-1} in velocity space away from each other. The presence of a relatively bright continuum source in the "redshifted" part of the velocity profile could support this interpretation. Ongoing observations of other emission lines such as He II, C IV (e.g., with MUSE), and Hα (e.g., with Keck/MOSFIRE) will help disentangling these two possibilities.

Other spectroscopic observations of radio-quiet nebulae (e.g., Martin et al. 2014; Prescott et al. 2015) have found signatures of coherent velocity fields over scales of several tens of kpc that have been interpreted as evidence of infall (e.g., Martin et al. 2014) or large-scale rotation in a disk (Prescott et al. 2015). In particular, the observations of Prescott et al. (2015) of coherent velocity shears of about $\sim$500 km s^{-1} in both Lyα and in non-resonant lines such as He II within the central 50 kpc of a 80 kpc-sized Lyα nebula at $z \sim 1.67$ provide evidences that the extended gas in this system is produced in situ by recombination radiation and settled in a rotating disk that is at least partially stable against collapse. In a larger spectroscopic study of eight small radio-quiet nebulae at $z \sim 2.3$ including rest-frame optical emission lines (mostly OIII and Hα), Yang et al. (2014) found instead no significant evidences for bulk motions such as inflow, rotation, or outflows and suggested that the gas should be "stationary" or slowly outflowing at speed less than 250 km s^{-1} with respect to the central galaxies in these systems.

The MUSE observations of about 17 bright radio-quiet quasars and 2 radio-loud quasars at $3 < z < 4$ presented by Borisova et al. (2016) provided the first large statistical sample of giant ($>$100 kpc) Lyα nebulae with full kinematical information from integral field spectroscopy over their full detectable extent. Figures 3 and 4 show, respectively, the maps of the first and second moment of the flux distribution, i.e., the flux-weighted velocity centroid shift and the dispersion relative to the peak of the integrated Lyα emission for each of the MUSE Quasar Nebulae (MQN) (figures taken from Borisova et al. 2016). While some systems, e.g., MQN15, show possible evidences of rotation in a disk-like structure with a velocity shear of about

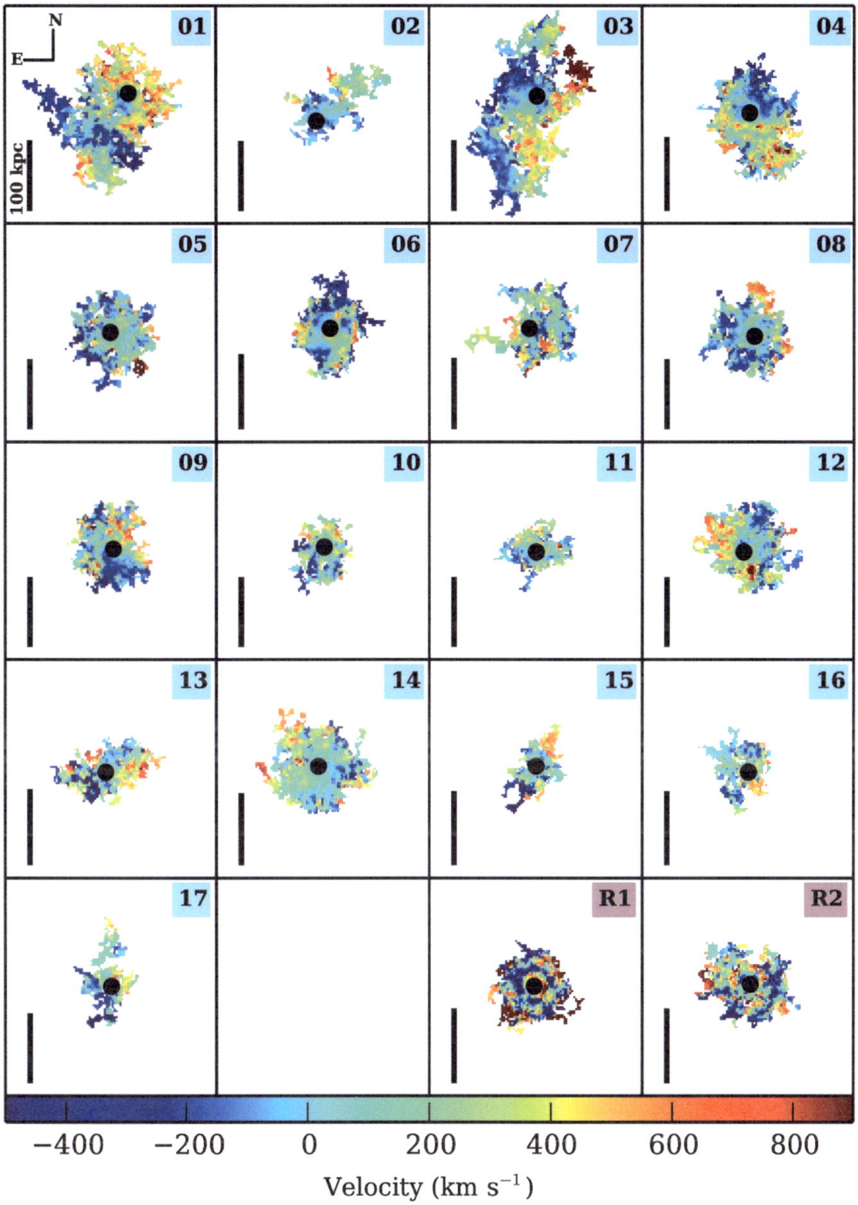

Fig. 3 "Velocity maps" of the MUSE Quasar Nebulae presented in Fig. 2 obtained from the first moment of the flux distribution (see Borisova et al. 2016 for details). This is the largest sample to date of kinematical maps of giant Lyα nebulae obtained with integral-field-spectroscopy. As discussed in Sect. 4, while some systems (e.g., MQN15) show possible evidences of rotation in a disk-like structure the majority of the nebulae do not show clear evidences of rotation or other ordered kinematic patterns. Several nebulae show instead coherent kinematical structures over scales as large as 100 kpc (e.g., MQN01 and MQN03) (Figure reproduced with permission from Borisova et al. 2016)

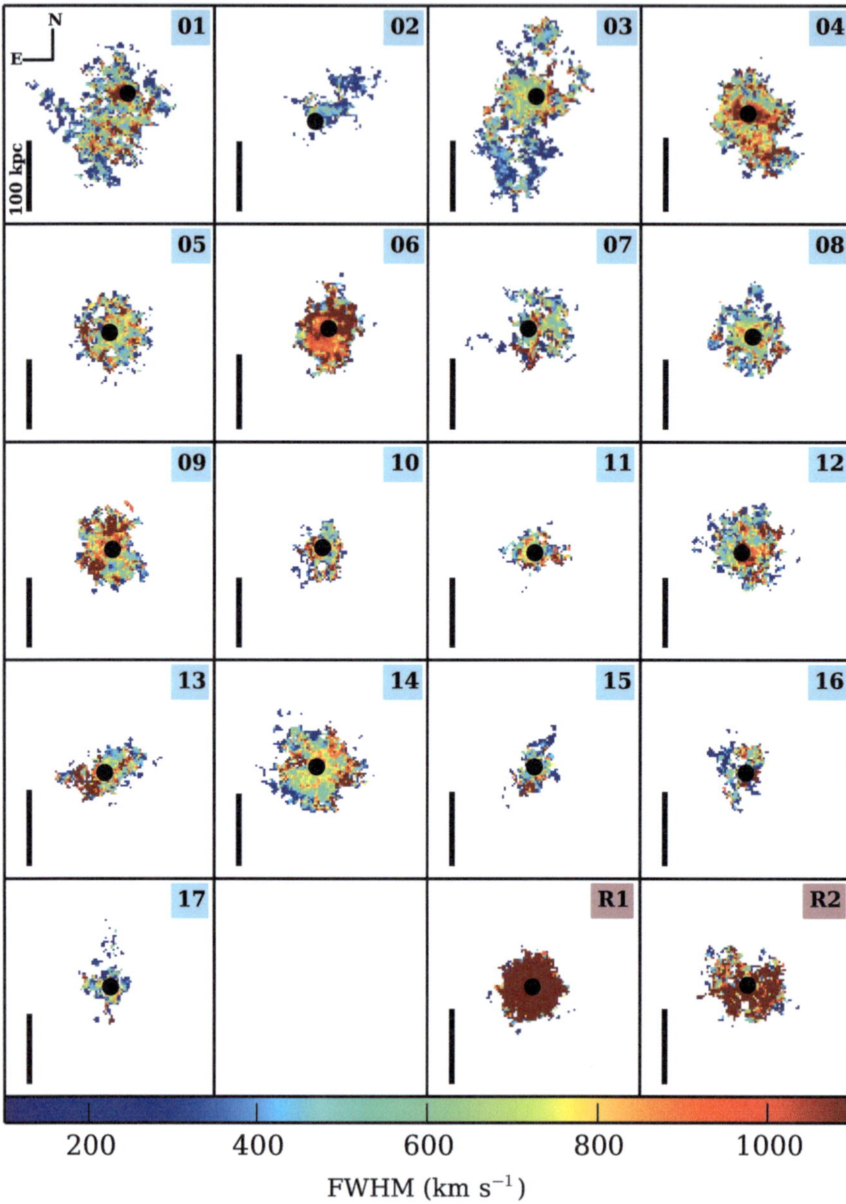

Fig. 4 "Velocity dispersion maps" of the MUSE Quasar Nebulae presented in Fig. 2 obtained from the second moment of the flux distribution and expressed in terms of Gaussian-equivalent FWHM (see Borisova et al. 2016 for details). This figure shows that the large majority of radio-quiet nebulae are narrower in Lyα emission (FWHM$\sim$ 500 km s^{-1}) than radio-loud systems, i.e., MQN-R1 and MQN-R2 (FWHM> 1000 km s^{-1}), in agreement with the overall kinematical results discussed in Sect. 4 (Figure reproduced with permission from Borisova et al. 2016)

800 km s^{-1}, the majority of the MQN do not show clear evidences of rotation or other ordered kinematic patterns. Several MQN, especially the largest ones, show instead coherent kinematical structures over scales as large as 100 kpc, e.g., MQN01 and MQN03. The velocity dispersions, expressed in terms of Gaussian-equivalent FWHM, clearly shows the main difference between radio-quiet and radio-loud systems: the large majority of radio-quiet nebulae are narrower in Lyα emission (FWHM$\sim$ 500 km s^{-1}) than radio-loud systems (FWHM$>$ 1000 km s^{-1}) in agreement with previous results discussed in this section. The only exception is MQN06 but this nebula is peculiar and more similar to radio-loud nebulae in terms of all the properties studied in Borisova et al. (2016), including the SB profiles and higher He II/Lyα and C IV/Lyα ratios with respect to the non-detection for radio-quiet systems.

The emerging picture from these observations seems to suggest therefore that kinematics in radio-loud nebulae may be dominated by ionized outflows of relatively cold and metal-enriched material within at least the inner 30–50 kpc from the AGN, while, on average, the ionized and clumpy gas in radio-quiet nebulae may be in a more "stationary" situation and in some cases settled in a possibly rotating structure. Clear evidences from Lyα emission for gas accretion into galaxies from these cold gas reservoirs are not currently detected, either because "washed-out" by Lyα radiative transfer effects or because their magnitude and projection effects could make them to small to be detected with current facilities. Future deep surveys using other bright, non-resonant lines such as hydrogen Hα or He II 1640 would be extremely helpful to search for small velocity shears and therefore for signature of cosmological gas accretion onto galaxies and AGN.

5 Summary

Several decades of observations and discoveries have produced an extensive literature of large and giant Lyα nebulae. Depending on the technique and original target of the observation, i.e., quasars, radio-galaxies, overdense regions, or "blank" fields, they have been classified in various ways and with different nomenclatures. For historical and practical reasons, I have divided them in quasar Lyα nebulae, radio-galaxy halos, and Lyα Blobs (LAB) but, as discussed in Sect. 2, their comparable volume densities, luminosities, and their almost invariable association with AGN or massively star forming galaxies suggest, however, that these nebulae could be just apparently different manifestations of the same phenomenon.

Among the three processes associated with the production of Lyα photons (recombination radiation, "scattering," and collisional excitation; see Sect. 3), recombination radiation currently appears as the most viable scenario to explain the observed Lyα luminosities and Surface Brightness for the large majority of the nebulae. If contribution from scattering and Lyα collisional excitation is negligible, this would imply that the emitting gas should be in the form of dense ($n > 1\,\mathrm{cm}^{-3}$), highly ionized, and cold ($T \sim 10^4$) structures ("clumps") with volume filling factors smaller than 10^{-3} or, analogously, with clumping factors larger than about a thousand on projected scales with size of about 5–8 kpc. The apparent ubiquity of giant Lyα nebulae around bright AGN, at least at $3 < z < 4$, and in overdense environment suggests that this cold gas should be a common occurrence within and around the halos of massive galaxies. Deep observations of non-resonant lines such as He II 1640 and hydrogen Hα from giant Lyα nebulae can better constrain the scattering and collisional contribution to the emission but they are currently lacking for the majority of the nebulae. Ongoing surveys (e.g., with MUSE, MOSFIRE, KMOS, and JWST in the future) will soon provide these data and therefore potentially help in refining our understanding of the physical properties of this cold gas.

Lyα integral-field and long-list spectroscopy shows that radio-loud nebulae (i.e., associated with radio-galaxies and radio-loud quasars) are almost invariably associated with larger velocity widths with respect to the majority of radio-quiet systems (see Sect. 4). Together with the analysis of the He II/Lyα and C IV/Lyα ratios, these observations seem to suggest that kinematics in radio-loud nebulae may be dominated by ionized outflows of relatively cold and metal-enriched material within at least the inner 30–50 kpc from the AGN. On the other hand, these observations suggest that ionized and clumpy gas in radio-quiet nebulae should be in a more "stationary" situation and only in some cases there are possible evidences that this gas is settled in rotating structures. Definitive evidences for the accretion of this cold gas into galaxies from Lyα emission are not clearly detected in the current data. However, it is important to notice that accretion signatures could be "washed-out" by Lyα radiative transfer effects or too small to be detected with current facilities because of their magnitude and possible projection effects. Again, future deep surveys using other bright, non-resonant lines such as hydrogen Hα or He II 1640 would be extremely helpful to search for small velocity shears and therefore for clearer signature of cosmological gas accretion onto galaxies and AGN.

Acknowledgements I gratefully acknowledge support from Swiss National Foundation grant PP00P2_163824

References

Antonucci, R. 1993, ARA&A, 31, 473
Antonucci, R. 2012, Astronomical and Astrophysical Transactions, 27, 557
Ao, Y., Matsuda, Y., Beelen, A., et al. 2015, A&A, 581, A132
Arrigoni Battaia, F., Hennawi, J. F., Prochaska, J. X., & Cantalupo, S. 2015, ApJ, 809, 163
Arrigoni Battaia, F., Hennawi, J. F., Cantalupo, S., & Prochaska, J. X. 2016, ApJ, 829, 3
Baker, J. G., & Menzel, D. H. 1938, ApJ, 88, 52
Barthel, P. D. 1989, ApJ, 336, 606
Basu-Zych, A., & Scharf, C. 2004, ApJ, 615, L85
Beck, M., Scarlata, C., Hayes, M., Dijkstra, M., & Jones, T. J. 2016, ApJ, 818, 138
Bergeron, J., Dennefeld, M., Boksenberg, A., & Tarenghi, M. 1983, MNRAS, 202, 125
Bergeron, J., Petitjean, P., Cristiani, S., et al. 1999, A&A, 343, L40

Borisova, E., Cantalupo, S., Lilly, S. J., et al. 2016, ApJ, 831, 39
Bremer, M. N., Fabian, A. C., Sargent, W. L. W., et al. 1992, MNRAS, 258, 23P
Bunker, A., Smith, J., Spinrad, H., Stern, D., & Warren, S. 2003, Ap&SS, 284, 357
Cai, Z., Fan, X., Yang, Y., et al. 2016, ApJ, in press (arXiv:1609.04021)
Cantalupo, S., Porciani, C., Lilly, S. J., & Miniati, F. 2005, ApJ, 628, 61
Cantalupo, S., Lilly, S. J., & Porciani, C. 2007, ApJ, 657, 135
Cantalupo, S., Porciani, C., & Lilly, S. J. 2008, ApJ, 672, 48–58
Cantalupo, S., Lilly, S. J., & Haehnelt, M. G. 2012, MNRAS, 425, 1992
Cantalupo, S., Arrigoni-Battaia, F., Prochaska, J. X., Hennawi, J. F., & Madau, P. 2014, Nature, 506, 63
Chapman, S. C., Lewis, G. F., Scott, D., et al. 2001, ApJ, 548, L17
Chapman, S. C., Scott, D., Windhorst, R. A., et al. 2004, ApJ, 606, 85
Dey, A., Bian, C., Soifer, B. T., et al. 2005, ApJ, 629, 654
Dekel, A., & Birnboim, Y. 2006, MNRAS, 368, 2
Dijkstra, M., & Loeb, A. 2008, MNRAS, 386, 492
Dijkstra, M., & Loeb, A. 2009, MNRAS, 400, 1109
Christensen, L., Jahnke, K., Wisotzki, L., & Sánchez, S. F. 2006, A&A, 459, 717
Courbin, F., North, P., Eigenbrod, A., & Chelouche, D. 2008, A&A, 488, 91
Djorgovski, S., Spinrad, H., McCarthy, P., & Strauss, M. A. 1985, ApJ, 299, L1
Eales, S. A., Rawlings, S., Dickinson, M., et al. 1993, ApJ, 409, 578
Fardal, M. A., Katz, N., Gardner, J. P., et al. 2001, ApJ, 562, 605
Faucher-Giguère, C.-A., Kereš, D., Dijkstra, M., Hernquist, L., & Zaldarriaga, M. 2010, ApJ, 725, 633
Francis, P. J., Woodgate, B. E., Warren, S. J., et al. 1996, ApJ, 457, 490
Francis, P. J., Williger, G. M., Collins, N. R., et al. 2001, ApJ, 554, 1001
Fried, J. W. 1998, A&A, 331, L73
Fumagalli, M., Cantalupo, S., Dekel, A., et al. 2016, MNRAS, 462, 197
Furlanetto, S. R., Schaye, J., Springel, V., & Hernquist, L. 2005, ApJ, 622, 7
Geach, J. E., Matsuda, Y., Smail, I., et al. 2005, MNRAS, 363, 1398
Geach, J. E., Alexander, D. M., Lehmer, B. D., et al. 2009, ApJ, 700, 1
Goerdt, T., Dekel, A., Sternberg, A., et al. 2010, MNRAS, 407, 613
Haiman, Z., & Rees, M. J. 2001, ApJ, 556, 87
Hayes, M., Scarlata, C., & Siana, B. 2011, Nature, 476, 304
Heckman, T. M., Baum, S. A., van Breugel, W. J. M., & McCarthy, P. 1989, ApJ, 338, 48
Heckman, T. M., Miley, G. K., Lehnert, M. D., & van Breugel, W. 1991a, ApJ, 370, 78
Heckman, T. M., Lehnert, M. D., Miley, G. K., & van Breugel, W. 1991b, ApJ, 381, 373
Hennawi, J. F., & Prochaska, J. X. 2013, ApJ, 766, 58
Hennawi, J. F., Prochaska, J. X., Cantalupo, S., & Arrigoni-Battaia, F. 2015, Science, 348, 779
Herenz, E. C., Wisotzki, L., Roth, M., & Anders, F. 2015, A&A, 576, A115
Hine, N. K., Geach, J. E., Matsuda, Y., et al. 2016, MNRAS, 460, 4075
Hu, E. M., & Cowie, L. L. 1987, ApJ, 317, L7
Hu, E. M., Songaila, A., Cowie, L. L., & Stockton, A. 1991, ApJ, 368, 28
Humphrey, A., Villar-Martín, M., Fosbury, R., Vernet, J., & di Serego Alighieri, S. 2006, MNRAS, 369, 1103
Humphrey, A., Villar-Martín, M., Fosbury, R., et al. 2007, MNRAS, 375, 705
Humphrey, A., Villar-Martín, M., Sánchez, S. F., et al. 2008, MNRAS, 390, 1505
Humphrey, A., Vernet, J., Villar-Martín, M., et al. 2013, ApJ, 768, L3
Keel, W. C., Cohen, S. H., Windhorst, R. A., & Waddington, I. 1999, AJ, 118, 2547
Kollmeier, J. A., Zheng, Z., Davé, R., et al. 2010, ApJ, 708, 1048
Kurk, J. D., Röttgering, H. J. A., Pentericci, L., et al. 2000, A&A, 358, L1
Leitherer, C., Schaerer, D., Goldader, J. D., et al. 1999, ApJS, 123, 3
Lilly, S. J., Cowie, L. L., & Gardner, J. P. 1991, ApJ, 369, 79
Lusso, E., Worseck, G., Hennawi, J. F., et al. 2015, MNRAS, 449, 4204
Martin, D. C., Chang, D., Matuszewski, M., et al. 2014, ApJ, 786, 106
Martin, D. C., Matuszewski, M., Morrissey, P., et al. 2015, Nature, 524, 192

Matsuda, Y., Yamada, T., Hayashino, T., et al. 2004, AJ, 128, 569
Matsuda, Y., Yamada, T., Hayashino, T., Yamauchi, R., & Nakamura, Y. 2006, ApJ, 640, L123
McCarthy, P. J., Spinrad, H., Djorgovski, S., et al. 1987, ApJ, 319, L39
McCarthy, P. J., Spinrad, H., Dickinson, M., et al. 1990, ApJ, 365, 487
McCarthy, P. J. 1993, ARA&A, 31, 639
McCarthy, P. J., Baum, S. A., & Spinrad, H. 1996, ApJS, 106, 281
Møller, P., Warren, S. J., Fall, S. M., Jakobsen, P., & Fynbo, J. U. 2000, The Messenger, 99, 33
Nesvadba, N. P. H., Lehnert, M. D., Eisenhauer, F., et al. 2006, ApJ, 650, 693
Nesvadba, N. P. H., Lehnert, M. D., De Breuck, C., Gilbert, A. M., & van Breugel, W. 2008, A&A, 491, 407
Nilsson, K. K., Fynbo, J. P. U., Møller, P., Sommer-Larsen, J., & Ledoux, C. 2006, A&A, 452, L23
North, P. L., Courbin, F., Eigenbrod, A., & Chelouche, D. 2012, A&A, 542, A91
Osterbrock, D. E. 1989, University Science Books, 1989, 422 p.,
Overzier, R. A., Nesvadba, N. P. H., Dijkstra, M., et al. 2013, ApJ, 771, 89
Palunas, P., Teplitz, H. I., Francis, P. J., Williger, G. M., & Woodgate, B. E. 2004, ApJ, 602, 545
Pentericci, L., Roettgering, H. J. A., Miley, G. K., Carilli, C. L., & McCarthy, P. 1997, A&A, 326, 580
Prescott, M. K. M., Dey, A., & Jannuzi, B. T. 2009, ApJ, 702, 554
Prescott, M. K. M., Smith, P. S., Schmidt, G. D., & Dey, A. 2011, ApJ, 730, L25
Prescott, M. K. M., Dey, A., & Jannuzi, B. T. 2013, ApJ, 762, 38
Prescott, M. K. M., Martin, C. L., & Dey, A. 2015, ApJ, 799, 62
Prochaska, J. X., Hennawi, J. F., Lee, K.-G., et al. 2013, ApJ, 776, 136
Reuland, M., van Breugel, W., Röttgering, H., et al. 2003, ApJ, 592, 755
Roche, N., Humphrey, A., & Binette, L. 2014, MNRAS, 443, 3795
Rosdahl, J., & Blaizot, J. 2012, MNRAS, 423, 344
Röttgering, H. J. A., Lacy, M., Miley, G. K., Chambers, K. C., & Saunders, R. 1994, A&AS, 108, 79
Saito, T., Shimasaku, K., Okamura, S., et al. 2006, ApJ, 648, 54
Sánchez, S. F., & Humphrey, A. 2009, A&A, 495, 471
Scarlata, C., Colbert, J., Teplitz, H. I., et al. 2009, ApJ, 706, 1241
Shen, Y., Strauss, M. A., Oguri, M., et al. 2007, AJ, 133, 2222
Schneider, D. P., Gunn, J. E., Turner, E. L., et al. 1987, AJ, 94, 12
Smith, D. J. B., Jarvis, M. J., Simpson, C., & Martínez-Sansigre, A. 2009, MNRAS, 393, 309
Steidel, C. C., Sargent, W. L. W., & Dickinson, M. 1991, AJ, 101, 1187
Steidel, C. C., Adelberger, K. L., Dickinson, M., et al. 1998, ApJ, 492, 428
Steidel, C. C., Adelberger, K. L., Shapley, A. E., et al. 2000, ApJ, 532, 170
Steidel, C. C., Bogosavljević, M., Shapley, A. E., et al. 2011, ApJ, 736, 160
Taniguchi, Y., & Shioya, Y. 2000, Apj, 532, L13
Trainor, R. F., & Steidel, C. C. 2012, ApJ, 752, 39
Trebitsch, M., Verhamme, A., Blaizot, J., & Rosdahl, J. 2016, A&A, 593, A122
Ueda, Y., Akiyama, M., Ohta, K., & Miyaji, T. 2003, ApJ, 598, 886
Valentino, F., Daddi, E., Finoguenov, A., et al. 2016, ApJ, 829, 53
van Ojik, R., Roettgering, H. J. A., Carilli, C. L., et al. 1996, A&A, 313, 25
van Ojik, R., Roettgering, H. J. A., Miley, G. K., & Hunstead, R. W. 1997, A&A, 317, 358
Veilleux, S., Shopbell, P. L., Rupke, D. S., Bland-Hawthorn, J., & Cecil, G. 2003, AJ, 126, 2185
Venemans, B. P., Röttgering, H. J. A., Miley, G. K., et al. 2007, A&A, 461, 823
Villar-Martín, M., Vernet, J., di Serego Alighieri, S., et al. 2003, MNRAS, 346, 273
Villar-Martín, M., Sánchez, S. F., Humphrey, A., et al. 2007, MNRAS, 378, 416
Weidinger, M., Møller, P., & Fynbo, J. P. U. 2004, Nature, 430, 999
Weidinger, M., Møller, P., Fynbo, J. P. U., & Thomsen, B. 2005, A&A, 436, 825
Wisotzki, L., Bacon, R., Blaizot, J., et al. 2016, A&A, 587, A98
Yang, Y., Zabludoff, A., Tremonti, C., Eisenstein, D., & Davé, R. 2009, ApJ, 693, 1579
Yang, Y., Zabludoff, A., Jahnke, K., & Davé, R. 2014, ApJ, 793, 114

Gas Accretion and Galactic Chemical Evolution: Theory and Observations

Kristian Finlator

This chapter reviews how galactic inflows influence galaxy metallicity. The goal is to discuss predictions from theoretical models, but particular emphasis is placed on the insights that result from using models to interpret observations. Even as the classical "G-dwarf problem" endures in the latest round of observational confirmation, a rich and tantalizing new phenomenology of relationships between M_*, Z, SFR, and gas fraction is emerging both in observations and in theoretical models. A consensus interpretation is emerging in which star-forming galaxies do most of their growing in a quiescent way that balances gas inflows and gas processing, and metal dilution with enrichment. Models that explicitly invoke this idea via equilibrium conditions can be used to infer inflow rates from observations, while models that do not assume equilibrium growth tend to recover it self-consistently. Mergers are an overall subdominant mechanism for delivering fresh gas to galaxies, but they trigger radial flows of previously accreted gas that flatten radial gas-phase metallicity gradients and temporarily suppress central metallicities. Radial gradients are generically expected to be steep at early times and then flattened by mergers and enriched inflows of recycled gas at late times. However, further theoretical work is required in order to understand how to interpret observations. Likewise, more observational work is needed in order to understand how metallicity gradients evolve to high redshifts.

K. Finlator (✉)
New Mexico State University, Las Cruces, NM, USA
e-mail: finlator@nmsu.edu

A. Fox, R. Davé (eds.), *Gas Accretion onto Galaxies*, Astrophysics and Space Science Library 430, DOI 10.1007/978-3-319-52512-9_10

1 Introduction

Theoretically, inflows happen. The classical and perhaps most famous motivation for this view is the failure of the "closed-box" model (Schmidt 1963; Tinsley 1980) to account for the observed paucity of low-metallicity G and K stars (Pagel and Patchett 1975; Casuso and Beckman 2004), a discrepancy that persists even with the most recent measurements of stars throughout the Milky Way disk (Schlesinger et al. 2012). The closed-box model has other problems such as its inability to account for the slow decline in galaxy gas fractions (Tacconi et al. 2013) and the cosmic abundance of neutral hydrogen (Wolfe et al. 2005). Likewise, it cannot account for the weak observed evolution in galaxy metallicities during the interval $z = 2 \rightarrow 0$ (Erb et al. 2006), an epoch during which most of the present-day stellar mass formed.

More importantly, however, galaxy growth *without* inflows is theoretically incompatible with the current ΛCDM paradigm. In this picture, galaxies are viewed as condensations of cold baryons within dark matter halos. The dark matter halos themselves grow via a sequence of mergers that is decoupled from baryon physics and straightforward to compute using either analytic (White and Rees 1978; White and Frenk 1991) or numerical methods (Springel et al. 2005). As the halos grow, they accrete gas directly from the intergalactic medium (IGM). The vast majority of this gas accretes in a smooth fashion; that is, it does not arrive having previously condensed into an interstellar medium (ISM; Nelson et al. 2013). Halos that are more massive than the cosmic Jeans mass are expected to acquire a mass of baryons that is of order $\frac{\Omega_b}{\Omega_M} \times M_{\mathrm{DM}}$, where M_{DM} is the halo's total mass (Gnedin 2000; Okamoto et al. 2008; Christensen et al. 2016). Roughly half of this material collapses from the halo onto the central galaxy (Christensen et al. 2016), driving further star formation.

The expected thermal history of collapsed gas prior to its arrival in the central galaxy remains a topic of active study. It was originally assumed that all gas is shock-heated to the virial temperature and then cools in a spherically symmetric way (White and Rees 1978). This was challenged a decade ago by numerical calculations, which found that much of the gas accretes directly onto the central galaxy without ever being heated, particularly at masses below $10^{12} M_\odot$ (Kereš et al. 2005; Dekel and Birnboim 2006). The most recent calculations that include significantly improved hydrodynamic solvers contradict those results, attributing the lack of shock-heating and the inefficient cooling of the hot gas in previous calculations to numerical problems (Nelson et al. 2013). The new calculations indicate that the majority of gas at all halo masses is heated to the virial temperature before accreting onto the halo. However, it does not accrete in a spherically symmetric fashion as originally envisioned (White and Rees 1978). Instead, it tends to concentrate in coherent structures that connect to large-scale intergalactic medium (IGM) filaments. The upshot is that, one way or another, gas readily accretes efficiently enough in ΛCDM to form the observed galaxy populations, with most gas arriving in the form of smooth inflows.

Once the gas condenses to densities of $\sim$1 atom per cm^{-3}, gravitational instability triggers the formation of molecular clouds and eventually stars. Feedback energy from the young stars limits the efficiency of star formation and regulates the ISM's structure in a number of ways. For our purposes, the most important of these is the generation of galactic outflows, which are inevitably observed wherever there is vigorous star formation (Veilleux et al. 2005). Theoretical models consistently predict that the mass of material that is ejected is comparable to or greater than the mass of stars that form (Murray et al. 2005; Muratov et al. 2015; Christensen et al. 2016). This enriched material then becomes available for re-accretion after a few dynamical times (Oppenheimer et al. 2010; Henriques et al. 2013; Christensen et al. 2016).

Outflows thus give rise to two conceptually distinct gas accretion channels, "Primordial Gas" and "Recycled Gas." Primordial gas dominates inflows at early times and low masses (Oppenheimer et al. 2010; Ma et al. 2016), and it dilutes galaxies' gas-phase metallicities. Recycled gas becomes increasingly important at late times and high masses. It is pre-enriched, and therefore less effective at dilution.

To summarize, in the era of ΛCDM, galaxy growth driven by ongoing inflows is *unavoidable*. The central conceit of this chapter is that measurements of galaxy metallicities may be used to test models of those inflows. To motivate our discussion of how they do so, we list the observational probes that have been deployed:

- Stellar metallicity distributions;
- The slope, normalization, and evolution of the mass-metallicity relation (MZR);
- Third-parameter dependencies of metallicity on SFR, gas fraction, redshift, and environment; and
- Radial metallicity gradients (chiefly of the gas).

Stellar metallicity distributions have historically been an important indicator that inflows occur, but they are only available for the Milky Way and a handful of its satellite galaxies (Kirby et al. 2011). For this reason, we will not discuss them further. Rather, we will focus on extragalactic diagnostics where larger samples are available. We also note that, throughout this discussion, we will focus on the oxygen metallicity as it is the most widely observed tracer of the overall gas-phase metallicity. In Sect. 2, we review the physical processes through which inflows modulate galaxy metallicities. In Sect. 3, we discuss the extent to which galaxy growth tracks the host halo growth. In Sect. 4, we introduce the Equilibrium Model, which is the simplest way for relating observables to inflows. In Sect. 5, we discuss departures from equilibrium growth. Finally, in Sect. 6 we summarize.

2 Physical Processes

Gas inflows impact galaxy metallicities on a wide range of timescales and spatial scales, and through a variety of processes that we associate either with Mergers, Galactic Fountains, or Environment. In this section, we summarize observational and theoretical progress in understanding these processes.

2.1 Mergers

Both observations and theory indicate that mergers are not the dominant way in which fresh gas is delivered to galaxies (Papovich et al. 2011; Behroozi et al. 2013). However, on timescales comparable to a dynamical time, mergers can cause relatively unenriched gas that has previously settled in a galaxy's outskirts to plummet into the central 100–1000 parsecs (pc), simultaneously suppressing the central gas metallicity and boosting the central star formation rate.

This effect arises naturally in hydrodynamic simulations of galaxy mergers. Torrey et al. (2012) found that interactions suppress the nuclear metallicity of gas-poor galaxies (gas fractions of $\leq$20%) by an amount that fluctuates depending on which merger stage the galaxies are observed in, but is typically $\approx$0.07 dex. Gas-rich galaxies, by contrast, can experience nuclear metallicity boosts during interactions although this effect is not expected in typical SDSS galaxies owing to their generally low gas fractions. A similar result was presented by Rupke et al. (2010a), who found that the suppressed metallicities were associated with the period between the first and second pericenter passages of a merger event.

Does this mean that metallicities can be used to detect mergers? In fact, the statistical signature of nuclear inflows has been identified observationally in galaxy pairs from the Sloan Digital Sky Survey (SDSS). In particular, Scudder et al. (2012, 2013) showed that star-forming galaxies have suppressed central metallicity (by 0.02–0.3 dex) and enhanced central SFR (by 60%) if they are separated from a paired galaxy by a projected separation of $\leq$80 kpc h^{-1} (see also Rupke et al. 2010b). It is also possible that galaxy mass and merger mass ratio modulate this effect: Splitting the merging SDSS galaxies by stellar mass, Michel-Dansac et al. (2008) have argued that mergers systematically boost the metallicity of low-mass galaxies while suppressing the metallicity of massive ones. The different behavior reflects a competition between dilution and enrichment, and is qualitatively expected based on the results presented by Rupke et al. (2010a) if low-mass galaxies are relatively gas-rich while massive ones are gas-poor.

While the ability of mergers to drive nuclear flows has been identified both in simulations and in observations, these flows have yet to be invoked as a rigorous constraint on galaxy evolution models. As a first step in this direction, Grønnow et al. (2015) asked whether a small but robust population of galaxies whose metallicities were lower than expected for their combination of stellar mass and star formation rate (Mannucci et al. 2010) could be interpreted as ongoing mergers or merger remnants. They developed a simple analytical model invoking the known halo merger rate along with the assumption that mergers boost SFR and suppress metallicity for a set time and showed that this model's parameters can be tuned to yield excellent agreement with the observed outlier population. They found that post-merger galaxies must exhibit diluted nuclear metallicities for $1.568^{+0.020}_{-0.027}$ Gyr, and the average dilution for mergers with mass ratios of 1:1–1:5 was 0.114 dex. Encouragingly, these inferences are within a factor of two of expectations from numerical simulations and observations.

A caveat to analyses of SDSS metallicities is that the SDSS spectroscopic fibers subtend a diameter of 3 arcsec, hence they capture only the central 5 kpc of galaxies at the typical redshift of SDSS galaxies ($z = 0.1$). While this makes them ideal for studying *nuclear* metallicities, it leaves unanswered the question of how mergers effect the metallicity at larger radii. Rupke et al. (2010a) showed that the same interaction-triggered flows that dilute central metallicities also flatten overall gas metallicity gradients. This effect has likewise been observed in ongoing mergers (Kewley et al. 2010). Unfortunately, it is difficult to use the physics of ongoing mergers to test galaxy formation models because results are so sensitive to details of the merger such as gas fractions, mass ratio, and orbital parameters (Torrey et al. 2012).

Fortunately, the tendency for mergers to flatten metallicity gradients and to build up central bulges does leave an observable signature: Galaxies that are more massive or have larger (classical) bulges should have flatter gas metallicity gradients. Fu et al. (2013) have shown that this effect is strong observationally, and can readily be reproduced by their semi-analytical model (their Fig. 12). However, they did not explore which of the assumptions underlying their merger model are required to yield the good agreement. Moreover, they address only local observations, and suggest that extending the work to higher redshifts would be a useful next step. We will return to this point in Sect. 4.2.

2.2 *Outflows and Galactic Fountains*

The way in which galactic inflows modulate metallicities and metallicity gradients depends critically on galactic fountains. Observations indicate that vigorous star formation is inevitably associated with outflows that eject gas from galaxies at a rate that is comparable to the star formation rate (Veilleux et al. 2005). Observations and models both suggest that they carry away most of the metals that are generated in core-collapse supernovae (Martin et al. 2002; Fu et al. 2013). The gas is expected to re-accrete on a dynamical timescale, particularly in massive galaxies ($\log(M_*/M_\odot) > 10.5$) and at relatively late times ($z < 1$; Oppenheimer et al. 2010; Henriques et al. 2013). The gas that a central galaxy re-accretes contains contributions both from its progenitor and from satellite systems, while satellite systems probably do not accrete much gas. Importantly, while the ejected gas has predominantly low angular momentum, it picks up angular momentum from the halo and re-accretes at generally larger radii than where it was ejected (Brook et al. 2012; Christensen et al. 2016). This has a number of observational consequences.

2.2.1 Galaxy-Averaged Quantities

First, re-accreted metals boost the central galaxy's metallicity with respect to models with unenriched inflows (or, viewed from a different perspective, they suppress dilution). If the metallicity of inflowing gas varies with mass, then recycled gas must affect the *slope* of the mass-metallicity relation. In particular, if low-mass galaxies accrete predominantly pristine material while inflows into massive galaxies are pre-enriched (Ma et al. 2016; Brook et al. 2014), then this differential pre-enrichment steepens the slope of the mass-metallicity relation.

If pre-enrichment levels vary with redshift, then they furthermore contribute to the evolution of the normalization of the mass-metallicity relation (Davé et al. 2011). For example, if inflows are pristine at early times and enriched to roughly the same level as the ISM at $z = 0$, then they are more effective at diluting gas reservoirs at high redshift, driving stronger evolution in the normalization of the mass-metallicity relation. The normalization of the MZR is observed to increase (Maiolino et al. 2008; Faist et al. 2016), which could be explained in this way. The observational challenge is to disentangle this factor from other influences that could also evolve with redshift (or mass) such as the initial mass function (van Dokkum 2008; Davé 2008).

It is instructive to compare this interpretation of how the MZR's normalization evolves upwards to that presented in Ma et al. (2016). In the latter work, high-resolution simulations were used to show that the relationship between gas metallicity and stellar mass fraction f_* (that is, the stellar mass divided by the baryonic mass) is nearly that of a closed-box throughout $z = 3 \rightarrow 0$ when averaged over the entire halo rather than just the central galaxy. In this view, growth in the MZR's normalization tracks growth in f_*. While this is a suggestive insight, it does not directly address growth in *galaxy* metallicities, which are far more readily observable. It is not hard to imagine, for example, that gas flows might hide metals within the halo at high redshifts but then shift progressively more of them into observability at late times. In other words, galaxy metallicities may evolve to become an increasingly unbiased probe of the halo metallicity, which in turn tracks closed-box expectations. A more detailed study of the evolving relationship between halo-averaged and ISM-averaged metallicity is probably indicated. For the present, however, it is clear that pre-enriched inflows have the potential to drive MZR evolution.

2.2.2 Radial Metallicity Gradients

If inflows deposit gas that has relatively uniform metallicity over a range of radii as expected theoretically (Brook et al. 2012; Christensen et al. 2016), then they flatten radial metallicity gradients because the metallicity at any point is driven by the inflowing gas, washing out other influences such as radial gradients in star formation efficiency or wind characteristics, or radial flows that escort low-metallicity gas in from the galaxy's outskirts. This raises the possibility of using observed metallicity gradients to constrain inflows.

At low redshift, it has long been known observationally that star-forming galaxies have slowly declining metallicity gradients (Zaritsky et al. 1994), and recent analyses have confirmed these results (Fu et al. 2013; Carton et al. 2015; Ho et al. 2015). Fu et al. (2013) used a semi-analytical model of galaxy formation to interpret observations. They showed that inflows can readily dominate metallicity gradients (their Fig. 6), while the effect of radial flows is probably relatively weak. They further found that it was necessary to assume that 80% of all newly formed metals are launched into the halo in order to match observed metallicity gradients, qualitatively consistent with inferences from X-ray observations of local outflows (Martin et al. 2002). In their model, the tendency for ejected metals to re-accrete over a range of radii makes the baryon cycle into an efficient method for redistributing metals.

Carton et al. (2015) interpreted locally observed metallicity gradients using a simpler model that assumes a local equilibrium between inflows, star formation, and outflows (Sect. 4.2). In their model, radial flows are ignored, and inflows are assumed to be uniform across the disk. Hence the observed metallicity gradient is driven by the radial dependence of the mass-loading factor (that is, the ratio of the outflow to star formation rates), with weaker outflows yielding higher metallicity in the center and stronger outflows suppressing it toward the disk edge. This model requires inflows in order to balance ongoing enrichment (Lilly et al. 2013), but that does not mean that the data require strong inflows. In fact, it is not even possible to use the Carton et al. (2015) model to measure inflow rates or metallicities owing to degeneracies. For example, a high observed metallicity could reflect a high metal yield and unenriched inflows, or a low metal yield and highly enriched inflows.

The model explored by Ho et al. (2015) leads to a different interpretation, tying locally observed metallicity gradients to gradients in gas fraction. These authors relax the assumption that all regions of a galaxy are in local enrichment equilibrium, but assume that the mass-loading factor and the ratio of inflow to star formation rates $\dot{M}_{g,\mathrm{out}}/\mathrm{SFR}$ are all constant. In this case, metallicity decreases with radius because more diffuse regions have higher gas fraction, implying that they are chemically less mature. They also find that their models fit observations if they assume that both inflows and outflows are weak, indicating nearly closed-box chemical evolution for systems with low gas fraction.

The Ho et al. (2015) and Carton et al. (2015) studies both leverage high-quality measurements of radial trends in metallicity and gas fraction, but their modeling efforts lead to different conclusions regarding the flow of gas into and out of the galaxy because they invoke different assumptions. Which model is more correct? More theoretical insight into how observables connect to the underlying physics would certainly help. At the same time, it is to be hoped that future studies that leverage measurements on halo metallicities or inflow rates (from, for example, absorption-line campaigns) will eventually break the underlying degeneracies.

Fu et al. (2013) indicated that the evolution of metallicity gradients to high redshift is a complementary constraint on star formation and gas flows. This echoes Pilkington et al. (2012), who analyzed how the radial and vertical gradients evolved in 25 cosmological simulations of Milky Way analogs from several groups as well as two independent analytical models. They found that, although all models roughly reproduce the Milky Way's current radial gradient, some predicted dramatically steeper gradients at earlier times while others did not. They concluded that steep gradients in metallicity reflect steep gradients in star formation efficiency and noted that strong feedback can wash out metallicity gradients. However, their discussion did not consider the possible role of inflows in balancing enrichment.

Observationally, the high-redshift story is far from clear. An early integral field study of three star-forming galaxies at $z \sim 3$ uncovered *inverted* metallicity gradients (i.e., the gas-phase metallicity is lower in the center; Cresci et al. 2010). The authors interpreted their findings as evidence that inflows deliver pristine gas preferentially to the center of high-redshift galaxies, although in fact this is not expected to produce inverted gradients generically (Pilkington et al. 2012).

Over the following years, detailed analyses of a few strongly lensed high-redshift systems yielded steeply declining metallicity gradients (Yuan et al. 2011; Jones et al. 2013) as predicted in Pilkington et al. (2012). These results seemed to indicate that strong inflows do not flatten or invert metallicity gradients at high redshifts.

Most recently, however, a study of a large sample of unlensed systems indicated that metallicity gradients are flatter than in the local Universe or even absent at higher redshifts (Wuyts et al. 2016). As noted in Wuyts et al. (2016), it is not obvious why lensed systems should show strong gradients while unlensed ones do not, particularly given that the samples overlap in stellar mass. Wuyts et al. (2016) consider a number of effects such as beam-smearing, AGN, or shocks that could artificially suppress the intrinsic metallicity gradient in unlensed systems, but conclude that they would have detected strong gradients if they were there. Hence while further work is needed to control biases, current observations support the idea that strong inflows flatten high-redshift metallicity gradients, as qualitatively suggested by some (but not all) models (Brook et al. 2012; Fu et al. 2013).

2.2.3 Future Work

At a technical level, the pioneering study of re-accretion presented in Oppenheimer et al. (2010) deserves to be re-visited in the context of more recent numerical models for two reasons. The less important of these is that galactic outflow models have grown more realistic owing to high-resolution simulations (Muratov et al. 2015) as well as to increasingly detailed comparison with measurements (Mitra et al. 2015). The more important reason is that simulations now incorporate significantly improved hydrodynamic solvers and dynamic range, which are critical for resolving the complicated interaction between outflows and the circumgalactic medium (Nelson et al. 2015). Relatedly, improved cross-fertilization of insights regarding outflows and enriched inflows between hydrodynamic simulations and semi-analytical models would be helpful both for distilling insight from the numerical models and for exploring its implications within a more flexible framework.

With an improved understanding of the baryon and metal cycles, a straightforward next step would be to review the hypothesis presented in Davé et al. (2011) that pre-enriched inflows drive the MZR's shape and evolution. A detailed budgeting of how galaxies distribute their baryons and metals within the ISM and halos would be necessary for this step, and would inform the next generation of measurements of the CGM, which has become a very active field over the past few years.

In the specific case of merger-induced inflows, it is not to soon to ask whether simulations can accommodate the rich set of observations of how interactions trigger nuclear flows driving star formation, gas dilution, as well as AGN activity (cf. the "Galaxy Pairs in the Sloan Digital Sky Survey" paper series by S. Ellison and collaborators); this would test the hypothesis that mergers can be identified based on their suppressed nuclear metallicities (Grønnow et al. 2015).

Finally, the effective use of metallicity gradients for constraining gas inflows in a cosmological context will require further development both in models and in observations. An improved understanding of what drives the metallicity gradients that are predicted by numerical simulations, a converged picture for the observed evolution of metallicity gradients to higher redshift, and the continued development of techniques for using observations to test models (for example, accounting for biases associated with sample selection and strong-line abundance indicators) will yield further insight into the baryon cycle.

2.3 Environment

At fixed stellar mass, galaxies that live in richer environments are observed to have slightly higher metallicities (Mouhcine et al. 2007; Cooper et al. 2008), although the effect is weak and not always observed (Hughes et al. 2013). A qualitatively similar relationship occurs in cosmological simulations (Davé et al. 2011), suggesting that it may in fact be real. If so, it could reflect systematic variation in the basic properties of the baryon cycle. For example, it is easy to imagine that weaker outflows, more efficient re-accretion of previously ejected gas, or enriched inflows could boost the metallicities in overdense regions. Indeed, analytical models have been used to argue that enriched inflows are a viable explanation for the offset (Peng and Maiolino 2014).

In reality, it may be that none of these is the correct explanation. A detailed study of galaxies in the ILLUSTRIS simulations (Genel 2016) recently traced the metallicity-environment correlation to two causes. First, at a given stellar mass, satellite galaxies (which dominate richer environments) tend to form earlier than centrals, draining their gas reservoirs and boosting their metallicities. Second, satellites' disks tend to be truncated and more centrally concentrated such that observations are weighted toward their metal-rich cores. Combining these effects essentially accounts for the entire simulated offset.

Genel (2016) notes that the CGM around satellite galaxies is more enriched than around centrals of the same stellar mass, and in fact the offset is comparable to offset in the galaxies' metallicities. However, he argues that this does not dominate the dependence of metallicity on environment because centrals and satellites with similar stellar mass and star formation history have nearly the same metallicity—despite the apparently higher metallicity of the gas around satellites.

Future work disentangling the metallicity of inflowing and outflowing gas may be needed in order to understand how a more enriched CGM does not drive higher galaxy metallicities. For the present, however, it seems that the tendency for metallicity to increase in rich environments does not relate to inflows.

3 Galaxy Growth and Halo Growth

The previous section drew attention to our imperfect understanding of how galaxies and halos grow together. A specific problem related to this question is the "dwarf galaxy conundrum," or the suite of observational clues that galaxies with stellar masses below $10^{10} M_{\odot}$ do not grow at the same specific rate as their host halos (for example, Weinmann et al. 2012). This behavior is very difficult for conventional galaxy evolution models to accommodate (White et al. 2015; Somerville and Davé 2015): generically, the strong outflows that seem required in order to bring the predicted stellar mass function at $z = 0$ into agreement with observations lead to metallicities and gas fractions that are under-predicted at low redshifts. Conversely, adopting weaker outflows at low masses in order to match MZR observations yields too many stars in low-mass halos at $z = 0$.

The problem clearly points to the need for a qualitatively new physical mechanism that retards gas processing in low-mass systems. This raises the question of whether that process operates in the IGM/CGM, or in the ISM. In a creative analytic model, Bouché et al. (2010) argued for a CGM-based solution, showing that forbidding halos less massive than $10^{11} M_{\odot}$ from cooling their gas onto galaxies improves agreement with measurements of star formation and stellar mass growth. They unfortunately did not address metallicities. Lilly et al. (2013) also addressed this problem. They found that, under the assumption that enrichment and dilution balance exactly (Sect. 4), the fraction of galaxy gas that is converted to stars can be inferred directly from the MZR, and the resulting scaling matches the requirement that is implied by the stellar mass function. In this argument, the dwarf galaxy conundrum can be resolved by processes that occur within the ISM, although inflows are required.

In order to assault the problem using a more comprehensive suite of observables, White et al. (2015) implement several galaxy-slowing mechanisms (both galaxy-based and CGM-based) into a semi-analytical model. They show that two changes to the fiducial model can improve agreement: in their "preferential reheating" model, low-mass galaxies eject systematically more gas in outflows per unit of stellar mass formed at higher redshift. In their "parking lot" model, ejected gas is not permitted to re-accrete for a delay time that depends on halo mass (see also Henriques et al. 2013). By manipulating the parameters governing these processes, they found improved agreement with measurements of cold gas fractions, specific star formation rates, and ISM metallicities

In the context of numerical simulations, Ma et al. (2016) have shown that accounting more realistically for the physical processes that occur within the interstellar medium (ISM) tends naturally to decouple the growth of low-mass galaxies from their host halo, leading to improved agreement with observations of the MZR evolution (see also Hopkins et al. 2014). Their findings are qualitatively consistent with the inferences that White et al. (2015) draw from semi-analytical models, but it is difficult to draw robust conclusions owing to the small number of halos that have been simulated at high resolution. For example, is the bottleneck that retards gas processing located in the ISM, in the CGM, or both? Ongoing efforts to distill their results into scalings that can be implemented into cosmological simulations and semi-analytical models will generate further insight (Davé et al. 2016).

4 The Equilibrium Model

Over the last decade, an increasing number of theoretical studies have arrived at the conclusion that reasonably massive star-forming galaxies ($M_* > 10^{10} M_\odot$) grow in a quasi-smooth fashion whereby gas is accreted from the CGM and processed at roughly the same rate. This is a generalization of the idea that star formation self-regulates to match the gas accretion rate (Tinsley and Larson 1978; Köppen and Edmunds 1999) in that it replaces "star formation" with "gas processing," where by the latter we mean that gas is processed either into stars *or* outflows. In this section, we introduce the concepts governing the equilibrium model and discuss inquiries into when and where it applies.

4.1 A Single Zone

We begin by considering a galaxy to be a single zone and define the growth rate of stellar mass in long-lived stars $\dot{M}_*$, the rate of change of the gas mass $\dot{M}_g$, the gas and star accretion rates $\dot{M}_{g,\rm in}$, $\dot{M}_{*,\rm in}$, and the gas outflow rate $\dot{M}_{g,\rm out}$. Generically, these are related by

$$\dot{M}_* + \dot{M}_g = \dot{M}_{g,\rm in} - \dot{M}_{g,\rm out} + \dot{M}_{*,\rm in} \tag{1}$$

Equation (1) is a tolerably comprehensive statement of the conservation of baryonic mass within the context of galaxy evolution. Note that star formation does not appear in this equation because it does not change the galaxy's baryonic mass.

To derive the Equilibrium Model, we begin by assuming that accretion of already-formed stars in actively star-forming galaxies is negligible; that is, $\dot{M}_{*,\rm in} = 0$. This implies that stellar mass growth owes predominantly to star formation, or $\dot{M}_* = \rm SFR$, which is supported observationally (Papovich et al. 2011; Behroozi et al. 2013). Next, we assume that the outflow rate equals the star formation rate of long-lived stars multiplied by a slowly varying "mass-loading factor" η; that is, $\dot{M}_{g,\rm out} = \eta \dot{M}_*$. Finally, we assume that the galaxy balances accretion exactly with gas processing so that $\dot{M}_g = 0$. This condition cannot, of course, be exactly correct, as it would imply that galaxies have no gas. However, high-resolution simulations suggest that, indeed, the mass of material that is processed and ejected by disks is comparable to the mass that is accreted (Christensen et al. 2016), justifying it as a first-order approximation. What is left is (Finlator et al. 2008):

$$\dot{M}_*(1+\eta) = \dot{M}_{g,\text{in}} \tag{2}$$

For clarity, we note that many authors define a total star formation rate (SFR), compute the "return fraction," R, or the fraction of mass that collapses into massive stars and is then returned to the ISM on a short timescale (Tinsley 1980), and then write down the growth of stellar mass in long-lived stars as $\dot{M}_* = \text{SFR}(1-R)$. In this case, we define η as the ratio of the outflow rate to the total star formation rate, and Eq. (2) is unchanged.

Equation (2) is a statement of *gas processing equilibrium*. The intuition is that, if the gas accretion rate temporarily exceeds the gas processing rate, then the gas density increases, boosting gas processing until the excess has been worked through. Conversely, if accretion temporarily sputters, then the gas density declines, choking off gas processing until new inflows are available. Equation (2) assumes that this balancing occurs on a timescale that is very short. In doing so, it departs conceptually from the traditional view that the star formation rate is fundamentally determined by the gas density. In an equilibrium scenario, the ISM adjusts itself so as to satisfy Eq. (2); the gas density merely reflects the details of how it achieves this.

The factor of $(1+\eta)$ is what distinguishes Eq. (2) from classical analytical models (Tinsley and Larson 1978; Köppen and Edmunds 1999). To see how important it is, consider the influences that determine a galaxy's gas-phase metallicity. At any given time, inflows from the CGM dilute the galaxy's metallicity. However, the inflows trigger continuing star formation, which simultaneously enriches the gas. The competition between dilution and enrichment gives rise to an *Equilibrium Metallicity* Z_{eq} (Finlator et al. 2008; Davé et al. 2012):

$$Z_{\text{eq}} \equiv Z_0 + \frac{y}{(1+\eta)} \tag{3}$$

Here, y is the metal yield, or the mass of metals ejected by core-collapse supernovae per mass of long-lived stars formed; and Z_0 is the metallicity of inflowing gas, which for convenience we may re-cast as $Z_0 \equiv \alpha_Z * Z$, where α_Z is the ratio of the metallicity of inflowing gas to the ISM metallicity and Z is the ISM metallicity. If η is redefined such that $\eta \equiv \text{SFR}/\dot{M}_{g,\text{out}}$ where SFR is the total star formation rate, then we replace $\eta \rightarrow \eta/(1-R)$ in Eq. (3). Note that we assume that outflows have the same metallicity as the ISM; this is supported by high-resolution simulations (for example, Fig. 11 of Ma et al. 2016).

Equation (3) is a statement of *enrichment equilibrium*. The intuition is that enrichment and dilution balance each other once the galaxy's metallicity reaches Z_{eq}. If Z somehow jumps above Z_{eq}, then inflows dilute it. If Z dips, then continued star formation boosts it back to Z_{eq}. If the SFR responds instantaneously to fluctuations in the inflow rate, then departures from Z_{eq} are erased on a *dilution time* $M_g/\dot{M}_{g,\text{in}}$, or the timescale for inflows to replenish the ISM completely.

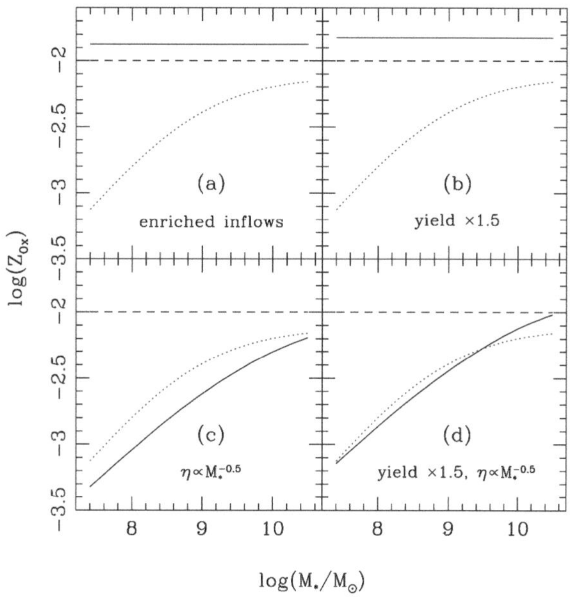

Fig. 1 Illustration of the parameters that control metallicity in Eq. (3). In each panel, the *black dashed line* indicates the same "base model" while the *blue dashed curve* is the observed trend (Andrews and Martini 2013). In panel **(a)**, inflows are enriched above the base model. In panel **(b)**, the assumed yield is boosted instead. In panel **(c)**, the mass-loading factor scales inversely with the square root of the stellar mass. Panel **(d)** shows that combining the models in panels **(b)** and **(c)** yields good agreement with observations

Equations (2)–(3) involve two parameters, η and Z_0. For galaxies that are massive enough to be in equilibrium, the slope and redshift dependence of the mass-metallicity relation may be expressed as evolution in these parameters.

In order to illustrate how powerful Eq. (3) is, we use it to interpret the $z = 0$ mass-metallicity relation (where by "metallicity" we mean the oxygen mass fraction Z_{ox}). In each panel of Fig. 1, the black dashed line indicates Z under the assumption of pristine inflows ($Z_0 = 0$), an oxygen yield of $y = 0.01$, and no outflows ($\eta = 0$). Meanwhile, the blue dotted curve is the observed trend at $z \sim 0.1$ (Andrews and Martini 2013), where we convert from units of $12 + \log(O/H)$ to oxygen mass fraction assuming a hydrogen mass fraction of 0.76.

With no outflows ($\eta = 0$), metallicity has no dependence on mass as shown by the dashed black segment in each panel of Fig. 1. Increasing the inflow metallicity Z_0 from 0 to 0.00333 boosts Z at all masses (panel a). Likewise, increasing the yield y from 0.01 to 0.015 increases Z at all masses (panel b). Assuming that the outflow rate η varies with stellar mass as $(10^{10} M_\odot / M_*)^{0.5}$ immediately yields a strong dependence of Z on M_* (panel c). Finally, restoring the assumption of pristine inflows and tuning both the yield and the outflows yields a plausible level of agreement with observations (panel d). We do not use a rigorous approach to tune our parameters because our goal is merely to illustrate its potential as a conceptual framework. However, the model has been generalized and tuned elsewhere, yielding promising insight into the nature of the baryon cycle (Mitra et al. 2015, 2016); we will discuss these efforts shortly.

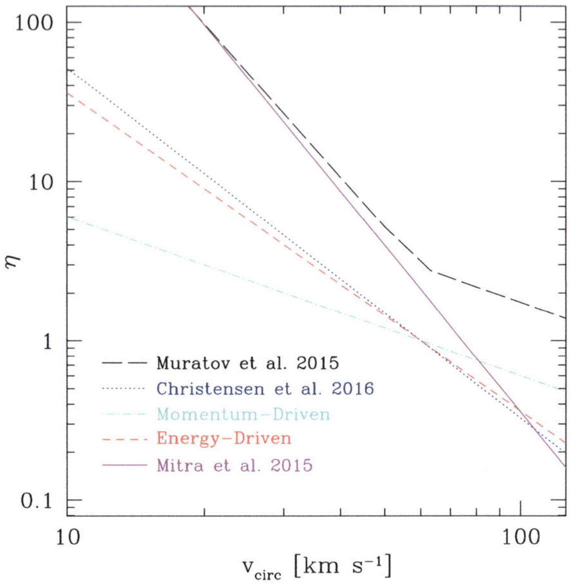

Fig. 2 Recent models for the ratio η of the gas outflow rate to the star formation rate (a.k.a. the mass-loading factor). The models of Christensen et al. (2016) and Muratov et al. (2015) are high-resolution ab-initio simulations. The Mitra et al. (2015) curve results from fitting an equilibrium model to observations. The momentum-driven and energy-driven scalings are arbitrarily normalized to $\eta = 1$ at $60\,\mathrm{km\,s^{-1}}$

This exercise shows that the MZR can easily be interpreted within the equilibrium scenario (Eqs. (2)–(3)). However, the interpretation's value is limited by model's simplicity. For example, η and Z_0 could vary with redshift, and Z_0 could additionally vary with mass. For improved realism, further insight from numerical models is required.

η is now relatively well-studied theoretically, with a consensus that it decreases with increasing mass and exceeds 1 in halos whose circular velocity v_{circ} lies below $60\,\mathrm{km\,s^{-1}}$ (Muratov et al. 2015; Christensen et al. 2016). In detail, there is an ongoing debate as to whether it follows expectations from momentum-driven or energy-driven outflows. Briefly, if the characteristic speed of galactic outflows is the halo circular velocity v_{circ} and if star formation deposits a constant amount of energy in outflows per unit of stellar mass formed, then η varies as v_{circ}^{-2}. Alternatively, if star formation deposits a constant amount of momentum in outflows per unit of stellar mass formed, then $\eta \propto v_{\mathrm{circ}}^{-1}$ (Murray et al. 2005).

We show three recent estimates for η in Fig. 2. Muratov et al. (2015) use high-resolution numerical simulations to find that $v_{\mathrm{circ}}^{-1.0}$ for $v_{\mathrm{circ}} > 60\,\mathrm{km\,s^{-1}}$ and $v_{\mathrm{circ}}^{-3.2}$ at lower masses. By contrast, Christensen et al. (2016) study a complementary suite of high-resolution simulations and derive $\eta \propto v_{\mathrm{circ}}^{-2.2}$ at all masses. These simulations model star formation and feedback in different ways, hence it is telling that they both predict $\eta > 1$ for $v_{\mathrm{circ}} < 60\,\mathrm{km\,s^{-1}}$ and $\eta > 4$ for $v_{\mathrm{circ}} = 30\,\mathrm{km\,s^{-1}}$ which is the lowest halo mass that can efficiently accrete baryons from a photoheated IGM (Gnedin 2000; Okamoto et al. 2008). The generally higher values of η found by Muratov et al. (2015) probably reflect additional feedback sources such as radiation pressure that are not considered in the Christensen et al. (2016) model.

Mitra et al. (2015) take a different approach and use a Bayesian Monte Carlo Markov Chain approach to infer η (among other things) directly from observations. In particular, they assume that galaxies obey Eqs. (2)–(3) at all times and then infer the parameters that are required in order to match simultaneously the observed stellar mass - halo mass relation, the MZR, and the stellar mass - star formation rate relation. They find that $\eta \propto M_h^{-1.16} \propto v_{\rm circ}^{-3.48}$, slightly steeper than the dependencies obtained from numerical simulations. When η is expressed as a function of stellar mass, however, the differences shrink: Muratov et al. (2015) find $\eta \propto M_*^{-0.35}$ while Mitra et al. (2015) find $\eta \propto M_*^{-0.5}$ for $M_* < 10^{10} M_\odot$ (note that this is the same scaling as in panel d of Fig. 1).

Given that η decreases with mass both in ab-initio calculations and in efforts to model observations, it assuredly drives the MZR's slope. As shown in Fig. 1, a large η removes more enriched gas, with the result that dilution by inflows is more efficient at suppressing the gas metallicity. This in turn leads to a positive dependence of Z on mass.

The way in which enriched inflows (that is, $Z_0 > 0$) affect the MZR has received considerably less study than η but is probably just as important. Davé et al. (2011) showed that, on average, α_Z increases with time, driving or at least contributing to the observed growth in the MZR's normalization. The analysis of metal flows in high-resolution simulations presented by Ma et al. (2016) qualitatively supports this view. Furthermore, it suggests that inflows into massive halos (those that grow to $10^{11} M_\odot$ by $z = 0$) have roughly one third the ISM's metallicity at all times, or $\alpha_Z \sim \frac{1}{3}$. Meanwhile, the α_Z in low-mass halos ($10^{10} M_\odot$ at $z = 0$) falls from $\sim\frac{1}{3}$ at early times to a few percent at $z = 0$.

In short, both η and Z_0 probably vary in a way that drives an MZR to increase with mass and time. The connection between this *differential dilution* and the MZR depends critically on the equilibrium model. This discussion then raises several points for future research.

Which galaxies are in equilibrium? As discussed in Sect. 3, observations suggest that low-mass halos depart from equilibrium expectations in the sense that they do not cool their gas onto galaxies at the same rate as they accrete it from the IGM. Is there a similar mass cutoff below which galaxies do not convert their stars into gas at the same rate as they accrete it from the CGM? Does it vary with redshift? If so, then one expects the MZR normalization to plummet above the redshift at which galaxies first achieved equilibrium. Indeed, observations have suggested that the MZR drops rapidly in normalization from $z = 2$–3 (Maiolino et al. 2008) and then flattens (that is, the mass dependence weakens) to $z = 5$ (Faist et al. 2016). This behavior could indicate that $z \geq 3$ is the "gas accumulation" epoch, when metal and dust production were swamped by inflows and η and α_Z did not yet govern metallicities. A detailed analysis of what is expected from hydrodynamic simulations in this regard would be illuminating.

What is equilibrium? For completeness, we note that, while some works define gas processing equilibrium as the condition that a galaxy's gas *mass* be constant (Finlator et al. 2008; Davé et al. 2011, 2012), it may be equally defensible to propose that a galaxy's gas *fraction* [defined as $\mu \equiv M_g/M_*$ or $M_g/(M_g + M_*)$]

is constant (Feldmann 2013). If true, then the equilibrium metallicity is slightly different from Eq. (3) (i.e., set $\dot{\mu} \equiv 0$ in Eq. (4)). It would be interesting to ask which of these conditions more accurately describes galaxy growth in cosmological simulations or semi-analytical models.

Can we close the loop? Once the mass and redshift range has been identified where galaxies are expected to be in equilibrium, it would be interesting to ask whether Eqs. (2)–(3) really do govern the observed evolution within simulations and models. Conceptually, this would be accomplished in several (rather involved) steps:

- Identify which galaxies obey Eq. (2) or an analogous statement of gas processing equilibrium such as $\dot{\mu} = 0$;
- Measure $\eta(M_*, z)$, $Z_0(M_*, z)$, and the MZR and its evolution;
- Ask whether Eq. (3) or an analogous statement of enrichment equilibrium correctly describes the predicted MZR evolution.

A more complete conceptual understanding of what governs the MZR in hydrodynamic simulations or semi-analytical models would inform the interpretation of existing and upcoming observations.

4.2 Multi-Zone Models

If the equilibrium model describes galaxies as a whole, then it is natural to ask whether it also describes individual regions within the galaxy. There is a lower limit of course; we would not apply the equilibrium model on spatial scales comparable to a molecular cloud (10–100 pc) where metals diffuse or are spread by individual supernovae. But what about a region of characteristic size ~1 kpc? To be specific, we may ask two questions:

1. What is the smallest region within a galaxy whose gas reservoir obeys Eqs. (2)–(3)? This is important because the relative roles of inflows and outflows are more difficult to assess (or even discuss) in systems that are not in equilibrium.
2. How does the balance between radial flows, outflows, and inflows vary within a star-forming galaxy?

Within the context of an analytical model, Fu et al. (2013) found that radial flows do not impact the overall metallicity gradient while inflows do. They did not study the extent to which individual annuli were in equilibrium, and the assumptions that were necessarily to make their model tractable were strong, motivating further insight from numerical models.

A year later, a different group took the bait and carried out a detailed study of high-resolution, one-dimensional simulations of star-forming disks (Forbes et al. 2014a). They found that, indeed, individual annuli do obey Eq. (2) locally. In detail, the way in which cosmological inflows, star formation, galactic outflows, and radial gas flows (triggered by gravitational instability-driven torques that occur wherever gas densities are high) balance varies with mass and with redshift. Annuli at large radii have low densities and low star formation rates, hence they do not efficiently process newly accreted gas or transport it to smaller radii. Moving inwards, gas densities increase to the point that star formation and radial transport of gas balance accretion. Near a galaxy's center, star formation and outflows are so efficient that they can only be balanced by radial transport of gas from larger radii; cosmological inflows are a subdominant gas source here. The slowly declining cosmological infall rate causes this balance to evolve in such a way that the region where gas is gravitationally unstable migrates outwards with time. By $z = 0$ it operates only in galaxies' outskirts, if at all.

The finding that high-redshift galaxies obey Eq. (2) locally as well as globally certainly raises the hope that Eq. (3) also applies, but Forbes et al. (2014a) unfortunately do not comment on the predicted metallicity gradients or their evolution. However, simple applications of this idea have been able to accommodate locally observed radial trends fairly well (Carton et al. 2015).

It would be interesting to compare high-resolution one-dimensional simulations (Forbes et al. 2014a) with lower-resolution cosmological simulations (Pilkington et al. 2012) in order to assess (1) Whether cosmological simulations resolve gravitational instabilities and their impact; (2) Whether there are effects in three-dimensional simulations (such as bars) that modulate gas flows and that cannot be captured in one-dimensional simulations; and most broadly (3) Whether the cosmological simulations predict the same level of equilibrium as the one-dimensional simulations. Likewise, it would be useful to distill insights from the Forbes et al. (2014a) study (such as the radial flow rate) into a form that can be used to improve the realism of semi-analytical models.

We conclude from these studies that the expected signature of cosmological inflows in galaxies' metallicity gradients remains unclear theoretically. There is room both for an improved understanding of observed metallicity gradients (Sect. 2.2), and of what they are telling us about inflows and gas processing.

5 Extensions to the Equilibrium Model

Observations indicate that there is more to galaxy metallicities than equilibrium. For example, at a fixed stellar mass, galaxies show finite scatter in gas-phase metallicity (Tremonti et al. 2004; Zahid et al. 2012; Guo et al. 2016), and in star formation rate (Whitaker et al. 2012). Do these observations reflect intrinsic scatter in η or Z_0, or do they indicate departures from equilibrium?

The observed level of scatter is intuitively consistent with the idea that star formation is at least slightly irregular or "bursty" at any mass, which could be associated with departures from equilibrium. For example, if star formation rate is driven by stochastic processes (Gerola et al. 1980) or if the ISM responds slowly to inflow fluctuations, then the galaxy would not be perfectly in equilibrium (Forbes et al. 2014b); this is in fact expected for dwarf galaxies (Lilly et al. 2013;

Hopkins et al. 2014; Wise et al. 2014). Alternatively, the burstiness could owe to irregular inflows, which are not themselves completely smooth. Indeed, it has been shown that, by modeling the bursty inflow rates realistically but retaining the assumption that galaxies are in perfect equilibrium with those bursty inflows (which is theoretically supported Rathaus and Sternberg 2016), the observed level of scatter can be accounted for (Mitra et al. 2016; Dutton et al. 2010). This means that the observed scatter need not reflect truly out-of-equilibrium behavior.

The most important handle on the nature of the observed scatter is the identification of *correlated departures* from equilibrium, in particular departures from mean-trend behavior in Z, SFR, and gas fraction at constant M_*, as quantified by mean scaling relations. These have prompted a range of generalizations to the Equilibrium Model. Before delving into the models, however, we review the observations.

5.1 The M_*-Z-SFR Relation: Observations and Intuition

Ellison et al. (2008) first pointed out that the SDSS MZR has a third-parameter dependence on SFR in the sense that, at a fixed M_*, galaxies with higher SFR have lower Z. Subsequently, other authors have found qualitatively similar results at a variety of redshifts (Mannucci et al. 2010; Lara-López et al. 2010; Cresci et al. 2012; Hunt et al. 2012; Henry et al. 2013; Stott et al. 2013; Cullen et al. 2014; Maier et al. 2014; Nakajima and Ouchi 2014; de los Reyes et al. 2015; Salim et al. 2015; Kacprzak et al. 2016). There have been suggestions that the SFR-dependence reflects observational systematics or disappears at high redshift (Sánchez et al. 2013; Steidel et al. 2014; Sanders et al. 2015). However, detailed analyses of possible biases have shown that it persists even if a variety of effects is taken into account both in the SDSS sample (Andrews and Martini 2013; Salim et al. 2014; Telford et al. 2016) and at high redshift (Salim et al. 2015). Moreover, early suggestions that the sign of the effect has a mass dependence in the sense that massive galaxies with enhanced SFR have enhanced metallicity (Yates et al. 2012) may simply reflect tricky selection biases (Salim et al. 2014, 2015). Initial results suggesting that the M_*-Z-SFR relation is redshift-independent gave rise to the idea that the M_*-Z-SFR relation is "fundamental" (Mannucci et al. 2010), which would be powerful if true. However, more recent results indicate that, while *sense* of the SFR dependence is unchanged throughout z =0–2.3, its *strength* evolves (Brown et al. 2016; Grasshorn Gebhardt et al. 2016), particularly at high stellar masses ($10^{11}M_\odot$; Salim et al. 2015).

While the dependence of metallicity on SFR was identified first, it has recently been argued that the dependence on gas fraction is just as strong, if not stronger (Hughes et al. 2013; Bothwell et al. 2013, 2016; Lara-López et al. 2013). This result has also been found in cosmological simulations (Lagos et al. 2016). These findings indicate that SFR perturbations really just serve as a proxy for gas supply perturbations. In other words, the M_*-Z-SFR relation is driven by bursty inflows rather than bursty star formation (Dutton et al. 2010; Mitra et al. 2016). SFR is much easier to measure for large samples, however, and will probably continue to be used as the *de facto* third parameter for some time.

It thus appears that the M_*-Z-SFR relation is real, if perhaps a bit elusive. Theoretical explanations generally invoke a picture along the following lines: suppose that a galaxy is growing in a quiescent way that obeys Eqs. (2)–(3). If, for some reason, the gas accretion rate increases, two things will happen. First, the gas mass will grow, likely suppressing the galaxy's metallicity. Second, the SFR will increase owing to the boosted gas density. Finally, the perturbations to Z and SFR subside once the extra gas is processed and the metallicity is diluted by further inflows (Dalcanton 2007).

Of course, one could also imagine the opposite effect: Dips in the gas accretion rate could lead to suppressed SFR and enhanced metallicity (because the metals generated by ongoing star formation would be less diluted by inflows). The problem with this is that, whereas accretion spikes can increase the size of the gas reservoir and boost SFR on a disk dynamical time ($\sim$100 Myr), accretion dips can only suppress the SFR on the timescale given by the gas mass divided by the SFR. This is, to within a factor of a few, the same as the halo specific growth rate, or 1–10 Gyr (see, for example, Fig. 3 of Lilly et al. 2013). It is perhaps for this reason that, observationally, the M_*-Z-SFR dependence is driven by systems with enhanced SFR (Salim et al. 2014; Telford et al. 2016), and evidence for short-term accretion dips is hard to come by.

A variety of theoretical approaches have been used to study the scatter in the M_*-Z-SFR relation and its possible use as a probe of gas inflows. We now summarize key insights gleaned from these efforts.

5.2 The M_*-Z-SFR Relation: Equilibrium Treatments

5.2.1 The Power of Assuming $\dot{Z} = \dot{M}_g = 0$: Davé et al. (2012)

Inspired by cosmological hydrodynamic simulations, Davé et al. (2012) devised an equilibrium model that relates galaxy growth to host halo growth under the condition that galaxies obey Eqs. (2)–(3). In order to connect halo growth with galaxy growth, they parameterized the impact of photoionization heating, quenching by accretion shocks, and re-accretion of previously ejected gas. They also studied the condition for galaxies to be in gas processing equilibrium, finding that there is a characteristic redshift above which galaxies cannot process their gas as rapidly as they accrete it. The precise redshift depends both on halo mass and on the outflow model because stronger outflows lead galaxies to reach equilibrium sooner. For momentum-driven outflow scalings, $10^{12}M_\odot$ halos achieve equilibrium at $z \sim 5$ while galaxies in lower-mass halos ($\leq 10^{11}M_\odot$) are always in equilibrium (because their outflows are strong). Hence star-forming galaxies are in or near equilibrium for most of cosmic time.

Equations (2)–(3) are too simple to permit a third-parameter dependence of Z on SFR, but Davé et al. (2012) did note that such a dependence arises naturally in hydrodynamic simulations and can readily be attributed to weak violations of the condition $\dot{M}_g = 0$.

The Davé et al. (2012) picture has since been generalized into an eight-parameter model and tuned to match the observed relationships between M_*, Z, SFR, and halo mass as well as their redshift dependence and scatter (Mitra et al. 2015, 2016). The resulting fits are excellent, supporting the model's realism. Notably, three of the eight parameters describe the rate at which ejected gas re-accretes and two describe "quenching," or the suppression of inflows into galaxies hosted by massive halos. Such quenching could be associated, for example, with active galactic nuclei or virial shocks. These studies show forcefully that, within a suitable framework, it is possible to use observations of Z and SFR to constrain the efficiency of inflows.

5.2.2 Relaxing Gas Processing Equilibrium: Lilly et al. (2013)

Initial efforts to interpret ensemble measures of galaxy evolution within an equilibrium framework assumed that the galaxy's gas reservoir does not grow (Finlator et al. 2008; Davé et al. 2011, 2012). However, it was soon realized that relaxing this assumption and allowing the gas fraction to fluctuate naturally gives rise to a M_*-Z-SFR relation that qualitatively resembles observations (Lilly et al. 2013). Re-deriving the condition for enrichment equilibrium while relaxing the assumption of gas processing equilibrium leads to a generalization of Eq. (3):

$$Z_{\rm eq} \equiv Z_0 + \frac{y}{(1+\eta) + \epsilon^{-1}\left[\frac{\rm SFR}{M_*} + (1-R)^{-1}\frac{d\ln\mu}{dt}\right]} \tag{4}$$

Here, ϵ^{-1} is the star formation timescale and $\mu \equiv M_g/M_*$ is the ratio of gas mass to stellar mass. Equation (4) beautifully connects Z, SFR, and M_*. If the timescale ϵ^{-1} for a galaxy to run through its gas reservoir is very short, then Z should not correlate with SFR because the ISM recovers quickly from perturbations (by processing gas efficiently). In this case, the gas mass is constant and Eq. (4) reduces to Eq. (3). On the other hand, if star formation is less efficient (ϵ^{-1} is longer), then the ISM can be out of gas processing equilibrium for some time. In this case, a high specific star formation rate SFR/M_* is associated with a low-metallicity as observed, irrespective of the relationship between the inflow and gas processing rates.

After presenting this model, Lilly et al. (2013) used it to draw a number of inferences from observations. In particular:

1. If the model is tuned to match the observed MZR at $z = 0$, then it immediately predicts the strong observed dependence of stellar mass on halo mass (Moster et al. 2010), suggesting that the latter trend may be governed entirely by

the details of feedback and gas processing that occur deep within the ISM. This does not rule out the possible importance of suppression or feedback processes that occur in the CGM, but it indicates that they are not required.

2. The model can readily be tuned to match observations of the M_*-Z-SFR dependence, and it predicts that this relation does not evolve with redshift unless the parameters governing gas processing (i.e., η, Z_0, and ϵ) also evolve. By contrast, it predicts that the MZR's normalization increases with time because high-redshift galaxies have systematically higher specific star formation rates.

The gas regulator model contains a number of approximations, chief among them the assumption of enrichment equilibrium. It is naturally of interest to ask whether the model describes what happens within a less approximate framework. Pipino et al. (2014) tested the Lilly et al. (2013) model using exact analytical expressions for the evolution of gas-phase metallicity, gas mass, and star formation rate, finding excellent agreement across a wide range of potential growth histories. This suggests that the gas regulator model is a powerful generalization of the equilibrium model introduced by Davé et al. (2012). Not surprisingly, it is now being tested and applied in a wide variety of contexts (for example, Birrer et al. 2014; Dekel and Mandelker 2014; Bouché and Blandin 2015; Wu et al. 2016).

To date, there has been no effort to determine whether the Lilly et al. (2013) model describes what happens within three-dimensional hydrodynamic simulations or semi-analytical models; this may be a promising avenue for future research.

5.3 *The M_*-Z-SFR Relation: Non-equilibrium Treatments*

5.3.1 An Analytical Model Without Equilibrium: Dayal et al. (2013)

Can the observed M_*-Z-SFR relation can be explained without any explicit assumption of equilibrium? In a purely analytical study, Dayal et al. (2013) presented a minimal model for galaxy growth in which the evolution of the stellar, gas, and metal reservoirs depends only on the unknown gas accretion and outflow rates. They assume that outflow and infall rates depend only on stellar mass, that the metal yield y is constant, and that inflows are pristine. They then show that they can fit the model's governing parameters using the observed M_*-Z-SFR relation (Mannucci et al. 2010) to achieve good agreement.

In common with Lilly et al. (2013), this model predicts no evolution in the overall M_*-Z-SFR relation for a given set of parameters, but it has the advantage that it assumes neither gas processing or enrichment equilibrium. Nonetheless, a mass-dependent balance between inflows and outflows emerges as a central prediction. Massive galaxies have weak winds and rapidly achieve enrichment equilibrium (similar to Eq. (3), although they do not impose this). By contrast, low-mass galaxies lose most of their metals through much stronger outflows and are hence more susceptible to dilution.

In fact, they show that, subject to their modeling assumptions, the infall rate and its mass dependence can be inferred directly from the observed M_*-Z-SFR relation. Intriguingly, the inferred outflow rate η varies with mass in a way that follows expectations from momentum-driven outflow models (Murray et al. 2005). The model can also accommodate high-redshift observations, with the result that gas fractions and inflow rates were previously higher while outflow efficiencies were essentially the same (Hunt et al. 2016).

Their model also naturally reproduces the observed relation between oxygen abundance and neutral gas fraction (Hughes et al. 2013). This finding reinforces the tight connection between metallicity and gas fraction that emerges in any plausible model of galaxy growth. It would be interesting to ask whether re-writing the Dayal et al. (2013) model to match the observed M_*-SFR-gas fraction relation automatically recovers the Z dependence. Likewise, it would be interesting to quantify precisely how close the galaxies are to enrichment and gas processing equilibrium in the Dayal et al. (2013) model once the inflow and outflow parameters are tuned to match observations.

5.3.2 Stochastic Accretion Histories and Scatter: Forbes et al. (2014b)

Observationally, there is residual scatter in Z even after the SFR dependence is removed (Salim et al. 2014, 2015). Can this scatter be used as a complementary probe of inflows and gas processing? This question is taken up by Forbes et al. (2014b), who propose that, even if individual galaxies are not in equilibrium, they may be close enough to it that ensemble statistics such as the slope of the Z-SFR relation are in equilibrium. They explore this idea by modeling galaxy growth as a connected sequence of intervals whose constant duration is given by the "coherence time" $t_{\rm coherence}$. At the start of each interval, a new accretion rate is drawn from a log-normal distribution with fixed median and scatter. The galaxy then uses star formation and outflows to respond to the new accretion rate on a "loss timescale" $t_{\rm loss}$, which incorporates gas losses to both outflows and star formation. If $t_{\rm loss}$ is short compared to other timescales, then galaxies are always in equilibrium. Conversely, if $t_{\rm loss}$ is long, then they retain a "memory" of previous accretion episodes.

The model's parameters can be tuned to yield qualitative agreement with the observed M_*-Z-SFR relations. This has two important implications:

1. Galaxy metallicities, when combined with measures of M_* and SFR, may probe both the mean and the distribution in gas accretion rates;
2. As in Dayal et al. (2013), observations do not require galaxies to obey either of the equilibrium relations [Eqs. (2)–(3)] explicitly, although in practice they may be nearly in equilibrium.

The price for relaxing the assumption of equilibrium is that this model is more complicated than the Lilly et al. (2013) or Davé et al. (2012) models. It would therefore be useful to ask how close galaxies are to equilibrium when this model is tuned to reproduce observations. Moreover, as the authors point out, the Forbes et al. (2014b) model for inflow fluctuations is extremely simple, invoking only a single coherence timescale. In reality, inflows fluctuate on a range of timescales. It would therefore be useful to generalize this aspect of the model for improved realism.

5.3.3 Cosmological Hydrodynamic Simulations

Cosmological hydrodynamic simulations and semi-analytical models make fewer assumptions regarding inflows and gas processing than analytical models (Somerville and Davé 2015), and are hence a particularly realistic framework for studying the M_*-Z-SFR relation. Reassuringly, they inevitably predict relations that qualitatively resemble observations (Davé et al. 2011; Yates et al. 2012; Obreja et al. 2014; de Rossi et al. 2015; Kacprzak et al. 2016; Lagos et al. 2016; Cousin et al. 2016).

In order to understand this prediction better, Lagos et al. (2016) apply a principal component analysis to the predicted ensemble distribution of global galaxy attributes in the EAGLE simulations (Schaye et al. 2015). In agreement with previous work, they identify a particularly thin plane in the space of M_*, SFR, and neutral+molecular gas fraction where simulated galaxies tend to live. Moreover, they show that this plane is observed. They interpret its existence as an indication that galaxies live in a slowly evolving balance between gas accretion and gas processing reminiscent of Eq. (2). Interestingly, the predicted scatter about this plane tightens if stellar feedback assumed in the model is stronger. This could readily reflect the fact that fresh inflows drive galaxies back to their equilibrium state more quickly if the gas fraction is lower.

Just as the statistics of the EAGLE galaxies imply a balance between gas inflows and processing, they also imply a balance between dilution and enrichment: the predicted metallicity is correlated with the gas fraction in the observed sense, namely that galaxies with high gas fractions have systematically low metallicities. Interestingly, however, it is found that Z contributes less to the overall variation in galaxy properties than M_*, SFR, and gas fraction: whereas the latter three attributes dominate the first principal component, Z only enters into the second. This qualitatively supports observational suggestions that gas fraction is more fundamental than metallicity (Hughes et al. 2013; Bothwell et al. 2013, 2016; Lara-López et al. 2013). Further theoretical work is needed in order to understand why gas fraction is more fundamental.

Unlike recent high-resolution simulations of individual galaxies (Christensen et al. 2016; Ma et al. 2016), the study presented in Lagos et al. (2016) stops short of directly measuring the flow of gas and metals through the simulated galaxies because it is based on a principal component analysis, which leverages the statistical advantage that lower-resolution cosmological simulations have over high-resolution "zoom" studies. This means that they do not ask whether galaxies keep gas mass or gas fraction constant, or how Z_0 varies with mass and redshift. It would therefore be interesting to merge the two approaches by quantifying the expected flow of gas and metals in cosmological simulations in order to understand in more detail how they give rise to the M_*-Z-SFR-gas fraction relation.

5.3.4 An Analytical Model Without Inflows: Magrini et al. (2012)

Thus far, all of the models for the M_*-Z-SFR relation that we have discussed invoke inflows, giving the impression that cosmological inflows of pristine gas are an unavoidable inference. A different perspective is offered by Magrini et al. (2012), who construct analytic, one-zone, closed-box evolutionary models that qualitatively match the observed M_*-Z-SFR relation. The key ingredient in their model is a distinction between "active" and "passive" modes of star formation: active-mode galaxies have high-density molecular clouds whose collisions trigger vigorous star formation while the clouds in passive-mode galaxies are more diffuse. When they confront their model with observations, they find that most high-redshift galaxies and the occasional, unusually active low-redshift ones can be identified with the active mode while quiescent systems can be identified with the passive mode.

While it is interesting that the model in Magrini et al. (2012) contains observations within its accessible parameter space, the discussion leaves several basic questions unaddressed. First, the model does not predict the distribution of active and passive-mode systems or its evolution. Hydrodynamic simulations attribute the prevalence of active-mode star-formers at high redshift and the scaling relations that they obey to high gas accretion rates and the detailed way in which galaxies achieve an equilibrium with those inflows (Forbes et al. 2014a; Lagos et al. 2016). By contrast (and similarly to Dayal et al. 2013), Magrini et al. (2012) attribute the observed scaling relationships to unexplained variations in initial conditions, evolutionary state, and ISM properties. Second, the lack of inflows means that the Magrini et al. (2012) predict stellar metallicity distributions that conflict with local observations (Sect. 1). Future work that includes inflows following cosmological expectations may reconcile this model with simulations.

6 Summary

Over the past 20 years, a marked dichotomy has emerged that distinguishes between how dark matter halos and galaxies grow. While halos do much or most of their growing by merging with other halos, star-forming galaxies with stellar masses below $\sim 10^{11} M_{\odot}$ acquire most of their baryons through relatively smooth gas flows. Theoretical models predict that these inflows leave a number of observable signatures in galaxies' gas-phase metallicities. Mergers leave a statistically detectable imprint both on galaxies' central gas-phase metallicities and on their radial metallicity gradients, but they are not the primary mechanism for delivering fresh gas to galaxies.

Radial metallicity gradients are a promising probe of inflows because they are expected to flatten once previously ejected gas begins to re-accrete. This is because metals that form and are ejected in a galaxy's core are "spun up" by the halo and re-distributed to large radii. To date, however, radial metallicity gradients in nearby galaxies have not been demonstrated to require inflows. Instead, they have been attributed either to radial gradients in ISM properties such as the strength of outflows, or to a radially varying "evolutionary state" in which annuli at large radii are simply less evolved. These conflicting interpretations point to the need for further theoretical inquiry into how low-redshift observations ought to be interpreted.

At high redshifts, simulations predict that metallicity gradients may have been stronger than in the local Universe, possibly because of the weaker role of recycled gas and mergers at early times. Meanwhile, observations seem split between lensed galaxies with generally strong gradients and unlensed galaxies with weak ones. Further observational work is evidently required in order to clarify the level of agreement with theoretical models.

Moving from radial gradients to ensemble statistics, the interplay between inflows and gas processing leads to a rich phenomenology in which M_*, Z, SFR, and gas fraction are tightly coupled. The resulting correlations are observed out to at least $z = 2$, and they are being disentangled via a wide variety of theoretical studies.

Analytical treatments have shown that the observed M_*-Z-SFR correlation follows naturally from the assumption that gas processing is nearly in equilibrium with inflows, and that enrichment is nearly in equilibrium with dilution. Models that do not explicitly assume equilibrium tend to recover it once their parameters are tuned to match observations. Consequently, they can be used to *infer* inflow rates (in addition to other physical parameters) from the M_*-Z-SFR relation. They have further shown that there is a tight connection between Z, SFR, and gas fraction in the sense that tuning the model to match one observable often yields agreement with another, free of charge. Detailed measurement of inflows and outflows in high-resolution numerical simulations support the importance of quasi-equilibrium behavior governed by outflows whose efficiency decreases with mass. Meanwhile, cosmological simulations (and at least one semi-analytical model) find that a realistic M_*-Z-SFR-gas fraction relation arises naturally within sufficiently realistic frameworks. This ensemble statistical behavior has been interpreted as support for predominantly quiescent, equilibrium-like galaxy growth.

The progress that has been made in understanding the M_*-Z-SFR relation, while promising, does not yet explain everything. Even after removing the dependence of Z on specific star formation rate SFR$/M_*$, the residual observed scatter in Z at a given M_* remains substantial (Salim et al. 2014, 2015). If the SFR (or gas fraction) dependence can largely be accounted for by relaxing the assumption of gas processing equilibrium, then does the residual scatter indicate departure from enrichment equilibrium or intrinsic scatter in η or Z_0? This will be a useful question to take up in future theoretical studies.

In summary, it is now widely recognized that star-forming galaxies do most of their growing in a quiescent mode where the global SFR, Z, and gas fraction constantly adjust on a relatively short timescale to reflect the influence of inflows. Further study of the flow of gas and metals in numerical simulations is required in order to improve our understanding of equilibrium growth, but its broad role is now beyond dispute. From the beginning, measurements of metallicities have driven the development of this paradigm. There is no doubt that they will continue to do so.

Acknowledgements The author thanks R. Davé and A. Fox for offering him the opportunity to contribute this chapter, and for their patience as he drafted it. Additional thanks go to A. Klypin for detailed and honest comments on an early version.

References

Andrews, B. H., & Martini, P. 2013, ApJ, 765, 140
Behroozi, P. S., Wechsler, R. H., & Conroy, C. 2013, ApJ, 770, 57
Birrer, S., Lilly, S., Amara, A., Paranjape, A., & Refregier, A. 2014, ApJ, 793, 12
Bothwell, M. S., Maiolino, R., Kennicutt, R., et al. 2013, MNRAS, 433, 1425
Bothwell, M. S., Maiolino, R., Cicone, C., Peng, Y., & Wagg, J. 2016, arXiv:1606.04102
Bouché, N., Dekel, A., Genzel, R., et al. 2010, ApJ, 718, 1001
Bouché, N., & Blandin, S. 2015, Revolution in Astronomy with ALMA: The Third Year, 499, 63
Brook, C. B., Stinson, G., Gibson, B. K., et al. 2012, MNRAS, 419, 771
Brook, C. B., Stinson, G., Gibson, B. K., et al. 2014, MNRAS, 443, 3809
Brown, J. S., Martini, P., & Andrews, B. H. 2016, MNRAS, 458, 1529
Carton, D., Brinchmann, J., Wang, J., et al. 2015, MNRAS, 451, 210
Casuso, E., & Beckman, J. E. 2004, A&A, 419, 181
Christensen, C. R., Davé, R., Governato, F., et al. 2016, ApJ, 824, 57
Cooper, M. C., Tremonti, C. A., Newman, J. A., & Zabludoff, A. I. 2008, MNRAS, 390, 245
Cousin, M., Buat, V., Boissier, S., et al. 2016, A&A, 589, A109
Cresci, G., Mannucci, F., Maiolino, R., et al. 2010, Nature, 467, 811
Cresci, G., Mannucci, F., Sommariva, V., et al. 2012, MNRAS, 421, 262
Cullen, F., Cirasuolo, M., McLure, R. J., Dunlop, J.S., & Bowler, R.A.A. 2014, MNRAS, 440, 2300
Dalcanton, J. J. 2007, ApJ, 658, 941
Davé, R. 2008, MNRAS, 385, 147
Davé, R., Finlator, K., & Oppenheimer, B. D. 2011, MNRAS, 416, 1354
Davé, R., Finlator, K., & Oppenheimer, B. D. 2012, MNRAS, 421, 98
Davé, R., Thompson, R. J., & Hopkins, P. F. 2016, arXiv:1604.01418
Dayal, P., Ferrara, A., & Dunlop, J. S. 2013, MNRAS, 430, 2891
Dekel, A., & Birnboim, Y. 2006, MNRAS, 368, 2
Dekel, A., & Mandelker, N. 2014, MNRAS, 444, 2071
de los Reyes, M. A., Ly, C., Lee, J. C., et al. 2015, AJ, 149, 79
De Rossi, M. E., Theuns, T., Font, A. S., & McCarthy, I. G. 2015, MNRAS, 452, 486
Dutton, A. A., van den Bosch, F. C., & Dekel, A. 2010, MNRAS, 405, 1690
Ellison, S. L., Patton, D. R., Simard, L., & McConnachie, A. W. 2008, ApJ, 672, L107
Erb, D. K., Shapley, A. E., Pettini, M., et al. 2006, ApJ, 644, 813
Feldmann, R. 2013, MNRAS, 433, 1910
Faisst, A. L., Capak, P. L., Davidzon, I., et al. 2016, ApJ, 822, 29
Finlator, K., & Davé, R. 2008, MNRAS, 385, 2181
Forbes, J. C., Krumholz, M. R., Burkert, A., & Dekel, A. 2014a, MNRAS, 438, 1552
Forbes, J. C., Krumholz, M. R., Burkert, A., & Dekel, A. 2014b, MNRAS, 443, 168

Fu, J., Kauffmann, G., Huang, M.-l., et al. 2013, MNRAS, 434, 1531
Genel, S. 2016, ApJ, 822, 107
Gerola, H., Seiden, P. E., & Schulman, L. S. 1980, ApJ, 242, 517
Grasshorn Gebhardt, H. S., Zeimann, G. R., Ciardullo, R., et al. 2016, ApJ, 817, 10
Grønnow, A. E., Finlator, K., & Christensen, L. 2015, MNRAS, 451, 4005
Gnedin, N. Y. 2000, ApJ, 542, 535
Guo, Y., Koo, D. C., Lu, Y., et al. 2016, ApJ, 822, 103
Henriques, B. M. B., White, S. D. M., Thomas, P. A., et al. 2013, MNRAS, 431, 3373
Henry, A., Martin, C. L., Finlator, K., & Dressler, A. 2013, ApJ, 769, 148
Ho, I.-T., Kudritzki, R.-P., Kewley, L. J., et al. 2015, MNRAS, 448, 2030
Hopkins, P. F., Kereš, D., Oñorbe, J., et al. 2014, MNRAS, 445, 581
Hughes, T. M., Cortese, L., Boselli, A., Gavazzi, G., & Davies, J. I. 2013, A&A, 550, A115
Hunt, L., Magrini, L., Galli, D., et al. 2012, MNRAS, 427, 906
Hunt, L., Dayal, P., Magrini, L., & Ferrara, A. 2016, MNRAS,
Jones, T., Ellis, R. S., Richard, J., & Jullo, E. 2013, ApJ, 765, 48
Kacprzak, G. G., van de Voort, F., Glazebrook, K., et al. 2016, ApJ, 826, L11
Kereš, D., Katz, N., Weinberg, D. H., & Davé, R. 2005, MNRAS, 363, 2
Kewley, L. J., Rupke, D., Zahid, H. J., Geller, M. J., & Barton, E. J. 2010, ApJ, 721, L48
Kirby, E. N., Cohen, J. G., Smith, G. H., et al. 2011, ApJ, 727, 79
Köppen, J., & Edmunds, M. G. 1999, MNRAS, 306, 317
Lagos, C. d. P., Theuns, T., Schaye, J., et al. 2016, MNRAS, 459, 2632
Lara-López, M. A., Cepa, J., Bongiovanni, A., et al. 2010, A&A, 521, L53
Lara-López, M. A., Hopkins, A. M., López-Sánchez, A. R., et al. 2013, MNRAS, 433, L35
Lilly, S. J., Carollo, C. M., Pipino, A., Renzini, A., & Peng, Y. 2013, ApJ, 772, 119
Ma, X., Hopkins, P. F., Faucher-Giguère, C.-A., et al. 2016, MNRAS, 456, 2140
Magrini, L., Hunt, L., Galli, D., et al. 2012, MNRAS, 427, 1075
Maier, C., Lilly, S. J., Ziegler, B. L., et al. 2014, ApJ, 792, 3
Maiolino, R., Nagao, T., Grazian, A., et al. 2008, A&A, 488, 463
Mannucci, F., Cresci, G., Maiolino, R., Marconi, A., & Gnerucci, A. 2010, MNRAS, 408, 2115
Martin, C. L., Kobulnicky, H. A., & Heckman, T. M. 2002, ApJ, 574, 663
Michel-Dansac, L., Lambas, D. G., Alonso, M. S., & Tissera, P. 2008, MNRAS, 386, L82
Mitra, S., Davé, R., & Finlator, K. 2015, MNRAS, 452, 1184
Mitra, S., Davé, R., Simha, V., & Finlator, K. 2016, arXiv:1606.07436
Moster, B. P., Somerville, R. S., Maulbetsch, C., et al. 2010, ApJ, 710, 903
Mouhcine, M., Baldry, I. K., & Bamford, S. P. 2007, MNRAS, 382, 801
Muratov, A. L., Kereš, D., Faucher-Giguère, C.-A., et al. 2015, MNRAS, 454, 2691
Murray, N., Quataert, E., & Thompson, T. A. 2005, ApJ, 618, 569
Nakajima, K., & Ouchi, M. 2014, MNRAS, 442, 900
Nelson, D., Vogelsberger, M., Genel, S., et al. 2013, MNRAS, 429, 3353
Nelson, D., Genel, S., Vogelsberger, M., et al. 2015, MNRAS, 448, 59
Obreja, A., Brook, C. B., Stinson, G., et al. 2014, MNRAS, 442, 1794
Okamoto, T., Gao, L., & Theuns, T. 2008, MNRAS, 390, 920
Oppenheimer, B. D., Davé, R., Kereš, D., et al. 2010, MNRAS, 406, 2325
Pagel, B. E. J., & Patchett, B. E. 1975, MNRAS, 172, 13
Papovich, C., Finkelstein, S. L., Ferguson, H. C., Lotz, J. M., & Giavalisco, M. 2011, MNRAS, 412, 1123
Peng, Y.-j., & Maiolino, R. 2014, MNRAS, 438, 262
Pilkington, K., Few, C. G., Gibson, B. K., et al. 2012, A&A, 540, A56
Pipino, A., Lilly, S. J., & Carollo, C. M. 2014, MNRAS, 441, 1444
Rathaus, B., & Sternberg, A. 2016, MNRAS, 458, 3168
Rupke, D. S. N., Kewley, L. J., & Chien, L.-H. 2010b, ApJ, 723, 1255
Rupke, D. S. N., Kewley, L. J., & Barnes, J. E. 2010a, ApJ, 710, L156
Salim, S., Lee, J. C., Ly, C., et al. 2014, ApJ, 797, 126
Salim, S., Lee, J. C., Davé, R., & Dickinson, M. 2015, ApJ, 808, 25

Sanders, R. L., Shapley, A. E., Kriek, M., et al. 2015, ApJ, 799, 138
Schaye, J., Crain, R. A., Bower, R. G., et al. 2015, MNRAS, 446, 521
Schmidt, M. 1963, ApJ, 137, 758
Schlesinger, K. J., Johnson, J. A., Rockosi, C. M., et al. 2012, ApJ, 761, 160
Sánchez, S. F., Rosales-Ortega, F. F., Jungwiert, B., et al. 2013, A&A, 554, A58
Scudder, J. M., Ellison, S. L., Torrey, P., Patton, D. R., & Mendel, J. T. 2012, MNRAS, 426, 549
Scudder, J. M., Patton, D. R., Ellison, S. L., Torrey, P., & Mendel, J. T. 2013, *Galaxy Mergers in an Evolving Universe*, ASP Conf. Ser. 477, San Francisco:ASP, p241
Somerville, R. S., & Davé, R. 2015, ARA&A, 53, 51
Springel, V., White, S. D. M., Jenkins, A., et al. 2005, Nature, 435, 629
Steidel, C. C., Rudie, G. C., Strom, A. L., et al. 2014, ApJ, 795, 165
Stott, J. P., Sobral, D., Bower, R., et al. 2013, MNRAS, 436, 1130
Tacconi, L. J., Neri, R., Genzel, R., et al. 2013, ApJ, 768, 74
Telford, O. G., Dalcanton, J. J., Skillman, E. D., & Conroy, C. 2016, ApJ, 827, 35
Tinsley, B. M., & Larson, R. B. 1978, ApJ, 221, 554
Tinsley, B. M. 1980, Fundamentals of Cosmic Physics, 5, 287
Torrey, P., Cox, T. J., Kewley, L., & Hernquist, L. 2012, ApJ, 746, 108
Tremonti, C. A., Heckman, T. M., Kauffmann, G., et al. 2004, ApJ, 613, 898
van Dokkum, P. G. 2008, ApJ, 674, 29–50
Veilleux, S., Cecil, G., & Bland-Hawthorn, J. 2005, ARA&A, 43, 769
Weinmann, S. M., Pasquali, A., Oppenheimer, B. D., et al. 2012, MNRAS, 426, 2797
Whitaker, K. E., van Dokkum, P. G., Brammer, G., & Franx, M. 2012, ApJ, 754, L29
White, S. D. M., & Frenk, C. S. 1991, ApJ, 379, 52
White, S. D. M., & Rees, M. J. 1978, MNRAS, 183, 341
White, C. E., Somerville, R. S., & Ferguson, H. C. 2015, ApJ, 799, 201
Wise, J. H., Demchenko, V. G., Halicek, M. T., et al. 2014, MNRAS, 442, 2560
Wolfe, A. M., Gawiser, E., & Prochaska, J. X. 2005, ARA&A, 43, 861
Wu, H.-Y., Doré, O., & Teyssier, R. 2016, arXiv:1607.02546
Wuyts, E., Wisnioski, E., Fossati, M., et al. 2016, ApJ, 827, 74
Yates, R. M., Kauffmann, G., & Guo, Q. 2012, MNRAS, 422, 215
Yuan, T.-T., Kewley, L. J., Swinbank, A. M., Richard, J., & Livermore, R. C. 2011, ApJ, 732, L14
Zahid, H. J., Bresolin, F., Kewley, L. J., Coil, A. L., & Davé, R. 2012, ApJ, 750, 120
Zaritsky, D., Kennicutt, R. C., Jr., & Huchra, J. P. 1994, ApJ, 420, 87

Gas Accretion and Angular Momentum

Kyle R. Stewart

1 Introduction

In the standard Lambda Cold Dark Matter (ΛCDM) paradigm, galaxies form at the center of extended dark matter halo. These halos grow hierarchically through halo mergers (including embedded galaxies) as well as by diffuse accretion of dark matter and gas from the cosmic web. Diffuse infalling gas is expected to shock-heat to the virial temperature of the halo, mixing within the halo until it virializes, with gas eventually cooling out of this hot gaseous halo, sinking to the center of the halo and onto the central galaxy (e.g., Binney 1977; Rees and Ostriker 1977; Silk 1977; White and Rees 1978; White and Frenk 1991; Maller and Bullock 2004). Under this (admittedly simplified) picture of galaxy formation, it is expected that the inflowing gas (and thus the virialized hot gaseous halo) should share the same angular momentum distribution as the inflowing dark matter. Conserving angular momentum, the galaxy that ultimately forms should also have specific angular momentum similar to that of the dark matter halo, resulting in a rotationally supported disk galaxy (in many cases) with a spin proportional to the dark matter halo (Mestel 1963; Fall and Efstathiou 1980; Mo et al. 1998), the statistical properties of which have been well-studied via dissipationless cosmological N-body simulations (e.g., Bullock et al. 2001; Vitvitska et al. 2002; Maller et al. 2002; Avila-Reese et al. 2005; D'Onghia and Navarro 2007; Bett et al. 2010; Muñoz-Cuartas et al. 2011; Ishiyama et al. 2013; Trowland et al. 2013; Kim et al. 2015; Zjupa and Springel 2016).

K.R. Stewart (✉)
Department of Mathematical Sciences, California Baptist University, 8432 Magnolia Ave., Riverside, CA 92504, USA
e-mail: kstewart@calbaptist.edu

A. Fox, R. Davé (eds.), *Gas Accretion onto Galaxies*, Astrophysics and Space Science Library 430, DOI 10.1007/978-3-319-52512-9_11

However, in recent years, advances in galaxy formation theory (both in analytic work and via cosmological hydrodynamic simulations) have begun to emphasize the importance of the filamentary nature of gas accretion onto massive galaxies, particularly at high redshift when cosmic filaments are significantly narrower and denser than in the local universe. Filamentary gas accretion, though diffuse, may be dense enough to allow cold streams to maintain cooling times shorter than the compression time to establish a stable shock, leading to what has been referred to as "cold flows" or "cold-mode" gas accretion that can quickly penetrate from the virial radius of a dark matter halo all the way to the inner galactic region of the halo (e.g., Kereš et al. 2005; Dekel and Birnboim 2006; Ocvirk et al. 2008; Brooks et al. 2009; Dekel et al. 2009; Faucher-Giguère and Kereš 2011; Faucher-Giguère et al. 2011; van de Voort et al. 2011, 2015; Hobbs et al. 2015).

While there has been some contention in recent years as to whether or not these cold streams are truly capable of delivering unshocked gas directly onto the galaxy, without heating in the inner regions of the halo (e.g., Torrey et al. 2012; Nelson et al. 2013, 2016), the importance of distinguishing between these dense filamentary forms of gas accretion to galaxy halos (verses isotropic "hot-mode" gas accretion) remains a crucial one for understanding galaxy formation. In particular—as it relates to this chapter—under this developing paradigm of filamentary versus isotropic gas accretion, halo gas (and particularly gas accreted in the cold mode) tends to show considerably higher specific angular momentum than the dark matter in the halo (Chen et al. 2003; Sharma and Steinmetz 2005; Kereš et al. 2009; Kereš and Hernquist 2009; Agertz et al. 2009; Brook et al. 2011; Stewart et al. 2011b, 2013, 2016; Kimm et al. 2011; Codis et al. 2015; Danovich et al. 2015; Prieto et al. 2015; Teklu et al. 2015; Tillson et al. 2015). In this picture, the resulting angular momentum of simulated stellar disks may be significantly different than that of the accreted gas, in part because feedback effects preferentially expel low angular momentum gas from galaxies (e.g., Maller and Dekel 2002; Governato et al. 2010; Brook et al. 2011; Guedes et al. 2011), such that the total cumulative spin of a growing galactic disk may not be expected to match the cumulative spin of accreted dark matter or gas to the virial radius of the halo.

As a result, this emerging picture of galaxy grown seems to be in tension with the canonical picture in which the spin of the accreted gas (and ultimately, the galaxy) mimics the dark matter. Thus, a modified picture of angular momentum acquisition that takes into account these distinctions between filamentary and isotropic gas accretion seems to be developing in the literature. While this picture is by no means fully developed, nor universally agreed upon, one of the main goals of this review is to serve as a synthesis for a number of recent theoretical works (primarily utilizing the results from hydrodynamic cosmological simulations) in order to present a consistent new picture for angular momentum acquisition to galaxies, as well as the expected angular momentum content of gaseous halos around galaxies.

The outline of this review is as follows. We will begin in Sect. 2 by briefly reviewing the origin of angular momentum in dark matter halos within the framework of Lambda Cold Dark Matter cosmology (LCDM), including a discussion of tidal torque theory (TTT), the role of mergers, and studies of dark matter halos from cosmological dissipationless N-body simulations. In Sect. 3 we will review the canonical model for galaxy formation in LCDM, which builds upon these properties of dark matter halos as a means of understanding the process by which gas within dark matter halos shock-heats, dissipates energy, and ultimately sinks to the center of the halo's gravitational potential to form stars (galaxies) there. We will also discuss some of the historic challenges in simulating galaxies with realistic angular momentum in hydrodynamic cosmological simulations. The most in-depth portion of this review is Sect. 4, which begins by briefly discussing some of the observational challenges to this canonical picture—namely, a number of recent observations of coherent co-rotation and high angular momentum gas in galaxy halos. This is followed by a deeper discussion of recent studies of hydrodynamic cosmological simulations that have begun to demonstrate a need to update the canonical picture of gas accretion onto galaxies (in large part by emphasizing the importance of "cold flow" filamentary gas accretion) and how these modifications are in better alignment with recent observations. We summarize and conclude in Sect. 5.

2 Angular Momentum of Dark Matter Halos

Before address the complications of gas dynamics on the angular momentum acquisition of galaxies, let us review the theoretical picture for the origin of angular momentum in dark matter halos in which galaxies are embedded. We will begin by briefly discussing some the general characteristics of the spin of dark matter halos as derived from cosmological dissipationless N-body simulations, before reviewing the tidal torque model in Sect. 2.1, and the role of major and minor mergers in angular momentum acquisition in Sect. 2.2.

Within the framework of LCDM, the universe forms hierarchically, with less massive dark matter halos forming first, and halos merging together to form more massive halos over time (e.g., Peebles 1982; Blumenthal et al. 1984; Davis et al. 1985). In this paradigm, since dark matter dominates the matter density of the universe, galaxies are expected to reside within the centers of massive dark matter halos. Thus, an important starting point for understanding the angular momentum of galaxies is to determine the angular momentum of dark matter halos in which galaxies reside, typically characterized by the dimensionless spin parameter (Peebles 1969),

$$\lambda_P \equiv \frac{J|E|^{1/2}}{GM^{5/2}} \tag{1}$$

or by the revised version of the spin parameter, first presented by Bullock et al. (2001), rewritten in terms of a dark matter halo's radius and virial velocity[1]:

$$\lambda = \frac{j}{\sqrt{2}V_{\rm vir}R_{\rm vir}} \tag{2}$$

where $j = J/M$ is the magnitude of the specific angular momentum of the halo, $V_{\rm vir} = \sqrt{GM_{\rm vir}/R_{\rm vir}}$ is the circular velocity at the halo virial radius, and $R_{\rm vir}$ and $M_{\rm vir}$ are the halo virial radius and virial mass, respectively.

Remarkably, a long history of studies of N-body simulations reveal that the spin parameters of dark matter halos do not show substantial trends with halo mass, redshift, or environment (e.g., Fall and Efstathiou 1980; Barnes and Efstathiou 1987; Bullock et al. 2001; Vitvitska et al. 2002; Bett et al. 2007, 2010; Macciò et al. 2007; Berta et al. 2008), which is a natural result of the expectations of Tidal Torque Theory as the origin of dark halo angular momentum (see Sect. 2.1, below). Instead, the distribution of halo spins from a number of simulations demonstrates a relatively good fit to a log-normal distribution:

$$P(\lambda) = \frac{1}{\lambda\sigma\sqrt{2\pi}} \exp\left(-\frac{\ln^2(\lambda/\lambda_0)}{2\sigma^2}\right) \tag{3}$$

with the distribution peaking at $\lambda_0 \simeq 0.035$ with a Gaussian width of $\sigma \simeq 0.5$. For comparison, examples of best-fit parameters to this log-normal distribution from Bullock et al. (2001), Bett et al. (2007), Knebe and Power (2008), Ishiyama et al. (2013) and Zjupa and Springel (2016),[2] across a wide range in dark matter halo masses, are shown in Fig. 1, showing remarkable agreement among simulations. The distributions of angular momentum within dark matter halos have also been shown to fit a universal two-parameter angular momentum profile (Bullock et al. 2001), based on the halo spin parameter as well as a halo shape parameter, though the underlying reason dark matter halos should be fit to a nearly universal angular momentum distribution profile is still not well understood.

[1]The revised spin parameter from Bullock et al. (2001) was first introduced as λ', but for the purposes of our discussions in this chapter, we will drop the prime and adopt this revised spin parameter as simply λ.

[2]Since Zjupa and Springel (2016) do not provide fitting parameters to the modified λ' from Bullock et al. (2001), we compare here with their SO–halo log-normal fit to λ_P from Peebles (1969), given in Eq. (1), which appears qualitatively similar to their results for the Bullock definition, based on their Fig. 9.

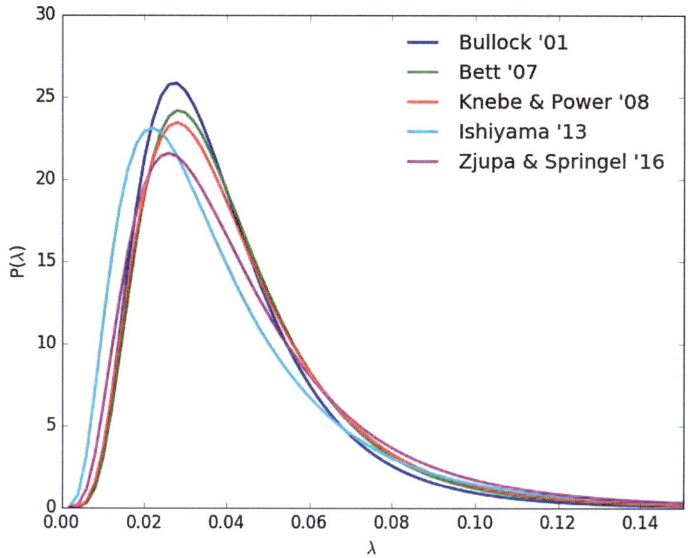

Fig. 1 Distributions of dark matter halo spin parameters from Bullock et al. (2001), Bett et al. (2007), Knebe and Power (2008), Ishiyama et al. (2013), and Zjupa and Springel (2016). Shown here are the best fits to the log-normal function defined in Eq. (3). Quantitative details vary among the simulation (based on, e.g., variations in the definition of the dark matter halo virial radius). However, there is good agreement among simulations, with best-fit parameters $\lambda_0 \simeq 0.035$ and $\sigma \simeq 0.5$

2.1 *Tidal Torque Theory*

The foundational picture of the origin of angular momentum in large-scale structures (such as galaxies and dark matter halo)s is provided by Tidal Torque Theory (TTT, e.g., Hoyle 1949; Peebles 1969; Sciama 1955; Doroshkevich 1970; White 1984; Barnes and Efstathiou 1987). In TTT (for a more thorough recent review, see Schäfer 2009), until the initial density perturbations in the universe reach maximum expansion (turnaround), their angular momentum grows linearly with cosmic time as a consequence of the torques exerted by the tidal gravitational fields from neighboring overdensities. After turnaround, structures decouple from the Hubble Flow and these overdensities collapse and virialize, conserving angular momentum in the process, such that the angular momentum of individual halos may be predicted (within an order of magnitude) by the initial large-scale gravitational tidal torques before maximum expansion. However, detailed analysis requires the inclusion of non-linear effects after turnaround, and thus is typically carried out by cosmological N-body simulations.

Under TTT, the total angular momentum acquired by a halo of a given mass at turnaround is expected to scale as $L \propto M^{5/3}$ (Peebles 1969), with more massive halos acquiring more angular momentum, in part because it takes a long time for more massive halos to reach turnaround and decouple from the Hubble expansion.

Taken in combination with the definition of the halo spin parameter in Eq. (2), in which $\lambda \propto J/MVR$, as well as the virial scaling relations $V \propto (M/R)^{1/2}$ and $R \propto M^{1/3}$, one straightforward expectation from TTT is that the halo spin parameter should be independent of halo mass (and redshift, as long as the redshift in question is after turnaround). This expectation of TTT has been well tested in a number of numerical studies using cosmological N-body simulations of large-scale structure, and is foundational to numerous semi-analytic models of galaxy formation (see Sect. 3.1), however, the importance of non-linear effects also lead to a number of quantitative disagreement between TTT and N-body simulations; for example, linear spatial correlations between the spins of halos on scales greater than about 1 Mpc are over-predicted in TTT by ~50–70% compared to simulations (e.g., Porciani et al. 2002a,b).

More recently, Codis et al. (2012, 2015) revisited the framework of TTT, emphasizing the importance of the anisotropic geometry of walls and filaments in the cosmic web. They found that the alignment direction of angular momentum in dark matter halos is dependent on the geometry of the cosmic web. The misalignments of accretion flows in the walls that collapse into filaments result in spin directions aligned *with* filaments for low mass halos. However, higher mass halos are strongly influenced by accretion (including mergers) that flow along these filaments, resulting in spin directions that are *perpendicular* to the filament. (This will be an important theoretical framework to keep in mind as we discuss revisions to the canonical picture of angular momentum acquisition onto galaxies in Sect. 4.)

2.2 Angular Momentum Acquisition via Mergers

An alternate way to study angular momentum acquisition in galaxy halos is by considering the impact of major and minor mergers during the hierarchical merger history of a given dark matter halo. In this model, the final angular momentum of a dark matter halo is determined by the sum of the orbital angular momentum of all merging satellites over the course of its accretion history, and was found to match the spin distribution of dark matter halos from cosmological N-body simulations (Vitvitska et al. 2002; Maller et al. 2002).

One useful way to conceptualize this approach in a complementary role to TTT (rather than a "competing" mechanism) is to consider that the initial large-scale tidal torques serve as a means of establishing the tangential velocities of infalling satellite halos. Thus, rather than attempting to fully understand the role of tidal torques in setting the ultimate angular momentum of a dark matter halo, one may instead focus on the particular kinematics of infalling satellites over the accretion history of the halo, which are in turn determined by the gravitational torques from the initial tidal field.

Under the merger model, galaxy halos tend to show larger variation in spin parameter than predicted in TTT. Larger mergers contribute substantial angular momentum based on their orbital motion, leading to significant spikes in the spin parameter of a post-merger halo, typically followed by a steady decline in spin parameter during epochs of gradual smooth accretion. Since major mergers can lead to a significant redistribution of angular momentum, under this model the high end of the angular momentum distribution of a dark matter halo is largely determined by the orbital angular momentum of its last major merger, with minor mergers (presumably from random infall directions) tending to contribute to the low end of the distribution.

3 The Angular Momentum of Galaxies

In the canonical picture of galaxy formation, the initial distribution of baryons in the universe matched that of the dark matter, such that when dark matter overdensities collapse and virialize, gas is also affected by the same large-scale tidal fields as the dark matter, resulting in a similar distribution of baryonic matter as that of the dark matter. However, since the halo gas is dissipational (unlike dark matter), it is capable of radiating away orbital and thermal energy, sinking to the center of the halo's gravitational potential, until it is ultimately cold and dense enough to form stars. Thus, all galaxies are thought to be embedded at the center of a massive dark matter halo, with the sizes, luminosities, morphologies, and angular momentum content of those galaxies owing to the details of their formation—which are likely to be correlated in some way with the formation of the dark matter halo. In this section, we will detail the canonical model for how this link between dark matter halo formation and galaxy formation is thought to operate, including the importance of this model in laying the foundation for semi-analytic models (SAMs) of galaxy formation, and discussing the levels of agreement between such models and observations. We will then give a brief review of the challenges and achievements in attempting to simulate galaxy formation directly with hydrodynamic cosmological simulations.

3.1 Modeling Gas Accretion onto Galaxies

The classic picture of galaxy formation (e.g., Fall and Efstathiou 1980; White and Frenk 1991; Mo et al. 1998) attempts to model the formation of galactic disks inside the hierarchical framework of LCDM by making a few assumptions about the relationship between the baryons and dark matter. In these relatively simple models it is possible to reproduce a number of observable properties of spiral galaxies (e.g., the slope and scatter of the Tully–Fisher relation) as well as damped Lyα absorbers, while making as few underlying assumptions as possible. For example, Mo et al. (1998) use the following fundamental assumptions:

1. As the halo forms, the gas initially relaxes into an isothermal distribution. Further gas accretion is shocked to the virial temperature of the halo. Virialized gas subsequently cools, conserving angular momentum.

2. The specific angular momenta of galaxy disks are thus similar to their parent halos, $j_d \simeq j$ (alternatively, $\lambda_d \simeq \lambda$). As a result, the total angular momenta of disks is expected to be a fixed fraction of that of the halo: $J_d/J \simeq M_d/M$.
3. Galaxy disks have masses that are a fixed fraction of roughly a few percent of the mass of their parent halos: $M_d/M \leq 0.05$
4. The resulting disk is assumed to be rotationally supported with an exponential surface density profile and $R_d \simeq \lambda R_{\rm vir}$.

Building on this approach, more recent semi-analytic models include additional physical models such as supernova feedback that expels gas from galaxies, black hole growth and feedback that heats gas in galaxy clusters, estimation of the cooling radius and cooling rate out of the hot halo, as well as effects of galaxy mergers such as starbursts and morphological transformation (recently, e.g., Cattaneo et al. 2006; Croton et al. 2006; Somerville et al. 2008, 2012; Dutton 2012). By tuning the input parameters of these models on certain observational constraints (e.g., tuning the chemical yield of supernovae to reproduce metallicities of stars in galaxies in Somerville et al. 2008), it is possible to produce modeled galaxy populations that reproduce a great number of physical properties of galaxies: e.g., cold gas fractions, stellar ages, specific star formation rates, stellar mass functions, etc.

While it is beyond the scope of this chapter to provide a more comprehensive review of semi-analytic models of galaxy formation, one important point for consideration (especially for our discussion in Sect. 4) is that while some more recent models do implement a distinction between "cold-mode" accretion that does not shock-heat to the virial temperature, versus "hot-mode" gas accretion which does shock-heat, the angular momentum of the halo gas (regardless of which "mode" is used) is still modeled by $\lambda_d = \lambda_{\rm gas} = \lambda_{\rm DM}$. However, hydrodynamic simulations suggest that galactic outflows may preferentially expel low angular momentum gas from the centers of galaxies, keeping galaxy formation inefficient and stopping forming galaxies from universally creating massive bulges at early times. In this case, even if galaxies initially form with $\lambda_d = \lambda_{\rm DM}$ at early times, this similarity would be expected to break over cosmic time, as outflows continue to preferentially remove low angular momentum gas from the galaxy (without similarly removing dark matter from the halo). Furthermore, while one might expect the overall spin of halo gas and dark matter to be in rough agreement, we will see in Sect. 4.2 that the detailed accretion geometry of different modes of gas accretion (particularly the contribution of dense filamentary "cold-mode" gas) results in a scenario where $\lambda_{\rm gas} \neq \lambda_{\rm DM}$.

3.2 *Hydrodynamic Simulations of Galaxy Formation*

Early work in cosmological hydrodynamic simulations showed great difficulty in successfully simulating disk-dominated galaxies. In what is often referred to as the "angular momentum catastrophe," simulations produced either spherical galaxies or disks with significantly lower angular momentum than the halo, with orbital angular momentum being transferred to the dark matter by dynamical friction before the baryons reach the center of the halo (e.g., Katz 1992; Navarro and White 1994; Sommer-Larsen et al. 1999; Steinmetz 1999; Navarro and Steinmetz 2000; D'Onghia et al. 2006). Not surprisingly, these simulated galaxies also produced unrealistic rotation curves and failed to match other observational constraints, such as the Tully–Fisher relation.

The alleviation of this problem seemed to be the inclusion of efficient star formation feedback, which preferentially removes low angular momentum gas (that would otherwise form stars) from the centers of galaxies during the formation process (e.g., Governato et al. 2007, 2010; Scannapieco et al. 2008; Brook et al. 2011; Guedes et al. 2011; Übler et al. 2014; Christensen et al. 2016). This feedback makes galaxies considerably less efficient at forming stars, also keeping them gas-rich for longer. This, in turn, also helps alleviate the tension between the observed abundance of disk-dominated galaxies (e.g., Weinmann et al. 2006) with the frequency of major mergers derived from *N*-body simulations (e.g., Stewart et al. 2008; Fakhouri et al. 2010), as both direct hydrodynamic simulation and semi-empirical galaxy formation models suggest that gas-rich major mergers may help *build* angular momentum supported disks from the surviving merger remnant, rather than transforming pre-existing disks into spheroids (Robertson et al. 2006; Stewart et al. 2009; Hopkins et al. 2009; Governato 2009). With these advances in star formation and feedback prescriptions (as well as more advance computational power), recent simulations have essentially eliminated the early angular momentum problem, allowing hydrodynamic simulations to successfully produce bulgeless exponential disk galaxies with properties quite similar to those observed in the real universe (e.g., Governato et al. 2010; Brook et al. 2011; Guedes et al. 2011).

Most importantly for our discussion of angular momentum acquisition in galaxies (and their halos), recent hydrodynamic simulations have also begun to place growing emphasis on the different "modes" of gas accretion onto galaxies, especially at high redshift. In what is labeled "hot-mode" accretion, gas continues to behave in the manner previously described, shock-heating to the virial temperature of the halo, mixing, and eventually cooling on the galaxy. However, the main mode of gas accretion for most galaxies is thought to be via "cold-mode" (or "cold flow") accretion—where the inflowing gas streams originating from filamentary accretion are dense enough at high redshift to have cooling times shorter than the shocking compression timescales, resulting in direct gas accretion from the cosmic web, through the galaxy halo, and onto the outskirts of the galaxy (e.g., Kereš et al. 2005; Dekel and Birnboim 2006; Brooks et al. 2009). As a result, this cold-mode gas does not necessarily mix with the existing gaseous halo, and so the specific angular momentum of gas that accretes onto the central galactic disk might *not* be well matched by that of the dark matter halo, as previously assumed. We will discuss possible implications of this dual mode of accretion for understanding angular momentum in galaxy halos in Sect. 4.2.

4 Angular Momentum of Gaseous Halos

The previously described model for galaxy formation (and angular momentum acquisition in particular) in Sect. 3.1 assumes that gaseous halos of galaxies should maintain a similar distribution of spin parameters to that of the dark matter (since the baryons and the dark matter both share the same initial tidal torques as an origin of their angular momentum). Thus, if one were to consider the spin parameter of the gas in a galaxy halo it should be the same distribution as the spin of the dark matter, which is well constrained from N-body simulations (Fig. 1), such that $\lambda_{gas} \simeq \lambda_{DM}$. This simple theoretical picture has provided reasonable agreement between theory and observations, in terms of matching distributions of galaxy sizes and luminosities, with characteristic galaxy sizes expected to be $R_d \sim \lambda R_{vir}$ (corresponding to ~10 kpc for a galaxy halo with R_{vir} ~300 kpc, e.g., Fall and Efstathiou 1980; Bullock et al. 2001; Dutton and van den Bosch 2009).

Having also reached the point where cosmological hydrodynamic simulations of galaxy formation (with properly tuned star formation and stellar feedback physics implemented) are able to produce galaxies with realistic disk scale lengths, bulge-to-disk ratios, and overall angular momentum content—resolving the angular momentum catastrophe—one might think that our picture of angular momentum acquisition is reasonably complete. However, recent observations have provided further complication to the issue of angular momentum acquisition, with numerous detections of baryons in galaxy halos with significantly *higher* spin than either the galaxy or expectations for dark matter halos. These frequent detections of high-spin baryons in galaxy halos would seem difficult to explain under the assumption that $\lambda_{gas} \simeq \lambda_{DM}$.

In this section, we will discuss the angular momentum content not of the dark matter halo, nor the baryons in the stellar content of the galaxy, but instead the baryons present in the gaseous halo of the galaxy—i.e., the circumgalactic medium (CGM). We begin in Sect. 4.1 by presenting the observational evidence for high angular momentum material in galaxy halos, challenging the classical picture of galaxy formation described above. We then discuss recent advances in our understanding of galaxy formation theory that may help explain these observations in Sect. 4.2.

4.1 Observations of High Angular Momentum Gas

In the local universe, some of these high angular momentum observations include detections of extended H I disks, XUV disks, and giant low surface brightness galaxies (e.g., Bothun et al. 1987; Matthews et al. 2001; Oosterloo et al. 2007; Christlein and Zaritsky 2008; Sancisi et al. 2008; Lemonias et al. 2011; Heald et al. 2011; Holwerda et al. 2012; Hagen et al. 2016), as well as low metallicity high angular momentum gas (presumably from fresh accretion) in polar ring galaxies

(Spavone et al. 2010). For example, observations of UGC 2082 from Heald et al. (2011) show a stellar disk diameter of $D_{25} = 24\,\text{kpc}$ (defined by a surface brightness of 25 mag/arcsec^2) but the H I disk (down to a minimum column density of $10^{20}\,\text{cm}^{-2}$) has significantly higher specific angular momentum—being larger by roughly a factor of ~ 2, $D_{\rm HI} = 44\,\text{kpc}$. In an even more extreme example, Oosterloo et al. (2007) detected H I disks as large as ~200 kpc in diameter around early type galaxies. As another example, the giant lower surface brightness galaxy UGC 1382 contains a low surface brightness stellar disk with a ~38 kpc radius, embedded in a ~110 kpc H I disk, residing in a $\sim 2\times10^{12} M_\odot$ halo (Heald et al. 2011). The high spin of such extended disk components is somewhat difficult to understand within the context of the canonical model, in which $j_d \simeq j_{\rm gas} \simeq j_{\rm DM}$. Furthermore, Courtois et al. (2015) also indicated that local extended H I disks may be dependent on the galaxy's filamentary environment, suggesting there may be a fundamental distinction between the angular momentum content of filamentary accretion versus isotropic accretion.

At moderate redshift ($z \sim 0.5$–1.5) there are a growing number of absorption line studies of the circumgalactic medium of galaxies that have begun to emphasize the bi-modal properties of absorbers (Kacprzak et al. 2010, 2012a,b; Bouché et al. 2012, 2013, 2016; Crighton et al. 2013; Nielsen et al. 2015; Diamond-Stanic et al. 2016; Bowen et al. 2016), where absorbers along a galaxy's major axis tend to show higher angular momentum inflows that are roughly co-rotating with the galactic disk and absorbers along a galaxy's minor axis tend to instead show observational signatures of outflowing gas. Increasingly, a number of absorption system observations seem to be in agreement with models that include massive, extended structures with inflowing disk-like kinematics (e.g., Bouché et al. 2016; Bowen et al. 2016).

At higher redshift ($z \sim 2$–3) kinematic studies of Lyα "blobs" have observed large-scale rotation that seems consistent with high angular momentum cold gas accretion (Martin et al. 2014; Prescott et al. 2015), and there have also been recent detections of massive protogalactic gaseous disks that are kinematically linked to gas inflow along a cosmic filaments (Martin et al. 2015, 2016). (We will show in Sect. 4.2.1 how these observations are strikingly similar to theoretical expectations for inspiraling cold streams from cosmological simulations.)

Taken together (though this is by no means an exhaustive list), such observations show growing evidence for the existence of coherent rotation with high angular momentum for cold halo gas, in stark contrast to the theoretical picture where halo gas should have specific angular momentum similar to that of the galaxy and the dark matter halo.

4.2 "Cold Flow" Gas Accretion and Angular Momentum

Recent advances in galaxy formation theory and cosmological simulations have also begun to complicate the picture of galaxy formation presented in Sect. 3, with growing emphasis on multiple modes of accretion to galaxy halos. While isotropic "hot-mode" accretion continues to behave in the manner previously assumed, it has also been shown that anisotropic "cold-mode" accretion along cosmic filaments may have cooling times shorter than the compression timescales for creating a stable shock (e.g., Binney 1977; Kereš et al. 2005; Dekel and Birnboim 2006; Ocvirk et al. 2008; Brooks et al. 2009; Dekel et al. 2009; Faucher-Giguère and Kereš 2011; Faucher-Giguère et al. 2011; Stewart et al. 2011a; van de Voort et al. 2011, 2015; Hobbs et al. 2015). As a result, this cold filamentary accretion does not spend sufficient time in the halo to become well mixed before sinking towards the central galaxy.[3]

In one of the seminal papers outlining the importance of "cold-mode" gas accretion to galaxies, Kereš et al. (2005) suggested that the angular momentum of filamentary cold gas accretion may be substantially different than that of isotropic "hot-mode" accretion, however, for a more thorough investigation of the angular momentum of these different types of gas accretion, we must look to more recent results (in part owing to the need for superior numerical resolution in simulations before an analysis of angular momentum in gas accretion could be considered reasonably robust).

Stewart et al. (2011b, 2013) studied 4 high resolution cosmological hydrodynamic simulations of roughly Milky Way size halos run to $z = 0$ (using the smooth particle hydrodynamics (SPH) code `Gasoline`, Wadsley et al. 2004), with particular emphasis on the distinction between "cold-mode" gas accretion (which is typically more filamentary and thus anisotropic) versus "hot-mode" gas accretion (typically more isotropic). They found that cold-mode gas in galaxy halos contains significantly higher specific angular momentum than the dark matter, $\lambda_{\rm cold} \sim 4\lambda_{\rm DM}$, and also has noticeably higher spin than the hot-mode accretion, $\lambda_{\rm cold} \sim 2\lambda_{\rm hot}$ (also see Fig. 2). As a preliminary look at origin of this discrepancy, they compared the spin of recently accreted isotropic versus anisotropic dark matter and found a qualitatively similar distinction.

They also found that fresh accretion (both gas and dark matter) contained $\sim$2 times higher angular momentum than the entire halo, with $\lambda \sim 0.1$ (rather than the canonical $\lambda \simeq 0.04$ for the entire halo). Furthermore, even at accretion to the virial radius, cold gas enters the halo with $\sim$70% more specific angular momentum than dark matter, and has a relatively short sinking time from the virial radius to the galactic disk (only $\sim$1–2 halo dynamical times). They argue that this combination naturally explains the high angular momentum nature of cold halo gas, compared to dark matter: dark matter in galactic halos represents a cumulative process of past accretion, while cold gas currently in a galactic halo both entered the halo with higher specific angular momentum than dark matter, and also represents recent accretion rather than a cumulative sum of all past accretion. As a result of the

[3]Insofar as the angular momentum of spheroids at low redshift are still thought to be most strongly correlated with the merger history of its dark matter halo, we note that the following discussion mostly pertains to the newfound importance of filamentary cold accretion to the growth of massive disk-dominated galaxies, or to the properties of gaseous halos of galaxies.

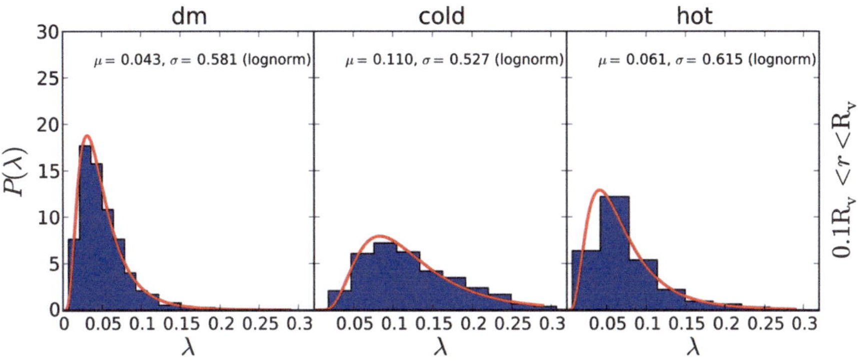

Fig. 2 Modified version of Fig. 6 from Danovich et al. (2015), showing distributions of halo spin parameters (for material in the halo but not the central $0.1R_{\rm vir}$) in dark matter (*left*), cold gas (*middle*), and hot gas (*right*) from 29 high resolution cosmological hydrodynamic "zoom-in" simulations between $z = 4 - 1.5$. *Red curves* represent best-fit log-normal fits to the data. The mean spin parameter of each best-fit curve ($\mu = \langle\lambda\rangle$) is given for each component, demonstrating that cold halo gas, $\langle\lambda_{\rm cold}\rangle = 0.110$, has significantly higher specific angular momentum than the dark matter in the same region: $\langle\lambda_{\rm dm}\rangle = 0.043$. This fundamental result is in close agreement with a number of other studies (Stewart et al. 2011b, 2013, 2016; Pichon et al. 2011; Kimm et al. 2011), which implement a variety of simulation codes and feedback implementations to study the spin of cold halo gas compared to that of the dark matter halo

coherent high nature of this cold inflow, they also reported the formation of transient "cold flow disk" structures in their simulations: massive extended planar structures of inflowing cold halo gas (not rotationally supported) that are often warped with respect to the central galaxy—aligned instead with the angular momentum of the inflowing cold filamentary gas.

In a set of companion papers, Pichon et al. (2011) and Kimm et al. (2011) analyzed a statistical sample of ~15,000 halos at $z > 1.5$ (from a somewhat lower resolution simulation) as well as ~900 intermediate resolution halos and 2 high resolution zoom-in simulations run to $z = 0$ (using the `Ramses` code, Teyssier 2002), finding that the ability of gas to radiative cool significantly alters the angular momentum transport of gas and dark matter into halos of a wide ranges of masses, primarily due to the dense filamentary nature of cold gas accretion. They reason that due to the asymmetry of cosmic voids, gas and dark matter flowing out of voids onto cosmic filaments gain a net transverse velocity, which acts as the seed of a halo's angular momentum as the material subsequently flows along the filament onto a nearby (gravitationally dominant) dark matter halo. In this picture, material initially farther away from the filament will gain a larger transverse velocity by the time it impacts the filament, naturally providing more angular momentum at later times, but with coherent direction (since the filament direction does not substantially change orientation over cosmic time).

While they find that both gas and dark matter tend to deliver similar specific angular momentum (as would be expected from TTT) the newly accreted material (at any given time) has significantly higher spin than that of the halo as a whole.

Specifically, at high redshift, a large discrepancy between the spin of gas and dark matter in the halo arises, in agreement with previous work (Stewart et al. 2011b), with $\lambda_{gas} \sim 2\text{–}4\lambda_{DM}$, depending on halo mass. This discrepancy is thought to be due to the coherent dense filamentary accretion of gas along cosmic filaments, without shock-heating to redistribute the angular momentum through the entire halo. Over time, both dark matter and baryons (including all gas and stars) lose a significant amount of angular momentum, primarily via vector cancellation, as freshly accreted material is never perfectly aligned with that of the halo as a whole.

Using 2 intermediate resolution simulations, Sales et al. (2012) studied 100 roughly Milky Way size halos run to $z = 0$ (using the `Ramses` code). They noted that the angular momentum of different gas components upon infall may not necessarily be indicative of how these modes contribute to galaxy type. They found that galaxy morphology was most strongly correlated with the *coherent alignment* of angular momentum over cosmic time. Specifically, they found that most disk-dominated galaxies formed their stars from hot-mode gas that shock-heated and eventually cooled onto the galaxy at later times, while more spheroids were formed from cold-mode gas that sinks onto the galaxy more quickly, forming stars at a much earlier time; as a result, later episodes of accretion may not be very well aligned with the galaxy, leading to a spheroidal morphology.

Danovich et al. (2015), building upon previous work (Danovich et al. 2012), analyzed 29 zoom-in simulations at $z > 1.5$ (using the `Art` code, Kravtsov et al. 1997; Kravtsov 2003), focusing on the angular momentum transport from the cosmic web onto massive galaxies, which they comprehensively detailed through four distinct phases, outlined below.

1. According to TTT, the spatial dependence of the angular momentum vector components, J_i, are given by the antisymmetric tensor product: $J_i \propto \epsilon_{ijk} T_{jl} I_{lk}$, where T_{jl} is the tidal tensor and I_{lk} is the inertial tensor. In the principle coordinates of the tidal tensor, the angular momentum is thus proportional to the difference in the corresponding eigenvalues of the inertial tensor: $J_1 \propto T_{23}(I_3 - I_2)$, $J_2 \propto T_{13}(I_3 - I_1)$, $J_3 \propto T_{12}(I_2 - I_1)$. Under the assumption that the underlying tidal tensor should be approximately the same for both dark matter and gas, any inherent differences in the specific angular momentum of dark matter versus gas (outside the virial radius, in the regime of TTT) should result from the difference between the inertial eigenvalues, $(I_j - I_k)$, as a proxy for the quadrupole moment. Focusing on dark matter and gas just outside the virial radius of the galaxy, $(1 < r/R_{vir} < 2)$, they found that the quadrupole moment is consistently higher for the cold gas than it is for the dark matter by a factor of ~1.5–2. Since these tidal torques may act prior to maximum, each stream may acquire a transverse velocity, so that it is no longer pointing directly at the center (in agreement with previous work by Pichon et al. 2011). The highest angular momentum streams thus have spin parameters as high as $\lambda \sim 0.3$ upon crossing the virial radius, however, misalignment between multiple streams typically lowers the net spin parameter of all cold gas entering the virial radius to a lower value of $\lambda \sim 0.1$ (in agreement with previous work by Stewart et al. 2013).

2. While inflowing dark matter virializes once inside the virial radius, the cold streams penetrate the halo quickly, without becoming well-mixed with the pre-existing gas in the halo, resulting in a higher spin parameter for cold gas in the halo ($0.1 < r/R_{\rm vir} < 1.0$) by a factor of $\sim$3 when compared to the dark matter in the same volume (see Fig. 2). The cold gas in the outer halo is also significantly more coherent than the dark matter, with a significantly smaller anti-rotating fraction.
3. The cold streams remain coherent, spiraling around the galaxy and sinking quickly towards the center of the halo. As the streams blend and mix together, they often form into what Danovich et al. (2015) refer to as "extended rings" of inflowing cold gas, the radius of which is typically set by the pericenter of the stream contributing the most angular momentum. These structures are typically warped with respect to the inner disk, in qualitative agreement with the "cold flow disk" structures reported previously by Stewart et al. (2011b, 2013). Similar structures have also been noted as areas of interest in previous simulations: for example, the "messy region" of Ceverino et al. (2010) or the "AM sphere" or Danovich et al. (2012). Angular momentum in these extended rings is ultimately lost as a result of strong tidal torques from the inner disk on timescales of roughly one orbital time, allowing the extended disk to gradually align with the inner disk.
4. The angular momentum lost by the inspiraling cold gas can ultimately be redistributed to both outflows and the dark matter. The inner disk is subject to angular momentum redistribution and violent disk instabilities.

Teklu et al. (2015) analyzed $\sim$600 intermediate resolution massive ($M_{\rm vir} > 5 \times 10^{10} M_\odot$) halos from the *Magneticum* simulation (Dolag et al. in preparation) over the redshift range $z = 2-0.1$. In agreement with previous results, they compared the spin parameter of all dark matter, stars, gas, cold gas, and hot gas in the virial radius (not cutting out the inner region of the halo where the galaxy resides), and found that the distribution of spins was well fit by log-normal distributions, with the gas (particularly the cold gas components) showing systematically higher spin than that of the dark matter, with the dark matter spin staying roughly constant with time, but the gas spin parameter growing with time: $\lambda_{\rm cold} \sim 2(3)\lambda_{\rm DM}$ at $z = 2(0.1)$. They also noticed a dichotomy in spin parameter with galaxy morphology: disk galaxies tend to populate halos with slightly higher spin parameters, and where there is better alignment between the angular momentum vector of the inner region of the dark matter halo versus that of the entire halo.

In an effort to test whether this changing picture of angular momentum acquisition is sensitive to simulation code architectures or specific feedback implementations, Stewart et al. (2016) carried out a code comparison of a single high resolution zoom-in simulation of a Milky Way sized halo (using common recent hydrodynamic/feedback implementations for each code, and utilizing identical analysis for each code) run with `Enzo` (Bryan et al. 2014), `Art` (Kravtsov et al. 1997; Kravtsov 2003), `Ramses` (Teyssier 2002), `Arepo` (Springel 2010), and `Gizmo-PSPH` (Hopkins 2015). While many quantitative differences were apparent among the codes, agreements included the spin of cold halo gas being $\sim$4 times

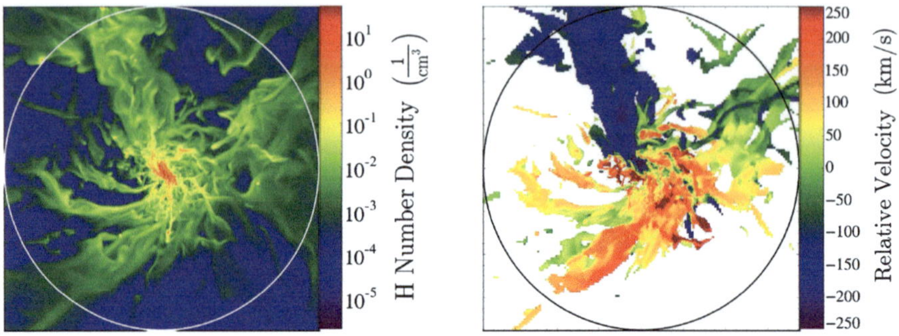

Fig. 3 Inspiraling cold streams in a galaxy simulation at $z = 3$. The halo virial radius is annotated by a *circle* in each panel. *Left panel* shows projected *H* number density, and the *right panel* shows density-weighted line of sight velocity for gas with a minimum density threshold of $n_{\rm H} > 3 \times 10^{-3}\,{\rm cm}^{-3}$ (approximately equivalent to $N_{\rm HI} \gtrsim 10^{17}\,{\rm cm}^{-2}$). The coherent bulk rotation of the inspiraling cold streams is apparent and should, in principle, represent an observable test of filamentary gas accretion in LCDM

higher than the dark matter in the halo (in agreement with previous work, e.g., Fig. 2, taken from Danovich et al. 2015), as well as the presence of inspiraling cold streams. These inspiraling cold streams often form extended transient structures of high angular momentum cold gas, co-rotating with the galaxy along a preferred plane that is kinematically linked to inflow via large-scale cosmic filaments (see Fig. 3 and discussion in Sect. 4.2.1). The agreement among disparate simulation codes and physics implementations suggests that these aspects (at minimum) are likely to be robust predictions of galaxy formation in the Lambda Cold Dark Matter paradigm.

4.2.1 Theoretical Predictions: High Angular Momentum, Co-rotation, and Inspiraling Cold Streams

The most direct observable predictions of this new picture, specifically as it relates to angular momentum is thus not likely to come from studies of galaxies themselves (which represent a complex cumulative history of past angular momentum acquisition—including mergers, stream misalignments, etc.—as well as effects from stellar feedback and outflows), but from observations of the circumgalactic medium. In the canonical picture (outlined in Sect. 3) cold gas in galaxy halos is thought to have cooled out of a virialized hot halo, and should have roughly the same angular momentum distribution as the dark matter. In this new picture, cold gas in galaxy halos should have ~4 times higher spin, and often form coplanar structures of coherent inflowing gas, fueled by filamentary gas accretion. In an attempt to find a middle ground between the different terms in the literature for such structures, we will refer more generally here to these phenomena as resulting from inspiraling cold streams, since the degree to which these structures resemble the disk-like (Stewart et al. 2011b, 2013) or ring-like (Danovich et al. 2015) morphologies from previous work may be sensitive to specific hydrodynamic codes, feedback implementations, and possibly the halo mass scale involved.

Figure 3 shows one example of inspiraling cold streams in a galaxy halos, taken from a cosmological hydrodynamic zoom-in simulation and visualized on the scale of the halo virial radius (denoted by the circle in each panel). The left panel shows the projected H number density and the right panel showing the projected density-weighted line of sight velocity for all sight lines that meet a minimum column density threshold[4] of $N_{\rm HI} \gtrsim 10^{17}\,{\rm cm}^{-2}$. The coherent rotational structure of the inspiraling cold streams (fueled and kinematically connected to the larger filamentary geometry) is apparent. Encouragingly, this type of extended rotational structure of inflowing gas bears a striking similarity to recent observations of giant protogalactic disks (Martin et al. 2015, 2016), which have been detected in $M_{\rm vir} \sim 5 \times 10^{12} M_{\odot}$ halos at $z \sim 2$–3. In qualitative agreement with simulations, these observed disk-like structures extend to diameters of ~100 kpc ($\sim R_{\rm vir}/2$), with rotational velocities of ~300 km/s that show a kinematic connection to an inflowing filament, and have very high angular momentum (estimated $\lambda \sim 0.1$–0.3), with orbital times comparable to the halo dynamical time.

In an effort to compare simulations to absorption line studies that match absorber kinematics to the rotation curve of the associated galaxy, Stewart et al. (2011b) also created mock absorption sightlines to infer that, for inflowing gas, ~90% of absorbers with $N_{\rm HI} \gtrsim 10^{16}\,{\rm cm}^{-2}$ should have line of sight velocities completely offset from the system velocity of the galaxy in a single direction (per sightline) by ~100 km/s, with most of these absorbers roughly co-rotating with the galactic disk. Again, the results from simulations are in encouraging agreement with recent absorption studies where the associated galaxy kinematics are known (e.g., Bouché et al. 2016; Bowen et al. 2016, also see Sect. 4.1), though larger statistical samples of both observations and high resolution zoom-in simulations will be important for characterizing the level of agreement in detail.

5 Summary and Conclusion

In this review of gas accretion and the angular momentum of galaxies and galaxy halos, we began (Sect. 2) by reviewing the origin of angular momentum in dark matter halos via Tidal Torque Theory, where large-scale tidal torques before halo turnaround set the initial angular momentum of a collapsing region based on the

[4]In detail, a minimum 3D hydrogen density cutoff of $n_{\rm H} > 3 \times 10^{-3}\,{\rm cm}^{-3}$ was implemented, however, this should correspond to a minimum hydrogen column density of $N_{\rm HI} \gtrsim 10^{17}\,{\rm cm}^{-2}$ (e.g., Altay et al. 2011; Schaye 2001)

structure of large-scale overdensities. This ultimately sets a distribution of halo spin parameters in dark matter halos, independent of halo mass, and with typical spins of $\lambda = j/\sqrt{2}V_{\rm vir}R_{\rm vir} \simeq 0.04$.

Under the canonical galaxy formation model (Sect. 3), it is presumed that the angular momentum of inflowing gas matches that of the dark matter, shock-heats to the virial temperature of the halo, where the gas becomes well-mixed, before ultimately cooling out of the halo while conserving angular momentum to form a rotationally supported disk galaxy at the halo center. Under this picture, the hot gas halo is expected to have roughly the same spin distribution of the dark matter ($\lambda_{\rm gas} = \lambda_{\rm DM}$), such that the disk galaxy that eventually forms should have a disk size of roughly $R_d \simeq \lambda R_{\rm vir}$. While this picture is a good approximation for estimating galactic disk sizes in practice, there are growing number of observations (outlined in Sect. 4.1) both in the local and distant universe that demonstrate the presence of significantly higher angular momentum material at large distances from the centers of galaxies, which might seem difficult to explain under the above scenario.

Our main emphasis in this review (Sect. 4.2) has been a summation of recent findings from hydrodynamic cosmological simulations that suggest a modified picture for angular momentum acquisition, particular as it relates to predictions for the circumgalactic medium (CGM) of massive galaxies at $z \gtrsim 1$. This emerging picture, found in qualitative agreement among a variety of simulations—including large statistical samples of intermediate resolution simulations, smaller statistical samples of high resolution zoom-in simulations, and a wide range of code architectures and feedback physics implementations—is summarized as follows:

- The dense filamentary nature of cosmic gas accretion plays an important role in the properties of the CGM, with the geometry of filamentary gas responding differently to the initial tidal torques that set angular momentum at early times (a factor of $\sim$1.7 enhancement versus the dark matter in the filament). This angular momentum translates to a net transverse velocity of the filament and subsequently a non-zero impact parameter as the filamentary gas enters the virial radius of the halo.
- Material (both gas and dark matter) in a cosmic void that is initially farther away from the filaments gains a larger transverse velocity by the time it impacts the filament. As a result, the specific angular momentum of material freshly accreted to the virial radius (for both dark matter and gas) increases with cosmic time. Compared to the cumulative average of the material that already exists in the halo, the spin of fresh accretion is also enhanced by a factor of $\sim$2.5.
- Filamentary gas tends to have cooling times shorter than compression times for stable shocks at high-z, even for massive galaxies. As a result, filamentary ("cold flow") gas accretion sinks quickly (in $\lesssim$2 times the freefall timescale) to the center of the halo upon entering the virial radius. Thus, cold gas currently in the CGM of a given galaxy shows an enhanced spin by a factor of $\sim$4 (compared to the dark matter). This is a combination of the previous two effects mentioned above: (1) the high intrinsic spin of cold filamentary gas and (2) the enhanced spin for recent accretion to the halo (when compared to the dark matter halo, which probes a cumulative total of all past accretion).

- One natural result of this high-spin inflow with short sinking times (and thus, often a coherent direction for the angular momentum vector of all cold gas currently in the CGM) is that the inspiraling cold streams in the halo often form massive extended structures of roughly coplanar cold CGM gas, showing coherent rotation (see Fig. 3) as the gas flows from the virial radius to the center of the halo. While the existence of these structures seems robust in a qualitative sense, the exact nature and prevalence of these structures (and their implications for galaxy formation) are still unknown. However, it is promising to note that recent observations of "protogalactic disks" seem to show qualitatively similar structures in the real universe.

It is important to note that much of emphasis in this modified picture for angular momentum acquisition has been on the process by which cold filamentary gas transitions from the cosmic web through the CGM on its way to the galaxy, particularly that this cold gas has significantly higher angular momentum while in the CGM than either the dark matter halo or the baryons in the galaxy. One important clarification is that this factor of $\sim$4 enhancement in cold halo gas spin versus dark matter is a result of the coupling of cold CGM gas properties (gas that is freshly accreted to the halo, and that is of a filamentary origin) working together to produce this enhancement. The same level of spin enhancement should *not* hold for the cold gas in the *galactic* region, as the baryons near the galactic center (or indeed within the galaxies themselves) are more likely to probe a prolonged accretion history from multiple "modes" of accretion, and is thus more likely to mimic the spin of the dark matter halo, as expected from the canonical picture of galaxy formation.

For example, Fig. 2 showed the spin parameter distribution for cold halo gas, but did not include the inner region (i.e., $0.1 < r/R_{\rm vir} < 1$). However, misalignment between the inspiraling cold streams and the baryons in the central region typically leads to significant vector cancellation, with a lower overall spin parameter for gas once the galactic region is included. Thus, while the mean spin parameter of cold halo gas shown in Fig. 2 is $\langle\lambda_{\rm cold}\rangle = 0.11$, in a complementary panel in the same figure (Fig. 6 from Danovich et al. 2015) the mean spin parameter for all cold gas within the virial radius of the halo ($r < R_{\rm vir}$) is noticeably reduced: $\langle\lambda_{\rm cold}\rangle = 0.086 \sim 2\langle\lambda_{\rm DM}\rangle$. Thus, in angular momentum studies where all material within the virial radius is included (including the galactic region), only this factor of $\sim$2 enhancement of cold gas versus dark matter is expected. For example, Teklu et al. (2015) compared of the spin of all cold gas within $r < R_{\rm vir}$ to that of the dark matter at $z = 2$, finding that $\lambda_{\rm cold} = 0.074 \sim 2\lambda_{\rm DM}$. Similarly, Zjupa and Springel (2016) recently studied the angular momentum of dark matter halos and their baryons for $\sim$320,000 moderately high resolution halos from the *Illustris* simulation (Vogelsberger et al. 2014)—comparing all gas within the virial radius ($r < R_{\rm vir}$) and not making any distinction between cold versus hot gas components, finding that $\lambda_{\rm gas} \simeq 0.1 \sim 2\lambda_{\rm DM}$, in agreement with other work reviewed here.

We also note that while the modifications suggested here for the standard picture of angular momentum acquisition in galaxy halos have strong implications for the angular momentum of baryons in the CGM, it is unclear at this time how this modified picture directly impacts of the angular momentum of the galaxies that form at the center of the halo. If cold gas accretion onto galaxies typically has higher spin than the dark matter, but that angular momentum is subsequently lost from the galactic disk by strong torques from inspiraling cold streams, or redistribution of angular momentum via subsequent mergers, diffuse accretion and/or outflows, it may be that the similar spins for galactic disks and dark matter halos are merely the result of coincidence. Alternatively, since the galaxy that ultimately forms at the center of a growing dark matter halo is the result of an extended, cumulative process, which must by its very nature account for misalignments in the angular momentum direction of accretion over cosmic timescales, it may not be surprising (or coincidental) that the specific angular momentum of galactic disks are similar to their dark matter halos. After all, the dark matter halo also probes the cumulative history of angular momentum acquisition over cosmic time.

Acknowledgements KRS thanks Ari Maller for providing useful comments and discussions.

References

Agertz, O., Teyssier, R., & Moore, B. 2009, MNRAS, 397, L64
Altay, G., Theuns, T., Schaye, J., et al. 2011, ApJ, 737, L37
Avila-Reese, V., Colín, P., Gottlöber, S., et al. 2005, ApJ, 634, 51
Barnes, J. & Efstathiou, G. 1987, ApJ, 319, 575
Berta, Z. K., Jimenez, R., Heavens, A. F., & Panter, B. 2008, MNRAS, 391, 197
Bett, P., Eke, V., Frenk, C. S., Jenkins, A., et al. 2007, MNRAS, 376, 215
Bett, P., Eke, V., Frenk, C. S., Jenkins, A., & Okamoto, T. 2010, MNRAS, 404, 1137
Binney, J. 1977, ApJ, 215, 483
Blumenthal, G. R., Faber, S. M., Primack, J. R., & Rees, M. J. 1984, Nature, 311, 517
Bothun, G. D., Impey, C. D., Malin, D. F., & Mould, J. R. 1987, AJ, 94, 23
Bouché, N., Hohensee, W., Vargas, R., et al. 2012, MNRAS, 426, 801
Bouché, N., Murphy, M. T., Kacprzak, G. G., et al. 2013, Science, 341, 50
Bouché, N., Finley, H., Schroetter, I., et al. 2016, ApJ, 820, 121
Bowen, D. V., Chelouche, D., Jenkins, E. B., et al. 2016, ApJ, 826, 50
Brook, C. B., Governato, F., Roškar, R., et al. 2011, MNRAS, 415, 1051
Brooks, A. M., Governato, F., Quinn, T., Brook, C. B., & Wadsley, J. 2009, ApJ, 694, 396
Bryan, G. L., Norman, M. L., O'Shea, B. W., et al. 2014, ApJS, 211, 19
Bullock, J. S., Dekel, A., Kolatt, T. S., et al. 2001, ApJ, 555, 240
Cattaneo, A., Dekel, A., Devriendt, J., et al. 2006, MNRAS, 370, 1651
Ceverino, D., Dekel, A., & Bournaud, F. 2010, MNRAS, 404, 2151
Chen, D. N., Jing, Y. P., & Yoshikaw, K. 2003, ApJ, 597, 35
Christensen, C. R., Davé, R., Governato, F., et al. 2016, ApJ, 824, 57
Christlein, D. & Zaritsky, D. 2008, ApJ, 680, 1053
Codis, S., Pichon, C., & Pogosyan, D. 2015, MNRAS, 452, 3369
Codis, S., Pichon, C., Devriendt, J., et al. 2012, MNRAS, 427, 3320
Courtois, H. M., Zaritsky, D., Sorce, J. G., Pomarède, D. 2015, MNRAS, 448, 1767
Crighton, N. H. M., Hennawi, J. F., & Prochaska, J. X. 2013, ApJ, 776, L18
Croton, D. J., Springer, V., White, S. D. M., et al. 2006, MNRAS, 365, 11
Danovich, M., Dekel, A., Hahn, O., et al. 2015, MNRAS, 449, 2087
Danovich, M., Dekel, A., Hahn, O., & Teyssier, R. 2012, MNRAS, 422, 1732

Davis, M., Efstathiou, G., Frenk, C. S., & White, S. D. M. 1985, ApJ, 292, 371
Dekel, A. & Birnboim, Y. 2006, MNRAS, 368, 2
Dekel, A., Birnboim, Y., Engel, G., et al. 2009, Nature, 457, 451
Diamond-Stanic, A. M., Coil, A. L., Moustakas, J., et al. 2016, ApJ, 824, 24
D'Onghia, E., Burkert, A., Murante, G., & Khochfar, S. 2006, MNRAS, 372, 1525
D'Onghia, E. & Navarro, J. F. 2007, MNRAS, 380, L58
Doroshkevich, A. G. 1970, Astrofizika, 6, 581
Dutton, A. A. 2012, MNRAS, 424, 3123
Dutton, A. A. & van den Bosch, F. C. 2009, MNRAS, 396, 141
Fakhouri, O., Ma, C., & Boylan-Kolchin, M. 2010, MNRAS, 406, 2267
Fall, S. M. & Efstathiou, G. 1980, MNRAS, 193, 189
Faucher-Giguère, C.-A. & Kereš, D. 2011, MNRAS, 412, L118
Faucher-Giguère, C.-A., Kereš, D., & Ma, C.-P. 2011, MNRAS, 417, 2982
Governato, F., Brook, C. B., Brooks, A. M., et al. 2009, MNRAS, 398, 312
Governato, F., Willman, B., Mayer, L., et al. 2007, MNRAS, 374, 1479
Governato, F., Brook, C., Mayer, L., et al. 2010, Nature, 463, 203
Guedes, J., Callegari, S., Madau, P., & Mayer, L. 2011, ApJ, 742, 76
Hagen, L. M. Z., Seibert, M., Hagen, A., et al. 2016, ApJ, 826, 210
Heald, G., Józsa, G., Serra, P., et al. 2011, A&A, 526, A118
Hobbs, A., Read, J., & Nicola, A. 2015, MNRAS, 452, 3593
Holwerda, B. W., Pirzkal, N., & Heiner, J. S. 2012, MNRAS, 427, 3159
Hopkins, P. F. 2015, MNRAS, 450, 53
Hopkins, P. F., Somerville, R. S., Cox, T. J., et al. 2009, MNRAS, 397, 802
Hoyle, F. 1949, MNRAS, 109, 365
Ishiyama, T., Rieder, S., Makino, J., et al. 2013, ApJ, 767, 146
Kacprzak, G. G., Churchill, C. W., Ceverino, D., et al. 2010, ApJ, 711, 533
Kacprzak, G. G., Churchill, C. W., & Nielsen, N. M. 2012a, ApJ, 760, L7
Kacprzak, G. G., Churchill, C. W., Steidel, C. C., et al. 2012b, MNRAS, 427, 3029
Katz, N. 1992, ApJ, 391, 502
Kereš, D. & Hernquist, L. 2009, ApJ, 700, L1
Kereš, D., Katz, N., Fardal, M., et al. 2009, MNRAS, 395, 160
Kereš, D., Katz, N., Weinberg, D. H., & Davé, R. 2005, MNRAS, 363, 2
Kim, J., Choi, Y.-Y., Kim, S. S., & Lee, J.-E. 2015, ApJS, 220, 4
Kimm, T., Devriendt, J., Slyz, A., et al. 2011, preprint, ArXiv:1106.0538
Knebe, A. & Power, C. 2008, ApJ, 678, 621
Kravtsov, A. V. 2003, ApJ, 590, L1
Kravtsov, A. V., Klypin, A. A., & Khokhlov, A. M. 1997, ApJS, 111, 73
Lemonias et al. 2011, ApJ, 733, 74
Macciò, A. V., Dutton, A. A., van den Bosch, F. C., et al. 2007, MNRAS, 378, 55
Maller, A. H. & Bullock, J. S. 2004, MNRAS, 355, 694
Maller, A. H. & Dekel, A. 2002, MNRAS, 335, 487
Maller, A. H., Dekel, A., & Somerville, R. 2002, MNRAS, 329, 423
Martin, D. C., Chang, D., Matuszewski, M., et al. 2014, ApJ, 786, 107
Martin, D. C., Matuszewski, M., Morrissey, P., et al. 2015, Nature, 524, 192
Martin, D. C., Matuszewski, M., Morrissey, P., et al. 2016, ApJ, 824, L5
Matthews, L. D., van Driel, W., & Monnier-Ragaigne, D. 2001, A&A, 365, 1
Mestel, L. 1963, MNRAS, 126, 553
Mo, H. J., Mao, S., & White, S. D. M. 1998, MNRAS, 295, 319
Muñoz-Cuartas, J. C., Macciò, A. V., Gottlöber, S., & Dutton, A. A. 2011, MNRAS, 411, 584
Navarro, J. F. & Steinmetz, M. 2000, ApJ, 538, 477
Navarro, J. F. & White, S. D. M. 1994, MNRAS, 267, 401
Nelson, D., Genel, S., Pillepich, A., et al. 2016, MNRAS
Nelson, D., Vogelsberger, M., Genel, S., et al. 2013, MNRAS, 429, 3353
Nielsen, N. M., Churchill, C. W., Kacprzak, G. G., Murphy, M.T., & Evans, J.L. 2015, ApJ, 812, 83

Ocvirk, P., Pichon, C., & Teyssier, R. 2008, MNRAS, 390, 1326
Oosterloo et al. 2007, A&A, 465, 787
Peebles, P. J. E. 1969, ApJ, 155, 393
—. 1982, ApJ, 263, L1
Pichon, C., Pogosyan, D., Kimm, T., et al. 2011, MNRAS, 418, 2493
Porciani, C., Dekel, A., & Hoffman, Y. 2002a, MNRAS, 332, 325
—. 2002b, MNRAS, 332, 339
Prescott, M. K. M., Martin, C. L., & Dey, A. 2015, ApJ, 799, 62
Prieto, J., Jimenez, R., Haiman, Z., & González, R. E. 2015, MNRAS, 452, 784
Rees, M. J. & Ostriker, J. P. 1977, MNRAS, 179, 541
Robertson, B., Bullock, J. S., Cox, T. J., et al. 2006, ApJ, 645, 986
Sales, L. V., Navarro, J. F., Theuns, T., et al. 2012, MNRAS, 423, 1544
Sancisi, R., Fraternali, F., et al. 2008, A&A Rev., 15, 189
Scannapieco, C., Tissera, P. B., et al. 2008, MNRAS, 389, 1137
Schäfer, B. M. 2009, International Journal of Modern Physics D, 18, 173
Schaye, J. 2001, ApJ, 562, L95
Sciama, D. W. 1955, MNRAS, 115, 3
Sharma, S. & Steinmetz, M. 2005, ApJ, 628, 21
Silk, J. 1977, ApJ, 211, 638
Somerville, R. S., Gilmore, R. C., Primack, J. R., & Domínguez, A. 2012, MNRAS, 423, 1992
Somerville, R. S., Hopkins, P. F., Cox, T. J., et al. 2008, MNRAS, 391, 481
Sommer-Larsen, J., Gelato, S., & Vedel, H. 1999, ApJ, 519, 501
Spavone, M., Iodice, E., Arnaboldi, M., et al. 2010, ApJ, 714, 1081
Springel, V. 2010, MNRAS, 401, 791
Steinmetz, M. 1999, Ap&SS, 269, 513
Stewart, K., Maller, A., Oñorbe, J., et al. 2016, ArXiv:1606.08542
Stewart, K. R., Bullock, J. S., Wechsler, R. H., & Maller, A. H. 2009, ApJ, 702, 307
Stewart, K. R., Bullock, J. S., Wechsler, R. H., Maller, A.H., & Zentner, A.R. 2008, ApJ, 683, 597
Stewart, K. R., Kauffmann, T., Bullock, J. S., et al. 2011a, ApJ, 735, L1
— 2011b, ApJ, 738, 39
Stewart, K. R., Brooks, A. M., Bullock, J. S., et al. 2013, ApJ, 769, 74
Teklu, A. F., Remus, R.-S., Dolag, K., Beck, A. M., et al. 2015, ApJ, 812, 29
Teyssier, R. 2002, A&A, 385, 337
Tillson, H., Devriendt, J., Slyz, A., et al. 2015, MNRAS, 449, 4363
Torrey, P., Vogelsberger, M., Sijacki, D., et al. 2012, MNRAS, 427, 2224
Trowland, H. E., Lewis, G. F., & Bland-Hawthorn, J. 2013, ApJ, 762, 72
Übler, H., Naab, T., Oser, L., et al. 2014, MNRAS, 443, 2092
van de Voort, F., Schaye, J., Booth, C. M., et al. 2011, MNRAS, 414, 2458
van de Voort, F., Davis, T. A., Kereš, D., et al. 2015, MNRAS, 451, 3269
Vitvitska, M., Klypin, A. A., Kravtsov, A. V., et al. 2002, ApJ, 581, 799
Vogelsberger, M., Genel, S., Springel, V., et al. 2014, MNRAS, 444, 1518
Wadsley et. al. 2004, New Astronomy, 9, 137
Weinmann, S. M., van den Bosch, F. C., Yang, X., & Mo, H. J. 2006, MNRAS, 366, 2
White, S. D. M. 1984, ApJ, 286, 38
White, S. D. M. & Frenk, C. S. 1991, ApJ, 379, 52
White, S. D. M. & Rees, M. J. 1978, MNRAS, 183, 341
Zjupa, J. & Springel, V. 2016, preprint, ArXiv:1608.01323

Observational Diagnostics of Gas Flows: Insights from Cosmological Simulations

Claude-André Faucher-Giguère

1 Introduction

In recent years, there has been a growing realization that the "cosmic baryon cycle" is both a primary driver and a primary regulator of galaxy formation. Continuous accretion of gas from the intergalactic medium (IGM) is necessary to sustain observed star formation rates (SFRs) over a Hubble time (e.g., Erb 2008; Prochaska and Wolfe 2009; Bauermeister et al. 2010). However, models in which the intergalactic gas accreted by galaxies is efficiently converted into stars produce galaxies with stellar masses that exceed observed ones by an order of magnitude or more (e.g., White and Frenk 1991; Navarro et al. 1995; Kereš et al. 2009a). In the latest generation of models, star formation-driven galactic winds regulate galaxy growth below $\sim L^{\star}$ by ejecting back into the IGM most of the accreted gas before is has time to turn into stars (see the review by Somerville and Davé 2015). Despite a broad consensus regarding the importance of inflows and outflows in galaxy evolution, many questions regarding their nature and effects remain at the forefront of current research.

For example, many cosmological simulations and semi-analytic models now suggest that wind recycling (the fallback of gas previously ejected in galactic winds) plays an important role in shaping the galaxy stellar mass function and setting the level of late-time galactic accretion (Oppenheimer et al. 2010; Henriques et al. 2013; Anglés-Alcàzar et al. 2016). While galactic accretion is a generic prediction

C.-A. Faucher-Giguère (✉)
Department of Physics and Astronomy and Center for Interdisciplinary Exploration and Research in Astrophysics (CIERA), Northwestern University, 2145 Sheridan Road, Evanston, IL 60208, USA
e-mail: cgiguere@northwestern.edu

A. Fox, R. Davé (eds.), *Gas Accretion onto Galaxies*, Astrophysics and Space Science Library 430, DOI 10.1007/978-3-319-52512-9_12

of cosmological simulations (e.g., Kereš et al. 2005, 2009b; Dekel et al. 2009; Brooks et al. 2009; Faucher-Giguère et al. 2011; van de Voort et al. 2011), its properties are subject to uncertainties in how the accretion flows are affected by shocks and hydrodynamical instabilities as they interact with galaxy halos (e.g., Birnboim and Dekel 2003; Nelson et al. 2013; Mandelker et al. 2016). Galactic winds are driven by feedback processes that operate on the scale of individual star-forming regions, which are generally not well resolved in current simulations. As a result, detailed properties such as their phase structure remain uncertain even in today's highest resolution zoom-in simulations of galaxy formation (e.g., Shen et al. 2013; Hopkins et al. 2014; Marinacci et al. 2014; Agertz and Kravtsov 2015). In large-volume cosmological simulations, it is not yet possible to resolve how galactic winds are launched so even the bulk properties of galactic winds in such simulations are typically tuned to match observables such as the galaxy stellar mass function (e.g., Davé et al. 2011; Vogelsberger et al. 2014; Schaye et al. 2015). Theoretical predictions for inflows and outflows are furthermore complicated by the fact that inflows and outflows inevitably interact with each other (e.g., van de Voort et al. 2011; Faucher-Giguère et al. 2011, 2015; Nelson et al. 2015).

The importance of inflows and outflows for galaxy evolution, as well as the significant theoretical uncertainties, imply that observations of these processes are critical to test and inform galaxy formation theories. Since observational techniques for probing inflows and outflows generally provide only fragmentary information about the physical nature of the observed gas (e.g., 1D skewers through galactic halos for typical quasar absorption line measurements), forward modeling using cosmological simulations and comparing the simulations with observations will likely continue to play a central role in disentangling these processes. In this chapter, we review the current status of using cosmological simulations to develop observational diagnostics of galactic accretion. Since the dynamics inflows and outflows are intertwined in the circum-galactic medium (CGM), this chapter will also cover relevant outflow diagnostics.

This chapter is largely organized around our group's research on the topic, but attempts to provide a broad review of theoretical research relevant to interpreting recent and upcoming observations. The chapter is divided into two main sections, one on absorption diagnostics (Sect. 2) and one on emission diagnostics (Sect. 3). Interspersed within our discussion of different observational diagnostics, we include some remarks on numerical uncertainties and the sensitivity of different predictions to the numerical method employed. We conclude in Sect. 4 with a synthesis of lessons from existing simulations of galactic accretion and comparisons with observations, and suggest some promising directions for future work. We focus on observational diagnostics applicable to galaxies other than the Milky Way.

2 Absorption Diagnostics

We divide our discussion of absorption diagnostics into different observational statistics.

2.1 *H I Covering Fractions*

Covering fractions of absorbers within different impact parameters from foreground galaxies have been extensively modeled using simulations and provide the most basic consistency test between simulations and observations. Over the last decade, large observational datasets on absorption by the CGM gas flows have been assembled using quasar spectra transverse to galaxies of different types and at different redshifts. For example, this technique has been applied at both low and high redshifts to foreground dwarf galaxies (e.g., Bordoloi et al. 2014b), damped Lyα absorbers (e.g., Rubin et al. 2015), luminous red galaxies (e.g., Gauthier et al. 2010), $\sim L^{\star}$ star-forming galaxies (e.g., Adelberger et al. 2003; Tumlinson et al. 2011; Rudie et al. 2012; Turner et al. 2014), and quasars (e.g., Hennawi et al. 2006; Prochaska et al. 2013). In a study of $z \sim 2$–3 Lyman break galaxies (LBGs), Steidel et al. (2010) also demonstrated that useful constraints on the CGM can be extracted from spectra of ordinary, fainter background galaxies (see also Bordoloi et al. 2011). With the advent of 30-m class ground-based telescopes in the next decade, spectroscopy of background galaxies will become increasingly powerful as it becomes generically possible to obtain spectra of multiple sight lines through the halos of individual foreground galaxies.

Lyman limit systems (LLSs; usually quantitatively defined as systems with HI column density $N_{\rm HI} \geq 10^{17.2}\,{\rm cm}^{-2}$) in particular are useful tracers of inflows and outflows, being dense enough to be closely associated with galaxy halos but not sufficiently dense to arise only in galactic disks.[1] Cosmological simulations show that LLSs are good tracers of cool filamentary accretion, especially at high redshift ($z \sim 2$–4) where these are most prevalent (e.g., Fumagalli et al. 2011, 2014; Faucher-Giguère and Kereš 2011; Kimm et al. 2011; Goerdt et al. 2012; Shen et al. 2013; Faucher-Giguère et al. 2015).

[1]Galactic disks are better traced by damped Lyα absorbers (DLAs; $N_{\rm HI} \geq 2 \times 10^{20}\,{\rm cm}^{-2}$; e.g., Wolfe et al. 2005; Neeleman et al. 2015). At very high redshift, the increased cosmic mean density and declining cosmic ultraviolet background (UVB) cause absorbers of fixed HI column to probe structures more closely associated with the low-density IGM with increasing redshift (e.g., McQuinn et al. 2011). As a result, LLSs become associated with structures such as intergalactic filaments and some DLAs may arise in the CGM. There rapid increase in LLS incidence observed at $z \gtrsim 3.5$ suggests that LLSs commonly arise outside galaxy halos at these redshifts (Fumagalli et al. 2013) while the rapid evolution of the DLA metallicity distribution at $z \gtrsim 5$ suggest that DLAs at these redshifts commonly arise outside galaxies (Rafelski et al. 2014).

Radiative transfer is important to properly model LLSs since these systems are optically thick at the Lyman limit by definition. However, until recently most cosmological simulations computed ionization balance assuming that all systems are optically thin. In early studies using simple approximations for the ionization state of the gas (e.g., Dekel et al. 2009 though see Kimm et al. 2011), the predicted covering fractions of cold accretion streams were well in excess of observational constraints (Steidel et al. 2010). Properly processing simulations with ionizing radiative transfer—thus allowing more accurate identification of strong HI absorbers—showed that the LLS covering fractions of cold accretion streams are in fact quite small in simulations. As a result, the predicted small LLS covering fractions of cold accretion streams are consistent with present observational constraints ($\lesssim 10\%$ for LBG-mass halos at $z \sim 2$; Faucher-Giguère and Kereš 2011; Fumagalli et al. 2011).

The more accurate treatments of radiative transfer actually revealed tension in the opposite direction. In an analysis of high-resolution quasar spectra transverse to $z \sim$2–2.5 LBGs, Rudie et al. (2012) measured an LLS covering fraction within a projected virial radius of $30 \pm 14\%$, at face value a factor $\sim$3 higher than cosmological simulations without strong galactic winds (Faucher-Giguère and Kereš 2011; Fumagalli et al. 2011). This discrepancy has been plausibly resolved in the latest generation of cosmological simulations with stronger stellar feedback, necessary to produce realistic galaxy stellar masses (Fumagalli et al. 2014; Faucher-Giguère et al. 2015). These simulations showed that $\gtrsim$50% of the cool halo gas giving rise to LLSs around $z \sim$ 2–3 LBGs arises not from IGM accretion but rather from galactic winds. Figure 1 shows two simulations of the same low-mass LBG halo, one with strong galactic winds and one without galactic winds, at $z =$ 2–4.

Galactic winds enhance LLS covering fractions in the simulations in two ways: (1) they eject cool interstellar gas into the CGM, and (2) they increase the cross section of inflows through hydrodynamic interactions. Importantly, it is not only the galactic wind from the central galaxy that interacts with infalling gas, but also outflows from nearby satellites. The latter effect is enhanced because satellites tend to be embedded in surrounding large-scale structure filaments. These filaments are "puffed up" by galactic winds from embedded galaxies. It is apparent from the example in Fig. 1 that, absent galactic winds, the $z \sim 2$ LLS covering fraction from accreting gas is very small and almost entirely limited to galaxies and their immediate vicinity (at higher redshifts, where the halos are less massive, the filamentary inflows give rise to more extended LLSs). This is because cool filamentary inflows tend to disappear in higher mass, lower redshift halos (e.g., Kereš et al. 2005; Dekel and Birnboim 2006). Most latest-generation cosmological simulations, including those from the EAGLE (Schaye et al. 2015), Illustris (Vogelsberger et al. 2014), and FIRE (Hopkins et al. 2014)[2] projects, implement on-the-fly approximations based

[2]See the FIRE project web site at http://fire.northwestern.edu.

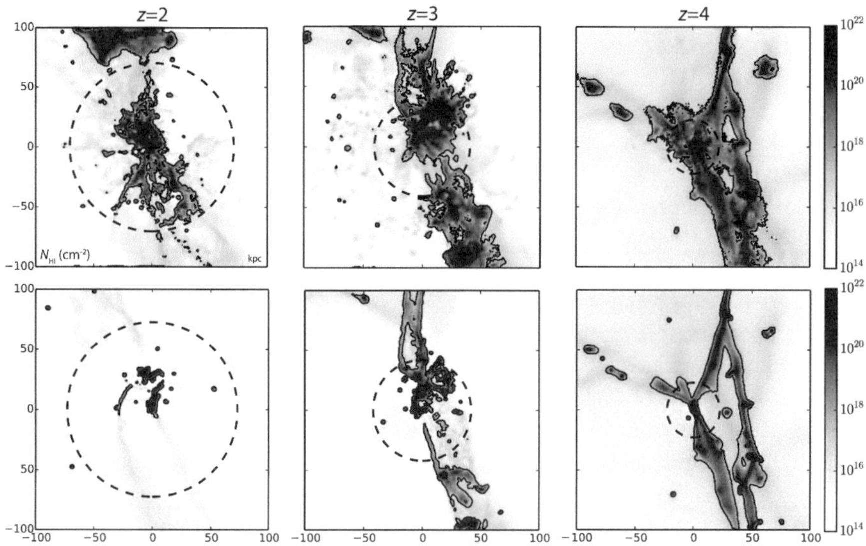

Fig. 1 *Top*: H I maps for a low-mass LBG simulation with stellar feedback from the FIRE project at $z = 2$, 3, and 4 ($M_h(z = 2) = 3 \times 10^{11}$ $M_\odot$). *Bottom*: Simulation from the same initial conditions but without galactic winds. The virial radius of the halo is indicated in each panel by the *dashed circles* and Lyman limit systems are indicated by *solid contours*. Stellar feedback increases the covering fractions in galaxy halos both by directly ejecting cool gas from galaxies and through the interaction of galactic winds with cosmological inflows. At $z = 2$, LLSs in this example are almost exclusively restricted to galaxies and their immediate vicinity absent galactic winds. Length scales are consistent across rows and columns. Adapted from Faucher-Giguère et al. (2015)

on local gas properties for self-shielding based on post-processing radiative transfer calculations (Faucher-Giguère et al. 2010; Fumagalli et al. 2011; Rahmati et al. 2013).

A current puzzle are the order unity LLS covering fractions measured in the halos of luminous quasars $z \sim$ 2–2.5. Prochaska et al. (2013) reported an LLS covering fraction $f_{cov}(> 10^{17.2}; < R_{vir}) \approx 0.64^{+0.06}_{-0.07}$ within a projected virial radius of $z \sim$ 2–2.5 quasars (see also Prochaska et al. 2014). This high LLS covering fraction should be compared to the lower fraction $f_{cov}(10^{17.2}; < R_{vir}) = 0.30 \pm 0.14$ measured by Rudie et al. (2012) around $z \sim$2–2.5 Lyman break galaxies (LBGs) in the Keck Baryonic Structure Survey (KBSS). The LBGs in KBSS reside in dark matter halos of characteristic mass $M_h \approx 10^{12}$ $M_\odot$ (Adelberger et al. 2005a; Trainor and Steidel 2012), a factor just $\sim 3\times$ lower than quasars. Using cosmological zoom-in simulations with stellar feedback but neglecting AGN feedback, Fumagalli et al. (2014) and Faucher-Giguère et al. (2015) found simulated LLS covering fractions consistent with those measured in LBG halos (see Shen et al. 2013). In both studies, however, the most massive simulated halos failed to explain the LLS covering fraction measured around quasars by a large factor, suggesting that the presence of a luminous AGN could affect the properties of CGM gas on $\sim$100 kpc scales.

More recent simulations by Rahmati et al. (2015) and Faucher-Giguère et al. (2016) were able to match the covering fractions observed by Prochaska et al. (2013) in quasar halos, but for different reasons. Recognizing that the distribution of halo masses probed by quasars is only crudely constrained by clustering measurements, Rahmati et al. (2015) made the optimistic assumption that all of Prochaska et al. (2013)'s quasars are hosted in halos of mass *greater than* the characteristic clustering mass $M_{\rm h} \approx 3 \times 10^{12}$ $M_{\odot}$. They then compared the quasar observations with the halos in the EAGLE simulation with mass above this threshold as a function of impact parameter in absolute units of proper distance. As a result, many of Rahmati et al. (2015)'s simulated LLSs are located at a smaller *fraction* of the virial radius than would be inferred assuming a constant virial radius corresponding to the characteristic quasar clustering halo mass (the assumption made in Prochaska et al. 2013). Since covering fractions decreases with increasing impact parameter, Rahmati et al. (2015)'s approach tends to boost the covering fractions, enough to bring them in agreement with those observed around real quasars. Rahmati et al. (2015)'s fiducial simulation included AGN feedback, but AGN feedback does not appear to play a significant role in explaining their results. Faucher-Giguère et al. (2016)'s simulations, on the other hand, are consistent with the $f_{\rm cov}(> 10^{17.2}; < R_{\rm vir})$ value reported by Prochaska et al. (2013). Faucher-Giguère et al. (2016)'s simulations, from the FIRE project, included strong stellar feedback but no AGN feedback. Relative to the analysis of Faucher-Giguère et al. (2015), which focused on LBG-mass halos, Faucher-Giguère et al. (2016) analyzed a much larger set of quasar-mass halos (15 vs. 1) and the new halos were simulated at order-of-magnitude better mass resolution than the previous quasar-mass halo.

Figure 2 summarizes how the simulated LLS covering fractions of Faucher-Giguère et al. (2015, 2016) compare with observed covering fractions at $z \sim 2$–2.5. For both LBG and quasar halos, the simulations rely on star formation-driven galactic winds to explain observations. Faucher-Giguère et al. (2016) performed a resolution convergence study of the covering fractions in quasar-mass halos, and found two important results. The first is that LLS covering fractions increase with increasing resolution. This is the primary reason why Faucher-Giguère et al. (2015)'s earlier quasar-mass run fell short of reproducing observations. The second is that much of the LLS material in quasar-mass halos is due to galactic winds not from central galaxies but instead from lower-mass satellite galaxies. This is illustrated in Fig. 3, which shows that the spatial distribution of LLSs in quasar-mass halos correlates strongly with the spatial distribution of satellites. As in lower-mass halos, galactic winds from satellites both eject gas into the CGM and increase the cross section of large-scale structure filaments. The velocity maps in Fig. 3 show that the LLS structures with embedded satellites are typically infalling, so these LLS are connected to galactic accretion, albeit somewhat indirectly.

The gas particle mass in Faucher-Giguère et al. (2016)'s simulations of quasar-mass halos is $m_{\rm b} \approx 3 \times 10^4$ $M_{\odot}$. At this resolution, Fig. 2 shows that the simulations

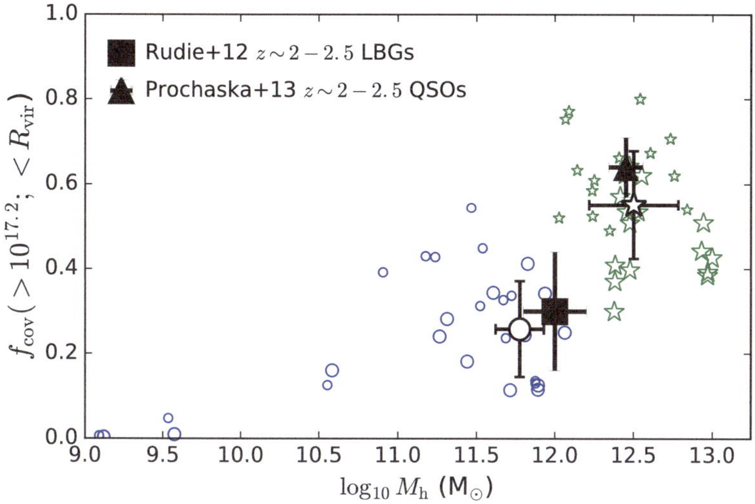

Fig. 2 *Blue circles*: Lyman limit system (LLS) covering fractions within a projected virial radius for the high-resolution simulated halos from the FIRE project analyzed in Faucher-Giguère et al. (2015). For each simulated halo, covering fractions for 25 snapshots over the redshift interval $z = 2 - 2.5$ are shown. The simulations are in good agreement with LLS covering fractions measured around LBGs in that redshift interval by Rudie et al. (2012) (*black square*). *Green stars*: Covering fractions at $z = 2$ (*large*) and $z = 2.5$ (*small*) for the quasar-mass halos analyzed in Faucher-Giguère et al. (2016). The quasar-mass simulated halos are compared to LLS measurements transverse to luminous quasars at $z \sim$ 2–2.5 by Prochaska et al. (2013) (*black triangle*). The *open black symbols* show averages over simulated LBG-mass halos and QSO-mass halos, with the error bars showing the standard deviations of the simulated data points included in the averages. Figure from Faucher-Giguère et al. (2016)

are in good agreement with observations of massive halos. Since these covering fractions are not far from unity, they are necessarily saturating with increasing resolution. However, a comparison with lower-resolution simulations indicates that the CGM properties of quasar-mass halos may not be fully converged even in the highest-resolution simulations presently available. The stringent resolution requirements are in part due to the necessity of resolving the generation of galactic winds from satellites. Since the high simulated covering fractions in Fu et al. (2016) do not require AGN feedback, one prediction is that similarly massive halos without a quasar—such as may be selected based on high stellar mass or SFR—should show similarly high covering fractions. In an analysis of three sight lines with impact parameter ~100–200 proper kpc from $z \sim$ 2–2.5 sub-millimeter galaxies (SMGs), Fu et al. (2016) did not find compelling evidence for LLS-strength absorbers. If the simulations are correct and this observational result persists when the SMG sample is increased, it would suggest that some SMGs are hosted by halos significantly less massive than luminous quasars at $z \sim 2$. A larger observational sample is, however, clearly needed to firm up the statistical significance of this observational result.

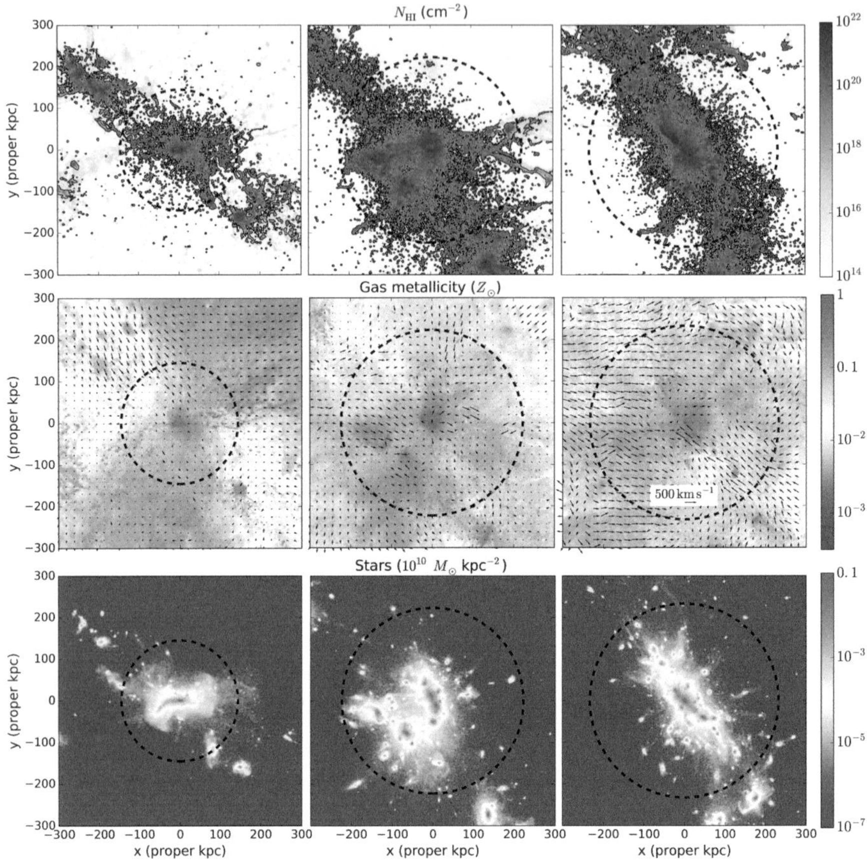

Fig. 3 H I column density (*top*), gas-phase metallicity (*middle*), and stellar mass surface density (*bottom*) maps for three representative quasar-mass halos from the FIRE project at $z = 2$ (*from left to right*: $M_{\rm h}(z = 2) = (2.4, 8.8, 9.9) \times 10^{12}$ $M_{\odot}$). The virial radius is indicated by *dashed circle* in each panel and *solid contours* indicate Lyman limit systems. The vectors on metallicity maps indicate projected mass-weighted velocities. The large-scale distribution of LLS gas correlates with the spatial distribution of satellite galaxies, indicating the importance of stellar feedback from satellites in producing large HI covering fractions. The velocity maps also show that the LLS structures with embedded satellites are typically infalling. Figure adapted from Faucher-Giguère et al. (2016)

The ability to develop robust observational diagnostics of galactic accretion depends critically on the ability of numerical codes to properly capture the hydrodynamics of gas accretion, so we briefly digress to comment on this issue. In the example shown in Fig. 1, cold streams (as traces by LLSs) disappear in a slightly lower-mass halo ($M_{\rm h}(z = 2) = 3 \times 10^{11}$ $M_{\odot}$) than predicted by some previous simulations (see, e.g., the simulations similar-mass halos in Faucher-Giguère and Kereš 2011) because older smooth particle hydrodynamics (SPH) simulations underestimated the destructive effects of shocks and hydrodynamical instabilities (Agertz et al. 2007; Sijacki et al. 2012; Nelson et al. 2013). Recent improvements to SPH algorithms (e.g., Read and Hayfield 2012; Saitoh and Makino 2013; Hopkins 2013; Hu et al. 2014) have greatly reduced the major historical differences with respect to grid-based codes, particular for fluid mixing instabilities. Overall, cold streams falling into galaxy halos tend to be more rapidly disrupted by hydrodynamic interactions with halo gas in codes of improved accuracy. The morphological differences in cold gas properties between hydrodynamic solvers are largest around the halo mass $M_{\rm h} \sim 3 \times 10^{11}$ $M_{\odot}$ above which quasi-static hot atmospheres start to develop. In lower-mass halos, cold streams are generically present in halos simulated using different numerical methods, at least at $z = 2$, which has been the subject of the most detailed simulation analyses. Interestingly, the most important differences overall for gas accretion between older SPH codes and grid-based codes are in the amount of "hot mode" accretion, i.e., the amount of hot gas that cools from hot atmospheres. Hot accretion is significantly more efficient in grid codes and updated SPH codes because spurious heating from the dissipation of turbulent energy on large scales prevents the same behavior in traditional SPH codes (Nelson et al. 2013). It is important to note, though, that most of the relevant direct code comparisons were performed on simplified cosmological simulations without strong galactic feedback. Observations of galaxy clusters clearly show that intra-cluster gas must be heated (likely by AGN feedback) to prevent a cooling catastrophe and avoid SFRs order-of-magnitude in excess of those observed in brightest cluster galaxies (e.g., McNamara and Nulsen 2007; Fabian 2012). This heating suppresses hot mode accretion. It is not yet clear how much different numerical methods for hot mode accretion diverge when realistic feedback is included. Comparing the predictions of simulations with observations will continue to play a critical role in identifying limitations of the simulations.

Before closing this section, we note that observations provide significantly more detail on the distribution of neutral hydrogen in galaxy halos than captured by the LLS covering fractions emphasized above, including better statistics on the incidence of (more numerous) lower-column absorbers and their line-of-sight velocity distributions (e.g., Rudie et al. 2012). More comprehensive comparisons with simulations will be necessary to fully exploit the discriminating power of these observations for galactic inflow and outflow models.

2.2 *Metal Absorption Systems Transverse to Galaxies*

Metal absorption is commonly observed out to $\sim$0.5–1 $R_{\rm vir}$ transverse to foreground galaxies of different types (e.g., Adelberger et al. 2005b; Steidel et al. 2010; Chen 2012; Werk et al. 2013; Bordoloi et al. 2014b; Liang and Chen 2014; Lau et al. 2016). However, gas that is first accreting from the IGM is expected to be relatively metal-poor (e.g., van de Voort and Schaye 2012). In cosmological simulations with relatively weak stellar feedback, Fumagalli et al. (2011) found that the mean metallicity of cold streams in $M_{\rm h} \sim 10^{10}$–10^{12} $M_{\odot}$ halos at

$z = 1.3$–4 is $\sim$0.01 $Z_{\odot}$, weakly dependent on halo mass and redshift. Similarly, Goerdt et al. (2012) concluded that cold streams will be challenging to detect in metal line absorption due to their low-metallicity and small covering fractions. The low-metallicity cold streams found in simulations severely under-predict the metal line equivalent widths observed around LBGs, strongly suggesting that most of the metal absorption observed transverse to LBGs originates from gas that has been processed by galaxies, such as galactic winds. Indeed, Turner et al. (2015) used photoionization modeling to argue that at least some of the metal-enriched gas ($\gtrsim 0.1 Z_{\odot}$) observed around $z \sim 2.3$ star-forming galaxies arises in the hot phase of galactic winds. In a few instances where abundance ratios have been measured, metal-rich CGM absorbers have abundance ratios consistent with either core collapse (e.g., Lau et al. 2016) or Type Ia supernovae (e.g., Zahedy et al. 2016). Cosmological simulations also convincingly demonstrate that star formation-driven galactic winds are necessary to explain metals observed in the CGM (e.g., Shen et al. 2012; Hummels et al. 2013; Suresh et al. 2015; Liang et al. 2016; Ford et al. 2016; Turner et al. 2016).

Despite the association between metal-rich gas and outflows, there is no clear cut metallicity division between inflows and outflows. In Sect. 2.1, we emphasized how galactic winds from infalling satellites can puff up large-scale structure filaments. Thus, a good fraction of the gas first accreting onto galaxies may come into contact with metal-enriched material. The extent to which this metal-enriched gas contaminates galactic accretion depends on how efficiently metals mix in the CGM. While simulations provide some indication of the expected mixing, observations of closely spaced sight lines toward gravitationally lensed quasars and photoionization modeling show that metal absorption systems are often compact and poorly mixed (e.g., Rauch et al. 1999, 2001; Simcoe et al. 2006; Schaye et al. 2007; Crighton et al. 2015). Overall, cool metal absorbers have inferred sizes ranging from $\sim$1 pc to $\gtrsim$1 kpc, with some evidence that typical size increases with increasing ionization state. Some clouds may be less than a solar mass in mass.

All numerical methods are limited in their ability to capture metal mixing near their resolution limit. While grid codes tend to over-mix metals due to diffusive errors at the grid scale,[3] standard SPH codes "lock" metals into SPH particles. Because of this, it is often assumed that SPH under-mixes metals. However, this is only true for metal clumps above the resolution limit: tiny metal clumps below the resolution limit will appear over-mixed in SPH codes because their metals will be spread over the gas mass of individual SPH particles. For reference, state-of-the-art zoom-in SPH simulations of Milky Way-mass galaxies have typical gas particle masses $\sim 10^4$–10^5 $M_{\odot}$ (Shen et al. 2013; Stinson et al. 2013; Hopkins et al. 2014). Thus, even state-of-the-art SPH simulations likely underestimate metal mixing in the CGM in at least some circumstances. In such circumstances, sub-resolution

[3]Such errors are mitigated in moving-mesh codes in which grid cells are advected with the flow, such as Arepo (Springel 2010), as well as in the "meshless finite mass" (MFM) method implemented in GIZMO (Hopkins 2015).

SPH models that attempt to model metal diffusion owing to unresolved turbulence (e.g., Shen et al. 2010) could vitiate rather than improve the solution. An important question for future work will be to identify the kinds of CGM absorbers that can be reliably resolved in cosmological simulations. If metal absorbers are compact because supernova ejecta take a long time to mix with ambient gas, then absorption by hydrogen and helium (elements synthesized in the Big Bang) may not suffer from the same clumpiness effects. Warm and hot gas phases, which tend to be more volume-filling, may also be easier to resolve in simulations. OVI, now routinely detected at both low and high redshift (e.g., Tumlinson et al. 2011; Lehner et al. 2014; Turner et al. 2015), stands out as one particular ion for which it will be important to determine the convergence properties of cosmological simulations.

Upcoming observations of multiple sight lines through the halos of individual galaxies will provide useful information regarding the size scales of different CGM absorbers. The best local laboratory to carry out such an experiment is M31. Lehner et al. (2015) recently analyzed 18 sight lines within $2R_{\rm vir} \approx 600\,{\rm kpc}$ of M31, thus producing a partial map of the multi-phase CGM around the galaxy. New HST/COS observations of quasars behind M31 as part of the AMIGA (Absorption Maps In the Gas of Andromeda) project (PI N. Lehner) will improve on this pilot analysis with a total of 25 sight lines within $1.1R_{\rm vir}$. While these observations will not constrain structure on the fine scales possible with gravitationally lensed quasars or photoionization modeling, the spatially resolved map of M31's CGM will help us better interpret observations of single sight lines through the halos of similar-mass galaxies, such as those from the COS-Halos program (Tumlinson et al. 2011). One caveat with drawing inferences based on detailed studies of a single system, however, is that simulations show that the CGM can be quite dynamic and time variable (e.g., Faucher-Giguère et al. 2015; Hafen et al. 2016). Thus, such observational analyses will be most powerful when combined with simulations that can inform how the observational inferences can be generalized to other halos. Another approach for observationally constraining the size scale of CGM structures is to quantify the fraction of the area of a background source absorbed by foreground CGM clouds. Quasar accretion disks have diameters $\sim$0.01 pc while galaxies generally have half-light radii $\gtrsim$1 kpc. Thus, background galaxies probe absorber size scales larger than background quasars (e.g., Diamond-Stanic et al. 2016).

2.3 Down-the-Barrel Metal Absorption Lines

Another observational approach to detect galactic accretion is to use single "down-the-barrel" spectra of galaxies (e.g., Steidel et al. 2010; Rubin et al. 2012; Martin et al. 2012). One advantage of the down-the-barrel observations, relative to galaxy pairs, is that gas that absorbs stellar light from a galaxy is known to be located between the galaxy and the observer. Thus, redshifted absorption can be unambiguously associated with gas with a radial velocity component in the direction of the galaxy.[4] Down-the-barrel observations, however, suffer from a different difficulty due to the fact that the typical infall velocity of IGM accretion is comparable to the halo velocity.[5] Thus, even for sight lines fortuitously aligned with infalling CGM cool gas, absorption from the infalling gas will typically overlap in velocity space with interstellar medium (ISM) gas. Since ISM gas is expected to be generally both denser and more enriched with metals than cold streams, down-the-barrel absorption by cold streams will typically appear as a minor perturbation to ISM absorption (Kimm et al. 2011).[6] In detailed analyses of down-the-barrel spectra, Rubin et al. (2012) and Martin et al. (2012) reported detections of infalling gas in a small fraction, $\sim$3–6%, of $z \sim 0.4$–1.4 galaxies. These detections were made using low-ionization metal absorption lines and thus likely trace relatively metal-rich gas, such as infalling dwarf galaxies on their way to merging or recycling wind gas, rather than gas accreting from the IGM for the first time. Unfortunately, there has been relatively little modeling of the inflow signatures expected in down-the-barrel observations. Despite the challenges in using this technique for probing IGM accretion, more modeling would very valuable given the very rich observational datasets now available, which may make it possible to extract even subtle signatures.

2.4 *Kinematic and Azimuthal Angle Diagnostics*

A simple toy physical picture for inflows and outflows is one in which inflows from the IGM bring in the angular momentum that creates rotating galactic disks and in which galactic winds have a bi-conical morphology due to collimation normal to the galactic plane. If this toy model were correct, it would suggest that absorption by gas normal to the plane of disk galaxies should arise primarily from galactic winds, while absorption in the disk plane may be commonly due to infalling gas. In this picture, the infalling gas would typically co-rotate with the disk as it approaches the galaxy. Thus, a combination of azimuthal angle and kinematic diagnostics would constitute a powerful probe of inflows and outflows. This is indeed a promising

[4]In galaxy pair experiments, an outflowing absorber located behind the foreground galaxy would also appear redshifted. This introduces a generic ambiguity in the interpretation of absorption lines transverse to foreground galaxies.

[5]Absent hydrodynamic interactions and angular momentum, the radial velocity would be simply equal to the velocity of free fall into halo. Current cosmological simulations indicate that asymptotic cold stream radial velocities are typically closer to half the halo circular velocity (Kereš et al. 2005; Goerdt and Ceverino 2015).

[6]If the stellar light of a galaxy were concentrated in a point source and the ISM were rotating in perfectly circular motion around the center, then the ISM would move purely tangentially with respect to the light source and could not mimic infalling gas. In real galaxies, stellar light is, however, spatially extended and the internal dynamics and morphology of the ISM can be quite complex. For example, at $z \sim 2$ the nebular line emission of galaxies is often very clumpy (e.g., Förster Schreiber et al. 2009). These effects could cause some ISM gas to appear as infalling.

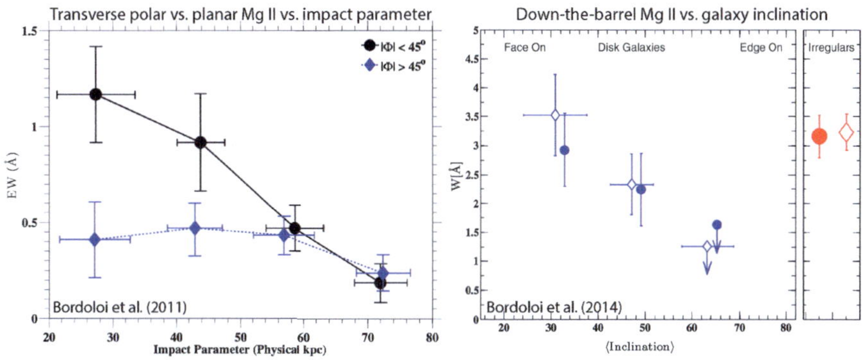

Fig. 4 Examples of how observed Mg II absorption varies as a function of azimuthal angle relative to the semi-minor axis of a galaxy and as a function of inclination angle in down-the-barrel spectra of galaxies. *Left*: Average Mg II equivalent width around $0.5 \leq z \leq 0.9$ zCOSMOS disk galaxies as a function of impact parameter along the disk axis ($|\phi| < 45°$) vs. near the disk plane ($|\phi| > 45°$; Bordoloi et al. 2011). *Right*: Average MgII equivalent width vs. inclination from co-added down-the-barrel spectra of zCOSMOS $1 \leq z \leq 1.5$ galaxies (Bordoloi et al. 2014a). The *circles* and *diamonds* correspond to two different ways of making the measurement. These observations are consistent with galactic winds preferentially expanding normal to the plane of disk galaxies but there has been so far relatively little modeling of these observations

avenue for identifying inflows and outflows, with some observational support for physical differences between planar and extra-planar absorbers in the CGM of galaxies. However, the latest high-resolution cosmological simulations indicate that the character of both galactic accretion and galactic winds change significantly with redshift and galaxy mass (e.g., Muratov et al. 2015; Hayward and Hopkins 2015). Observations also show that galactic winds become significant weaker as star formation activity in galaxies declines from its peak at $z \sim 2$ (e.g., Steidel et al. 2010) to the present (e.g., Heckman and Borthakur 2016). Thus, it is likely that the toy model outlined above is too simple, and this is an area where more detailed and systematic modeling is likely to prove critical.

2.4.1 Azimuthal Angle Diagnostics

Observationally, there is support for galactic winds preferentially expanding normal to the plane of disk galaxies from spectroscopic observations transverse to foreground galaxies (expressed in terms of azimuthal angle relative to the semi-minor axis of the galaxy projected on the sky; Bordoloi et al. 2011; Bouché et al. 2012; Kacprzak et al. 2012) and from down-the-barrel spectra of galaxies as a function of inclination angle of the disk (Kornei et al. 2012; Rubin et al. 2012, 2014; Martin et al. 2012; Bordoloi et al. 2014a). Figure 4 shows observations of each type for $z \sim 0.5$–1.5 galaxies in zCOSMOS. The larger average Mg II absorption equivalent widths along the poles of disk galaxies suggest that strong polar Mg II absorbers trace galactic winds. Because down-the-barrel sight lines toward low-inclination disk galaxies probe regions of the CGM similar to transverse spectra at small azimuthal (polar) angles, we will refer to both types of observations as probing the azimuthal angle dependence of CGM gas in what follows.[7] In Mg II absorption, the observed azimuthal angle dependence appears to be stronger for systems with high rest equivalent widths $W_0 \gtrsim 1$ Å. This is consistent with an origin of strong MgII absorbers in galactic winds, which is also supported by other observations (e.g., Ménard and Chelouche 2009; Nestor et al. 2011; Matejek and Simcoe 2012). Recently, Kacprzak et al. (2015) reported evidence that OVI absorbers in the CGM of $0.08 \leq z \leq 0.67$ galaxies arise primarily either along their minor axis or their major axis, with stronger absorbers being preferentially found along the minor axis.

Most existing observations of azimuthal angle dependence are at low to intermediate redshift, $z \lesssim 1.5$. This is in part because these observations require high-resolution imaging of the foreground galaxy in order to measure inclination or azimuthal angle. At high redshift, this requires long integrations with either HST or adaptive optics. In an analysis of the rest-frame optical morphological properties of $z \sim 2$–3 star-forming galaxies, Law et al. (2012) concluded that in contrast to galaxies at lower redshifts, there is no evidence for a correlation between outflow velocity and galaxy inclination. On the other hand, Newman et al. (2012) found 3σ evidence in a sample of 27 $z \sim 2$ star-forming galaxies with spatially resolved spectroscopic data that the mass loading factors of galactic winds are higher in face-on galaxies. It will be interesting to expand studies of azimuthal angle dependence in this redshift regime.

From the theoretical standpoint, it is possible that any azimuthal dependence present at lower redshift will be either weaker or absent in the CGM of $z \gtrsim 2$ galaxies. First, $z \gtrsim 2$ star-forming galaxies are often observed to have clumpy morphologies, especially in the UV light that traces star formation (e.g., Förster Schreiber et al. 2011), and galactic winds are to a large extent driven by outflows from prominent star-forming clumps (e.g., Genzel et al. 2011; Bordoloi et al. 2016). Chaotic and clumpy galaxy morphologies at high redshift followed by the gradual emergence of stable disks are also a generic finding of recent cosmological simulations (e.g., Agertz et al. 2009; Ceverino et al. 2010, 2016; Hopkins et al. 2014; Oklopcic et al. 2016; Ma et al. 2016b). Thus, there may simply often not be well-defined gaseous disks to neatly collimate galactic winds at high redshift. Furthermore, at $z \sim 2$ SFRs can be elevated relative to the local Universe by up to ~ 2 orders of magnitude, and some simulations suggest that galactic wind bursts may be sufficiently powerful to expel most of the ISM from galaxies (e.g., Muratov et al. 2015). In that case, even if it were present, a gaseous disk may not offer enough resistance to significantly collimate the galactic wind.

[7]Note, however, that the two types of observations are not equivalent since down-the-barrel spectra are always sensitive to high-density material near (or within) the target galaxy, while transverse spectra only probe the CGM at distances from the galaxy equal to the impact parameter or greater.

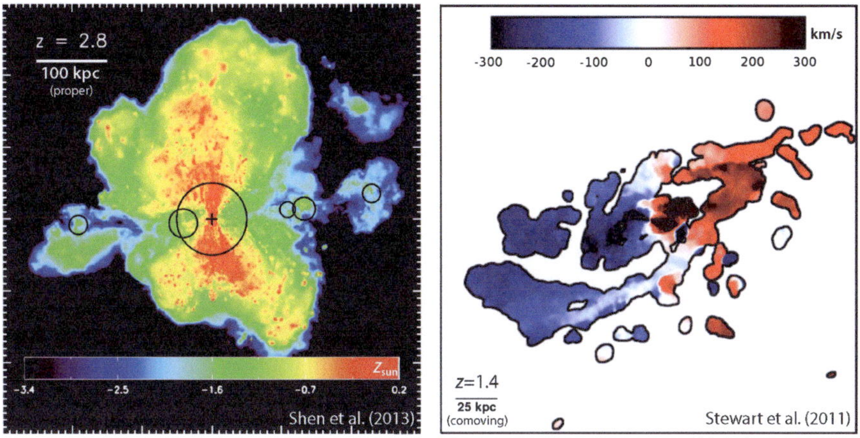

Fig. 5 Simulations that suggest the potential of azimuthal angle and kinematic diagnostics of galactic accretion. *Left*: Projected gas metallicity in the Eris2 simulation at $z = 2.8$. In this image, the stellar disk is nearly edge-on and the metal-enriched gas entrained by the galactic wind forms a rough bi-cone above and below the disk. The *circles* show virial radii, including of the five most massive nearby dwarf galaxies, which are aligned in the direction of metal-poor infalling filaments (adapted from Shen et al. 2013). *Right*: Line-of-sight velocity map of cool halo gas ($N_{\rm HI} > 10^{16}\,{\rm cm}^{-2}$) for a simulated Milky Way-mass halo at $z = 1.4$. The cool halo gas tends to rotate in the same direction as cool halo gas out to $R \gtrsim 40$ proper kpc (100 co-moving kpc). Both results shown here are based on zoom-in simulations of single halos, so more work is needed to determine which aspects generalize to different halo masses, assembly histories, redshifts, and details of how the baryonic physics is modeled. Adapted from Stewart et al. (2011)

Overall, azimuthal angle dependence is a promising approach for separating inflows and outflows statistically, but a more systematic analysis of the predictions of galaxy formation simulations will be needed to inform when a significant azimuthal angle dependence is expected. The left panel of Fig. 5 shows an example of how the metallicity of CGM gas varies with azimuthal angle in the Eris2 simulation of a Milky Way progenitor at $z = 2.8$ (Shen et al. 2013). In this example, gas is substantially more metal enriched above and below the galaxy due to the effects of galactic winds, but it remains to be shown whether this simulation is representative.

2.4.2 Kinematic Diagnostics

Weaker Mg II absorbers can arise from either inflows or outflows, but kinematics can potentially distinguish different origins. Several simulations indicate that accreting cool gas preferentially joins galactic disks in a co-rotating structure, consistent with galactic disks acquiring angular momentum from the accreting gas (Kereš and Hernquist 2009; Brooks et al. 2009; Stewart et al. 2011, 2013, 2016; Kauffmann et al. 2016).[8] Because the infalling gas co-rotates with the galaxy, its distribution is expected to be at least mildly flattened along the galactic semi-major axis. The right panel shows an example of a zoom-in simulation of a Milky Way-mass halo at $z = 1.4$ in which the radial velocity profile of the cool halo gas with HI column $N_{\rm HI} > 10^{16}\,{\rm cm}^{-2}$ indicates co-rotation with the growing galactic disk out to $R \gtrsim 40$ proper kpc (Stewart et al. 2011).

Observationally, there are several tentative detections of co-rotating MgII absorbers at low and high redshifts (e.g., Steidel et al. 2002; Kacprzak et al. 2010; Bouché et al. 2013, 2016), but current observational samples are small and conflicting results have been reported (e.g., Kacprzak et al. 2011). The simple picture of co-rotating MgII absorbers tracing galactic accretion is no doubt complicated by the different possible origins of MgII absorbers, including outflows, which can also carry angular momentum imparted as they are launched from a rotating disk. Over the next several years, kinematic diagnostics of halo gas will become increasingly interesting with the advent of a new generation of integral field surveys, including MaNGA (Bundy et al. 2015), KMOS^{3D} (Wisnioski et al. 2015), and with MUSE (e.g., Bacon et al. 2015), which will provide a new handle on the internal kinematics of galaxies. As with azimuthal angle diagnostics, systematically analyzing simulation predictions for the kinematics of gas galactic accretion relative to the orientation and internal kinematics of galaxies for a wide range of redshifts and galaxy properties will be critical to make progress. Currently, our theoretical expectations are limited by the small number of simulated halos for which kinematic relationships have been analyzed in detail, with existing studies being typically limited to a one or a few zoom-in simulations.

2.5 Cosmological Absorber Statistics

All the observational diagnostics discussed so far rely on an association between absorbing gas and a galaxy. However, quasar absorption spectra contain a wealth of information on intergalactic absorbers without known galaxy associations. Nevertheless, many of the stronger absorption systems in the spectra of arbitrarily selected quasars arise in the CGM of foreground galaxies and thus provide important statistical constraints on galactic accretion.

[8]This is not to say that infalling cool gas solely determines the angular momentum of disk galaxies. In an analysis of the Illustris simulation, Zjupa and Springel (2017) identify the important roles of specific angular momentum transfer from dark matter onto gas during mergers and from feedback expelling low specific angular momentum gas from halos.

2.5.1 The HI Column Density Distribution

We saw in Sect. 2.1 that LLSs in galaxy halos arise from a mix of inflows and outflows. Thus, LLSs from galactic accretion contribute to the observed HI column density distribution. Using a simulation from the OWLS project post-processed with radiative transfer, van de Voort et al. (2012) quantified the contribution of cold accretion flows to the observed $z = 3$ HI column density distribution. The simulation analyzed by van de Voort et al. (2012) reproduces the observed HI column density distribution over ten orders of magnitude in $N_{\rm HI}$ (Altay et al. 2011). In this simulation most LLSs arise within galaxy halos and most of these are infalling toward a nearby galaxy. On this basis, van de Voort et al. (2012) concluded that cold accretion flows predicted by cosmological simulations have been statistically detected in the observed HI column density distribution at $z = 3$. The argument is compelling, though there are some caveats that future simulations and observational analyses should attempt to address to firm up the conclusion. Observationally, measurements of the column density distribution are relatively uncertain in the LLS regime (e.g., Prochaska et al. 2010), in part because LLSs are on the flat part of the curve of growth. Theoretically, the simulation results summarized in Sect. 2.1 indicate that galactic winds can contribute comparably to—or even dominate over—cold accretion streams in explaining LLSs in galaxy halos at $z \sim 2$–2.5. Since the properties of galactic winds are uncertain, it is plausible that reasonable agreement with the observed column density distribution could be obtained absent cold streams for some wind models. Finally, a combination of resolution effects and numerical limitations of different hydrodynamic solvers introduces additional uncertainties in the theoretical predictions (Bird et al. 2013; Nelson et al. 2013, 2016).

2.5.2 The Metallicity Distribution of LLSs

Recently, Lehner et al. (2013) and Wotta et al. (2016) reported evidence that the metallicity distribution of randomly selected LLSs at $z < 1$ is bimodal, with dearth of LLSs with metallicity of about 10% solar.[9] These authors interpreted the high-metallicity branch as arising in outflows, recycling winds, and tidally stripped gas around galaxies, while the low-metallicity branch may trace cool, dense accreting gas. If this interpretation is correct, then LLS metallicity would be an extremely powerful way to identify cool galactic accretion at $z < 1$. At $z > 2$, the observational analyses of Fumagalli et al. (2016b), Lehner et al. (2016), and Glidden et al. (2016) indicate instead a broad unimodal distribution of LLS metallicities.

Motivated by the low-redshift observations of Lehner et al. (2013) and Wotta et al. (2016), Hafen et al. (2016) analyzed the LLS metallicity distribution at

[9]In the updated $z \leq 1$ metallicity analysis of Wotta et al. (2016), the statistical evidence for a bimodality is strongest for a subsample restricted to partial LLSs, with $16.2 \leq \log N_{\rm HI} \leq 17.2$.

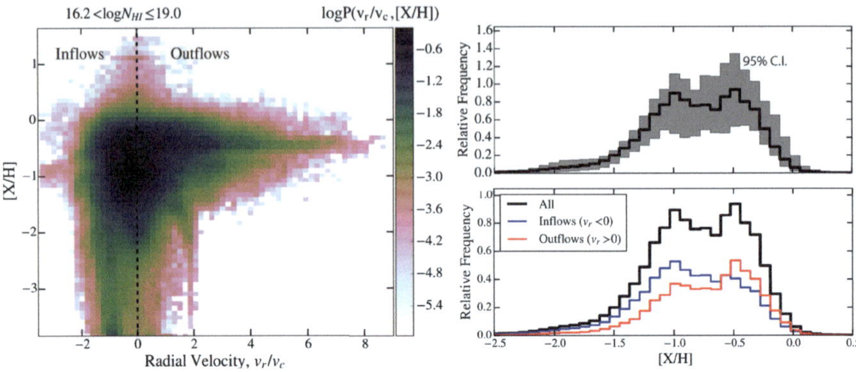

Fig. 6 Properties of the cosmological distribution of (randomly selected) LLSs at $0 < z < 1$ computed by convolving a suite of zoom-in simulations from the FIRE project with the dark matter halo mass function. *Left*: Relative incidence (logarithmic units) of LLSs with $10^{16.2} < N_{\rm HI} \leq 10^{19}\,{\rm cm}^{-2}$ in the metallicity vs. radial kinematics plane (radial velocity $v_{\rm r}$ defined relative to the central galaxy of the halo hosting the LLS and expressed in units of halo circular velocity). High-velocity outflows (with radial velocity exceeding the circular velocity by a factor $\gtrsim 2$) tend to have higher metallicities ([X/H] ~ -0.5) while very low-metallicity LLSs ([X/H] < -2) are typically associated with gas infalling from the IGM. *Right*: The corresponding overall LLS metallicity distribution. In the *top panel*, the *gray region* shows the 95% confidence interval resulting from the limited number of zoom-in simulations included in the analysis. There is no significant evidence for multiple modes in the simulated metallicity distribution. In the *bottom panel*, the total metallicity distribution is divided between gas elements that are instantaneously inflowing ($v_{\rm r} < 0$) and outflowing ($v_{\rm r} > 0$) relative to their central galaxy. The inflowing and outflowing distributions overlap strongly in part because wind recycling is efficient at low redshift, so metal-enriched outflows are later identified as instantaneous inflows. Adapted from Hafen et al. (2016)

$z < 1$ using a sample of zoom-in simulations from the FIRE project. To model the cosmological distribution from a sample of zoom-in simulations, Hafen et al. (2016) convolved the LLS properties for individual halos with the dark matter halo mass function. In these simulations, LLSs are concentrated close to galaxies at $z < 1$ so this halo-based approach should capture the majority of LLSs; Hafen et al. (2016) showed that it reproduces the LLS cosmological incidence measured by Ribaudo et al. (2011). Figure 6 summarizes the key results from Hafen et al. (2016) regarding the LLS metallicity distribution, and the relationship of LLS metallicity with inflows and outflows defined based on instantaneous radial kinematic relative to central galaxies. The analysis indicates that high-velocity outflows (with radial velocity exceeding the halo circular velocity by a factor $\gtrsim$2) tend to have higher metallicities ([X/H] ~ -0.5) while very low-metallicity LLSs ([X/H] < -2) are typically associated with IGM inflows. However, most LLSs occupy an intermediate region in metallicity-radial velocity space. Overall, the simulated LLS metallicity distribution does not show significant evidence for bimodality. The strong overlap between instantaneous inflows and outflows for intermediate metallicity systems is in part due to the prevalence of wind recycling in the FIRE simulations at $z < 1$, which causes metal-rich galactic wind ejecta to later fall back onto galaxies (Anglés-Alcàzar et al. 2016). The lack of a clean bimodality in the simulated LLS metallicity distribution is also due to the fact that halos covering the broad mass range $M_h \sim 10^{10}$–10^{12} $M_\odot$ contribute significantly to the distribution. Since the ISM and CGM metallicities both increase with galaxy mass in the simulations (Ma et al. 2016a; Muratov et al. 2016), any narrow feature in the metallicity distribution is likely to be washed out in the cosmological average. One effect that could cause simulations to miss features in the metallicity distribution is the mixing of metals on small scales. As mentioned above, some observations indicate that metals can be locked in compact clumps that will not be resolvable in cosmological simulations for the foreseeable future (e.g., Schaye et al. 2007). Going forward, it will be useful to address this issue by supplementing cosmological simulations with higher-resolution calculations better suited to understand small-scale mixing. The analysis of Hafen et al. (2016) included only 14 simulated main halos, so it will also be important to firm up the statistical significance of the results. Furthermore, it will be interesting to use simulations to study in more detail the redshift evolution of LLS metallicities, as well as how the metallicity distribution changes with HI column, e.g., from LLSs to DLAs. Cooper et al. (2015) analyzed the $z = 3.5$ LLS metallicity distribution in a full-volume cosmological simulation with the Illustris galaxy formation model (Vogelsberger et al. 2013; Torrey et al. 2014) and also found a broad unimodal distribution.

3 Emission Diagnostics

CGM emission is a probe of galactic accretion complementary to absorption measurements. The principal advantage of emission measurements is that they provide spatial resolution in individual halos, which can be used to identify galactic accretion using morphological signatures, such as filaments. At present, the main challenge with emission observations is that circum-galactic gas is typically very faint, so emission measurements are currently only possible for dense gas relatively close to galaxies, or in halos with a luminous quasar that can power CGM emission out to larger radii. High-quality CGM emission observations will, however, become increasingly common over the next few years as a number of optical integral field spectrographs (IFS) with the capacity to detect low surface brightness, redshifted rest-UV CGM emission have recently been commissioned or are planned for the near future. The Palomar Cosmic Web Imager (PCWI; Matuszewski et al. 2010) started taking data in 2009 and the first science results on luminous spatially extended Lyα sources at $z \sim 2$–3 have been reported (Martin et al. 2014a,b). Its successor, the Keck Cosmic Web Imager (KCWI, Martin et al. 2010), to be mounted on the Keck II telescope, is currently being developed. The Multi-Unit Spectroscopic Explorer (MUSE; Bacon et al. 2010) on the Very Large Telescope (VLT) completed its commissioning in August 2014 and early science results are being reported (e.g., Fumagalli et al. 2016a; Wisotzki et al. 2016; Borisova et al.

2016). These IFSs provide kinematic information not available with narrowband imaging, and their spectroscopic resolution also enables more accurate background subtraction for line emission.

In this section, we provide a brief summary of simulation results regarding Lyα emission (Sect. 3.1), UV metal line emission (Sect. 3.2), and X-ray emission from the CGM (Sect. 3.3), as well as a summary of the observational status for each.

3.1 Lyα Emission from the CGM

Lyα emission is typically the brightest emission line from the CGM. Our first glimpses of CGM emission have indeed come from spatially extended Lyα sources known as "Lyα blobs" (LABs). The classical LABs have line luminosities up to $\sim 10^{44}\,\mathrm{erg\,s^{-1}}$ and spatial extents sometimes exceeding 100 proper kpc (Steidel et al. 2000; Matsuda et al. 2004; Yang et al. 2009). The physical nature of LABs is not yet well understood, but several studies suggested that they could be powered by the conversion of gravitational potential into Lyα photons as gas accretes onto halos or galaxies ("cooling radiation"). In this model, weak shocks continuously heat cold accreting gas to temperatures $T \sim 10^4\,\mathrm{K}$ and this energy is efficiently converted into Lyα emission via collisional excitation of HI (Haiman et al. 2000; Fardal et al. 2001; Dijkstra and Loeb 2009; Goerdt et al. 2010; Rosdahl and Blaizot 2012). However, a major hurdle in identifying diffuse Lyα radiation with this process is that the expected luminosity remains uncertain at the order-of-magnitude level. Fundamentally, this is because the Lyα collisional excitation emissivity is an exponentially steep function of temperature near $T = 10^4\,\mathrm{K}$, so that small errors in the thermodynamic history of the accreting cold gas can result in large differences in the predicted Lyα luminosity (e.g., Faucher-Giguère et al. 2010). There are two sources of error that can affect the thermodynamic history of accreting gas in cosmological simulations.

The first is the accuracy of the hydrodynamics, which must be able to correctly capture the properties of both weak *and* strong shocks experienced by accreting gas. The latter point regarding strong shocks is also important for the identification of cold accretion flows in simulations, and is worth expanding on. In both particle-based and grid-based hydrodynamic codes, shocks are often broadened across several resolution elements, which can lead to "in-shock cooling" (e.g., Hutchings and Thomas 2000; Creasey et al. 2011). This problem arises, for example, when a strong shock should produce hot $T \gtrsim 10^6\,\mathrm{K}$ gas with a long cooling time but in the code the gas cools artificially as it passes through the numerically broadened shock and encounters the peak of the cooling function at $T \sim 10^5\,\mathrm{K}$. In such circumstances, the hydrodynamic solver can overestimate radiative energy losses via low-energy processes, including Lyα. A specific situation where this effect likely occurs in cosmological simulations is when cool accreting gas impacts a galaxy, where cooling times can be very short owing to the relatively high local gas densities. In this case, not only will there be an error in the predicted Lyα emission, but also artifacts can be introduced in simple algorithms for identifying cold mode accretion based on the maximum temperature to which gas is heated (e.g., Kereš et al. 2005, 2009b; van de Voort et al. 2011; Nelson et al. 2013).

The second reason for the large uncertainties in predicted Lyα cooling luminosities is the treatment of ionizing radiative transfer. As discussed in Sect. 2.1, cold accretion streams are traced by LLSs, which by definition are optically thick to ionizing photons at the Lyman edge. Since most cosmological simulations to date do not include self-consistent ionizing radiative transfer, they do not accurately capture photoheating in dense self-shielded gas (but see Rosdahl and Blaizot 2012). In particular, simulations run with a uniform cosmic UVB and assuming optically thin ionization balance overestimate the amount of photoheating within cold streams. Faucher-Giguère et al. (2010) tested the sensitivity of their predictions to the treatment of dense gas and found that different assumptions produced Lyα luminosities differing by up to ~ 2 orders of magnitude.

Even if a significant fraction of the Lyα emission in LABs comes from cooling radiation, it is difficult to observationally separate cooling radiation from Lyα photons produced by other processes, such as star formation or AGN activity in galaxies. One reason is that bright cooling radiation requires high galactic accretion rates, which lead to SFRs (or AGN activity) sufficient to power most observed diffuse Lyα halos. Diffuse Lyα halos are now in fact generically observed around ordinary star-forming galaxies (e.g., Steidel et al. 2011) and these observations are consistent with the Lyα photons being produced by star formation inside galaxies. There are several ways in which star formation or AGN power can mimic the spatially extended emission expected from galactic accretion. Lyα photons produced inside galaxies can result in diffuse halos due to scattering of the Lyα photons in the CGM (e.g., Dijkstra and Kramer 2012). Ionizing photons that escape galaxies but are absorbed in the CGM can also produce fluorescent Lyα emission (e.g., Gould and Weinberg 1996; Cantalupo et al. 2005; Kollmeier et al. 2010). Alternatively, energy injected in the CGM as galactic winds driven by stellar or AGN feedback encounter halo gas can also power diffuse Lyα emission (Taniguchi and Shioya 2000; Taniguchi et al. 2001). Since Lyα photons typically scatter a large number of times before escaping the CGM, the apparent Lyα spatial and velocity extents are not necessarily representative of the gas producing the Lyα photons.

A more promising avenue for using Lyα emission to identify galactic accretion is to simply use the Lyα photons as a tracers of CGM gas at last scattering. For example, many Lyα sources have a filamentary morphology reminiscent of cosmic web filaments and their extensions into galactic halos as cold streams (e.g., Rauch et al. 2011, 2013, 2016; Cantalupo et al. 2014; Martin et al. 2014a,b). Of course, one must be careful not attribute every filamentary features to a cold accretion stream, since other phenomena such as tidally stripped gas can appear elongated. Nevertheless, a statistical study of the morphological properties of spatially extended Lyα, along with a comparison to the incidence rate of accreting filaments predicted by cosmological simulations, can test simulation predictions for galactic accretion. In at least one case with detailed spatially resolved kinematic observations (the Lyα image shown in Fig. 7), there is evidence that the filamentary structure traced by Lyα emission smoothly connects to a large, rotating proto-galactic disk (Martin et al. 2015).

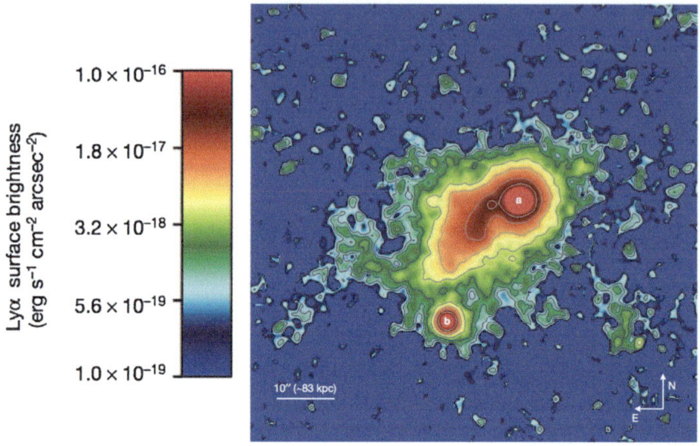

Fig. 7 Lyα image of the nebula around the UM 287 quasar ("*a*") at $z \approx 2.3$ (adapted from Cantalupo et al. 2014). The second bright spot labeled "*b*" marks the location of another, optically faint quasar at the same redshift. The extended filamentary morphology suggests that the Lyα emission traces a cold accretion flow. Follow-up integral field observations suggest a smooth kinematic profile consistent with a giant, rotating proto-galactic disk for the brightest portion of the filament, which appears to connect smoothly to the cosmic web (Martin et al. 2015)

Observations of particularly extended and luminous Lyα nebulae at high redshift provide further evidence for compact dense clumps in the gaseous halos of massive galaxies (see Sect. 2.2 for evidence from absorption measurements). Even if a luminous quasar can in principle power the observed Lyα luminosity through reprocessing of its ionizing radiation in the CGM, the integrated recombination rate over the nebula ($\propto \int dV\alpha(T)n_e n_{\rm HII}$, where α is the hydrogen recombination coefficient) must be sufficiently high to account for the Lyα luminosity attributed to fluorescence. Recent detailed analyses of giant Lyα nebulae indicate that in at least some cases the Lyα-emitting gas must be highly clumped and reach densities $\sim$1 cm^{-3} (more typical of ISM gas than CGM gas) over spatial scales $\sim$100 kpc (e.g., Cantalupo et al. 2014; Hennawi et al. 2015). If giant proto-galactic disks are relatively common at high redshift, one possibility is that much of the observed Lyα luminosity in giant nebulae originates in fluorescence due to a luminous quasar shining on such a disk rather than CGM gas (e.g., Martin et al. 2015).

The contribution by S. Cantalupo in this volume provides a more exhaustive review of recent results on spatially extended Lyα sources.

3.2 UV Metal Line Emission from the CGM

Because metals are not as abundant, metal line emission is generally significantly fainter than Lyα. Metal lines can, however, provide very useful complementary information on the physical conditions in the CGM. Since most metal lines are

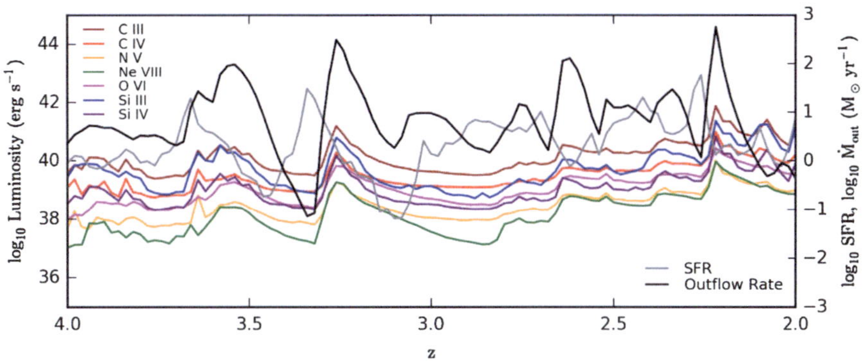

Fig. 8 Example simulation of the UV metal line emission from the CGM of an LBG-mass halo as a function of redshift. This simulation, from the FIRE project, includes strong stellar feedback. *Colored lines* show UV metal line luminosities within 1 $R_{\rm vir}$ but excluding the inner 10 proper kpc (a proxy for central galaxies). Star formation rates within 1 $R_{\rm vir}$ and gas mass outflow rates at 0.25 $R_{\rm vir}$ are plotted as *gray* and *black lines*, respectively. The UV metal line luminosities, star formation, and mass outflow rates are all strongly time variable and correlated. Peaks in CGM luminosity correspond more closely with peaks in mass outflow rates, which typically follow peaks of star formation with a time delay, indicating that energy injected by galactic winds is the primary source of CGM UV metal line emission. Adapted from Sravan et al. (2016)

optically thin, they are not subject to photon scattering effects and therefore more directly probe the spatial distribution and kinematics of the emitting gas. Furthermore, different metal ions probe different temperature regimes (e.g., Frank et al. 2012; van de Voort and Schaye 2013; Corlies and Schiminovich 2016). On the other hand, since metal emission preferentially probes metal-enriched gas, it is at present typically more useful as a general probe of the conditions in the CGM rather than of galactic accretion directly. For example, in an analysis of the UV metal line emission from the CGM of $z = 2 - 4$ simulated LBGs from the FIRE project, Sravan et al. (2016) showed the UV metal line emission arises primarily from gas collisionally excited by galactic winds (see Fig. 8).

3.3 X-ray Emission from Hot Halo Gas

Finally, we comment on the use of X-ray observations for probing galactic accretion. In galaxy clusters, it is well established that the hot intra-cluster medium (ICM) is primarily the result of gas shocked heated during the cluster assembly and that the ICM cooling observed in X-rays drives accretion onto galaxies (albeit with a strong apparent suppression of the cooling flows in many clusters, tentatively due to AGN feedback, e.g., McNamara and Nulsen 2007). But what processes do X-rays probe in lower-mass halos (e.g., Mulchaey and Jeltema 2010; Anderson et al. 2013; Li et al. 2016)?

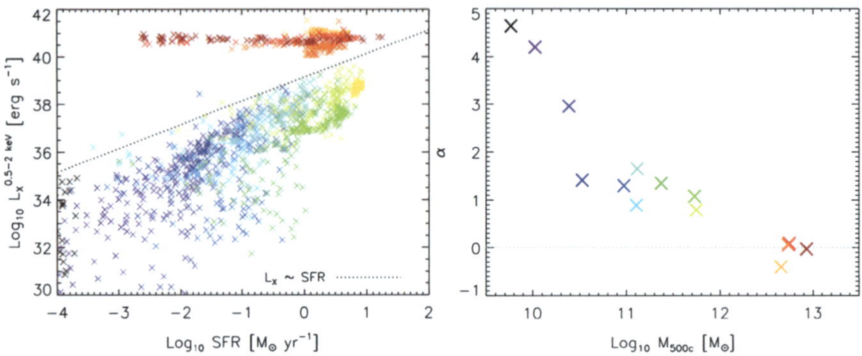

Fig. 9 *Left*: Soft X-ray luminosity L_X (0.5–2 keV) at $z = 0 - 0.5$ as a function of SFR (averaged over 100 Myr) for zoom-in cosmological simulations with stellar feedback from the FIRE project. *Crosses* with the same color belong to the same galaxy at different times (halo masses can be read off from the panel on the *right*). *Right*: The power, α, of the correlation between L_X and SFR ($L_X \propto SFR^{\alpha}$) for individual halos as a function of halo mass. The X-ray emission around dwarf galaxies is a strong function of their SFR, while halos with $M_{500c} \approx 10^{11-12}$ $M_{\odot}$ exhibit a correlation close to linear. There is no correlation between L_X and SFR for the most massive halos, because hot, virialized halo gas produces more X-rays than star formation-powered winds in those halos. Thus, X-ray emission is sensitive to gas accretion onto non-dwarf halos at low redshift (including Milky Way-mass halos, galaxy groups, and galaxy clusters) but primarily probes galactic winds in dwarfs. Adapted from van de Voort et al. (2016)

van de Voort et al. (2016) analyzed the X-ray emission in simulated halos from the FIRE project. As for the other FIRE simulations mentioned in this review, these simulations included stellar feedback but no AGN feedback. Figure 9 summarizes how the soft X-ray emission depends on SFR at $z < 0.5$, for different halo masses. The X-ray emission around dwarf galaxies is a strong function of their SFR but there is no correlation between L_X around massive galaxies or galaxy groups ($M_{500c} > 10^{12}$ $M_{\odot}$). In intermediate-mass halos ($M_{500c} \approx 10^{11-12}$ $M_{\odot}$), there is a close to linear relation between X-ray luminosity and SFR. These results indicate that diffuse X-rays primarily probe star formation-driven galactic winds in low-mass halos (see also the analytic wind models of Zhang et al. 2014).

4 Conclusions and Outlook

A common thread of this review is that there is no silver bullet in the quest to test models of galactic accretion. Each observational diagnostic that we discussed has not only some advantages, but also some ambiguities. There is in general likely no simple criterion that can be used to robustly identify galactic accretion in individual measurements. This is because the CGM is a complex and dynamic environment in which galactic accretion interacts with galactic winds, satellite galaxies, as well as more quiescent ambient gas. Furthermore, simulations predict that—like galaxies—

the properties of the CGM evolve significantly with redshift and halo mass. This complexity underscores the crucial role that simulations will continue to play in testing models of galactic accretion and feedback. Indeed, our current understanding points toward systematic, statistical comparisons with full-physics cosmological simulations as a necessary step to test the models.

We conclude with a brief list of general areas in which progress is likely to be particularly fruitful over the next few years:

1. Since cosmological simulations cannot resolve some of the fine-scale structure apparent in CGM observations (see Sect. 2.2), it will be important to clarify which observational tracers can be robustly compared with simulation predictions. For example, are the main observable characteristics of the more massive and volume-filling CGM phases reasonably converged?
2. Relatedly, a better understanding of how metals returned by stellar evolution mix with ambient gas (both inside galaxies and after being ejected into the CGM) will ultimately be essential to make robust predictions for the metallicity distribution of CGM gas. Current sub-grid models for metal mixing due to unresolved turbulence rely on simplified schemes and do not account for the fact that metals may be injected as compact clumps well below the resolution limit.
3. Non-ideal hydrodynamic effects, such as magnetic fields and thermal conduction, affect the survival and phase structure of CGM clouds, and should therefore be investigated.
4. With the exception of a few recent studies, most previous simulations used to study diagnostics of galactic accretion either neglected galactic winds or used stellar feedback too weak to reproduce observed outflows and galaxy stellar masses. To develop reliable accretion diagnostics, it is critical to use feedback models that reproduce observed galaxy properties.
5. Some promising diagnostics of inflows and outflows (e.g., azimuthal angle and kinematic diagnostics; Sect. 2.4) have so far only been studied using small samples of simulated halos and for limited redshift ranges. Since galaxies and their CGM evolve strongly with redshift and mass, it will be necessary to analyze larger simulation samples that systematically cover relevant mass and redshift ranges to quantify the statistical robustness and limitations of the diagnostics.
6. Given the ambiguities of different inflow/outflow diagnostics when applied in isolation, quantifying how different diagnostics (e.g., metallicity, azimuthal angle, and kinematics) could be *jointly* used to distinguish between inflows and outflows would be very useful.
7. Differential studies comparing observations at epochs where inflows/outflows are predicted to be more/less prominent could also help in breaking degeneracies in single observations.

We note that most of these issues are not specific to the CGM, but of general importance to galaxy formation. It is thus clear that studies of the CGM will remain a very active area at the forefront of research on galaxy evolution for the foreseeable future.

Acknowledgements We are grateful to many colleagues and collaborators who have helped shape our views on galactic accretion, including: Chuck Steidel, Gwen Rudie, Alice Shapley, Xavier Prochaska, Joe Hennawi, Michele Fumagalli, Nicolas Lehner, Chris Howk, Lars Hernquist, Joop Schaye, Freeke van de Voort, Andrey Kravtsov, Cameron Liang, Mark Dijkstra, Norm Murray, Eliot Quataert, Dusan Kereš, Phil Hopkins, Alexander Muratov, Daniel Anglés-Alcázar, and Zach Hafen. Our research on galactic accretion has been supported by NSF and NASA.

References

Adelberger, K. L., Erb, D. K., Steidel, C. C., et al. 2005a, ApJ, 620, L75

Adelberger, K. L., Shapley, A. E., Steidel, C. C., et al. 2005b, ApJ, 629, 636

Adelberger, K. L., Steidel, C. C., Shapley, A. E., & Pettini, M. 2003, ApJ, 584, 45

Agertz, O., & Kravtsov, A. V. 2015, ApJ, 804, 18

Agertz, O., Moore, B., Stadel, J., et al. 2007, MNRAS, 380, 963

Agertz, O., Teyssier, R., & Moore, B. 2009, MNRAS, 397, L64

Altay, G., Theuns, T., Schaye, J., Crighton, N. H. M., & Dalla Vecchia, C. 2011, ApJ, 737, L37

Anglés-Alcàzar, D., Faucher-Giguère, C.-A., Kereš, D., Hopkins, P. F., Quataert, E., Murray, N. 2016, MNRAS, submitted, arXiv:1610.08523

Anderson, M. E., Bregman, J. N., & Dai, X. 2013, ApJ, 762, 106

Bacon, R., Brinchmann, J., Richard, J., et al. 2015, A&A, 575, A75

Bacon, R., Accardo, M., Adjali, L., et al. 2010, SPIE, 7735, 8

Bauermeister, A., Blitz, L., & Ma, C. 2010, ApJ, 717, 323

Bird, S., Vogelsberger, M., Sijacki, D., et al. 2013, MNRAS, 429, 3341

Birnboim, Y., & Dekel, A. 2003, MNRAS, 345, 349

Bordoloi, R., Lilly, S. J., Hardmeier, E., et al. 2014a, ApJ, 794, 130

Bordoloi, R., Lilly, S. J., Knobel, C., et al. 2011, ApJ, 743, 10

Bordoloi, R., Rigby, J. R., Tumlinson, J., et al. 2016, MNRAS, 458, 1891

Bordoloi, R., Tumlinson, J., Werk, J. K., et al. 2014b, ApJ, 796, 136

Borisova, E., Cantalupo, S., Lilly, S. J., et al. 2016, ApJ, 831, 39

Bouché, N., Finley, H., Schroetter, I., et al. 2016, ApJ, 820, 121

Bouché, N., Hohensee, W., Vargas, R., et al. 2012, MNRAS, 426, 801

Bouché, N., Murphy, M. T., Kacprzak, G. G., et al. 2013, Science, 341, 50

Brooks, A. M., Governato, F., Quinn, T., Brook, C. B., & Wadsley, J. 2009, ApJ, 694, 396

Bundy, K., Bershady, M. A., Law, D. R., et al. 2015, ApJ, 798, 7

Cantalupo, S., Arrigoni-Battaia, F., Prochaska, J. X., Hennawi, J. F., & Madau, P. 2014, Nature, 506, 63

Cantalupo, S., Porciani, C., Lilly, S. J., & Miniati, F. 2005, ApJ, 628, 61

Ceverino, D., Dekel, A., & Bournaud, F. 2010, MNRAS, 404, 2151

Ceverino, D., Primack, J., Dekel, A., & Kassin, S. A. 2016, MNRAS, submitted (arXiv:1608.02114)

Chen, H.-W. 2012, MNRAS, 427, 1238

Cooper, T. J., Simcoe, R. A., Cooksey, K. L., O'Meara, J. M., & Torrey, P. 2015, ApJ, 812, 58

Corlies, L., & Schiminovich, D. 2016, ApJ, 827, 148

Creasey, P., Theuns, T., Bower, R. G., & Lacey, C. G. 2011, MNRAS, 415, 3706

Crighton, N. H. M., Hennawi, J. F., Simcoe, R. A., et al. 2015, MNRAS, 446, 18

Davé, R., Oppenheimer, B. D., & Finlator, K. 2011, MNRAS, 415, 11

Dekel, A., & Birnboim, Y. 2006, MNRAS, 368, 2

Dekel, A., Birnboim, Y., Engel, G., et al. 2009, Nature, 457, 451

Diamond-Stanic, A. M., Coil, A. L., Moustakas, J., et al. 2016, ApJ, 824, 24

Dijkstra, M., & Kramer, R. 2012, MNRAS, 424, 1672

Dijkstra, M., & Loeb, A. 2009, MNRAS, 400, 1109

Erb, D. K. 2008, ApJ, 674, 151

Fabian, A. C. 2012, ARA&A, 50, 455

Fardal, M. A., Katz, N., Gardner, J. P., et al. 2001, ApJ, 562, 605

Faucher-Giguère, C.-A., Kereš, D., Dijkstra, M., Hernquist, L., & Zaldarriaga, M. 2010, ApJ, 725, 633

Faucher-Giguère, C.-A., Feldmann, R., Quataert, E., et al. 2016, MNRAS, 461, L32

Faucher-Giguère, C.-A., Hopkins, P. F., Kereš, D., et al. 2015, MNRAS, 449, 987

Faucher-Giguère, C.-A., & Kereš, D. 2011, MNRAS, 412, L118

Faucher-Giguère, C.-A., Kereš, D., & Ma, C.-P. 2011, MNRAS, 417, 2982

Ford, A. B., Werk, J. K., Davé, R., et al. 2016, MNRAS, 459, 1745

Förster Schreiber, N. M., Genzel, R., Bouché, N., et al. 2009, ApJ, 706, 1364

Förster Schreiber, N. M., Shapley, A. E., Erb, D. K., et al. 2011, ApJ, 731, 65

Frank, S., Rasera, Y., Vibert, D., et al. 2012, MNRAS, 420, 1731

Fu, H., Hennawi, J. F., Prochaska, J. X., Mutel, R., Casey, C., Cooray, A., Kereš, D., Zhang, Z.-Y., Clements, D., Isbell, J., Lang, C., McGinnis, D., Michałowski, M. J., Mooley, K., Perley, D., Stockton, A., & Thompson, D. 2016, The circumgalactic medium of submillimeter galaxies. I. First results from a radio-identified sample. ApJ, 832, 52. arXiv:1607.00016. doi:10.3847/0004-637X/832/1/52. http://adsabs.harvard.edu/abs/2016ApJ...832...52F

Fumagalli, M., Cantalupo, S., Dekel, A., et al. 2016a, MNRAS, 462, 1978

Fumagalli, M., Hennawi, J. F., Prochaska, J. X., et al. 2014, ApJ, 780, 74

Fumagalli, M., O'Meara, J. M., & Prochaska, J. X. 2016b, MNRAS, 455, 4100

Fumagalli, M., O'Meara, J. M., Prochaska, J. X., & Worseck, G. 2013, ApJ, 775, 78

Fumagalli, M., Prochaska, J. X., Kasen, D., et al. 2011, MNRAS, 418, 1796

Gauthier, J.-R., Chen, H.-W., & Tinker, J. L. 2010, ApJ, 716, 1263

Genzel, R., Newman, S., Jones, T., et al. 2011, ApJ, 733, 101

Glidden, A., Cooper, T. J., Cooksey, K. L., Simcoe, R. A., & O'Meara, J. M. 2016, Predominantly low metallicities measured in a stratified sample of Lyman limit systems at $Z = 3.7$. ApJ, 833, 270. arXiv:1604.02144. doi:10.3847/1538-4357/833/2/270. http://adsabs.harvard.edu/abs/2016ApJ...833..270G

Goerdt, T., & Ceverino, D. 2015, MNRAS, 450, 3359

Goerdt, T., Dekel, A., Sternberg, A., et al. 2010, MNRAS, 407, 613

Goerdt, T., Dekel, A., Sternberg, A., Gnat, O., & Ceverino, D. 2012, MNRAS, 424, 2292

Gould, A., & Weinberg, D. H. 1996, ApJ, 468, 462

Hafen, Z., Faucher-Giguere, C.-A., Angles-Alcazar, D., et al. 2016, preprint (arXiv:1608.05712)

Haiman, Z., Spaans, M., & Quataert, E. 2000, ApJ, 537, L5

Hayward, C. C., & Hopkins, P. F. 2015, preprint (arXiv:1510.05650)

Heckman, T. M., & Borthakur, S. 2016, ApJ, 822, 9

Hennawi, J. F., Prochaska, J. X., Burles, S., et al. 2006, ApJ, 651, 61

Hennawi, J. F., Prochaska, J. X., Cantalupo, S., & Arrigoni-Battaia, F. 2015, Science, 348, 779

Henriques, B. M. B., White, S. D. M., Thomas, P. A., et al. 2013, MNRAS, 431, 3373

Hopkins, P. F. 2013, MNRAS, 428, 2840

—. 2015, MNRAS, 450, 53

Hopkins, P. F., Kereš, D., Oñorbe, J., et al. 2014, MNRAS, 445, 581

Hu, C.-Y., Naab, T., Walch, S., Moster, B. P., & Oser, L. 2014, MNRAS, 443, 1173

Hummels, C. B., Bryan, G. L., Smith, B. D., & Turk, M. J. 2013, MNRAS, 430, 1548

Hutchings, R. M., & Thomas, P. A. 2000, MNRAS, 319, 721

Kacprzak, G. G., Churchill, C. W., Barton, E. J., & Cooke, J. 2011, ApJ, 733, 105

Kacprzak, G. G., Churchill, C. W., Ceverino, D., et al. 2010, ApJ, 711, 533

Kacprzak, G. G., Churchill, C. W., & Nielsen, N. M. 2012, ApJ, 760, L7

Kacprzak, G. G., Muzahid, S., Churchill, C. W., Nielsen, N. M., & Charlton, J. C. 2015, ApJ, 815, 22

Kauffmann, G., Borthakur, S., & Nelson, D. 2016, MNRAS, 462, 3751

Kereš, D., & Hernquist, L. 2009, ApJ, 700, L1

Kereš, D., Katz, N., Davé, R., Fardal, M., & Weinberg, D. H. 2009a, MNRAS, 396, 2332

Kereš, D., Katz, N., Fardal, M., Davé, R., & Weinberg, D. H. 2009b, MNRAS, 395, 160
Kereš, D., Katz, N., Weinberg, D. H., & Davé, R. 2005, MNRAS, 363, 2
Kimm, T., Slyz, A., Devriendt, J., & Pichon, C. 2011, MNRAS, 413, L51
Kollmeier, J. A., Zheng, Z., Davé, R., et al. 2010, ApJ, 708, 1048
Kornei, K. A., Shapley, A. E., Martin, C. L., et al. 2012, ApJ, 758, 135
Lau, M. W., Prochaska, J. X., & Hennawi, J. F. 2016, ApJS, 226, 25
Law, D. R., Steidel, C. C., Shapley, A. E., et al. 2012, ApJ, 759, 29
Lehner, N., Howk, J. C., Tripp, T. M., et al. 2013, ApJ, 770, 138
Lehner, N., Howk, J. C., & Wakker, B. P. 2015, ApJ, 804, 79
Lehner, N., O'Meara, J. M., Fox, A. J., et al. 2014, ApJ, 788, 119
Lehner, N., O'Meara, J. M., Howk, J. C., Prochaska, J. X., & Fumagalli, M. 2016, The cosmic evolution of the metallicity distribution of ionized gas traced By Lyman Limit systems. ApJ, 833, 283. arXiv:1608.02588. doi:10.3847/1538-4357/833/2/283. http://adsabs.harvard.edu/abs/2016ApJ...833..283L
Li, J.-T., Bregman, J. N., Wang, Q. D., Crain, R. A., & Anderson, M. E. 2016, ApJ, 830, 134
Liang, C. J., & Chen, H.-W. 2014, MNRAS, 445, 2061
Liang, C. J., Kravtsov, A. V., & Agertz, O. 2016, MNRAS, 458, 1164
Ma, X., Hopkins, P. F., Faucher-Giguère, C.-A., et al. 2016a, MNRAS, 456, 2140
Ma, X., Hopkins, P. F., Wetzel, A. R., et al. 2016b, MNRAS, submitted (arXiv:1608.04133)
Mandelker, N., Padnos, D., Dekel, A., et al. 2016, MNRAS, 463, 3921
Marinacci, F., Pakmor, R., Springel, V., & Simpson, C. M. 2014, MNRAS, 442, 3745
Martin, C., Moore, A., Morrissey, P., et al. 2010, SPIE, 7735, 21
Martin, C. L., Shapley, A. E., Coil, A. L., et al. 2012, ApJ, 760, 127
Martin, D. C., Chang, D., Matuszewski, M., et al. 2014a, ApJ, 786, 106
Martin, D. C., Chang, D., Matuszewski, M., et al. 2014b, ApJ, 786, 107
Martin, D. C., Matuszewski, M., Morrissey, P., et al. 2015, Nature, 524, 192
Matejek, M. S., & Simcoe, R. A. 2012, ApJ, 761, 112
Matsuda, Y., Yamada, T., Hayashino, T. et al., 2004, AJ, 128, 569
Matuszewski, M., Chang, D., Crabill, R. M., et al. 2010, SPIE, 7735, 24
McNamara, B. R., & Nulsen, P. E. J. 2007, ARA&A, 45, 117
McQuinn, M., Oh, S. P., & Faucher-Giguère, C.-A. 2011, ApJ, 743, 82
Ménard, B., & Chelouche, D. 2009, MNRAS, 393, 808
Mulchaey, J. S., & Jeltema, T. E. 2010, ApJ, 715, L1
Muratov, A. L., Kereš, D., Faucher-Giguére, C.-A., et al. 2016, MNRAS, submitted (arXiv:1606.09252)
Muratov, A. L., Kereš, D., Faucher-Giguère, C.-A., et al. 2015, MNRAS, 454, 2691
Navarro, J. F., Frenk, C. S., & White, S. D. M. 1995, MNRAS, 275, 56
Neeleman, M., Prochaska, J. X., & Wolfe, A. M. 2015, ApJ, 800, 7
Nelson, D., Genel, S., Pillepich, A., et al. 2016, MNRAS, 460, 2881
Nelson, D., Genel, S., Vogelsberger, M., Springel, V., et al. 2015, MNRAS, 448, 59
Nelson, D., Vogelsberger, M., Genel, S., et al. 2013, MNRAS, 429, 3353
Nestor, D. B., Johnson, B. D., Wild, V., et al. 2011, MNRAS, 412, 1559
Newman, S. F., Genzel, R., Förster-Schreiber, N. M., et al. 2012, ApJ, 761, 43
Oklopčić, A., Hopkins, P. F., Feldmann, R., Kereš, D., Faucher-Giguère, C.-A., & Murray, N. 2017, Giant clumps in the FIRE simulations: a case study of a massive high-redshift galaxy. MNRAS, 465, 952-969. arXiv:1603.03778. doi:10.1093/mnras/stw2754. http://adsabs.harvard.edu/abs/2017MNRAS.465..952O
Oppenheimer, B. D., Davé, R., Kereš, D., et al. 2010, MNRAS, 406, 2325
Prochaska, J. X., Hennawi, J. F., & Simcoe, R. A. 2013, ApJ, 762, L19
Prochaska, J. X., Lau, M. W., & Hennawi, J. F. 2014, ApJ, 796, 140
Prochaska, J. X., O'Meara, J. M., & Worseck, G. 2010, ApJ, 718, 392
Prochaska, J. X., & Wolfe, A. M. 2009, ApJ, 696, 1543
Rafelski, M., Neeleman, M., Fumagalli, M., Wolfe, A. M., & Prochaska, J. X. 2014, ApJ, 782, L29
Rahmati, A., Pawlik, A. H., Raicevic, M., & Schaye, J. 2013, MNRAS, 430, 2427

Rahmati, A., Schaye, J., Bower, R. G., et al. 2015, MNRAS, 452, 2034
Rauch, M., Becker, G. D., & Haehnelt, M. G. 2016, MNRAS, 455, 3991
Rauch, M., Becker, G. D., Haehnelt, M. G., et al. 2011, MNRAS, 418, 1115
Rauch, M., Becker, G. D., Haehnelt, M. G., Gauthier, J.-R., & Sargent, W. L. W. 2013, MNRAS, 429, 429
Rauch, M., Sargent, W. L. W., & Barlow, T. A. 1999, ApJ, 515, 500
—. 2001, ApJ, 554, 823
Read, J. I., & Hayfield, T. 2012, MNRAS, 422, 3037
Ribaudo, J., Lehner, N., & Howk, J. C. 2011, ApJ, 736, 42
Rosdahl, J., & Blaizot, J. 2012, MNRAS, 423, 344
Rubin, K. H. R., Hennawi, J. F., Prochaska, J. X., et al. 2015, ApJ, 808, 38
Rubin, K. H. R., Prochaska, J. X., Koo, D. C., & Phillips, A. C. 2012, ApJ, 747, L26
Rubin, K. H. R., Prochaska, J. X., Koo, D. C., et al. 2014, ApJ, 794, 156
Rudie, G. C., Steidel, C. C., Trainor, R. F., et al. 2012, ApJ, 750, 67
Saitoh, T. R., & Makino, J. 2013, ApJ, 768, 44
Schaye, J., Carswell, R. F., & Kim, T.-S. 2007, MNRAS, 379, 1169
Schaye, J., Crain, R. A., Bower, R. G., et al. 2015, MNRAS, 446, 521
Shen, S., Madau, P., Aguirre, A., et al. 2012, ApJ, 760, 50
Shen, S., Madau, P., Guedes, J., et al. 2013, ApJ, 765, 89
Shen, S., Wadsley, J., & Stinson, G. 2010, MNRAS, 407, 1581
Sijacki, D., Vogelsberger, M., Kereš, D., Springel, V., & Hernquist, L. 2012, MNRAS, 424, 2999
Simcoe, R. A., Sargent, W. L. W., Rauch, M., & Becker, G. 2006, ApJ, 637, 648
Somerville, R. S., & Davé, R. 2015, ARA&A, 53, 51
Springel, V. 2010, MNRAS, 401, 791
Sravan, N., Faucher-Giguere, C.-A., van de Voort, F., et al. 2016, MNRAS, 463, 120
Steidel, C. C., Adelberger, K. L., Shapley, A. E., et al. 2000, ApJ, 532, 170
Steidel, C. C., Bogosavljević, M., Shapley, A. E., et al. 2011, ApJ, 736, 160
Steidel, C. C., Erb, D. K., Shapley, A. E., et al. 2010, ApJ, 717, 289
Steidel, C. C., Kollmeier, J. A., Shapley, A. E., et al. 2002, ApJ, 570, 526
Stewart, K., Maller, A., Oñorbe, J., et al. 2016, ApJ, submitted (arXiv:1606.08542)
Stewart, K. R., Brooks, A. M., Bullock, J. S., et al. 2013, ApJ, 769, 74
Stewart, K. R., Kaufmann, T., Bullock, J. S., et al. 2011, ApJ, 738, 39
Stinson, G. S., Brook, C., Macciò, A. V., et al. 2013, MNRAS, 428, 129
Suresh, J., Bird, S., Vogelsberger, M., et al. 2015, MNRAS, 448, 895
Taniguchi, Y., & Shioya, Y. 2000, ApJ, 532, L13
Taniguchi, Y., Shioya, Y., & Kakazu, Y. 2001, ApJ, 562, L15
Torrey, P., Vogelsberger, M., Genel, S., et al. 2014, MNRAS, 438, 1985
Trainor, R. F., & Steidel, C. C. 2012, ApJ, 752, 39
Tumlinson, J., Thom, C., Werk, J. K., et al. 2011, Science, 334, 948
Turner, M. L., Schaye, J., Crain, R. A., Theuns, T., & Wendt, M. 2016, MNRAS, 462, 2440
Turner, M. L., Schaye, J., Steidel, C. C., Rudie, G. C., & Strom, A. L. 2014, MNRAS, 445, 794
—. 2015, MNRAS, 450, 2067
van de Voort, F., Quataert, E., Hopkins, P. F., et al. 2016, MNRAS, 463, 4533
van de Voort, F., & Schaye, J. 2012, MNRAS, 423, 2991
—. 2013, MNRAS, 430, 2688
van de Voort, F., Schaye, J., Altay, G., & Theuns, T. 2012, MNRAS, 421, 2809
van de Voort, F., Schaye, J., Booth, C. M., Haas, M. R., & Dalla Vecchia, C. 2011, MNRAS, 414, 2458
Vogelsberger, M., Genel, S., Sijacki, D., et al. 2013, MNRAS, 436, 3031
Vogelsberger, M., Genel, S., Springel, V., et al. 2014, MNRAS, 444, 1518
Werk, J. K., Prochaska, J. X., Thom, C., et al. 2013, ApJS, 204, 17
White, S. D. M., & Frenk, C. S. 1991, ApJ, 379, 52
Wisnioski, E., Förster Schreiber, N. M., Wuyts, S., et al. 2015, ApJ, 799, 209
Wisotzki, L., Bacon, R., Blaizot, J., et al. 2016, A&A, 587, A98

Wolfe, A. M., Gawiser, E., & Prochaska, J. X. 2005, ARA&A, 43, 861
Wotta, C. B., Lehner, N., Howk, J. C., O'Meara, J. M., & Prochaska, J. X. 2016, Low-metallicity absorbers account for half of the dense circumgalactic gas at z $\sim$ 1. ApJ, 831, 95. arXiv:1608.02584. doi:10.3847/0004-637X/831/1/95. http://adsabs.harvard.edu/abs/2016ApJ...831...95W
Yang, Y., Zabludoff, A., Tremonti, C., Eisenstein, D., & Davé, R. 2009, ApJ, 693, 1579
Zahedy, F. S., Chen, H.-W., Rauch, M., Wilson, M. L., & Zabludoff, A. 2016, MNRAS, 458, 2423
Zhang, D., Thompson, T. A., Murray, N., & Quataert, E. 2014, ApJ, 784, 93
Zjupa, J., & Springel, V. 2017, Angular momentum properties of haloes and their baryon content in the Illustris simulation. MNRAS, 466, 1625–1647. arXiv:1608.01323 doi:10.1093/mnras/stw2945. http://adsabs.harvard.edu/abs/2017MNRAS.466.1625Z

The Effect of Galactic Feedback on Gas Accretion and Wind Recycling

Freeke van de Voort

1 Introduction

Structure formation in the Universe is governed by gravity. Observations of the cosmic microwave background radiation allow us to measure small fluctuations in temperature, which correspond to small fluctuations in density. These fluctuations grow through gravitational collapse, forming bound structures like galaxies and stars. In the standard cosmological constant cold dark matter (ΛCDM) model, mass assembles hierarchically, with the smallest structures forming first. As the matter in the Universe collapses, it forms a network of sheets and filaments, the so-called cosmic web. While the collapse of dark matter halts as it reaches virial equilibrium in near-spherical halos, baryons can radiate away their binding energy, allowing them to collapse further and fragment into smaller structures. Galaxies form in the densest regions and the most massive galaxies form where filaments intersect. The evolution of galaxies is critically linked to the cosmic web and gas flows in their halos.

The formation of galaxies and, specifically, the accretion of gas onto galaxies are complex problems. There is a variety of evidence showing that galaxies do not sit passively in the densest regions of the cosmic web, but instead have a profound impact on their surroundings. The intergalactic medium (IGM) is enriched with

F. van de Voort (✉)
Theoretical Astrophysics, Heidelberg Institute for Theoretical Studies, Schloss-Wolfsbrunnenweg 35, 69118 Heidelberg, Germany

Academia Sinica Institute of Astronomy and Astrophysics, P.O. Box 23-141, Taipei 10617, Taiwan

Heidelberg Institute for Theoretical Studies, Schloss-Wolfsbrunnenweg 35, 69118 Heidelberg, Germany

Astronomy Department, Yale University, PO Box 208101, New Haven, CT 06520-8101, USA
e-mail: freeke.vandevoort@h-its.org

A. Fox, R. Davé (eds.), *Gas Accretion onto Galaxies*, Astrophysics and Space Science Library 430, DOI 10.1007/978-3-319-52512-9_13

metals, as shown by observations of metal absorption lines in quasar spectra (e.g., Péroux et al. 2005; Turner et al. 2016). The baryon fraction in galaxy groups, as determined from X-ray observations of the hot gas around these galaxies, is much lower than the cosmic mean baryon fraction (e.g., Sun et al. 2009; Gonzalez et al. 2013). The observed cosmic star formation rate (SFR) density is much below the one predicted by cosmological, hydrodynamical simulations without galactic winds (e.g., Schaye et al. 2010).

This indicates that winds emanating from galaxies remove gas from the interstellar medium (ISM) and change the properties of the circumgalactic medium (CGM) and IGM by polluting it with metals and potentially heating it. However, gas inflow and outflow cannot be viewed separately. They interact with each other gravitationally and, more importantly, hydrodynamically. Outflows will plow into static or inflowing gas and sweep up extra material on their way out. This inevitably slows down the outflow until it either joins the warm-hot intergalactic medium (Davé et al. 2001) or forms cold gas clouds through efficient cooling. It can then possibly reverse its trajectory and re-accrete onto the galaxy. This is known as wind recycling and sometimes referred to as a halo fountain or intergalactic fountain (Oppenheimer and Davé 2008; Oppenheimer et al. 2010; Kereš et al. 2009). There is no clear distinction between wind recycling and galactic fountains, but in the case of wind recycling the gas reaches larger distances, out in the halo, before rejoining the galaxy. Another form of gas recycling happens through stellar mass loss, where gas that was previously locked up in stars becomes part of a stellar wind, rejoins the ISM, and can collapse again to form stars (see, e.g., Segers et al. 2016, and the references therein). This chapter, however, is concerned with wind recycling between galaxies and their halos.

Figure 1 shows the gas density in a $(450\,\mathrm{kpc})^3$ volume in and around a $10^{12}\,\mathrm{M}_\odot$ halo at $z = 2$. The halo shown is being fed primarily by dense, clumpy filaments and as well as by external galaxies embedded in the filaments. Gas outside the halos is, in general, moving towards the halo and relatively unenriched. Within the virial radius, however, most of the underdense regions are filled with diffuse, outflowing gas. The inflowing gas is generally associated with the dense, cold streams, which can penetrate the halo and feed the central disc, although the hot, diffuse gas can also cool and accrete onto the galaxy (e.g., Kereš et al. 2005; van de Voort et al. 2011a). At lower redshift, filaments are less pronounced and cooling from the hot halo is fractionally more important. The virial radius only approximately marks the boundary between the unaffected IGM and the enriched and heated CGM and many studies find effects of galaxy formation out to $2R_{\mathrm{vir}}$ or beyond (e.g., Wetzel et al. 2012; Rasmussen et al. 2012; Bahé et al. 2013).

Satellites contribute directly to the growth of the central galaxy through merging. Gas previously processed in and ejected or stripped from companion galaxies can also accrete onto the main galaxy. It is a matter of definition whether or not to consider this gas wind recycling. One argument in favor of including it is that the gas has had its properties altered through feedback. Not including it may be a better choice when interested in the return of gas to the same galaxy in order to quantify the efficiency of galactic winds.

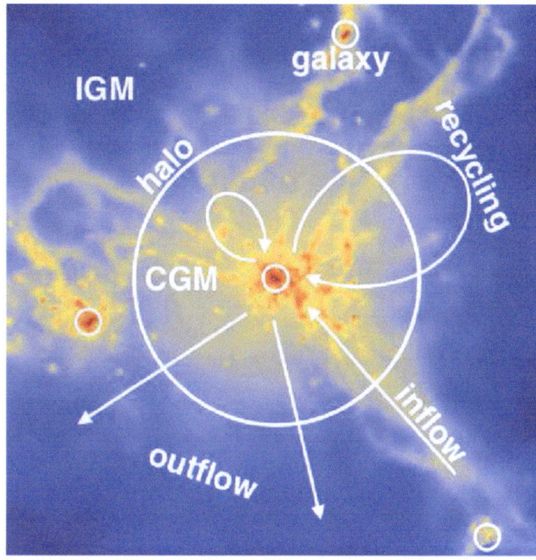

Fig. 1 Gas density in a $(450\,\text{kpc})^3$ volume centered on a $10^{12}\,\text{M}_\odot$ halo at $z = 2$. The *large white circle* indicates the virial radius and the *small white circles* show the locations of several galaxies. Gas inflow happens primarily via dense filaments connected to the cosmic web. Outflows escape along the path of least resistance in underdense regions. Wind recycling can take place within the halo or the ejected gas can escape the halo and re-accrete onto the galaxy on a longer time-scale

This chapter discusses our current understanding of how gas accretes and re-accretes onto galaxies and how feedback affects the accretion rate and changes the properties of the galaxy and its CGM. Section 2 discusses some theoretical background of galaxy and halo formation that is necessary to understand gas accretion. We discuss three methods in Sect. 3 that are used to study galaxy formation, semi-analytic models, equilibrium models, and hydrodynamical simulations. We focus on hydrodynamical simulations in Sect. 4 and describe how feedback affects the accretion rate onto galaxies and properties of the accreting gas and how important wind recycling is. Finally, we compare and discuss the results from the literature and conclude in Sect. 5.

2 Virial Relations

The following discusses a highly idealized, but nonetheless useful case of spherical collapse in a flat universe, where the radiation density parameter Ω_γ, the matter density parameter Ω_m, and the cosmological constant or vacuum density parameter Ω_Λ add up to one, $\Omega_\gamma + \Omega_\text{m} + \Omega_\Lambda = \Omega = 1$.

$$\Omega \equiv \frac{\langle\rho\rangle}{\rho_\text{crit}} = \frac{8\pi G\langle\rho\rangle}{3H^2}, \tag{1}$$

with $\rho_{\rm crit}$ the redshift-dependent density at which the universe is flat, or "critically bound." G is the gravitational constant, $\langle\rho\rangle$ the mean density of the universe, and H the Hubble constant. The mean cosmic matter density is

$$\langle\rho_m\rangle = \Omega_{\rm m}\rho_{\rm crit} = \Omega_{\rm m}(1+z)^3\rho_{\rm crit,0} = \Omega_{\rm m}(1+z)^3\frac{3H_0^2}{8\pi G}, \tag{2}$$

where the subscript 0 is used to indicate the present-day value. The density at which objects collapse or virialize can be calculated analytically for an Einstein-de Sitter universe (Padmanabhan 2002), but it holds for more general cosmologies (Bryan and Norman 1998).

$$\rho_{\rm coll} \approx 18\pi^2\langle\rho_m\rangle \tag{3}$$

Assuming spherical collapse, the mass of this halo then becomes

$$M_{\rm halo} = \frac{4\pi}{3}\rho_{\rm coll}R_{\rm vir}^3. \tag{4}$$

This leads to an expression for the virial radius

$$R_{\rm vir} \approx \left(\frac{2GM_{\rm halo}}{H_0^2\Omega_{\rm m}18\pi^2}\right)^{1/3}\frac{1}{1+z} \tag{5}$$

or

$$R_{\rm vir} \approx 3.4\times10^2\ {\rm kpc}\left(\frac{M_{\rm halo}}{10^{12}\ {\rm M}_\odot}\right)^{1/3}\frac{1}{1+z}. \tag{6}$$

$R_{\rm vir}$ approximately marks the location of an accretion shock, within which the shocked gas is in virial equilibrium with the dark matter.

The virial theorem states that the total kinetic energy, K, of a system is equal to minus one half of the average total potential energy, U.

$$K = -\frac{1}{2}U \tag{7}$$

Taking $K = \frac{1}{2}mv_{\rm c}^2$ and $U = -\frac{GM_{\rm halo}m}{R_{\rm vir}}$, with m the mass of a test particle, the circular velocity, $v_{\rm c}$, of the system is

$$v_{\rm c}^2 = \frac{GM_{\rm halo}}{R_{\rm vir}}. \tag{8}$$

In a monatomic ideal gas, the thermal energy of the particles is equal to their kinetic energy

$$\frac{3}{2}k_{\rm B}T = \frac{1}{2}\mu m_{\rm H} v_{\rm c}^2, \tag{9}$$

where $k_{\rm B}$ is Boltzmann's constant and $m_{\rm H}$ the mass of a proton. For a fully ionized plasma with primordial composition, the mean molecular weight $\mu \approx 0.59$. Adding heavier elements does not change this quantity significantly. Combining Eqs. (7) and (9) gives an expression for the virial temperature of a monatomic ideal gas

$$T_{\rm vir} = \frac{\mu m_{\rm H} G M_{\rm halo}}{3 k_{\rm B} R_{\rm vir}}. \tag{10}$$

Filling in the expression for the virial radius [Eq. (5)] one gets

$$T_{\rm vir} = \left(\frac{G^2 H_0^2 \Omega_{\rm m} 18\pi^2}{54}\right)^{1/3} \frac{\mu m_{\rm H}}{k_{\rm B}} M_{\rm halo}^{2/3}(1+z) \tag{11}$$

or

$$T_{\rm vir} \approx 3.0 \cdot 10^5\ {\rm K}\ \left(\frac{\mu}{0.59}\right)\left(\frac{M_{\rm halo}}{10^{12}\ {\rm M}_\odot}\right)^{2/3}(1+z)\,. \tag{12}$$

There is an uncertainty of a factor of a few because of the assumption we made for the average total potential energy. The virial temperature of a halo is completely determined by its mass and redshift, for given cosmological parameters. Due to adiabatic compression, the temperature of the hot, quasi-hydrostatic halo gas can increase for $R < R_{\rm vir}$, but this is only a minor effect. Feedback from supernovae or supermassive black holes can also heat the gas to above $T_{\rm vir}$ (e.g., van de Voort et al. 2016; Fielding et al. 2016).

The gas density, ρ, increases by orders of magnitude from $R_{\rm vir}$ towards the center of the halo. Assuming that the density profile is approximately self-similar for different halo masses, it follows that the pressure of the gas also increases roughly as $M_{\rm halo}^{2/3}$ at fixed $R/R_{\rm vir}$, since $P = k_{\rm B}T\rho/\mu m_{\rm H}$. At fixed radius R, the pressure scales linearly with $M_{\rm halo}$. It is therefore expected that galactic winds, if launched at the same velocity, slow down more quickly in more massive halos, partially because of their deeper potential wells, but also because of the much higher ram pressure forces.

2.1 Cooling Time

Gas falling towards a galaxy gains kinetic energy and in order for it to join the galaxy's ISM, it has to lose this excess kinetic energy. If the gas' velocity is supersonic when it reaches the hydrostatic halo, it will experience a shock and heat to the virial temperature of the halo. According to the simplest picture of spherical collapse, all gas in a dark matter halo is heated to the virial temperature of that halo, reaching a quasi-static equilibrium supported by the pressure of the hot gas.

Within the so-called cooling radius, the cooling time of the gas will, however, be shorter than the age of the Universe. If the cooling radius lies well inside the halo, which is the case for high-mass halos, a quasi-static, hot atmosphere will indeed form. Accretion onto the galaxy is then regulated by the cooling function. If, on the other hand, this radius is larger than the virial radius, then there will be no hot halo and the gas will not go through an accretion shock at the virial radius (Rees and Ostriker 1977; White and Rees 1978; Birnboim and Dekel 2003; Kereš et al. 2005). The accretion rate onto the central galaxy then depends on the infall rate, but not on the cooling rate (White and Frenk 1991). A clumpy or filamentary structure of the IGM and CGM will change this simplified picture and result in only part of the infalling gas shock heating at $R_{\rm vir}$ (Kereš et al. 2005; Dekel and Birnboim 2006).

Gas with a temperature below about 10^6 K can cool down rapidly, through line emission. Cooling occurs mostly through particle collisions and the cooling rate therefore depends on the number densities of both the colliding particle species. These number densities depend on the density of the gas. The cooling time depends on the density via

$$t_{\rm cool} = \frac{u}{\Lambda} \propto \frac{\rho}{\rho^2} = \frac{1}{\rho} \tag{13}$$

where $t_{\rm cool}$ is the cooling time, Λ the cooling function, and u the thermal energy density, equal to $\frac{3}{2}nk_{\rm B}T$ and therefore proportional to the density of the gas. The cooling time is thus inversely proportional to the density of the cooling gas.

The dependence of the cooling function on temperature and metallicity is shown in Fig. 2. The curves show the cooling function for an optically thin gas with hydrogen number density $n_{\rm H} = 0.001\,{\rm cm}^{-3}$ in photo-ionization equilibrium with the Haardt and Madau (2001) UV background. Different curves have different metallicities, primordial (solid), 10% of the solar metallicity (dotted), and solar metallicity ($Z_\odot$; dashed). At $Z_\odot$ cooling through metal-line emission dominates except at $T \approx 10^4$ K, where hydrogen lines are most important, and at $T > 10^7$ K, where Bremsstrahlung takes over. Cooling through metal lines becomes less important with decreasing metallicity and at $0.1Z_\odot$ changes Λ by less than a factor of two at most temperatures. At higher redshift, the density of the CGM is higher and the average cooling time is therefore shorter. Because galactic feedback changes the temperature, density, and metallicity of the CGM, it has a strong effect on the

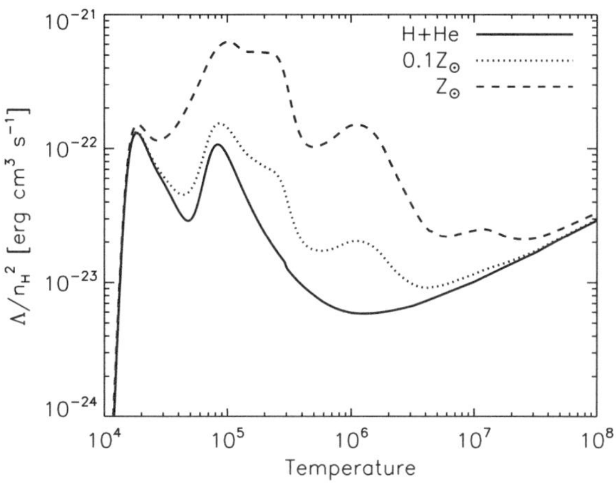

Fig. 2 Cooling function for optically thin gas with hydrogen number density $n_{\rm H} = 0.001\,{\rm cm}^{-3}$ in photo-ionization equilibrium with the Haardt and Madau (2001) UV background. The *solid*, *dotted*, and *dashed curves* show the cooling function for gas with primordial composition, 10% solar metallicity, and solar metallicity, respectively. Adapted from Wiersma et al. (2009)

cooling of the halo gas. Because gas that is ejected from a galaxy has a higher metallicity than average, it will cool more efficiently (at fixed temperature and density) and possibly re-accrete onto the galaxy.

3 Methods to Probe the Gas Cycle

Observations of groups of galaxies have revealed that their gas and baryon fractions are substantially below the cosmic mean baryon fraction, whereas clusters of galaxies have fractions much closer to the cosmic mean (e.g., Sun et al. 2009; Vikhlinin et al. 2009; Gonzalez et al. 2013). The reason for this is likely that feedback from galaxies removes a substantial fraction of gas from group-sized halos. Simulations predict that the baryon fraction decreases further towards lower halo mass, where it is easier for stellar feedback to expel gas from the halo (e.g., Chan et al. 2015; Christensen et al. 2016; van de Voort et al. 2016). The gas and baryon fractions are also lower at lower redshift, after powerful outflows remove a large amount of gas from the halo (Muratov et al. 2015; van de Voort et al. 2016). This results in halos of different mass not being self-similar and also directly effects observables such as the X-ray luminosity and the magnitude of the Sunyaev-Zel'dovich effect (Crain et al. 2010; Le Brun et al. 2014; van de Voort et al. 2016).

3.1 Semi-Analytic Models

The technique of modeling galaxy evolution on top of a dark matter-only N-body simulation is known as "semi-analytic." Semi-analytic models combine dark matter halo merger trees with analytic recipes for baryonic processes. A clear advantage over doing full hydrodynamical simulations is their lower computational cost. Galaxy formation is too complex to construct an ab initio model. Therefore, it is possible to conduct a large parameter study in order to test how observations of, for example, the galaxy stellar mass function can best be reproduced.

Even the earliest semi-analytic models included feedback from supernovae in order to suppress the efficiency of galaxy formation and flatten the low-mass end of the luminosity function. The details of how winds are driven and stalled and thus the time-scale over which the gas ejected from the galaxy can return has varied. Some models assume the ejected ISM gas becomes part of the hot halo, where it immediately becomes available for cooling and re-accreting onto the galaxy, whereas others assume it gets ejected into an external IGM reservoir for a certain amount of time, before it rejoins the hot halo (e.g., De Lucia et al. 2004). There are also models that assume that the gas that escapes from the halo and reaches the Hubble flow never accretes onto the halo or galaxy again (Bertone et al. 2007).

Although these models do not directly trace wind recycling back into the galaxy, it can be informative to take a look at the time it takes before the ejected gas is reincorporated into the hot halo. Henriques et al. (2013) used Monte Carlo Markov Chain methods to find a redshift-independent reincorporation time-scale that is inversely proportional to halo mass: $t_{\rm reinc} = 1.8 \times 10^{10}\,(10^{10}\,{\rm M}_{\odot}/M_{\rm halo})$ yr. This means that dwarf galaxies with $M_{\rm halo} \lesssim 10^{10}\,{\rm M}_{\odot}$ retain few of their ejected baryons and therefore have little wind recycling, whereas massive galaxies with $M_{\rm halo} \gtrsim 10^{12}\,{\rm M}_{\odot}$ have very short reincorporation times and the halo baryon fraction increases with halo mass, as observed (e.g., Lin et al. 2012). Even though there is no explicit redshift dependence, ejection is most efficient at high redshift for halos at fixed mass, because cooling, star formation, and feedback are stronger on average. This results in a suppression of the formation of low-mass galaxies at early times and a more efficient build-up at later times, as observed via the galaxy stellar mass function. We will compare this time-scale to those derived from other models in Sect. 5.

3.2 Equilibrium Models

The interpretation of semi-analytic models and cosmological, hydrodynamical simulations can be difficult, because of their high level of complexity. An alternative approach is to employ simplified models based on the idea that galaxies self-regulate their growth, in a way that gas inflow and outflow approximately balance (e.g., Finlator and Davé 2008; Bouché et al. 2010; Davé et al. 2012; Lilly et al. 2013; Dekel and Mandelker 2014). This is in part motivated by the observation that star-forming galaxies lie on a fairly tight relation between their SFR and stellar mass, which means that there appears to be an approximate equilibrium between gas supply and gas removal (e.g., Noeske et al. 2007; Lee et al. 2015).

Equilibrium models start with the assumption that the equilibrium between accretion, star formation, and galactic winds evolves slowly. Galaxy mergers are generally not taken into account, but could set the scatter in galaxy scaling relations (Mitra et al. 2017). The accretion rates are based on those derived at the virial radius from dark matter-only simulations and modified by a mass-dependent parameter that quantifies the fraction of this gas that accretes onto a galaxy. Gas ejected from the ISM is usually parameterized as a mass loading factor in units of the galaxy's SFR. Recycling can be taken into account by either reducing the mass loading factor (e.g., Dekel and Mandelker 2014) or adding an additional term to the accretion rate (e.g., Mitra et al. 2015). The recycling material is assumed to have the same metallicity as the gas in the ISM.

Matching the model parameters to reproduce stellar masses, SFRs, or metallicities in observations or simulations can give a handle on the importance of wind recycling. Finlator and Davé (2008) find that roughly half of the accreting gas has gone through wind recycling at $z = 2$. Dekel and Mandelker (2014) find that even with their most efficient wind recycling, SFRs at $z = 2$ are still under-predicted and argue for a more accurate treatment of the re-accretion of gas. Mitra et al. (2015) find that they can match observables well with parameters that depend on halo mass and redshift. They find that the gas recycling time-scale decreases with increasing halo mass (or stellar mass) and increasing redshift, but with a relatively weak dependence, $t_{\rm recycle} \propto M_{\rm halo}^{-0.45}$. We will discuss this time-scale further in Sect. 5.

3.3 Hydrodynamical Simulations

Full hydrodynamical, cosmological simulations broadly exist in two categories. One in which the volume is discretized (mesh-based simulations) and one in which the mass is discretized (particle-based simulations). Particle-based simulations have the ability to trace the trajectory of gas parcels and can therefore determine unambiguously when gas returns to a galaxy after having been ejected. Simulations show an increase in the gas fraction within $R_{\rm vir}$ towards more massive halos (e.g., Crain et al. 2010; McCarthy et al. 2016; van de Voort et al. 2016). This is a direct consequence of feedback removing gas more efficiently from low-mass halos. This already provides evidence for the increase of wind recycling in more massive galaxies.

Cosmological simulations have limited resolution and therefore have to employ sub-grid models in order to simulate the large-scale effects of much smaller scale phenomena, such as supernovae or active galactic nuclei (AGN). The parameters of these models are not known a priori, but can be tuned in order to reproduce certain observables, such as the galaxy stellar mass function (e.g., Vogelsberger et al. 2014; Schaye et al. 2015), although there may be many degeneracies. The importance of wind recycling depends strongly on the chosen sub-grid feedback model. Winds that are generated with a higher velocity, momentum, or temperature will travel farther away from the galaxy. Winds that are experiencing full hydrodynamical forces, such as ram pressure, will slow down more than those that have been "decoupled" and travel ballistically for some distance or time (e.g., Oppenheimer et al. 2010). These differences likely result in the variety of recycling times found in the literature, as discussed in Sect. 5.

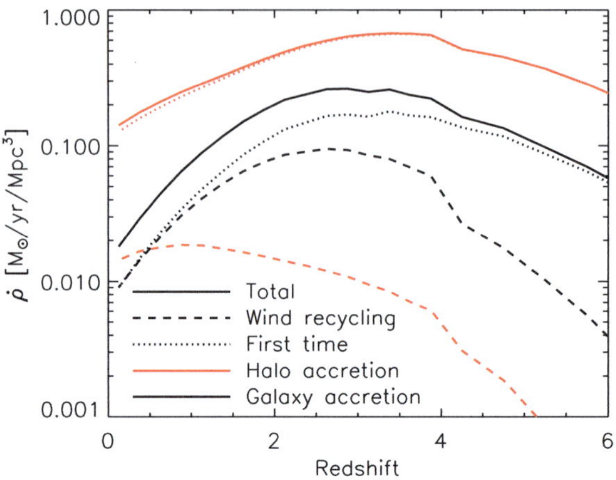

Fig. 3 Evolution of the global accretion rate densities in a (137 Mpc)^3 cosmological simulation with supernova and AGN feedback. The *solid*, *black* (*red*) *curves* show global accretion rate densities onto all galaxies (halos). *Dashed curves* show accretion rate densities resulting from wind recycling, *dotted curved* from gas that was never before in a galaxy's ISM ("first time"). The small "step" visible at $z \approx 4$ is caused by the sudden increase in the time resolution for determining accretion (the time between snapshots decreases by a factor of 2 at $z = 4$). The fraction of accretion onto halos due to wind recycling increases to almost 10% at $z = 0$ and is therefore clearly subdominant. At high redshift, wind recycling is also negligible for gas accretion onto galaxies. However, because of outflows driven by supernovae and AGN, its importance increases over time and rivals first time accretion at $z \approx 0$. Adapted from van de Voort et al. (2011b)

Another interesting question arises when we consider gas that was near the center of the halo and would have accreted onto the galaxy, but was pushed out by galactic outflows. This gas can later accrete, but since it was never actually part of any galaxy, it is not considered wind recycling. However, one could argue that the difference between "true" wind recycling that originates from a galaxy and the swept-up gas around the galaxy that accretes at a later time is somewhat artificial. However, the gas will be more enriched if it is ejected from the ISM and therefore have a different impact on galaxy formation.

4 The Importance of Feedback and Wind Recycling

Using hydrodynamical simulations it is possible to follow a parcel of gas as it flows into and out of a galaxy. As mentioned above, some studies consider gas to be wind recycling when they accrete onto the same galaxy for the second time, whereas others consider this the case when they re-accrete onto any galaxy. Usually studies that use zoom-in simulations to focus on individual galaxies follow the former approach, whereas full cosmological simulations tend to study wind recycling statistically and therefore follow the latter approach.

Figure 3 shows the global accretion rate density onto halos (solid, red curve) and galaxies (solid, black curve), i.e., the gas mass accreting onto halos and galaxies, respectively, per year and per comoving Mpc^3. These global accretion rate densities are averaged over the time interval between two snapshots. The global accretion rate onto halos is substantially higher than the accretion rate onto galaxies, indicating that most of the gas that accretes onto halos never participates in star formation.

We split the global accretion rate into separate contributions from gas that was never in a galaxy before (here labeled 'first time) and gas that was previously in a galaxy's ISM, but was ejected ("wind recycling"). Note that although the wind recycling gas is re-accreting, this could be onto a different galaxy than the one it was ejected from. Accretion onto halos is dominated by first time accreting gas at all times. Although accretion through wind recycling does become more important towards lower redshift, its accretion rate density is still 1 dex lower at $z = 0$. This means that the direct effect of feedback and gas processing in galaxies are only of minor importance for accretion onto halos. At very high redshift, the first time accretion rate density onto galaxies is about an order of magnitude higher than the wind recycling accretion rate density. This difference decreases gradually and vanishes by $z = 0$, even though first time accretion still dominates the growth of halos by 1 dex. This means that accretion onto galaxies through wind recycling happens efficiently and the majority returns to the galaxy without first leaving the halo.

Once the gas is in the ISM, wind recycling and first time accreted gas are treated equally. The global SFR density therefore follows the galaxy accretion rate density, although with a delay resulting from the time it takes to convert the interstellar gas into stars. The gas can, however, be removed from the ISM by supernova and AGN feedback, as well as by dynamical processes. This is why the overall normalization of the SFR is generally lower than the ISM accretion rate.

Figure 4 shows the evolution of the global SFR densities for a model without (solid, black curve) and with (dashed, black curve) efficient stellar feedback (Oppenheimer et al. 2010). Green curves show the global SFR density resulting from gas that has been accreted multiple times (wind recycling). Blue (red) curves show the global SFR density from gas that has only accreted once with a maximum past temperature of less (more) than 2.5×10^5 K. Wind recycling becomes more important at later times, in agreement with Fig. 3. It dominates the global SFR after $z = 1.7$ and is about an order of magnitude higher than first time accretion at $z = 0$. This simulation leads to the conclusion that the vast majority of the late-time gas accretion onto galaxies (and resulting star formation) is gas that was previously ejected from a galaxy. This difference with Fig. 3 is due to a different treatment of the sub-grid feedback and the subsequent evolution of the ejected gas. Compared with the observed SFR density (Hopkins and Beacom 2006) (gray dashed lines), the feedback model in Fig. 4 over-predicts the amount of star formation after $z \approx 1$, which is primarily a result of excessive star formation in massive galaxies. Suppressing some of the low redshift gas accretion and star formation with more efficient feedback could reduce the discrepancy with Fig. 3, although it is unlikely

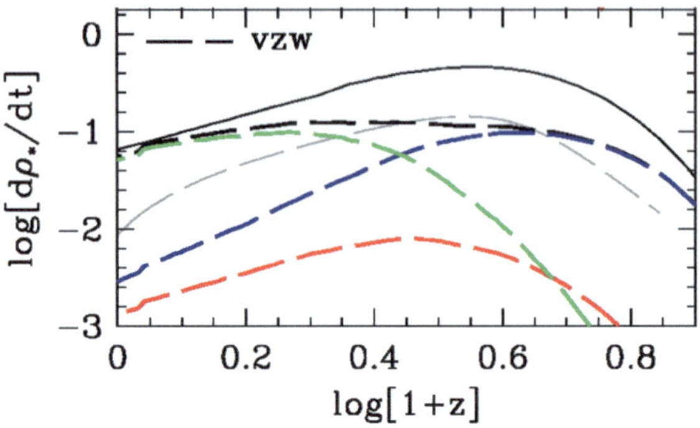

Fig. 4 Evolution of the global SFR densities in a $(69\,\mathrm{Mpc})^3$ cosmological simulation with (*dashed, black curve*) and without (*solid, black curve*) efficient stellar feedback (named "vzw"). The *dashed gray curve* shows the global SFR rate density derived from observations (Hopkins and Beacom 2006). *Colored curves* add up to the *dashed, black curve*. *Green curves* show the global SFR density resulting from gas that has been accreted multiple times (wind recycling). *Blue* (*red*) *curves* show the global SFR density from gas that has only accreted once with a maximum past temperature of less (more) than 2.5×10^5 K. At high redshift, most stars form from gas accreted for the first time, as also seen in Fig. 3. The importance of wind recycling increases with decreasing redshift and it dominates by about an order of magnitude at $z = 0$. This difference with Fig. 3 is due to a different treatment of the sub-grid feedback and the subsequent evolution of the ejected gas. Adapted from Oppenheimer et al. (2010)

to resolve it completely and shows how sensitive the quantitative results are to feedback implementations.

One can also pose the reverse question, not what fraction of accreted gas accretes through wind recycling, but what fraction of gas ejected from galaxies and never re-accreted. Christensen et al. (2016) find that this is about 50% of the mass in their zoom-in simulations (with $M_{\mathrm{halo}} < 10^{12}\,\mathrm{M}_{\odot}$).

4.1 Ejective and Preventive Feedback

Naively one would expect feedback from stars and black holes to reduce the amount of gas in the ISM and potentially also reduce the gas accretion rate (through the interaction between inflowing and outflowing gas). However, this is not necessarily the case because as gas is ejected or prevented from accreting at high redshift, it is still available for feeding galaxies at low redshift. In the absence of galactic winds, most baryons remain in the galaxy they first accreted onto. This means that a lot of baryons are already "locked up" at high redshift in relatively low-mass galaxies. However, if this gas is ejected from the galaxies (or never accreted in the first place) it remains available to fuel massive galaxies at low redshift (van de Voort et al. 2011a; Faucher-Giguère et al. 2011, e.g.,).

In Fig. 5 the specific accretion rate onto central galaxies as shown as a function of halo mass for four different simulations at $z = 2$ (black curves) and $z = 0$

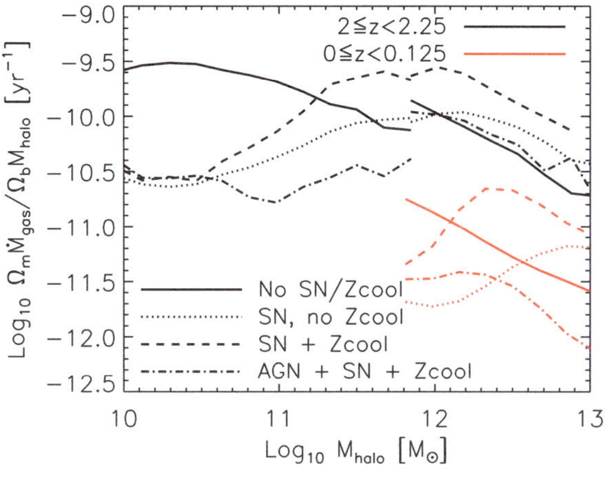

Fig. 5 Specific gas accretion rates onto galaxies against halo mass for different simulations at $z = 2$ (*black curves*) and $z = 0$ (*red curves*). The *curves* at low (high) halo masses are derived from simulations with higher (lower) resolution in smaller (larger) volumes. In the absence of supernova feedback (No SN/Zcool; *solid curves*), the specific accretion rate onto galaxies declines with halo mass, indicating that gas accretes less efficiently onto galaxies in higher mass halos. Efficient supernova feedback (SN, no Zcool; *dotted curves*) reduces the accretion rates by up to an order of magnitude for galaxies in low-mass halos, resulting in a peak in the specific gas accretion rate at $M_{\rm halo} \approx 10^{12}\,\rm M_{\odot}$. It increases the accretion rates at high masses, because less gas has been locked up in smaller galaxies at higher redshift and is still available for accretion at low redshift. Including metal-line cooling (SN + Zcool; *dotted curves*) increases the accretion rate at all masses, but most strongly for $M_{\rm halo} \approx 10^{12}\,\rm M_{\odot}$ ($T_{\rm vir} \approx 10^6$ K). Efficient AGN feedback (AGN + SN + Zcool; *dot-dashed curves*) reduces the accretion rates for the highest halo masses, also by up to an order of magnitude. Adapted from van de Voort et al. (2011a)

(red curves). The curves at low (high) halo masses are derived from simulations with higher (lower) resolution in smaller (larger) volumes. The solid curves show simulations without supernova feedback and metal-line cooling, the dotted curves include supernova feedback, but no metal-line cooling, the dashed curves include both supernova feedback and metal-line cooling, and the dot-dashed curves use simulations that additionally include AGN feedback (as in Fig. 3).

The specific accretion rates onto galaxies decrease with halo mass in the absence of galactic winds (even though the specific accretion rates onto halos stay roughly constant), because cooling becomes less efficient at higher temperatures (in more massive halos). Supernova feedback reduces the galaxy accretion rate onto low-mass galaxies strongly by up to an order of magnitude. On the other hand, it actually increases the accretion rate onto massive galaxies, because less gas has been locked up in smaller galaxies at earlier times. The inclusion of metal-line cooling enhances the gas accretion rate, showing that, besides gravity, cooling sets the rate at which gas can accrete. AGN feedback reduces the galaxy accretion rate onto high-mass halos strongly, again by up to an order of magnitude. In these feedback models, the specific galaxy accretion rate peaks around $10^{12}\,\rm M_{\odot}$, where galaxy formation proceeds most efficiently.

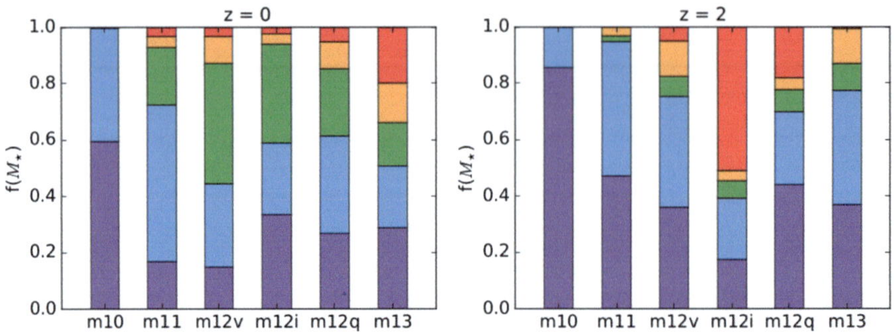

Fig. 6 Fraction of stellar mass at $z = 0$ (*left*) and $z = 2$ (*right*) contributed by gas accreted only once (*purple*), gas re-accreting onto the same galaxy (*blue*), gas re-accreted onto a different galaxy (*green*), gas accreted via galaxy mergers (*orange*), and stars accreted via galaxy mergers (*red*) for 6 simulated, central galaxies in order of increasing halo mass at $z = 0$ (from 7.8×10^9 to 1.4×10^{11} to 6.3×10^{11} to 1.1×10^{12} to 1.2×10^{12} to $6.1 \times 10^{12}\,M_{\odot}$). Gas accreted for the first time is responsible for less than half the stellar mass for all galaxies except the lowest mass dwarf. The re-accretion of gas becomes more important towards lower redshift, especially gas processed and ejected in external galaxies (mostly satellites) before accreting onto the central and forming stars. Adapted from Anglés-Alcázar et al. (2016)

Internal feedback processes are not the only ones that can prevent gas accretion onto galaxies. External processes can do this as well. The environment of a galaxy is thought to be primarily determined by properties of the dark matter-dominated halo it lives in (e.g., Crain et al. 2009; Blanton and Berlind 2007; Wilman et al. 2010). The removal of a satellite galaxy's gaseous halo can result in the suppression of gas accretion (and a resulting decline of the SFR), also known as "starvation" or "strangulation" (e.g., Larson et al. 1980; Balogh et al. 2000). Using hydrodynamical simulations, Simha et al. (2009) indeed found lower gas accretion rates for satellites in massive halos compared to centrals of the same mass. The suppression of the gas accretion rate increases with host halo mass (van de Voort et al. 2017). ISM gas lost through ram pressure stripping or galactic winds will also be less likely to re-accrete onto satellites, but is an important fraction of the total gas accretion onto central galaxies at low redshift (Anglés-Alcázar et al. 2016).

Figure 6 shows the fraction of the stellar mass formed in different ways for individual central galaxies from cosmological zoom-in simulations at $z = 0$ (left) and $z = 2$ (right). From bottom to top: gas that accreted for the first time is shown in purple, gas that re-accreted onto the galaxy after being ejected from the same galaxy is shown in blue, gas that accreted onto the central after having been ejected from surrounding (satellite) galaxies is shown in green, gas that is brought in through galaxy mergers is shown in orange, and stars formed in other galaxies and accreted through mergers is shown in red. Gas previously ejected from external galaxies becomes more important with decreasing redshift, while the fraction of gas accreted only once decreases. The fraction of gas that accreted onto the same galaxy is an important path through which galaxies grow at both high and low redshift.

4.2 The Effect of Feedback on the Properties of Accreting Gas

Gas accretion is difficult to observe directly. Besides correctly reproduced galaxy masses and SFRs, one major test of galaxy formation models is whether or not they can reproduce the metallicity distribution in the Universe. This is one of the main reasons why wind recycling is of great importance for understanding galaxy evolution. The distance out to which ejected gas travels determines the metallicity of the CGM and IGM. The time-scale on which it re-accretes onto galaxies plays an important role in setting the metallicity of the ISM and of the resulting stars that form. For example, a decrease of the infall of pristine gas can explain the fact that the metallicity is observed to increase with decreasing SFR at fixed stellar mass (Mannucci et al. 2010; Davé et al. 2011; Peng et al. 2015; Lagos et al. 2016; Kacprzak et al. 2016; Bahé et al. 2017).

Gas around galaxies has been detected in observations in absorption in quasar sightlines and in emission in hydrogen Lyman-α, O VI, and soft X-rays (e.g., Péroux et al. 2005; Turner et al. 2016; Steidel et al. 2011; Hayes et al. 2016; Anderson et al. 2015). However, it is difficult to determine whether this gas is accreting, outflowing, or static. Comparing results to hydrodynamical simulations is essential for their interpretation. van de Voort et al. (2012) used simulations to show that the vast majority of high-column density H I absorbers ($N_{\rm HI} \gtrsim 10^{17}\,{\rm cm}^{-2}$) at $z = 3$ are associated with galaxy halos. 60–70% of this neutral gas accretes onto a galaxy by $z = 2$, so within a Gyr. However, only roughly half of the accreting gas did so for the first time. Accreting gas will therefore not necessarily be metal-poor and it can be difficult to separate inflows and outflows based on their metallicity, except in the most extreme cases (Hafen et al. 2016).

Ford et al. (2014) find that only 3% of the low-redshift CGM and IGM is re-accreting onto galaxies after having been ejected from a galaxy, yet this recycling material contains 35% of the metal mass. The rest of the metal mass is primarily contained in previously ejected gas that is not accreting onto galaxies. The former accounts for most of the low-ionization metal-line absorbers, such as Mg II, while the latter accounts for the high-ionization absorbers, since it generally has a higher temperature. This suggests that accreting gas has likely already been detected via low-ionization absorption and wind recycling gas via low-ionization metal-line absorption (Ford et al. 2014). This identification can perhaps only be achieved statistically, because disentangling these from cool outflows is difficult. However, there are observations that show enhanced Mg II absorption along the major and minor axis of galaxies, potentially showing evidence for biconical outflows along the minor axis and recycled wind accretion along the major axis (Kacprzak et al. 2012).

Figure 7 shows the flux-weighted median gas overdensity (top, left panel), temperature (top, right panel), metallicity (bottom, left panel), and mean accretion rate (bottom, right panel) of all inflowing gas, i.e., gas with negative radial velocity with respect to the central galaxy, as a function of radius for halos with $10^{11.5} < M_{\rm halo} < 10^{12.5}\,{\rm M}_{\odot}$ at $z = 0$. The solid curves show the median values for all

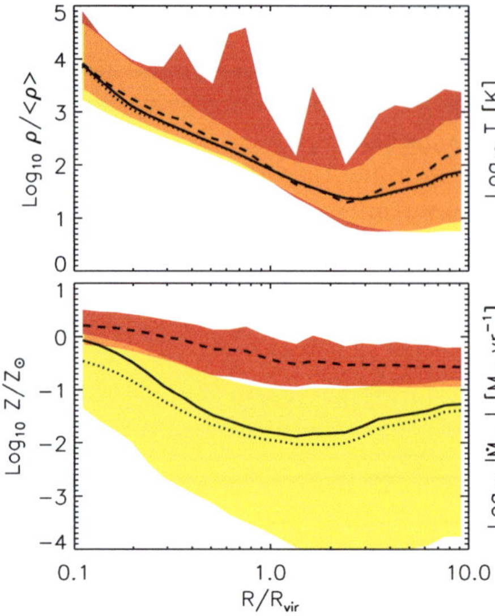

Fig. 7 Properties of gas in halos with $10^{11.5} < M_{\mathrm{halo}} < 10^{12.5}\,\mathrm{M}_\odot$ at $z = 0$ as a function of radius for all inflowing gas (*solid curves*), gas flowing in for the first time (*dotted curves*), and wind recycling gas (*dashed curves*). The *curves* show the median values, except for the *bottom, right panel* which shows the mean value. *Shaded regions* show values within the 16th and 84th percentiles, i.e., the $\pm 1\sigma$ scatter around the median, for first time accreting and wind recycling gas. *From the top-left to the bottom-right*, the different panels show the flux-weighted median gas overdensity, temperature, metallicity, and the mean accretion rate. A small fraction of the wind recycling gas has higher densities than average and most of it has a substantially lower temperature inside R_{vir}. Wind recycling gas has much higher metallicity than gas accreting for the first time, especially around R_{vir}. First time accreting gas dominates at all radii, but wind recycling contributes substantially in the centers of these halos and therefore to the accretion rate onto galaxies. Adapted from van de Voort and Schaye (2012)

accreting gas, except for the bottom, right panel which shows the mean value. The dashed (dotted) curves show the properties for wind recycling (first time accreting) gas, i.e., gas which was (not) previously in the ISM of a galaxy. The shaded regions show values within the 16th and 84th percentiles. Wind recycling gas at radii larger than about $2R_{\mathrm{vir}}$ is dominated by gas associated with external galaxies and gas flowing towards our selected Milky Way-mass galaxies are likely flowing out of other galaxies.

At and beyond the virial radius, only about 10% of accretion is due to wind recycling. This reflects the cosmic average fraction of gas ejected from galaxies by $z = 0$. However, it becomes more important at smaller radii until it reaches about 50% at $0.1R_{\mathrm{vir}}$, consistent with the global accretion rate density in Fig. 3. The gas density is similar for accreting and re-accreting gas, although the latter has some high-density peaks of gas, which has likely recently been ejected from satellite galaxies. The differences are much larger for the gas temperatures. Inflowing wind recycling gas has a higher temperature outside R_{vir}, likely because it has recently been heated by feedback from other galaxies. Inside R_{vir} the median temperature drops below the virial temperature and the difference between the temperature of gas flowing in for the first time and wind recycling gas is more than an order of magnitude around $0.4R_{vir}$. Since the densities are similar, the temperature difference is primarily a result of the metallicity difference. At all radii, the metallicity of wind recycling gas is substantially higher than that of gas accreting for the first time. The difference is largest around R_{vir} ($\approx$1.5 dex), but is still large ($\approx$0.7 dex at $0.1R_{vir}$). This shows that wind recycling is essential for accretion onto galaxies as well as for the temperature and metallicity structure of the CGM.

The inclusion of galactic feedback can result in more realistic galaxies and enhance disc formation. Galactic outflows due to stellar winds originate from the sites of star formation. Since the SFR is highest in the centers of galaxies, ejected gas preferentially comes from the center as well and thus has low angular momentum. Disc formation can then proceed in one of two ways. Either low angular momentum gas is removed from the galaxy completely, or the gas that re-accretes does so with higher angular momentum than when it was ejected. No consensus has been reached on which is more important. Übler et al. (2014) find no change in angular momentum for re-accreting gas from the time it was ejected. In their simulations, feedback promotes disc formation by removing low angular momentum gas at high redshift, allowing galaxies to be dominated by late-time gas accretion with high angular momentum. However, Brook et al. (2012) and Christensen et al. (2016) find that gas returns with significantly higher angular momentum than when it was ejected, because the ejected gas absorbs angular momentum from the CGM.

5 Discussion and Conclusions

This chapter reviews the effect of feedback on the accretion of gas, how it ejects material from the ISM and simultaneously prevents gas in the CGM (and IGM) from accreting onto the galaxy (and halo). Besides this negative effect on gas accretion, simulations show a positive effect as well, especially at late times and for massive galaxies. Gas that would have otherwise been locked up in galaxies becomes available for accretion at a later time, because of feedback. Furthermore, metal enrichment also enhances cooling, which further leads to enhanced gas accretion. We further focus on the accretion of gas that has previously been ejected from a galaxy and therefore has different properties than the gas that accretes directly from the IGM.

Mixing between hot outflows and the cooler CGM may be artificially suppressed in particle-based simulations and artificially enhanced in grid-based simulations or semi-analytic models. In reality, instabilities at the interfaces between the volume filling heated gas and embedded cool "clouds" can mix the different gas phases (e.g., Veilleux et al. 2005). Properly resolving these dynamics is challenging for simulations of isolated galaxies or idealized patches of galactic discs (e.g., Martizzi et al. 2016), let alone for cosmological simulations. The inclusion of magnetic fields and/or thermal conduction may also be critical for capturing the correct dynamics at the interface between the cool and hot phases (e.g., McCourt et al. 2015; Brüggen and Scannapieco 2016). In most particle-based simulations, metals are locked into gas particles. Either grid-based calculations or (particle-based simulations with explicit and realistic metal diffusion) would be useful for determining the extent to which small-scale mixing modifies the temperature, density, and, especially, metallicity of the ejected gas and thus its re-accretion. In the absence of such mixing the gas metallicity cannot converge at low metallicity. Thus, it is possible that the re-accretion of gas would change at much higher resolution and with the inclusion of additional physics.

Feedback is essential for bringing stellar masses into agreement with observations and for producing realistic galaxies. All methods discussed in Sect. 3 (semi-analytic models, equilibrium models, and hydrodynamical simulations) show that the re-accretion of gas after its ejection from a galaxy's ISM is an important ingredient for the growth of galaxies and for their metallicity evolution. However, there are large differences between estimates for its fractional contribution to galaxy growth and the time-scale for wind recycling. This can be understood by realizing that in order to reduce the ISM and stellar masses, one can either prevent more gas accretion onto the ISM or eject more gas from the ISM. Additionally, gas can be removed for a long time after a single ejection or be ejected many times if it re-accretes on a short time-scale.

Figure 8 shows a compilation of wind recycling time-scales from the literature Oppenheimer et al. (2010), Henriques et al. (2013), Mitra et al. (2015), Christensen et al. (2016), and Anglés-Alcázar et al. (2016). Different studies indeed find different time-scales for the re-accretion of gas onto galaxies, which scale differently with galaxy or halo mass. At the high-mass end, there is more agreement that gas re-accretes on a fairly short time-scale. At the low-mass end, however, there is a wide variety of predicted time-scales, some similar to those at the high-mass end and other similar to the age of the Universe. Each of the studies shown employs a different definition for wind recycling and differences may therefore be somewhat smaller in reality. However, even with the same definition, these discrepancies will likely remain, because of the very different feedback implementations.

Although these different feedback implementations may result in similar galaxy masses, other galaxy and CGM properties can be very different. For example, gas cycles on short time-scales can affect the gravitational potential. The collisionless dark matter and stars in the halo center/galaxy respond to this changing potential, which results in dark matter cores and an increase in the effective radius of dwarf galaxies (Pontzen and Governato 2012; El-Badry et al. 2016). Suarez et al. (2016) show in 2D grid simulations that the amount of cool gas formed in an outflow depends strongly on the feedback cycle as well. Shorter, stronger bursts produced by a rapidly fluctuating SFR generally lead to a larger fraction of cool gas forming in the outflow, although long feedback cycles (of a Gyr or more) also allow cool gas to form. This also clearly affects the structure of the CGM and the importance of re-accretion. Alternative feedback processes, not yet included in studies of wind recycling, such as cosmic ray feedback (e.g., Uhlig et al. 2012; Booth et al. 2013;

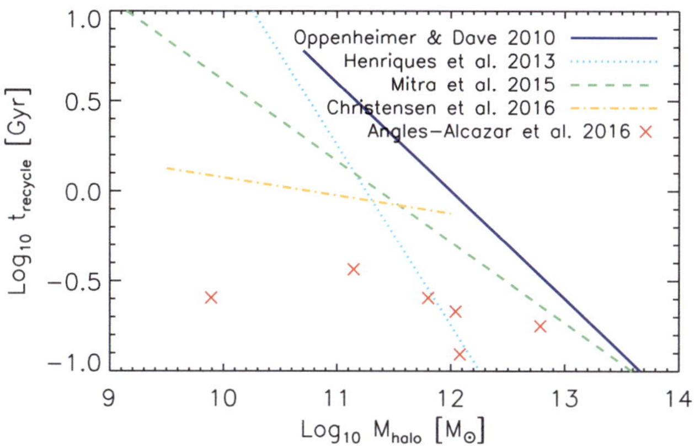

Fig. 8 Literature compilation of recycling time-scales as a function of halo mass from Oppenheimer et al. (2010), Henriques et al. (2013), Mitra et al. (2015), Christensen et al. (2016), and Anglés-Alcázar et al. (2016), as indicated by the legend. Most studies find relatively low recycling time-scales for massive galaxies, which means that ejected gas gets re-accreted fairly rapidly. Differences between studies are very large at the low-mass end, with some predicting that gas returns on a time-scale comparable to the age of the Universe, whereas others find time-scales similar to those at the high-mass end. Each of the studies shown employs a different definition for wind recycling (see also Sect. 3). In reality, differences may therefore be somewhat smaller, but are unlikely to disappear

Salem et al. 2014) could be important in driving outflows. For instance, cosmic rays may accelerate gas more smoothly than in the simulations described here, where feedback is more violent, and drive galactic winds with qualitatively different properties.

This chapter aimed to show how outflows driven by galactic feedback change the rates and properties of gas accretion onto galaxies. Although there are still large uncertainties, it is clear that wind recycling is an important component of the growth of galaxies, especially at late times. If we wish to understand galaxy evolution, we need to understand how outflows and inflows interact.

References

Anderson M. E., Gaspari M., White S. D. M., Wang W., Dai X., 2015, MNRAS, 449, 3806
Anglés-Alcázar D., Faucher-Giguère C.-A., Kereš D., et al. 2016, preprint, arXiv:1610.08523
Bahé Y. M., McCarthy I. G., Balogh M. L., Font A. S., 2013, MNRAS, 430, 3017
Bahé Y. M., Schaye J., Crain R. A., et al. 2017, MNRAS, 464, 508
Balogh M. L., Navarro J. F., Morris S. L., 2000, ApJ, 540, 113
Bertone S., De Lucia G., Thomas P. A., 2007, MNRAS, 379, 1143
Birnboim Y., Dekel A., 2003, MNRAS, 345, 349
Blanton M. R., Berlind A. A., 2007, ApJ, 664, 791
Booth C. M., Agertz O., Kravtsov A. V., Gnedin N. Y., 2013, ApJL, 777, L16
Bouché N., Dekel, A., Genzel, R., et al., 2010, ApJ, 718, 1001

Brook C. B., Stinson G., Gibson B. K., et al. 2012, MNRAS, 419, 771
Brüggen M., Scannapieco E., 2016, ApJ, 822, 31
Bryan G. L., Norman M. L., 1998, ApJ, 495, 80
Chan T. K., Kereš D., Oñorbe J., et al. 2015, MNRAS, 454, 2981
Christensen C. R., Davé R., Governato F., et al. 2016, ApJ, 824, 57
Crain R. A., Theuns T., Dalla Vecchia C., et al. 2009, MNRAS, 399, 1773
Crain R. A., McCarthy I. G., Frenk C. S., Theuns T., Schaye J., 2010, MNRAS, 407, 1403
Davé R., Cen R., Ostriker J. P., et al. 2001, ApJ, 552, 473
Davé R., Oppenheimer B. D., Finlator K., 2011, MNRAS, 415, 11
Davé R., Finlator K., Oppenheimer B. D., 2012, MNRAS, 421, 98
De Lucia G., Kauffmann G., White S. D. M., 2004, MNRAS, 349, 1101
Dekel A., Birnboim Y., 2006, MNRAS, 368, 2
Dekel A., Mandelker N., 2014, MNRAS, 444, 2071
El-Badry K., Wetzel A., Geha M., et al. 2016, ApJ, 820, 131
Faucher-Giguère C.-A., Kereš D., Ma C.-P., 2011, MNRAS, 417, 2982
Fielding D., Quataert E., McCourt M., Thompson T. A., 2016, preprint, arXiv:1606.06734
Finlator K., Davé R., 2008, MNRAS, 385, 2181
Ford A. B., Davé R., Oppenheimer B. D., et al. 2014, MNRAS, 444, 1260
Gonzalez A. H., Sivanandam S., Zabludoff A. I., Zaritsky D., 2013, ApJ, 778, 14
Haardt F., Madau P., 2001, in Neumann D. M., Tran J. T. V., eds, *Clusters of Galaxies and the High Redshift Universe Observed in X-rays*
Hafen Z., Faucher-Giguere C.-A., Angles-Alcazar D., et al. 2016, preprint, arXiv:1608.05712
Hayes M., Melinder J., Östlin G., et al. 2016, ApJ, 828, 49
Henriques B. M. B., White S. D. M., Thomas P. A., et al. 2013, MNRAS, 431, 3373
Hopkins A. M., Beacom J. F., 2006, ApJ, 651, 142
Kacprzak G. G., Churchill C. W., Nielsen N. M., 2012, ApJL, 760, L7
Kacprzak G. G., van de Voort F., Glazebrook K., et al. 2016, ApJL, 826, L11
Kereš D., Katz N., Weinberg D. H., Davé R., 2005, MNRAS, 363, 2
Kereš D., Katz N., Davé R., Fardal M., Weinberg D. H., 2009, MNRAS, 396, 2332
Lagos C. d. P., Theuns T., Schaye J., et al. 2016, MNRAS, 459, 2632
Larson R. B., Tinsley B. M., Caldwell C. N., 1980, ApJ, 237, 692
Le Brun A. M. C., McCarthy I. G., Schaye J., Ponman T. J., 2014, MNRAS, 441, 1270
Lee N., Sanders D. B., Casey C. M., et al. 2015, ApJ, 801, 80
Lilly S. J., Carollo C. M., Pipino A., Renzini A., Peng Y., 2013, ApJ, 772, 119
Lin Y.-T., Stanford S. A., Eisenhardt P. R. M., et al. 2012, ApJL, 745, L3
Mannucci F., Cresci G., Maiolino R., Marconi A., Gnerucci A., 2010, MNRAS, 408, 2115
Martizzi D., Fielding D., Faucher-Giguère C.-A., Quataert E., 2016, MNRAS, 459, 2311
McCarthy I. G., Schaye J., Bird S., Le Brun A. M. C., 2017, MNRAS, 465, 2936
McCourt M., O'Leary R. M., Madigan A.-M., Quataert E., 2015, MNRAS, 449, 2
Mitra S., Davé R., Finlator K., 2015, MNRAS, 452, 1184
Mitra S., Davé R., Simha V., Finlator K., 2017, MNRAS, 464, 2766
Muratov A. L., Kereš D., Faucher-Giguère C.-A., et al. 2015, MNRAS, 454, 2691
Noeske K. G., Weiner B. J., Faber S. M., et al. 2007, ApJL, 660, L43
Oppenheimer B. D., Davé R., 2008, MNRAS, 387, 577
Oppenheimer B. D., Davé R., Kereš D., et al. 2010, MNRAS, 406, 2325
Padmanabhan T. 2002, *Theoretical Astrophysics, Volume III: Galaxies and Cosmology*, pp. 638, Cambridge, UK: Cambridge University Press
Peng Y., Maiolino R., Cochrane R., 2015, Nature, 521, 192
Péroux C., Dessauges-Zavadsky M., D'Odorico S., Kim T.S., McMahon R. G., 2005, MNRAS, 363, 479
Pontzen A., Governato F., 2012, MNRAS, 421, 3464
Rasmussen J., Mulchaey J. S., Bai L., et al. 2012, ApJ, 757, 122
Rees M. J., Ostriker J. P., 1977, MNRAS, 179, 541
Salem M., Bryan G. L., Hummels C., 2014, ApJL, 797, L18

Schaye J., Dalla Vecchia C., Booth C. M., et al., 2010, MNRAS, 402, 1536
Schaye J., Crain R. A., Bower R. G., et al. 2015, MNRAS, 446, 521
Segers M. C., Crain R. A., Schaye J., et al. 2016, MNRAS, 456, 1235
Simha V., Weinberg D. H., Davé R., et al. 2009, MNRAS, 399, 650
Steidel C. C., Bogosavljević M., Shapley A. E., et al. 2011, ApJ, 736, 160
Suarez T., Pontzen A., Peiris H. V., Slyz A., Devriendt J., 2016, MNRAS, 462, 994
Sun M., Voit G. M., Donahue M., et al. 2009, ApJ, 693, 1142
Turner M. L., Schaye J., Crain R. A., Theuns T., Wendt M., 2016, MNRAS, 462, 2440
Übler H., Naab T., Oser L., et al. 2014, MNRAS, 443, 2092
Uhlig M., Pfrommer C., Sharma M., et al. 2012, MNRAS, 423, 2374
van de Voort F., Schaye J., 2012, MNRAS, 423, 2991
van de Voort F., Schaye J., Booth C. M., Haas M. R., Dalla Vecchia C., 2011a, MNRAS, 414, 2458
van de Voort F., Schaye J., Booth C. M., Dalla Vecchia C., 2011b, MNRAS, 415, 2782
van de Voort F., Schaye J., Altay G., Theuns T., 2012, MNRAS, 421, 2809
van de Voort F., Quataert E., Hopkins P. F., et al. 2016, MNRAS, 463, 4533
van de Voort F., Bahé Y. M., Bower R. G., et al. 2017, MNRAS, 466, 3460
Veilleux S., Cecil G., Bland-Hawthorn J., 2005, ARA&A, 43, 769
Vikhlinin A., Burenin R. A., Ebeling H., et al. 2009, ApJ, 692, 1033
Vogelsberger M., Genel S., Springel V., et al. 2014, MNRAS, 444, 1518
Wetzel A. R., Tinker J. L., Conroy C., 2012, MNRAS, 424, 232
White S. D. M., Frenk C. S., 1991, ApJ, 379, 52
White S. D. M., Rees M. J., 1978, MNRAS, 183, 341
Wiersma R. P. C., Schaye J., Smith B. D., 2009, MNRAS, 393, 99
Wilman D. J., Zibetti S., Budavári T., 2010, MNRAS, 406, 1701

Gas Accretion via Condensation and Fountains

Filippo Fraternali

1 Introduction

Galaxies grow through the continuous acquisition of gas from the surrounding intergalactic environment. This fundamental process of gas accretion is the key to understand galaxy formation and evolution and it has been described in all its most important aspects in the previous chapters of this book. Here, we focus on a mechanism that appears to work well for the Milky Way and similar galaxies today but has the potential to be extended to other types of galaxies and epochs. In this scheme, disc galaxies accrete gas from the surrounding massive hot corona, which is expected to extend out to the virial radius (Fukugita and Peebles 2004). The cooling and accretion occurs very close to the galactic disc and it is triggered and regulated by stellar feedback. Very strong and efficient feedback, like in starburst galaxies where we observe strong galactic winds, may have the effect of inhibiting accretion and heating the corona (Sokołowska et al. 2016). However, more common low-levels of feedback, typical of a galactic fountain regime in main-sequence galaxies, have the opposite effect of stimulating gas accretion. This overlooked effect of supernova feedback is the subject of this chapter.

Stellar winds and supernovae produce large expanding bubbles that bring large amounts of material from the disc into the halo (Mac Low et al. 1989). These phenomena set up a circulation (galactic fountain Shapiro and Field 1976; Fraternali and Binney 2006) that continuously lifts disc gas into the halo causing it to mix with the hot ($T \gtrsim 10^6$ K) gas of the corona. The disc gas is more metal-rich and colder

F. Fraternali (✉)
Department of Physics and Astronomy, University of Bologna, via Berti Pichat 6/2, 40127 Bologna, Italy

Kapteyn Astronomical Institute, University of Groningen, Postbus 800, 9700 AV Groningen, The Netherlands
e-mail: filippo.fraternali@unibo.it

A. Fox, R. Davé (eds.), *Gas Accretion onto Galaxies*, Astrophysics and Space Science Library 430, DOI 10.1007/978-3-319-52512-9_14

than the hot coronal gas and their mixing dramatically reduces the cooling time of the hot gas leading to the condensation and accretion of a significant fraction of the lower corona. This mechanism leaves a clear signature (lagging rotation) on the kinematics of the extraplanar cold/warm ($T \lesssim 10^5$ K) gas in disc galaxies, which is observed in local galaxies. The modelling of these data leads to the estimate of the amount of gas accretion via condensation and fountains. Remarkably, this turns out to be of the order of the star formation rate (SFR), providing a viable solution of the long-standing problem of the feeding of star formation in disc galaxies. In the literature, this mechanism is referred to as *supernova-driven accretion* or fountain-driven accretion, here we simply call it *fountain accretion*. Before embarking on the description of the physical phenomenon and its observational evidence, we review the main lines of thought that brought about the formulation of this theory and the key ingredients that are required to make it work.

A number of lessons can be indirectly learned about gas accretion by studying the stellar and gaseous components of galaxies and their star formation (Sánchez Almeida et al. 2014); we briefly summarize them here.

1. Star-forming galaxies have SFRs that are typically not sustainable in isolation for a Hubble time, i.e., the ratio between their gas mass and their SFR (*depletion time*) is of the order of a Gigayear for every galaxy at every redshift (e.g., Saintonge et al. 2013). This clearly points to the need of continuous gas accretion (Fraternali and Tomassetti 2012).
2. Galaxies, especially when going through starburst phases, have strong winds that can remove gas (Veilleux et al. 2005) making their need for accretion even more severe.
3. Galaxy discs grow inside-out, even today (Pezzulli et al. 2015), in the sense that the size of their stellar disc increases with time. This points to late accretion of gas preferentially in the outer parts and therefore at high-angular momentum (Goldbaum et al. 2015).
4. Chemical evolution models require late accretion of low-metallicity gas (e.g. Tosi 1988) in order to explain the chemistry of stars in the disc of the Milky Way and the radial metallicity gradients (Chiappini et al. 2001; Schönrich and Binney 2009; Kubryk et al. 2013).
5. Galaxies progressively evolve from the blue cloud to the red sequence with no evidence of significant going back (e.g., Peng et al. 2010). About 80% of red-sequence galaxies are not detected in deep *H I* surveys (Catinella et al. 2012).
6. Finally, galaxies have much less (factor 5 or more) baryonic matter in the form of cold gas and stars than we would expect given their dark matter masses (Papastergis et al. 2012; van Uitert et al. 2016). These *missing baryons* are at least partially located in the hot metal-poor coronae and thus being hard to detected (Bregman 2007). Only rich galaxy clusters contain the expected amount of baryons mostly in the form of hot gas (e.g., McGaugh et al. 2010).

The picture that emerges from above is that as long as they keep forming stars at the observed levels, disc galaxies must continuously gather low-metallicity and relatively high-angular momentum gas from the surrounding medium. This gas accretion proceeds until some form of star formation quenching occurs or the gas supply is somehow cut off (e.g., Lilly et al. 2013). This phenomenon is clearly more efficient in dense environments where the causes for gas removal are better understood, e.g., ram pressure stripping and tidal interactions (e.g., Fasano et al. 2000). After quenching either the gas reservoir is not available anymore or the galaxy loses the ability to harvest material from its reservoir and thereafter evolves passively as a red-sequence galaxy.

The above indirect evidence must be combined with direct observations of gas around galaxies to get a consistent picture of the census of baryons around galaxies and their physical state. These observations have been described in detail in the first part of this book, here we summarize those most relevant for fountain accretion.

a. Cold ($T \lesssim 10^4$ K) and dense gas (visible in *H I* emission) is only observed in large quantities very close (few kpc) to galaxy discs. This is the so-called *extraplanar gas* that we describe in some detail in Sect. 2. At larger distances, some *H I* clouds are observed that could be part of the same medium (see Sect. 6). In the Milky Way, they are called *High-Velocity Clouds* (HVCs Wakker and van Woerden 1997) and typically extend to ~10 kpc from the Galactic disc. The Galactic HVCs have relatively low metallicity (van Woerden et al. 2004) and appear to be falling onto the disc, however, their masses are typically 1–2 orders of magnitude smaller than what would be needed to feed the star formation (Putman et al. 2012). Further out, there is no evidence of large "free floating" *H I*-emitting clouds down to masses of $M_{\rm HI} \sim 10^6\, M_\odot$ (e.g., Pisano et al. 2011).
b. Large masses of neutral gas away from galaxies are only observed in the presence of strong interactions (encounters/mergers) or in tight groups (Yun et al. 1994; Putman et al. 2003; Lewis et al. 2013). Here we do not discuss these sporadic phenomena, which may also have relevance in promoting gas accretion from one galaxy to another (Nichols et al. 2011; Fox et al. 2014). Estimates of neutral gas accretion from minor mergers today return values that are orders of magnitude below the requirements to sustain star formation (Di Teodoro and Fraternali 2014). Moreover, the nearly constant amount of *H I* in galaxies across the whole Hubble time (e.g., Zafar et al. 2013) strongly suggests that most gas accretion must come from a gas phase different from *H I*.
c. Warm/ionized ($10^4 < T < 10^6$ K) gas is ubiquitously detected both around the Milky Way (Sembach et al. 2003; Shull et al. 2009; Lehner et al. 2012; Richter et al. 2013) and around external galaxies (Werk et al. 2013, 2014) via absorption towards distant quasars. This component has too low temperature to be in hydrostatic equilibrium in the potential of Milky Way-like galaxies, thus it is likely infalling or outflowing. The mass contained in this component is hard to estimate but it could provide accretion rates of the order of $\sim 1\, M_\odot\, {\rm yr}^{-1}$ (Lehner and Howk 2011).
d. HVCs both neutral and ionized must be confined by a hot medium generally referred to as the *galactic corona*. Recent observations in the X-rays have revealed the presence of this hot corona both around the Milky Way (Miller and Bregman 2015) and around external massive spiral galaxies (Dai et al. 2012;

Bogdán et al. 2013). The corona should extend out to the virial radius of the dark matter halo and contain a mass at least as large as the baryonic mass in the galaxy disc (Gatto et al. 2013).

In what follows we try to incorporate most of the direct and indirect evidence into a coherent picture. We first give a general description of the properties of the extraplanar gas (Sect. 2), the explanation of which generates problems for standard galactic fountain models pointing to the importance of interaction between disc and coronal gas at the disc-halo interface (Sect. 3). We describe the hydrodynamical simulations of disc-corona mixing that predict the condensation of the coronal gas (Sect. 4) and this leads us to the description of the fountain-driven accretion model (Sect. 5). We then give a full account of the comparison of the predictions of this model with the observations (Sect. 6) and finally, we discuss the potential relevance of this model for galaxy evolution focusing in particular on the onset of metallicity gradients and star formation quenching (Sect. 7).

2 Extraplanar Gas: Life at the Disc-Halo Interface

The discs of nearby spiral galaxies with sufficient star formation are surrounded by thick layers of neutral and ionized gas that extend kiloparsecs from the plane (Fraternali et al. 2002; Matthews and Wood 2003; Oosterloo et al. 2007; Zschaechner et al. 2015). This layer is also observed in the Milky Way (Lockman 2002; Marasco and Fraternali 2011). The HI in the extraplanar gas is dense enough to be observed in emission in deep observations (Sancisi et al. 2008) as is the ionized gas (Rossa and Dettmar 2003; Heald et al. 2007; Kamphuis 2008). This is of fundamental importance for what follows because these observations can be used to study the detailed kinematics of the gas at the disc-halo interface and understand its origin.

The extraplanar gas in disc galaxies has three main properties: (1) it is located considerably above the standard scale-height expected for the HI layer (of order 100 pc) and makes up about 10–30% of the total HI of a galaxy; (2) it shows a distinctive kinematics with respect to the disc gas, in particular it rotates more slowly (Fraternali et al. 2005) and it shows non-circular motions (Fraternali et al. 2001); (3) it is located mostly in the star-forming inner part of the galaxies. The first detection of a widespread layer of extraplanar gas with coherent kinematics (low rotation) was reported for the edge-on galaxy NGC 891 (Swaters et al. 1997). Then, thanks to its peculiar kinematics it has been detected and studied in galaxies not seen edge-on, in particular in NGC 2403 (Schaap et al. 2000; Fraternali et al. 2001). In these intermediate-inclination galaxies, it shows up as a smooth emission at lower rotation velocities with respect to the disc emission, in particular along the major axis of the galaxy (see position-velocity plot in Fig. 1). The peculiar kinematics of the extraplanar gas also allowed it to be identified in the Milky Way (Marasco and Fraternali 2011).

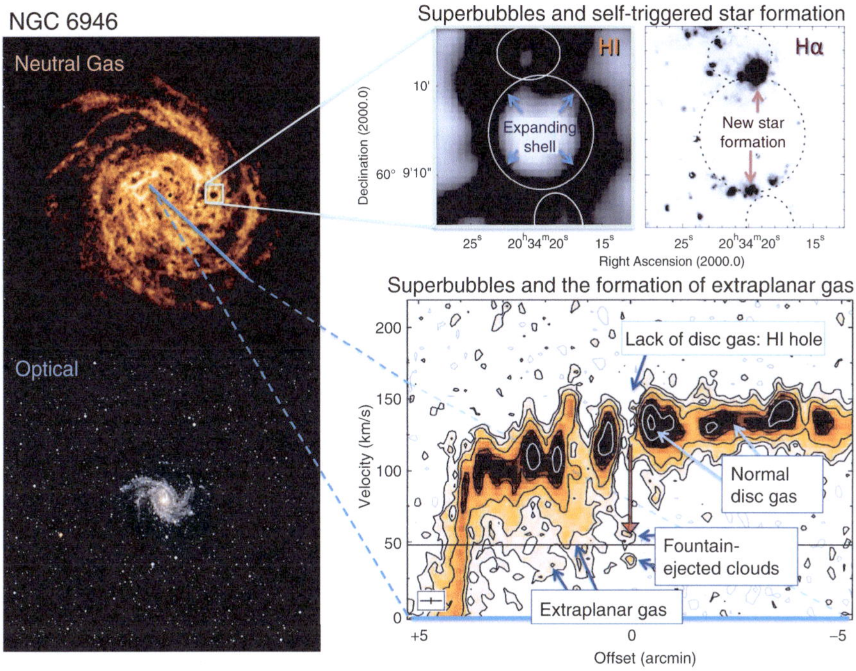

Fig. 1 The formation of the extraplanar gas layer from superbubble blowout in the nearby disc galaxy NGC 6946. *Left panels*: comparison between the extent of the *H I* disc in this galaxy and the bright optical disc (adapted from Boomsma et al. 2008). *Top right*: the expansion of a large superbubble has produced a hole in the *H I* distribution in the disc and caused the compression of the surrounding gas triggering a new generation of stars, testified by the intense Hα emission observed at the edges of the hole. *Bottom right*: position-velocity plot along the line shown overlaid on the top-left total *H I* map. In regions of depressed emission in the disc (*H I* holes) fast-moving clouds are observed leaving the disc and joining the diffuse material that makes up the extraplanar gas layer. Note how the extraplanar gas is located at velocities between the normal disc gas and the systemic velocity of the galaxy (*horizontal line*), i.e., it rotates more slowly that the disc gas

As we explain in Sect. 3, the origin of a significant fraction ($\sim$80%) of the extraplanar gas is likely to be stellar feedback. The combined action of stellar winds and supernovae in large star clusters (OB associations) produces the formation and expansion of large bubbles called *superbubbles* (Mac Low and McCray 1988). Unlike normal wind/supernova bubbles, superbubbles can easily reach sizes that exceed the typical thickness of the disc of cold gas. When the size of a superbubble shell (supershell) becomes of the order of the disc thickness, the expansion accelerates vertically and a blowout occurs (Mac Low et al. 1989). The blowout has two important consequences. First, it leaves *holes* in the distribution of the cold neutral gas. A large number of these *H I* holes are observed in nearby nearly face-on galaxies (Puche et al. 1992; Boomsma et al. 2008). Second, it ejects gas from the disc into the halo. This gas is both *cold* ($T \sim 10^4$ K) coming from the supershell and *hot* ($T \gtrsim 10^6$ K) from the bubble interior. Figure 1 illustrates these phenomena in

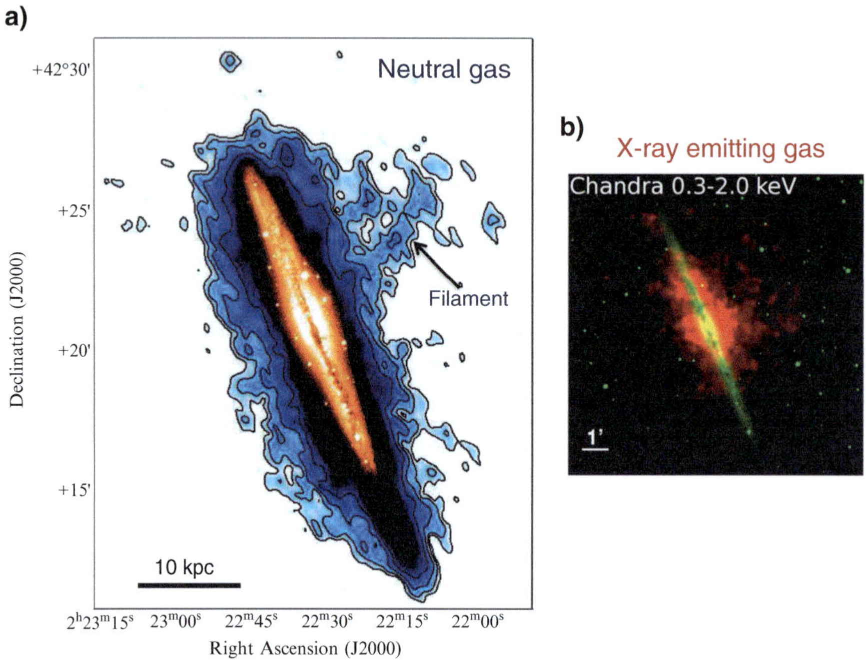

Fig. 2 Comparison between cold and hot gas in the halo region close to the disc of the edge-on galaxy NGC 891, the two images are on the same scale. (**a**) total *H I* map from the deep observations obtained with the Westerbork Synthesis Radio Telescope (WSRT) (Oosterloo et al. 2007) overlaid on an optical image (*orange*). The neutral cold ($T < 2 \times 10^4$ K) gas extends up to ~8 kpc everywhere above and below the inner disc with a feature (filament) reaching up to 20 kpc. The total *H I* mass extraplanar gas is of about $1.2 \times 10^9 M_\odot$ ($1.6 \times 10^9 M_\odot$ taking helium into account). (**b**) Chandra X-ray observations of NGC 891 overlaid on an optical image (from Hodges-Kluck and Bregman 2013). The X-ray emitting hot ($T = 4 \times 10^6$ K) gas appears more centrally concentrated and rounder in shape than the *H I* emission. Its total mass (assuming a cylindrical halo with scale-height 5 kpc) is $1–3 \times 10^8 M_\odot$. Note that this is only the very inner part of the galactic corona, which likely extends, albeit with much lower densities (and therefore undetectable with these observations), out to the virial radius (see text)

one galaxy, NGC 6946, seen at the relatively low inclination of 30° with respect to the line of sight. *H I* clouds ejected from the disc of this galaxy, often at the locations of *H I* holes are sustaining a layer of extraplanar neutral gas.

A somewhat more obvious way to visualize the extraplanar gas is to observe edge-on galaxies. Figure 2 shows deep *H I* observations (blue shade and contours) of the edge-on spiral NGC 891 overlaid on an optical image (Oosterloo et al. 2007). The *H I* layer extends above the plane to remarkable distances with a scale-height $h \simeq 2.2$ kpc. The total mass of *H I* located outside the disc is about $1.2 \times 10^9 M_\odot$, which corresponds to almost 30% of the total *H I* mass. This galaxy is likely to be on the high end of the spectrum of extraplanar gas properties, in particular for its large extraplanar *H I* mass. Preliminary results obtained by the HALOGAS survey

(Heald et al. 2011) show several other galaxies with relatively high SFR have typical extraplanar $H\,I$ masses of order 10–20% of their total $H\,I$ masses (Fraternali et al., in prep.).

3 Galactic Fountains and the Origin of Extraplanar Gas

Since its discovery, three models have been proposed to explain the cold extraplanar gas: hydrostatic equilibrium, accretion/inflow from the circumgalactic medium, and galactic fountain. The hypothesis that the extraplanar gas could be in equilibrium has been investigated with two different approaches, to which we can refer as hydrodynamical and dynamical. In the hydrodynamical model the gas layer is a baroclinic fluid[1] in hydrostatic equilibrium with the galactic potential (Barnabè et al. 2006). Despite the success in reproducing the extraplanar kinematics, this model predicts temperatures for the gas larger than 10^5 K, which are incompatible with the neutral gas phase. The second attempt assumed that the extraplanar layer consists of a large number of gas clouds characterized by large relative velocities, i.e., a velocity dispersion much higher than that of the disc $H\,I$. The equilibrium of such a medium was studied using the Jeans equations and their predictions compared in the 3D space (position on the sky and line-of-sight velocity) to the observed $H\,I$ emission of NGC 891 (Marinacci et al. 2010). It was found that the data are incompatible with a system of clouds with an isotropic velocity dispersion and the only configuration that resembles the observations is one with a strong vertical anisotropy ($\sigma_z \gg \sigma_\phi \sim \sigma_R$). This is akin to the anisotropy that one finds in galactic fountain models.

The possibility that the extraplanar gas is produced exclusively by extragalactic inflow has been studied using hydrodynamical simulations of galaxy formation (Kaufmann et al. 2006). A model of this kind reproduces the kinematics of the extraplanar gas, in particular the rotation curves of NGC 891 (Fraternali et al. 2005). This model is, however, unrealistic for a number of reasons. First, the cold clouds that formed in these early SPH simulations are now known to be numerical artifacts and not the product of thermal instabilities as initially thought (Kaufmann et al. 2009; Read and Hayfield 2012). In fact, the classical Field instability should not be at work at all in the coronae of these galaxies because it is counteracted by buoyancy and suppressed by thermal conduction (Binney et al. 2009; Joung et al. 2012). Moreover, pure infall models would give totally unrealistic infall rates, for example, of the order of $30\,M_\odot\,\mathrm{yr}^{-1}$ in NGC 891, an order of magnitude larger than the galaxy's SFR.

The third possibility that we can consider is that the extraplanar gas is the product of a widespread galactic fountain. As mentioned in Sect. 2, the blowout

[1] We recall that this is a fluid with an equation of state where pressure does not depend on density alone. This is in contrast to a barotropic fluid where $P = P(\rho)$.

Table 1 Property of the extraplanar gas in four representative galaxies

Galaxy	M_{HI} ($10^8\,M_\odot$)	$\frac{M_{HI}}{M_{HI,\,\rm disc}}$ (%)	Gradient (km s^{-1} kpc^{-1})	$M_{\rm warm}$ ($10^8\,M_\odot$)	$T_{\rm Xray}$ (10^6 K)	$M_{\rm Xray}$[a] ($10^8\,M_\odot$)	$Z_{\rm Xray}$ ($Z_\odot$)	Ref.
NGC 891	12	30	15	Few	3–4	1–3	0.1	(a, b, c)
Milky Way	3[b]	10	15	1.1	3	–[d]	0.3	(d, e, f)
NGC 2403	3[b]	10	10–15	Yes	2–8	∼0.1	–	(g, h, i)
NGC 6946	3–10[b]	4–15	–	Yes	4	0.4–0.8[c]	–	(j, l)

[a]This is only the mass of the X-ray emitting gas in the region close to the disc, which roughly overlaps with the region of the extraplanar neutral gas. [b]These masses may be underestimated because some extraplanar gas emission overlaps with the disc gas in these datacubes. [c]Assuming a range of scale-heights between 1 and 5 kpc (see Schlegel et al. 2003). [d]The density profile used for the hot gas in Miller and Bregman (2015) does not allow a reliable estimate of the inner mass

References: (a) Oosterloo et al. (2007), (b) Howk and Savage (2000), (c) Hodges-Kluck and Bregman (2013), (d) Marasco and Fraternali (2011), (e) Lehner and Howk (2011), (f) Miller and Bregman (2015), (g) Fraternali et al. (2002), (h) Fraternali et al. (2004), (i) Fraternali et al. (2002), (j) Boomsma et al. (2008), (l) Schlegel et al. (2003)

of superbubbles is expected to eject a large mass of gas out of the plane of the disc. The lack of this gas in the disc is observed in the form of HI holes and, in fact, the mass *missing* from the holes is comparable to the mass of the extraplanar gas layer (Boomsma et al. 2008). In the first models of galactic fountain, the general idea was that the gas is ejected hot from the disc and then it eventually cools down to produce a *rain* of HI clouds (Shapiro and Field 1976; Bregman 1980). However, current observations show that most of the gas in the region close to the galactic disc is in fact at relatively low temperatures. The example of NGC 891 is very instructive. Figure 2 shows the total HI map of this galaxy compared with the deepest X-ray observations with Chandra (Hodges-Kluck and Bregman 2013), on the same scale. Clearly, the neutral gas appears more widespread and even more extended in the vertical direction than the hot X-ray emitting gas. The mass of neutral extraplanar gas is $M_{\rm cold} \simeq 1.6 \times 10^9\,M_\odot$ (Oosterloo et al. 2007) including the correction for helium abundance. In roughly the same region, the mass of hot gas is instead at most a few times $10^8\,M_\odot$ (see Table 1), about an order of magnitude lower. This simple evidence shows that most of the circulation of gas in the extraplanar layer occurs at relatively low temperatures, either because the gas is ejected cold (Fraternali 2012; Hodges-Kluck and Bregman 2013) or because it cools rapidly (Houck and Bregman 1990). The same is true for other galaxies in which extraplanar gas has been studied in detail. In Table 1 we report the properties of four representative galaxies. A comprehensive compilation of the properties of the extraplanar gas in galaxies is currently lacking, only partial lists exist (Sancisi et al. 2008; Fraternali 2009; Zschaechner et al. 2015).

The most straightforward model of a galactic fountain is the so-called *ballistic model* (Collins et al. 2002; Spitoni et al. 2008). In this formulation, fountain clouds are treated like test particles shot up nearly vertically from the disc by stellar feedback. They follow trajectories governed only by gravity and return to another location in the disc where they join the local interstellar medium. This treatment is

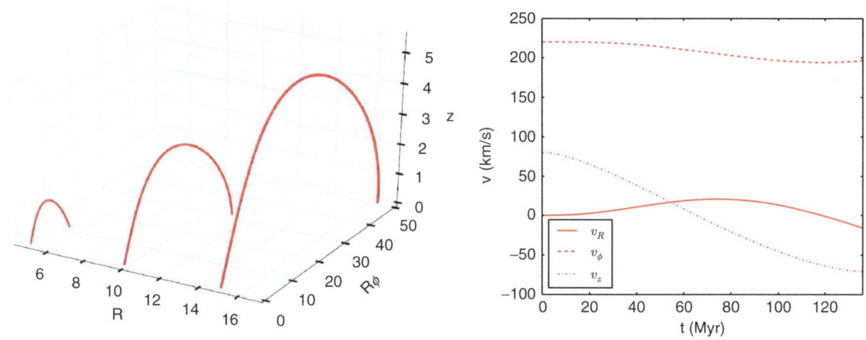

Fig. 3 *Left*: Orbits of galactic fountain particles in the potential of the Milky Way shot vertically upwards with $v_{\rm kick} = 75\,{\rm km\,s^{-1}}$ from three radii $R = 5, 10, 15$ kpc. All particles fall back to the disc at slightly larger radii with respect to the kick radii. Note how the maximum height reached varies with radius due to the decrease of the vertical force. Only gravity is included in this calculation. *Right*: Cylindrical components of the velocity of the fountain particle ejected from $R = 10$ kpc

clearly limited as it neglects the hydrodynamical nature of the fountain gas but it is a very useful base model to study the physical phenomenon (Fraternali and Binney 2006).

Once the galactic potential is fixed, the most important parameter of the model is the *kick velocity* of the fountain particles. Luckily, some of the observable properties strongly depend on the kick velocity and so its value can be accurately constrained. First, as one may expect, the higher the kick velocity the larger the scale-height of the extraplanar layer produced by the fountain. This led to an estimate of the average values of the kick velocity of about $v_{\rm kick} = 70$–$80\,{\rm km\,s^{-1}}$ for NGC 891 (Fraternali and Binney 2006) and the Milky Way (Marasco et al. 2012). These values are fully compatible with the velocities of the material in the shell of a superbubble at blowout coming from hydrodynamical simulations (Mac Low et al. 1989; Melioli et al. 2008). In practice, in a galactic fountain model, clouds are ejected at different radii with a characteristic velocity $v_{\rm kick}$ that is usually kept constant with radius. Figure 3 (left) shows the orbits that one obtains by shooting fountain particles vertically straight up in the potential of the Milky Way starting from three representative radii. Generally, the particles tend to move outward. This happens because, as a consequence of the kick, the particle acquires a velocity that always exceeds its circular velocity $\mathbf{v} = \mathbf{v}_{\rm c} + \mathbf{v}_{\rm kick} > \mathbf{v}_{\rm c}$ and thus it finds itself at the "pericenter" of the orbit (in the approximation of a nearly spherical potential). The maximum height reached varies substantially with radius reaching distances of up to several kpc if the ejection occurs at large radii.

Figure 3 (right) shows the three cylindrical components of the velocity of the particle ejected at $R = 10$ kpc. As one may expect for ballistic trajectories, v_z has a sinusoidal shape with the particle falling back to the disc at a speed roughly equal to the ejection velocity. The radial component $v_{\rm R}$ is mostly positive, given that the particle moves outward. Of great interest is the azimuthal component v_ϕ, which decreases slightly. This decrease is explained by angular momentum conservation. It can be shown (Marinacci et al. 2010) that the value of v_ϕ at each location along the orbit is similar to the "circular velocity" at that location:

$$v_c(R, z) = \left[R \frac{\partial \Phi(R, z')}{\partial R} \right]^{1/2}_{(z'=z)} . \tag{1}$$

where Φ is the axisymmetric galactic potential and the derivative is calculated at height z. This velocity provides a reference for the comparison between the prediction of the models and the rotation of the extraplanar gas.

The global model of a galactic fountain is constructed by integrating a large number of orbits like the ones described above. The first versions of the model were fully axisymmetric (but see end of Sect. 6). The clouds are shot with a Gaussian distribution of velocities with dispersion $v_{\rm kick}$ and a relatively small opening angle (estimated directly from observations; Fraternali and Binney 2006) with respect to the normal direction. The number of clouds ejected at each radius is proportional to the star formation rate surface density or, equivalently, the supernova rate density (Fraternali and Binney 2006). Once a suitable number of orbits have been integrated in the meridional plane (R, z), they are randomly distributed in azimuth. The 21-cm emission from each of these "clouds" is generated by assuming that they are partially or totally neutral at each time step along their orbits. A fraction of the material escaping from the superbubbles is likely to be photoionized given the strong radiation field—see, for instance, the Ophiucus superbubble in the Milky Way (Pidopryhora et al. 2007). To take this into account one introduces a second parameter: the *ionized fraction* ($f_{\rm ion}$) of the ascending fountain clouds. The descending clouds are usually considered neutral. If $f_{\rm ion} = 0$ the ascending clouds are always visible in *H I*, otherwise a fraction of their orbit will be invisible. Both $v_{\rm kick}$ and $f_{\rm ion}$ are parameters that can be fitted to observations (Sect. 6, Table 2).

Finally, the contribution of all the clouds in the simulation described above is recorded in a mock observation (*H I pseudo-datacube*) of the model galaxy. The comparison with real data is then carried out by building several models and by varying the parameters to minimize the data-model residuals. Figure 4a shows the total *H I* map produced by one such galactic fountain model for the edge on galaxy NGC 891 (see also Fig. 2). The model was produced by ejecting particles with $v_{\rm kick} \simeq 80\,{\rm km\,s^{-1}}$. The distribution and the extent of the model extraplanar gas layer appear quite similar to the data (see Fraternali and Binney 2006, for details). However, the kinematics of the extraplanar gas predicted by this type of fountain model is not compatible with the data. This can be clearly seen from Fig. 4b where the observed rotation curve of the extraplanar gas at $z \sim 5$ kpc (points) is compared to the prediction of a ballistic fountain model (red curve). The actual comparison was done much more thoroughly by simulating the whole datacube (Fraternali and Binney 2006), but Fig. 4 makes the point quite straightforwardly. The rotation of the

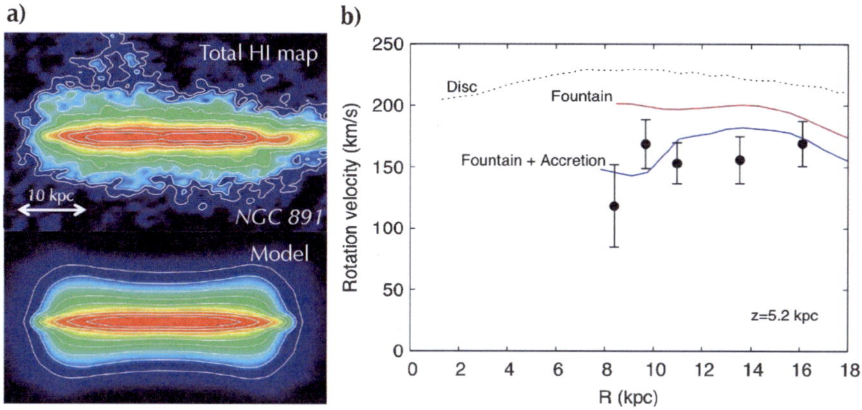

Fig. 4 (**a**) Comparison between the total H I map of NGC 891 (rotated with its major axis along the x-axis) and the mock map produced by a galactic fountain model (figure adapted from Fraternali and Binney 2006). The distribution of the gas in the extraplanar layer is reproduced by ejecting clouds from the disc with an average kick velocity of $v_{\rm kick} = 80$ km s^{-1}. (**b**) Comparison between the rotation above the plane of the extraplanar gas in NGC 891 (at $z \sim 5$ kpc) with a purely ballistic fountain (*red curve*) model and a fountain accretion (*blue curve*) model (Fraternali and Binney 2008), see text

extraplanar gas predicted by the fountain model is systematically higher than the observed rotation.

This comparison concludes this description of the ballistic galactic fountain model. The main feature of this model is that the particles conserve angular momentum throughout their trajectory. However, Fig. 4 shows that their rotation velocities should decrease, very likely as a consequence of hydrodynamical effects that are not included in the ballistic treatment. Unfortunately, a full hydrodynamical treatment of the problem is impractical because of the impossibility to fully cover the parameter space but it is possible to incorporate hydrodynamical effects on the motion of individual fountain clouds (Sect. 5). These effects allow us to reconcile the galactic fountain model with the data (blue curve in Fig. 4) and constitute the key evidence for gas accretion within the fountain cycle. In Sect. 4, we describe the tailored high-resolution hydrodynamical simulations that allowed us to go beyond the ballistic model.

4 Hydrodynamical Simulations of Disc-Corona Mixing

The interface between galaxy discs and the surrounding environment is a very complex multi-phase region that must be studied using hydrodynamical simulations at very high resolution. The standard approach is to simulated the whole galaxy disc or some fraction of it, either in isolation (Melioli et al. 2009; Marasco et al.

2015) or in a full cosmological box (Marinacci et al. 2014; Schaye et al. 2015). These studies allow us to study the global effect of feedback in the redistribution of gas (Brook et al. 2012), the removal of gas from the galaxy (Davé et al. 2011; Pontzen and Governato 2012), and the flow to and from the circumgalactic medium (Shen et al. 2013; Hobbs et al. 2013). However, they suffer from the choice of the feedback recipes and the prescriptions used for a number of sub-grid phenomena that one must take into account. Different groups and codes use various feedback prescriptions, often to maximize its efficiency, which are not in agreement with each other and can lead to very different results (Scannapieco et al. 2012). Moreover, the importance of phenomena often neglected in simulations like cosmic rays (Girichidis et al. 2016) and thermal conduction (Keller et al. 2014) is currently largely unconstrained.

A complementary approach is to use targeted simulations at very high (parsec scale) resolution to study the gas mixing at the disc-corona interface (Marinacci et al. 2010, 2011). With this approach, one does not aim to reproduce the entire phenomenon of supernova feedback but focuses on the physics of the interaction between disc and corona material. In these simulations, the disc material ejected by the supernova feedback is represented by a *cold* ($T \lesssim 10^4$ K) cloud that travels through an ambient medium with properties typical of the galactic corona. The cloud is typically at high metallicity (Solar or nearly so), while the corona is at lower metallicity usually 0.1 Solar, as observed in nearby galaxies (Hodges-Kluck and Bregman 2013; Bogdán et al. 2013). Clouds and corona are initially in pressure equilibrium, but this assumption is not too stringent and experiments show that if one starts with different conditions, the equilibrium is promptly reached in times $\lesssim$10 Myr. Figure 5 shows temperature snapshots of one such simulation (Marinacci et al. 2010). Here, a cold cloud with a mass of $\sim 10^4\,M_\odot$ is ejected at a small angle from the normal to the galactic disc and travels through the corona. The latter is a medium at $T = 2 \times 10^6$ K in hydrostatic equilibrium with a vertically stratified galactic potential.

The fundamental result obtained with simulations like the one shown in Fig. 5a is that the amount of cold gas in the box increases as a function of time (see Fig. 5b). This happens because part of the coronal gas condenses in the turbulent wake of the cloud. This gas will then follow the cloud back to the disc providing fresh gas accretion, this is the essence of fountain-driven (or supernova-driven) accretion.

The condensation in the cloud's wake occurs for the following reasons. The Milky Way's corona is at a temperature ($T \simeq 2 \times 10^6$ K) and density ($n \simeq 10^{-3}\,\mathrm{cm}^{-3}$, with n total gas density; Anderson and Bregman 2010) at which the cooling time is relatively long:

$$t_{\mathrm{cool}} \simeq 2.6 \times 10^9 \left(\frac{T}{2 \times 10^6\,\mathrm{K}}\right) \left(\frac{n}{10^{-3}\,\mathrm{cm}^{-3}}\right)^{-1} \left[\frac{\Lambda(T, Z)}{\Lambda(2 \times 10^6\,\mathrm{K}, 0.1\,Z_\odot)}\right]^{-1} \mathrm{yr} \tag{2}$$

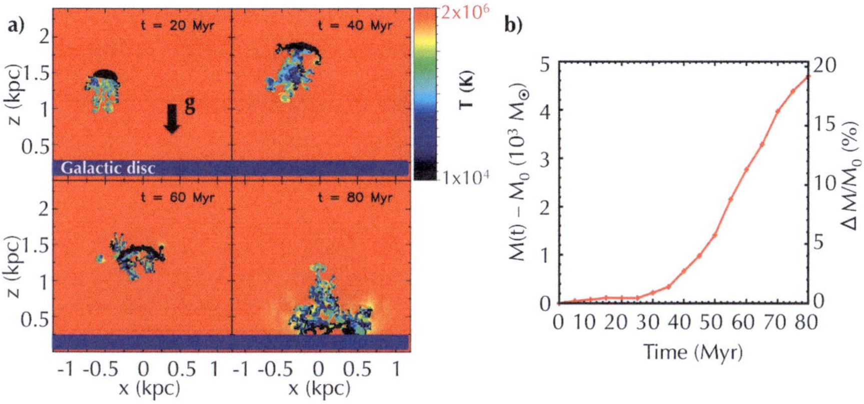

Fig. 5 (**a**) Temperature snapshots extracted at different times (see *top right corners*) from a 2D hydrodynamical simulation of a cold ($T = 10^4$ K) fountain cloud ejected upward from the Galactic plane and traveling through the hot ($T = 2 \times 10^6$ K) and metal-poor corona. (**b**) Behavior of the mass of cold/warm ($T < 10^5$ K) gas in the simulation box. The mass increases with time because part of the corona condenses in the turbulent wake of the cloud. This simulation is performed in 2D with the code ECHO++ (Marinacci et al. 2010) and with a grid size of 2 pc × 2 pc

where $\Lambda(T, Z)$ is the standard cooling function for collisional-ionization equilibrium (Sutherland and Dopita 1993) and we have used an electron density[2] $n_e \simeq 0.54n$. The value of $\Lambda(T, Z)$ at $T = 2 \times 10^6$ K is low but it increases by almost two orders of magnitude from there to $T \simeq 2 \times 10^5$ K, roughly the peak of the cooling function. In the wakes of the clouds, hot gas from the corona mixes with cooler gas from the disc thus $\Lambda(T, Z)$ moves progressively towards its peak. Moreover, the disc gas is metal-rich and this also increases the cooling rate of the mixture by up to another order of magnitude between 0.1 Solar and Solar. In conclusion, the cooling time of the coronal material entrained in the turbulent wake decreases by orders of magnitudes making its condensation possible within the fountain cycle ($t_{\rm fount} \sim 100$ Myr), see Fig. 5b.

The main consequence of condensation is a steady transfer of material at the disc-corona interface from the hot to the cold phase, which then follows nearly ballistic orbits and rains down to the disc. The average properties of the cold phase are significantly modified with respect to fountain orbits without condensation. The first modification is the increase in mass: the added mass (Fig. 5b) is made up exclusively of low-metallicity coronal material therefore providing significant accretion of metal-poor gas (∼10–20% of the initial mass of the fountain clouds). A second modification is introduced by the fact that the corona and the fountain gas

[2]Note that this is strictly valid only for $T > 10^5$ K.

have different angular momenta. To visualize this, consider a corona in equilibrium that rotates with a velocity different from that of a typical fountain cloud:

$$v_{\phi,\,\mathrm{hot}} \neq v_{\phi,\,\mathrm{cloud}} \simeq v_{\mathrm{c}}(R, z) \qquad (3)$$

where the equality is valid for a purely ballistic fountain [Sect. 3 and Eq. (1)]. Whenever T_{hot} is close to the virial temperature T_{vir}, the velocity of the corona will be lower than the circular velocity ($v_{\phi,\,\mathrm{hot}} < v_{\phi,\,\mathrm{cloud}}$). This can be intuitively understood by considering that such a corona is nearly sustained by pressure and it must give away part of its rotational support. The condensation of material from a slowly rotating corona has then the crucial consequence of decreasing the global rotation velocity of the cold extraplanar gas and reconciles it with the observations (see Fig. 4). We see now how one can derive from the simulations the momentum transfer to be introduced in the fountain model.

Marinacci et al. (2011) used simulations like the one shown in Fig. 7 but without a gravitational field. They set up fountain clouds of masses $M_{\mathrm{cold}} = \mathrm{few} \times 10^4\, M_{\odot}$ moving through a hot medium with constant density. The neglect of gravity makes little difference in the property of the wake and the condensation while greatly facilitating the interpretation of the momentum transfer. The stratification of the corona in z, induced by a gravity field, is very mild in the region where the fountain operates and the use of a constant density is also fully justified. We note that fountain clouds have typical masses (see Sect. 5) that make them fully pressure confined and self-gravity plays no role (Armillotta et al. 2016). Figure 6 (left panels) shows the velocity $v(t)$ of the cold ($T < 3 \times 10^4$ K) gas in the simulation box as a function of time starting from initial relative velocities $v_0 = 100\,\mathrm{km\,s^{-1}}$ and $v_0 = 75\,\mathrm{km\,s^{-1}}$. The velocity decreases because of two main effects: *drag* and *condensation*. The drag can be easily parameterized as described in Sect. 5 and its effect is shown by the black curve. This simple prescription, strictly valid for solids, works in fact quite well in predicting $v(t)$ in the adiabatic simulation. However, when radiative cooling is turned on, part of the coronal material condenses in the cloud's wake and further reduces the velocity of the cold gas phase. We discuss the parameterization of this condensation in Sect. 5.

The right panels of Fig. 6 show the remarkable effect that condensation has on the hot ($T > 1 \times 10^6$ K) phase. Here we see the momentum acquired by the hot gas as a function of time. Consider first the simulations without cooling, in this case the mass of the cold phase is slightly decreasing with time (not shown here) and so is its velocity, as a consequence the cold gas loses momentum. This momentum is transferred to the hot gas, which had initially null momentum. Only a negligible fraction of gas in these simulations occupies the regime at intermediate temperatures ($3 \times 10^4 < T < 10^6$ K) and thus this does not contribute to the momentum budget. This situation changes when the radiative cooling is included. The velocity of the cold gas decreases much more rapidly but, as mentioned, its mass increases (Fig. 5b). As a consequence, its momentum decreases less than in the case without cooling (Marinacci et al. 2011). This can be seen by looking at the momentum gained by the hot gas, which is significantly less than in the adiabatic

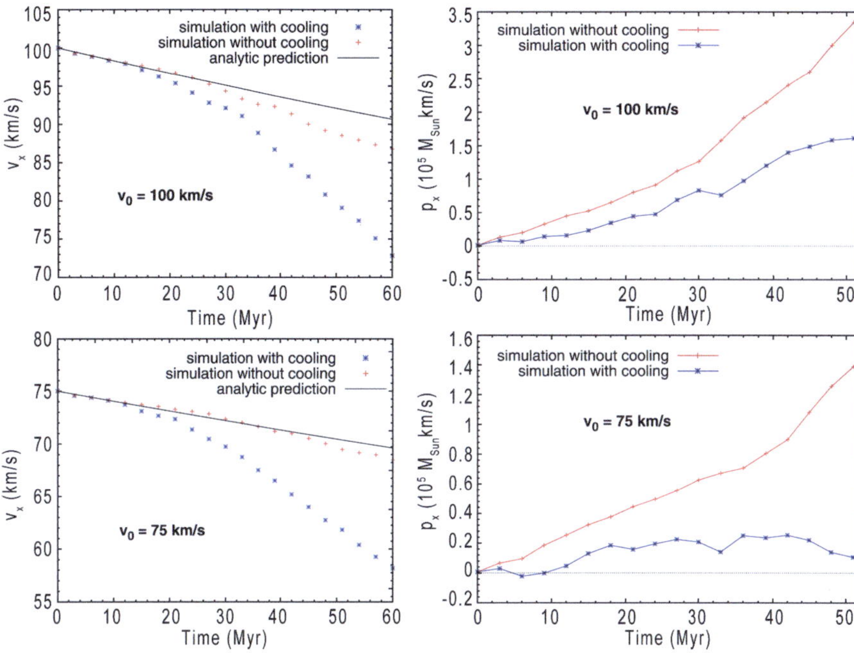

Fig. 6 Results of momentum transfer in simulations of fountain clouds that travel through a hot corona with initial relative velocities of $v = 100\,\mathrm{km\,s^{-1}}$ (*top panels*) and $v = 75\,\mathrm{km\,s^{-1}}$ (*bottom panels*). The *left panels* show the velocity of the cold ($T < 3 \times 10^4$ K) gas as a function of time, while the *right panels* show the momentum acquired by the hot gas. *Red* and *blue curves* refer to simulations in which the radiative cooling is, respectively, switched off (adiabatic) and on

case. Moreover, now part of the momentum lost by the cold gas is transferred to material at intermediate temperatures, generated by the cooling of the corona. All these effects conspire to have an approximately null transfer of momentum to the corona when the relative velocity between the cold and the hot gas falls below a threshold $v_{\rm thres} \simeq 75\,\mathrm{km\,s^{-1}}$ (Fig. 6, bottom right).

The above findings have important consequences. A problem that was identified in previous papers (Fraternali and Binney 2008) was that, given that the cold extraplanar gas has a mass larger than the lower corona (i.e., the hot gas within a few kpc from the disc, see Table 1), the transfer of momentum from the cold to the hot phase should spin the corona up to velocities so high that it would shoot away like a liquid in a centrifuge. This is obviously not happening because we observe the corona in the central parts of the halo region (Fig. 2b). The explanation is offered by the above result: there is a threshold in the velocity between extraplanar cold gas and the corona below which the hot gas does not gain momentum, therefore halting its spin-up. In other words, the momentum transferred to the hot phase through ram pressure is given back to the cold phase by the cooling of the corona. Furthermore, this process furnishes a prediction of the rotational speed that we can expect the corona to have. As long as the orbits of the fountain material are similar to the ones in Fig. 3, the largest velocity difference between corona and the fountain gas is expected to be in the azimuthal direction. The vertical velocity contributes by about $\langle v_z \rangle \simeq 40\,\mathrm{km\,s^{-1}}$ on average, while the radial velocity can be safely neglected. Thus, we can turn the momentum of the simulations in Fig. 6 into *angular momentum* and its velocity into azimuthal relative velocity $v_{\mathrm{lag}} = \sqrt{v_{\mathrm{thres}}^2 - \langle v_z \rangle^2} \simeq 63\,\mathrm{km\,s^{-1}}$. The threshold velocity changes slightly with the corona density, so, for instance, a density of $n = 0.5 \times 10^{-3}\,\mathrm{cm^{-3}}$ ($n = 2 \times 10^{-3}\,\mathrm{cm^{-3}}$) gives a $v_{\mathrm{thres}} \simeq 50\,\mathrm{km\,s^{-1}}$ ($v_{\mathrm{thres}} \simeq 85\,\mathrm{km\,s^{-1}}$) and $v_{\mathrm{lag}} \simeq 30\,\mathrm{km\,s^{-1}}$ ($v_{\mathrm{lag}} \simeq 75\,\mathrm{km\,s^{-1}}$). These values are to be considered with respect to the extraplanar cold gas, which is already lagging by about $15\,\mathrm{km\,s^{-1}\,kpc^{-1}}$ (Oosterloo et al. 2007; Marasco and Fraternali 2011). In the end, we expect the lower corona (say at $z = 1$–$2\,\mathrm{kpc}$) to rotate with a lag of $v_{\mathrm{lag}} = 45$–$105\,\mathrm{km\,s^{-1}}$ with respect to the disc rotation.[3] The rotation of the Milky Way's corona has been recently measured using O VII absorption lines and it was found to be $v_{\phi,\mathrm{hot}} = 183 \pm 41\,\mathrm{km\,s^{-1}}$ if the disc rotates at $240\,\mathrm{km\,s^{-1}}$ and thus with a $v_{\mathrm{lag}} \simeq 57\,\mathrm{km\,s^{-1}}$ (Hodges-Kluck et al. 2016), a value remarkably in line with the above prediction.

We conclude by remarking that, apart from predicting the rotation of the hot gas, the transfer of momentum from the cold to the hot phases and of mass from the hot to the cold phases has the major consequence of lowering the rotational velocity of the cold extraplanar gas. This is exactly what it is needed to reconcile the fountain model with the observations (see Sect. 3 and Fig. 4). After a short digression on the role of thermal conduction, we see in Sect. 5 how this can be implemented in large-scale fountain models allowing us to estimate the gas accretion from the corona.

An important physical process that was missing from previous simulations (Marinacci et al. 2010, 2011) is thermal conduction. Thermal (or heat) conduction causes the direct transfer of heat from regions at different temperatures and it is particularly efficient where strong temperature gradients are present in small regions of space, something that occurs very often in the disc-corona interface. In order to account for this process, Armillotta et al. (2016) used the fixed-grid hydrodynamical code *Athena* (Stone et al. 2008) and modified it to include implicit and explicit treatments of thermal conduction. They also include a standard radiative cooling function (Sutherland and Dopita 1993) and two metallicities for the different fluids. Here, we briefly discuss the effect of thermal conduction on coronal condensation.

Thermal conduction is implemented by introducing a heat flux that is described by the standard Spitzer formula (Spitzer 1962):

$$\mathbf{F}_{\mathrm{cond}} = -f\,\kappa_{\mathrm{Sp}}\,T^{5/2}\,\nabla T \tag{4}$$

[3]Note that the value of 80–$120\,\mathrm{km\,s^{-1}}$ quoted in Marinacci et al. (2011) was obtained by considering only a density of $n = 10^{-3}\,\mathrm{cm^{-3}}$ and a corona extending up to $z = 4\,\mathrm{kpc}$.

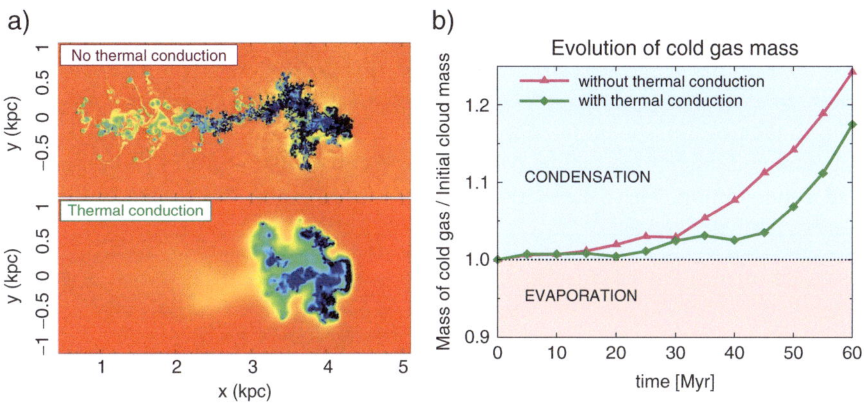

Fig. 7 (**a**) Temperature snapshots extracted after 60 Myr from a hydrodynamical simulation of a cold (10^4 K) fountain cloud moving (along the x-axis from left to right) through a hot ($T = 2 \times 10^6$ K) and metal-poor ($Z = 0.1\,Z_\odot$) medium, typical of the Milky Way's corona. The *top panel* shows a simulation with radiative cooling akin to the one in Fig. 5, while in the *bottom panel* thermal conduction is also included. (**b**) Behavior of the mass of cold ($T \leq 2 \times 10^4$ K) gas in the simulation box. The mass increases with time because of the condensation of the corona, even in the presence of thermal conduction. These simulations were performed with the code Athena on a 2D grid with cell size 2 pc × 2 pc

where T is the temperature at a given position, $\kappa_{\rm Sp} = 1.84 \times 10^{-5} \ln \Psi\, {\rm erg\, s^{-1}\, K^{-1} cm^{-1}}$, and the Coulomb logarithm is $\ln \Psi \simeq 30$ with a small dependence on temperature. The factor f in front of Eq. (4) is set to $f = 0.1$ and takes into account the suppression of conduction by magnetic fields (see Armillotta et al. 2016, for details) for details. Whenever the temperature scale length $(T/\nabla T)$ becomes shorter than the free mean paths of the electrons, the Spitzer formula is not valid any longer thus a prescription for the so-called saturated conduction (Dalton and Balbus 1993) was also implemented.

The general effect of thermal conduction is to smooth the temperature gradients at the interface between cold (fountain) gas and the hot medium of the corona. This has two main consequences. First, it makes the cloud more compact and resilient to destruction via Kelvin–Helmholtz instabilities (Vieser and Hensler 2007). Second, small cloudlets extracted from the main body of the cloud evaporate quite easily. As a result, the cloud's wake is mostly made of material at intermediate temperatures rather than very cold knots. This can be seen in Fig. 7a where we show temperature snapshots of simulations of clouds moving through the corona along the horizontal axis. These snapshots are taken after 60 Myr and apart from the switching on/off of thermal conduction they share the same initial setup. The right panel of Fig. 7 shows the mass of cold ($T \leq 2 \times 10^4$ K) gas in the simulation box as a function of time. The rate of condensation is lower in the simulation with thermal conduction. However, in the time of a typical fountain cycle (see Sect. 3), the mass of cold gas still increases substantially and thus we should expect significant gas accretion from the corona to occur in the region where the fountain operates.

5 Galactic Fountain with Accretion: Beyond the Ballistic Model

The simulations presented in Sect. 4 show that the interaction between the galactic fountain and the hot gas can be described as a composition of drag (ram pressure) and condensation. The deceleration of the fountain cloud due to the drag can be easily parameterized as:

$$\mathbf{a}_{\rm drag} = -\frac{v^2}{v_0 t_{\rm drag}} \frac{\mathbf{v}}{v} \tag{5}$$

where v is the relative velocity between the cloud and the hot gas, v_0 is the initial relative velocity. The *drag time* is:

$$t_{\rm drag} = \frac{M_{\rm cloud}}{C \rho_{\rm hot} \sigma_{\rm cloud} v_0} \tag{6}$$

where $M_{\rm cloud}$ and $\sigma_{\rm cloud}$ are the mass and the cross section of the cloud, $\rho_{\rm hot}$ is the density of the corona, and C is a dimensionless constant of order unity to account for the geometry of the cloud. Assuming that mass and cross section do not change during the cloud motion, the velocity of the cold gas decreases as follows:

$$v(t) = \frac{v_0}{1 + t/t_{\rm drag}}. \tag{7}$$

Equation (7) gives the black curves shown in the left panels of Fig. 6 and it is a good prediction for simulations without radiative cooling.

The accretion of coronal gas into the fountain wake has been described both as an exponential or a power-law growth (Fraternali and Binney 2008; Fraternali et al. 2015). Here we use an exponential growth for simplicity, thus the mass (M) of the cold phase grows as:

$$\dot{M} = \alpha M \tag{8}$$

where α is a *condensation parameter* to be determined, inverse of the condensation timescale. At each infinitesimal time step (δt) a mass δM from the corona joins the cold phase. The phenomenon can be described as an inelastic collision and the conservation of momentum gives a velocity for the cold gas at $t + \delta t$ (Fraternali and Binney 2008):

$$v(t + \delta t) = \frac{M(t)v(t) + \delta M v_{\rm hot}}{M(t) + \delta M} \tag{9}$$

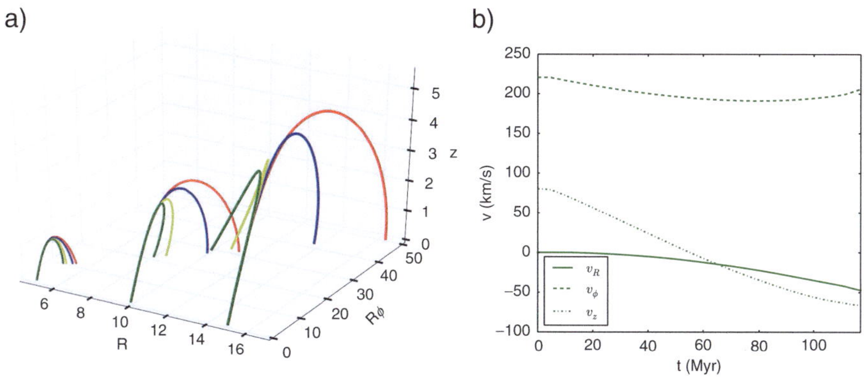

Fig. 8 (**a**) Orbits of galactic fountain particles in the potential of the Milky Way shot vertically upwards with $v_{\rm kick} = 75\,{\rm km\,s^{-1}}$ from three radii $R = 5, 10, 15$ kpc. The various effects of drag alone (*blue*), condensation/accretion alone (*pea green*), and drag+accretion (*dark green*) are shown and compared with the pure fountain (*red*, see also Fig. 3). (**b**) Cylindrical components of the velocity of the fountain particle ejected from $R = 10$ kpc when drag and accretion are taken into account

where $v_{\rm hot} = 0$ if we are considering relative velocities. Then, given that from Eq. (8) $\delta M = \alpha M \delta t$, the deceleration due to condensation simply becomes:

$$\mathbf{a}_{\rm accr} = -\alpha \mathbf{v}. \tag{10}$$

and thus v decreases exponentially at the same rate at which the mass increases. Figure 8 shows the effect of the inclusion of drag and condensation/accretion in the orbits of fountain clouds in the Milky Way's potential. The clouds are shot vertically from three radii ($R = 5, 10, 15$ kpc). The main effect of drag and accretion is the reduction of the specific angular momentum (and vertical momentum) of the cold gas. This has the following consequences: (1) the cold gas reaches lower heights (if shot with the same $v_{\rm kick}$), (2) its v_ϕ decreases more rapidly than in a pure fountain, and (3) it tends to move inward thus acquiring a negative radial velocity. As we see in Sect. 6, these effects are easily detectable in the kinematics of the extraplanar *H I* and therefore the values of the parameters of the models can be accurately determined.

In conclusion, the parameters of a galactic fountain model that includes hydrodynamical effects of the interaction between the fountain gas and the corona can be reduced to the following: (1) the characteristic kick velocity $v_{\rm kick}$, (2) the ionized fraction in the ascending part of the orbit $f_{\rm ion}$ (Sect. 3), (3) the drag time $t_{\rm drag}$, (4) the condensation/accretion rate α. In principle, all these parameters can have spatial variations although in most models until now they have been kept constant for simplicity. The drag time can be reasonably fixed by considering a typical mass for the fountain clouds of $M_{\rm cloud} > 1 \times 10^4\, M_\odot$ (Marinacci et al. 2010). Indeed Galactic Intermediate-Velocity Clouds (IVCs) and HVCs have typical massed between a few

$\times 10^4\, M_\odot$ and a few $\times 10^6\, M_\odot$, whereas smaller masses are unlikely to survive for long time in the corona (Heitsch and Putman 2009; Kwak et al. 2011). If the density does not change significantly within the cloud, its size can be simply estimated as:

$$R_{\rm cloud} = \left(\frac{3}{4\pi} \frac{T_{\rm cloud}\, \mu_{\rm hot}\, M_{\rm cloud}}{T_{\rm hot}\, \mu_{\rm cloud}\, \rho_{\rm hot}} \right)^{1/3} \tag{11}$$

where $\mu_{\rm cloud}$ and $\mu_{\rm hot}$ are the atomic weights and we have imposed pressure equilibrium between cloud and corona ($n_{\rm hot} T_{\rm hot} = n_{\rm cloud} T_{\rm cloud}$). The drag time becomes:

$$t_{\rm drag} \simeq 570 \left(\frac{M_{\rm cloud}}{10^4\, M_\odot} \right)^{1/3} \left(\frac{n_{\rm hot}}{10^{-3}\, {\rm cm}^{-3}} \right)^{-1/3} \left(\frac{T_{\rm hot}/T_{\rm cloud}}{200} \right)^{2/3} \left(\frac{v_0}{75\, {\rm km\, s}^{-1}} \right)^{-1} \text{Myr.} \tag{12}$$

for a neutral cloud, a fully ionized hot medium and assuming $C = 1$. As mentioned, we can expect that for most clouds $M_{\rm cloud} > 10^4\, M_\odot$ and $t_{\rm drag}$ in Eq. (12) to be a lower limit. Thus in general, the drag time is larger than the orbital time ($t_{\rm fount} \sim$ 100 Myr) and we can conclude that, as shown by simulations (Fig. 6), the main effect on the orbit is due to the condensation of coronal gas.

6 Observational Evidence of Fountain-Driven Accretion

The observational benchmark for the fountain accretion model described in this chapter has been the Milky Way, where several tracers have been used to test its validity and estimate the gas accretion from corona condensation. In this section we describe these successful tests in some detail. We first briefly mention early studies carried out by Fraternali and Binney (2006, 2008) on the two external galaxies NGC 891 and NGC 2403 (Fraternali and Binney 2006, 2008). The main feature of the extraplanar gas of NGC 891 is the lag in rotation that is not reproduced by a simple galactic fountain model (Fig. 4b). The introduction of accretion onto fountain clouds following Eq. (8) produces the nice match with the data shown by the blue curve in Fig. 4. This fit to the data requires an accretion of $\dot{M}_{\rm acc} = 3\, M_\odot\, {\rm yr}^{-1}$, slightly lower than the galaxy's SFR of $3.8\, M_\odot\, {\rm yr}^{-1}$ (Popescu et al. 2004).

The second galaxy that was considered in these early works is NGC 2403, a smaller galaxy with relatively high SFR=$1.2\, M_\odot\, {\rm yr}^{-1}$ (Kennicutt et al. 2003) and large amounts of extraplanar gas (Fraternali et al. 2002). The kinematics of the extraplanar HI is characterized by a lagging rotation and a global inflow towards the center. Remarkably, the fountain accretion models are able to reproduce this pattern very well (Fraternali and Binney 2008). The inflow is essentially due to the acquisition of low angular momentum from the ambient gas (Fig. 8). In order to reproduce the observed kinematics of the extraplanar gas, accretion at a rate of about $\dot{M}_{\rm acc} \simeq 0.8\, M_\odot\, {\rm yr}^{-1}$ is required. We stress that these accretion rates are estimated through the kinematics of the extraplanar gas that has no direct relation with the

SFR except through the fountain model. Thus, the fact that the accretion rates are found be similar the SFRs is a strong confirmation of the soundness of the fountain accretion model (or an extraordinary coincidence).

In more recent years, the fountain accretion model was used to reproduce the *H I* emission of the extraplanar neutral gas in the Milky Way (Marasco and Fraternali 2011). This component had been studied in several papers before mostly in the form of individual clouds, IVCs or HVCs (Richter et al. 2001; van Woerden et al. 2004). The general consensus is that IVCs, being at relatively small anomalous velocities and at nearly Solar metallicity, are fountain clouds, whereas HVCs are more likely external clouds. This separation is now rather debated and it may be less neat than previously thought. IVCs are just a local manifestation of a much broader component that extends above and below the plane and constitutes the extraplanar layer of the Milky Way (see Table 1) (Marasco and Fraternali 2011). HVCs, to some extent, may be produced by extreme and rare events of the same kind that produced IVCs (see below). Clearly, the first question to address in this context is whether the extraplanar *H I* is reproduced by a fountain model and whether its kinematics requires interactions with the galactic corona.

To answer this question, Marasco et al. (2012) built a model of galactic fountain that included the effect of drag and condensation (see Sect. 5). This model predicts the *H I* emission from fountain clouds that can be compared with the datacube from the Leiden-Argentina-Bonn (LAB) survey (Kalberla et al. 2005). The comparison is carried out by building a suite of all-sky pseudo-datacubes with different values of the main parameters and calculating the residuals between models and data in the 3D-space (position on the sky and line-of-sight velocity). Only locations where the extraplanar emission is present are used for the minimization (i.e., the Galactic disc is excluded). Figure 9a shows the comparison between models with and without coronal condensation and the *H I* data. Data and models are shown as longitude-velocity plots extracted at two representative latitudes above and below the plane ($b = \pm 30°$). At several locations, one can appreciate that the inclusion of gas accretion from the corona improves the fit substantially (see arrows). These are, in particular, locations where radial inflow motions, typical of an interacting fountain (Fig. 8), are present.

Figure 9b shows the values of the residuals between data and models obtained varying the two main parameters of the galactic fountain: the condensation parameter α and the ionization fraction $f_{\rm ion}$. A clear minimum is found for $f_{\rm ion} = 0.3$ (the gas is ionized for the initial 30% of the ascending part of the orbits) and $\alpha = 6.3\,{\rm Gyr}^{-1}$. This latter corresponds to a condensation/accretion timescale of $t_{\rm acc} = 160\,{\rm Myr}$, which then translates into a global accretion rate of $\dot{M}_{\rm acc} \simeq 2\,M_{\odot}\,{\rm yr}^{-1}$ including helium. In performing this fit the condensation timescale was not fixed to the value given by hydrodynamical simulations (Fig. 6). However, it remarkably turned out to be compatible with what the simulations predict (details in Marasco et al. 2012). Note also that, as expected, it is shorter than the drag time [Eq. (12)]. The model predicts that nearly 20–25% of the extraplanar *H I* in the Milky Way is made up of gas condensed from the corona, while the other 75–80% is fountain material.

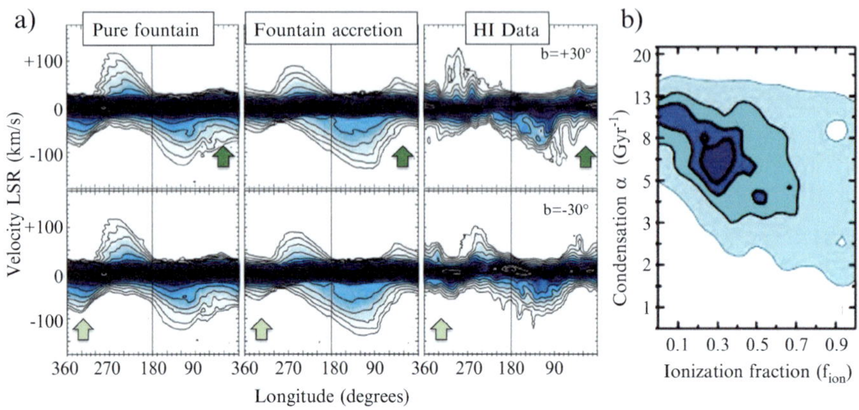

Fig. 9 (**a**) Comparison between the H I emission predicted by two fountain models and the LAB data (*third column*) in longitude-velocity plots extracted at two latitudes (*top*: $b = 30°$; *bottom*: $b = -30°$). The *first column* shows the best-fit pure galactic fountain and the *second* the best-fit fountain accretion model. The *arrows* indicate regions where the accretion model reproduces the data better than the pure fountain. (**b**) Confidence levels of residuals between models with corona condensation and data evaluated in the parameter space (α,$f_{\rm ion}$) for $v_{\rm kick} = 70\,{\rm km\,s^{-1}}$. The minimum corresponds to a global *H I* accretion rate onto the Milky Way's disc $\dot{M}_{\rm acc} = 1.6\,M_\odot\,{\rm yr}^{-1}$

Table 2 Observational evidence of fountain-driven accretion

Object	Tracer	DATA	SFR ($M_\odot\,{\rm yr}^{-1}$)	$v_{\rm kick}$ (${\rm km\,s^{-1}}$)	$f_{\rm ion}$	Accretion rate ($M_\odot\,{\rm yr}^{-1}$)	Ref.
NGC 891[a]	Extraplanar *H I*	WSRT	3.8	80	0–1	∼3	(a)
NGC 2403[a]	Extraplanar *H I*	VLA	1.2	55	0	0.8	(a)
Milky Way	Extraplanar *H I*	LAB	1–3	70	0.3	∼2	(b)
Milky Way	Metal absorbers	HST/COS	1–3	70[b]	0.3[b]	∼1	(c)
HVC complex C	*H I*+metals	LAB+COS	–	210	–	∼0.25[c]	(d)

[a]In these early works the ambient accreting medium was modeled as infalling and non-rotating. [b]These parameters were not fit to the COS data but kept fixed to the values obtained from the *H I* fit. [c]This is the total accretion rate if complex C keeps condensing at an exponential rate and all its warm/cold gas makes it back to the disc

References: (a) Fraternali and Binney (2008); (b) Marasco et al. (2012); (c) Fraternali et al. (2013); (d) Fraternali et al. (2015)

Individual clouds (IVCs/HVCs) may have different ratios depending on their past trajectories.

The application of the fountain accretion model to the Milky Way is successful at reproducing the detailed kinematics of the extraplanar *H I* and it returns a global accretion rate of the order of the Galactic SFR (Table 2). We now discuss an interesting prediction of this model. Figure 10 shows a comparison between the SFR surface density in the Milky Way (left) and the gas accretion surface density (right) predicted by the fountain accretion model. Clearly, most of the accretion tends to occur at larger radii with respect to the peak of the SFR density. This is

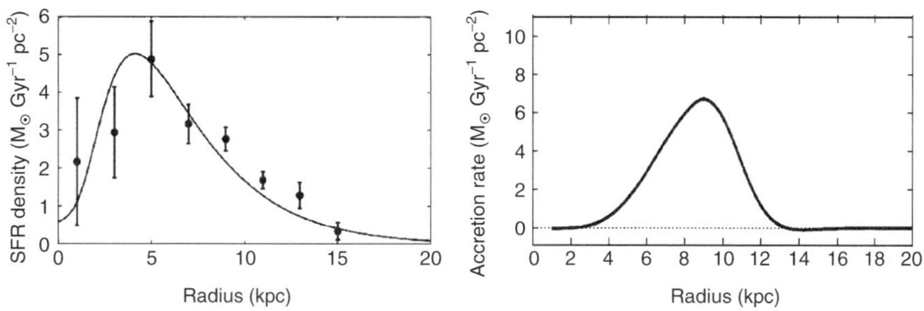

Fig. 10 *Left*: Star formation rate surface density in the Milky Way from supernova remnants (Case and Bhattacharya 1998); figure from Pezzulli and Fraternali (2016). *Right*: Accretion rate surface density from corona condensation predicted by the fountain-driven accretion model; figure from Marasco et al. (2012)

a consequence of the way the fountain operates: on the one hand, the orbital paths increase with radius (see Fig. 8) and the gas mass accreted from the corona increases with the orbital path [Eq. (8)]. On the other hand, the fountain flow (ejection rate density) has a shape proportional to the SFR density. The combined effect of the above is the accretion rate in Fig. 10 (right panel). The fate of the gas accreted from the corona at large radii is to move inward (before forming stars) because it has a rotation velocity lower than the local circular velocity [Eq. (3)]. We discuss this further in Sect. 7.

The cooling of a significant fraction of the Galactic corona in the region close to the disc must also produce a large amount of gas at temperatures intermediate between the virial temperature and that of the cold *H I* gas. Material at these temperatures is indeed observed surrounding the Milky Way's disc, thanks to the detection of absorption features towards quasars (Sembach et al. 2003; Shull et al. 2009). Recently, these absorbers have been identified also in the spectra of halo stars, which have the advantage to give an upper limit to the distance of the gas (the estimated distance of the star Lehner and Howk 2011). Fraternali et al. (2013) compared the predictions of the fountain accretion model with observations from the HST Cosmic Origins Spectrograph (COS) of low-ionization absorbers like Si III, Si IV, C II, C III and C IV, which encompass the temperature range $4.3 < \log(T/K) < 5.3$ (see also Marasco et al. 2013). Figure 11a shows a sketch of the physical process. The intermediate-temperature material arises in the turbulent wakes of fountain clouds, where coronal condensation occurs. The properties of the intermediate-temperature material with respect to the *H I* (velocity lag and relative density) were derived from hydrodynamical simulations like those in Sect. 4. Figure 11b shows the areas in the longitude-velocity space where the fountain model predicts the presence of intermediate-temperature gas compared to the data (points Lehner et al. 2012). A statistical test shows that 94% of the detected low-ionization absorbers are consistent with being produced in fountain wakes. The model also predicts the observed number of features along a typical line of sight and explains the large line broadening as due to turbulence in the wake. We stress that

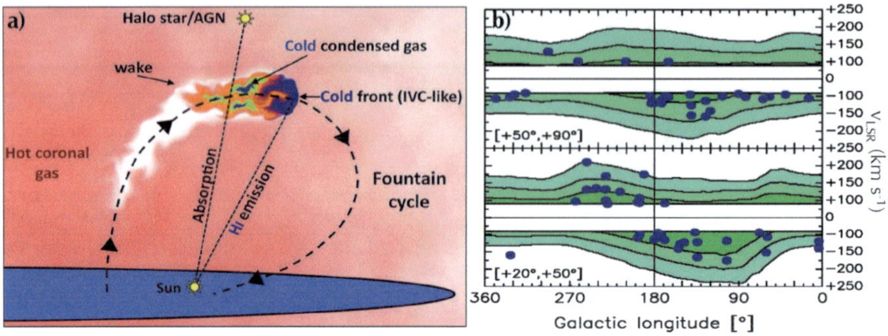

Fig. 11 (**a**) A sketch of a fountain cloud, ejected from the disc by supernova feedback and interacting with the hot gas in the corona. In the turbulent wake, coronal gas mixes with high-metallicity disc material triggering the condensation of a fraction of the corona, which is then accreted onto the disc. An observer looking towards a background source intercepting the wake detects absorption lines from the ionized intermediate-temperature material. An observer looking towards the cold front detects *H I* emission at velocities typical of IVCs and, occasionally, HVCs. (**b**) Predicted locations (*darker shades* show higher probability regions) in longitude and velocity of the intermediate-temperature material based on the fountain accretion model. The points show the HST/COS absorption features detected in the temperature range $4.3 < \log(T/K) < 5.3$. The vast majority (94%) of this ionized gas is consistent with being generated in wakes of fountain clouds

to obtain these results, the parameters of the fountain accretion model were not fit to the COS data but kept fixed to those (see Table 2) that reproduce the kinematics of the *H I* (Fig. 9).

We now turn to the Galactic HVCs (Wakker and van Woerden 1997), whose origin has been debated since their discovery with the two competing scenarios being either gas accreting from the intergalactic medium into the Milky Way (Oort 1970) or galactic fountain material (Bregman 1980). The estimate of a low metallicity for the prototypical cloud complex C (Wakker et al. 1999) pointed strongly at an external origin but the source of the accretion remained unknown. With the exclusion of spontaneous thermal instabilities in the corona (Binney et al. 2009), other possibilities are gas from satellites (Olano 2008) or accretion from cosmological filaments (Fernández et al. 2012). Fraternali et al. (2015) investigated whether complex C could have originated by the blowout of a powerful superbubble in the disc of the Milky Way that seeded the condensation of a large fraction of the lower corona. They calculated orbits of fountain clouds ejected from regions in the Milky Way's spiral arms and run an MCMC minimization between the predicted *H I* emission from these clouds and the LAB data of complex C. The results are shown in Fig. 12 and Table 2.

The most massive and metal poor of all Galactic HVCs appears to be the result of a superbubble blowout in the Cygnus-Outer arm occurred 150 Myr ago. The ejection of $\sim 3\times 10^6\, M_\odot$ of fountain gas has triggered the subsequent condensation of a similar amount of coronal material. This lowered the rotation velocity and the metallicity of the fountain material. The model is fitted only to the *H I* emission

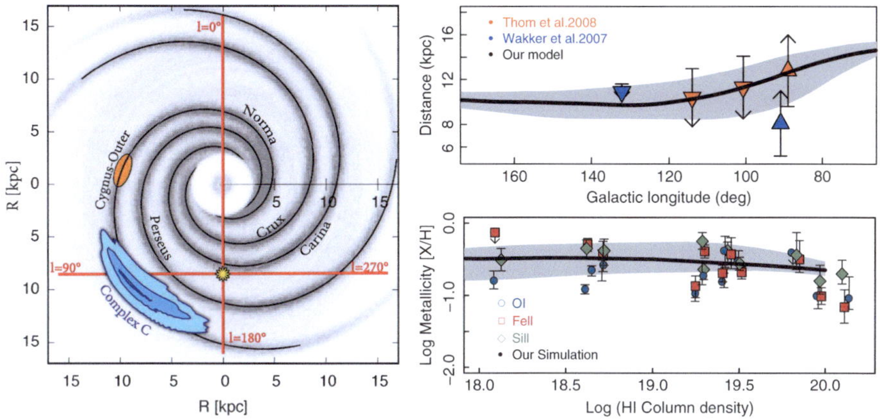

Fig. 12 *Left*: Location of the HVC complex C seen from a face-on view. Contours are obtained from a fountain accretion model and correspond to *H I* column densities of 1, 2, and 3 $\times 10^{19}$ cm^{-2} (Fraternali et al. 2015). The spiral arms, the position of the Sun, and the location of the superbubble blowout (*orange ellipse*) of the seed material are shown. *Right*: Comparison between the distance from the Sun as a function of longitude (*top*) and the abundances of oxygen, iron, and silicon as a function of *H I* column density (*bottom*) in the data (points and limits) and in hydrodynamical simulations of the complex C (*black curve*). The *gray bands* show the 16 and 84 percentiles. Adapted from Fraternali et al. (2015)

but it reproduces also the distance of complex C and its metallicity (see Fig. 12, right). The event that formed complex C requires an energy of $E \simeq 2 \times 10^{54}$ erg, which can be achieved by $\sim 1 \times 10^4$ supernovae assuming 20% efficiency. This is probably a rare event although ejections of similar *H I* masses are observed in spiral galaxies (Boomsma et al. 2008). Most superbubble blowouts likely produce material at lower anomalous velocities that end up being observed as IVCs. However, some (perhaps all) HVCs can be produced in this way. The recent discovery of a fountain-like metallicity for the Smith Cloud indeed suggests a similar formation mechanism (Fox et al. 2016; Marasco and Fraternali 2017).

7 Galaxy Evolution with Fountain Accretion

In Sect. 6 we have seen a number of successful comparisons between the predictions of the fountain accretion model and a variety of datasets. In summary, the model reproduces the kinematics of the cold/warm gas at the interface of the galactic corona in great detail (Marasco et al. 2012; Fraternali et al. 2013) and predicts the lag of the corona with respect to the disc recently measured in the Milky Way (Marinacci et al. 2011; Hodges-Kluck et al. 2016). Most important of all, it accounts for accretion of low-metallicity material at a rate that is comparable to the SFR, providing an elegant solution to the gas accretion problem in local spiral galaxies.

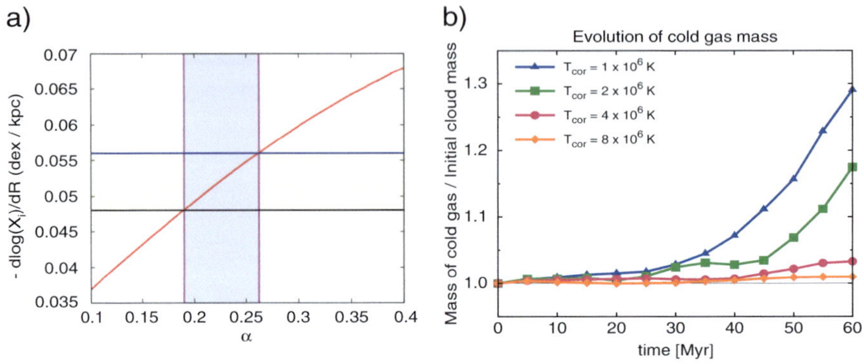

Fig. 13 (**a**) Abundance gradient (*red curve*) predicted for the Milky Way's disc as a function of the average rotation-velocity mismatch between the accreting and the disc gas ($\alpha = 1 - \frac{v_{\rm acc}}{v_{\rm disc}}$). The *shaded areas* show the regions allowed by the data (Luck and Lambert 2011), which correspond to material accreting with an average rotational velocity $v_{\phi,\,\rm acc} \simeq 170$–$195\,\rm km\,s^{-1}$ in agreement with recent observations (Hodges-Kluck et al. 2016); figure from Pezzulli and Fraternali (2016). (**b**) Corona condensation in the wakes of fountain clouds at the disc-halo interface as a function of the temperature of the corona. The condensation is very efficient for temperatures similar or lower than that of the Milky Way ($T \simeq 2 \times 10^6$ K) and it vanishes for temperatures above 4×10^6 K; figure from Armillotta et al. (2016)

In the future, much should be done to test the model against other observables and incorporate it into a self-consistent theory of galaxy evolution. In this section, we briefly describe two recent investigations in this direction that hint at the role of fountain accretion in the onset of metallicity gradients and the quenching of star formation.

Pezzulli and Fraternali (2016) built a model of disc evolution in the presence of a generic gas accretion (not necessarily fountain-driven) characterized by a rotation velocity different from the local circular velocity. In this scenario, which is very likely if the accreting gas comes from the condensation of the corona (Sect. 5), the accreted material lands in the disc with a deficit of specific angular momentum. The natural consequence is to induce inward radial flows of gas and the strength of these flows is regulated by the angular momentum mismatch between disc and accreting gas. These flows may be of great importance to bring gas from the outer regions of galaxies into the inner star-forming regions. Luckily, they have a strong impact on abundance gradients, providing an easily observable test of the theory (Bilitewski and Schönrich 2012). General analytic solutions for this problem have been found both in the absence or presence of inside-out growth of the disc (Pezzulli and Fraternali 2016). The current abundance gradient is steeper the higher the angular momentum mismatch. Figure 13a (red curve) shows the predicted gradient as a function of the average rotation-velocity mismatch of the accreting material for the Milky Way. The observed gradients are compatible with accretion at about 70–80 % of the local circular speed, i.e., with a $v_{\rm lag} \simeq 45$–$70\,\rm km\,s^{-1}$ with respect to the disc rotation. This value is very similar to the rotation of the corona that is required to fit the kinematics of the extraplanar gas in the Milky Way (Sect. 6). Thus, it appears that metallicity gradients as steep as those observed in local galaxies are compatible with a supernova-driven prescription for gas accretion.

In a second work, Armillotta et al. (2016) investigated the efficiency of condensation as a function of the coronal temperature using simulations that include thermal conduction and are akin to those shown in Fig. 7. The main result is shown in the right panel of Fig. 13, where we see the amount of condensation with respect to the initial mass of the cold gas as a function of time. Coronae with temperatures similar or lower than that of the Milky Way ($T \simeq 2 \times 10^6$ K) cool very efficiently but this trend stops for temperatures above $\sim 4 \times 10^6$ K. If we assume that the coronal temperature is of order of the virial temperature, then we can relate it to the virial mass of the dark matter halo. We conclude that for virial masses larger than the Milky Way fountain-driven condensation and cooling of the corona become less and less efficient and these galaxies will struggle to gather new gas from their environment. Gas accretion stops altogether for masses $M \gtrsim 10^{13} M_\odot$.

These findings may have important implications for the quenching of star formation in galaxies. As the virial mass of a galaxy grows because of dark matter assembly, its ability to cool the surrounding corona decreases and it may end up consuming all the available gas and thus quench. Moreover, when the galaxy enters a dense environment (group or cluster), its extended corona is extremely vulnerable to stripping and the galaxy will naturally become surrounded by gas at much higher-temperature. In these conditions, corona condensation stops and the star formation is starved of fuel. Thus, both mass and environmental quenching (Knobel et al. 2015) could be accommodated into the fountain accretion mechanism.

8 Concluding Remarks

In this chapter, we have presented the physical principles of the fountain-driven (or supernova-driven) accretion model and its successes in reproducing observational data for local galaxies. This model has been developed using tailored hydrodynamical simulations of the disc-corona interface. Physically motivated prescriptions from simulations have also been incorporated into large-scale galactic fountain models. The comparison with the data is done by building artificial observations (in particular *H I* pseudo-datacubes) and minimizing the residuals between these and the actual data. A general prediction is that, in the physical conditions typical of star-forming galaxies, significant accretion occurs from the cooling of the lower corona at the disc-halo interface. The cooling (condensation) is regulated by supernova feedback which can stimulate gas accretion at rates comparable with the galaxy SFR.

Cosmological simulations predict that galaxies with current virial masses above $10^{12} M_\odot$, after an initial phase of *cold-mode* (filamentary) accretion (Dekel et al. 2009), which lasts until $z \simeq 1$–2, move to a so-called *hot-mode* accretion phase (Kereš et al. 2009; Nelson et al. 2013). This accretion builds and feeds extended galactic coronae, which are unable to cool efficiently except at their very center where they can trigger episodic feedback from the central black hole (Binney and Fraternali 2012). Paradoxically, it is after $z \simeq 1$–2 that a galaxy like the Milky Way needs accretion of cold gas as the majority of its stars (at least the stars of the thin disk) are formed after $z = 1$ (Snaith et al. 2015). Fountain-driven accretion provides an explanation for this apparent contradiction, as it triggers continuous cooling of the lower corona across the whole star-forming disc. This mechanism could be the way star-forming galaxies extend their lives beyond the gas depletion time in the long epoch of hot-mode accretion.

We can also speculate that the shift of galaxies from the blue cloud to the red sequence is a consequence of a drop in the efficiency in gathering new gas. This can happen because the corona becomes too hot, the galaxy falls into a larger halo (Sect. 7) or because it simply loses its gaseous disc, e.g., due to mergers or gas stripping. In the fountain accretion scenario, the key ingredient for efficient accretion is the presence of a star-forming disc of cold gas, which acts as the "refrigerator" of the corona. If a galaxy loses this disc it may take long time to reform it (or it may never happen) and it will permanently become a red early type galaxy.

Acknowledgements I thank Lucia Armillotta, James Binney, Antonino Marasco, Federico Marinacci, and Gabriele Pezzulli for comments on the manuscript and Cecilia Bacchini for providing Figs. 3 and 8.

References

Anderson, M. E., & Bregman, J. N. 2010, ApJ, 714, 320
Armillotta, L., Fraternali, F., & Marinacci, F. 2016, MNRAS, 462, 4157
Barnabè, M., Ciotti, L., Fraternali, F., & Sancisi, R. 2006, A&A, 446, 61
Bilitewski, T., & Schönrich, R. 2012, Radial flows and angular momentum conservation in Galactic chemical evolution. MNRAS, 426, 2266–2282. arXiv:1208.0003. doi:10.1111/j.1365-2966.2012.21827.x. http://adsabs.harvard.edu/abs/2012MNRAS.426.2266B
Binney, J. & Fraternali, F. 2012, in European Physical Journal Web of Conferences, 19, 8001
Binney, J., Nipoti, C., & Fraternali, F. 2009, MNRAS, 397, 1804
Bogdán, Á., Forman, W. R., Vogelsberger, et al. 2013, ApJ, 772, 97
Boomsma, R., Oosterloo, T. A., Fraternali, F., van der Hulst, J. M., & Sancisi, R. 2008, A&A, 490, 555
Bregman, J. N. 1980, ApJ, 236, 577
—. 2007, ARA&A, 45, 221
Brook, C. B., Stinson, G., Gibson, B. K., et al. 2012, MNRAS, 419, 771
Case, G. L. & Bhattacharya, D. 1998, ApJ, 504, 761
Catinella, B., Schiminovich, D., Kauffmann, G., et al. 2012, A&A, 544, A65
Chiappini, C., Matteucci, F., & Romano, D. 2001, ApJ, 554, 1044
Collins, J. A., Benjamin, R. A., & Rand, R. J. 2002, ApJ, 578, 98
Dai, X., Anderson, M. E., Bregman, J. N., & Miller, J. M. 2012, ApJ, 755, 107
Dalton, W. W. & Balbus, S. A. 1993, ApJ, 404, 625
Davé, R., Oppenheimer, B. D., & Finlator, K. 2011, MNRAS, 415, 11
Dekel, A., Birnboim, Y., Engel, G., et al. 2009, Nature, 457, 451

Di Teodoro, E. M. & Fraternali, F. 2014, A&A, 567, A68
Fasano, G., Poggianti, B. M., Couch, W. J., et al. 2000, ApJ, 542, 673
Fernández, X., Joung, M. R., & Putman, M. E. 2012, ApJ, 749, 181
Fox, A. J., Lehner, N., Lockman, F. J., et al. 2016, ApJ, 816, L11
Fox, A. J., Wakker, B. P., Barger, K. A., et al. 2014, ApJ, 787, 147
Fraternali, F. 2009, in IAU Symposium Vol. 254, ed. J. Andersen, J. Bland-Hawthorn, & B. Nordström, 255
Fraternali, F. 2012, in EAS Publications Series, Vol. 56, ed. M. A. de Avillez, 355
Fraternali, F. & Binney, J. J. 2006, MNRAS, 366, 449
—. 2008, MNRAS, 386, 935
Fraternali, F., Marasco, A., Armillotta, L., & Marinacci, F. 2015, MNRAS, 447, L70
Fraternali, F., Marasco, A., Marinacci, F., & Binney, J. 2013, ApJ, 764, L21
Fraternali, F., Oosterloo, T., & Sancisi, R. 2004, A&A, 424, 485
Fraternali, F., Oosterloo, T., Sancisi, R., & van Moorsel, G. 2001, ApJ, 562, L47
Fraternali, F., Oosterloo, T. A., Sancisi, R., & Swaters, R. 2005, in ASP Conf. Ser. Vol. 331, *Extra-Planar Gas*, ed. R. Braun, 239
Fraternali, F. & Tomassetti, M. 2012, MNRAS, 426, 2166
Fraternali, F., van Moorsel, G., Sancisi, R., & Oosterloo, T. 2002, AJ, 123, 3124
Fukugita, M. & Peebles, P. J. E. 2004, ApJ, 616, 643
Gatto, A., Fraternali, F., Read, J. I., et al. 2013, MNRAS
Girichidis, P., Naab, T., Walch, S., et al. 2016, ApJ, 816, L19
Goldbaum, N. J., Krumholz, M. R., & Forbes, J. C. 2015, ApJ, 814, 131
Heald, G., Józsa, G., Serra, P., et al. 2011, A&A, 526, A118
Heald, G. H., Rand, R. J., Benjamin, R. A., & Bershady, M. A. 2007, ApJ, 663, 933
Heitsch, F. & Putman, M. E. 2009, ApJ, 698, 1485
Hobbs, A., Read, J., Power, C., & Cole, D. 2013, MNRAS, 434, 1849
Hodges-Kluck, E. J. & Bregman, J. N. 2013, ApJ, 762, 12
Hodges-Kluck, E. J., Miller, M. J., & Bregman, J. N. 2016, ApJ, 822, 21
Houck, J. C. & Bregman, J. N. 1990, ApJ, 352, 506
Howk, J. C. & Savage, B. D. 2000, AJ, 119, 644
Joung, M. R., Bryan, G. L., & Putman, M. E. 2012, ApJ, 745, 148
Kalberla, P. M. W., Burton, W. B., Hartmann, D., & et al. 2005, A&A, 440, 775
Kamphuis, P. 2008, PhD thesis, University of Groningen
Kaufmann, T., Bullock, J. S., Maller, A. H., Fang, T., & Wadsley, J. 2009, MNRAS, 396, 191
Kaufmann, T., Mayer, L., Wadsley, J., Stadel, J., & Moore, B. 2006, MNRAS, 370, 1612
Keller, B. W., Wadsley, J., Benincasa, S. M., & Couchman, H. M. P. 2014, MNRAS, 442, 3013
Kennicutt, Jr., R. C., Armus, L., Bendo, G., et al. 2003, PASP, 115, 928
Kereš, D., Katz, N., Fardal, M., Davé, R., & Weinberg, D. H. 2009, MNRAS, 395, 160
Knobel, C., Lilly, S. J., Woo, J., & Kovač, K. 2015, ApJ, 800, 24
Kubryk, M., Prantzos, N., & Athanassoula, E. 2013, MNRAS, 436, 1479
Kwak, K., Henley, D. B., & Shelton, R. L. 2011, ApJ, 739, 30
Lehner, N. & Howk, J. C. 2011, Science, 334, 955
Lehner, N., Howk, J. C., Thom, C., et al. 2012, MNRAS, 424, 2896
Lewis, G. F., Braun, R., McConnachie, A. W., et al. 2013, ApJ, 763, 4
Lilly, S. J., Carollo, C. M., Pipino, A., Renzini, A., & Peng, Y. 2013, ApJ, 772, 119
Lockman, F. J. 2002, ApJ, 580, L47
Luck, R. E. & Lambert, D. L. 2011, AJ, 142, 136
Mac Low, M., McCray, R., & Norman, M. L. 1989, ApJ, 337, 141
Mac Low, M.-M. & McCray, R. 1988, ApJ, 324, 776
Marasco, A., Debattista, V. P., Fraternali, F., et al.
Marasco, A. & Fraternali, F. 2011, A&A, 525, A134
— 2017, MNRAS, 464, L100
Marasco, A., Fraternali, F., & Binney, J. J. 2012, MNRAS, 419, 1107
Marasco, A., Marinacci, F., & Fraternali, F. 2013, MNRAS, 433, 1634

Marinacci, F., Binney, J., Fraternali, F., et al. 2010, MNRAS, 404, 1464
Marinacci, F., Fraternali, F., Nipoti, C., et al. 2011, MNRAS, 415, 1534
Marinacci, F., Pakmor, R., & Springel, V. 2014, MNRAS, 437, 1750
Matthews, L. D. & Wood, K. 2003, ApJ, 593, 721
McGaugh, S. S., Schombert, J. M., de Blok, W. J. G., & Zagursky, M. J. 2010, ApJ, 708, L14
Melioli, C., Brighenti, F., D'Ercole, A., & de Gouveia Dal Pino, E. M. 2008, MNRAS, 388, 573
—. 2009, MNRAS, 399, 1089
Miller, M. J. & Bregman, J. N. 2015, ApJ, 800, 14
Nelson, D., Vogelsberger, M., Genel, S., et al. 2013, MNRAS, 429, 3353
Nichols, M., Colless, J., Colless, M., & Bland-Hawthorn, J. 2011, ApJ, 742, 110
Olano, C. A. 2008, A&A, 485, 457
Oort, J. H. 1970, A&A, 7, 381
Oosterloo, T., Fraternali, F., & Sancisi, R. 2007, AJ, 134, 1019
Papastergis, E., Cattaneo, A., Huang, S., Giovanelli, R., & Haynes, M. P. 2012, ApJ, 759, 138
Peng, Y.-j., Lilly, S. J., Kovač, K., et al. 2010, ApJ, 721, 193
Pezzulli, G. & Fraternali, F. 2016, MNRAS, 455, 2308
Pezzulli, G., Fraternali, F., Boissier, S., & Muñoz-Mateos, J. C. 2015, MNRAS, 451, 2324
Pidopryhora, Y., Lockman, F. J., & Shields, J. C. 2007, ApJ, 656, 928
Pisano, D. J., Barnes, D. G., Staveley-Smith, L., et al. 2011, ApJS, 197, 28
Pontzen, A. & Governato, F. 2012, MNRAS, 421, 3464
Popescu, C. C., Tuffs, R. J., Kylafis, N. D., & Madore, B. F. 2004, A&A, 414, 45
Puche, D., Westpfahl, D., Brinks, E., & Roy, J.-R. 1992, AJ, 103, 1841
Putman, M. E., Peek, J. E. G., & Joung, M. R. 2012, ARA&A, 50, 491
Putman, M. E., Staveley-Smith, L., Freeman, K. C., Gibson, B. K., & Barnes, D. G. 2003, ApJ, 586, 170
Read, J. I. & Hayfield, T. 2012, MNRAS, 422, 3037
Richter, P., Fox, A. J., Wakker, B. P., et al. 2013, ApJ, 772, 111
Richter, P., Sembach, K. R., Wakker, B. P., & et al. 2001, ApJ, 559, 318
Rossa, J. & Dettmar, R.-J. 2003, A&A, 406, 505
Saintonge, A., Lutz, D., Genzel, R., et al. 2013, ApJ, 778, 2
Sánchez Almeida, J., Elmegreen, B. G., Muñoz-Tuñón, C., & Elmegreen, D. M. 2014, A&A Rev., 22, 71
Sancisi, R., Fraternali, F., Oosterloo, T., & van der Hulst, T. 2008, A&A Rev., 15, 189
Scannapieco, C., Wadepuhl, M., Parry, O. H., et al. 2012, MNRAS, 423, 1726
Schaap, W. E., Sancisi, R., & Swaters, R. A. 2000, A&A, 356, L49
Schaye, J., Crain, R. A., Bower, et al. 2015, MNRAS, 446, 521
Schlegel, E. M., Holt, S. S., & Petre, R. 2003, ApJ, 598, 982
Schönrich, R. & Binney, J. 2009, MNRAS, 396, 203
Sembach, K. R., Wakker, B. P., Savage, B. D., et al. 2003, ApJS, 146, 165
Shapiro, P. R. & Field, G. B. 1976, ApJ, 205, 762
Shen, S., Madau, P., Guedes, J., et al. 2013, ApJ, 765, 89
Shull, J. M., Jones, J. R., Danforth, C. W., & Collins, J. A. 2009, ApJ, 699, 754
Snaith, O., Haywood, M., Di Matteo, P., et al. 2015, A&A, 578, A87
Sokołowska, A., Mayer, L., Babul, A., Madau, P., & Shen, S. 2016, ApJ, 819, 21
Spitoni, E., Recchi, S., & Matteucci, F. 2008, A&A, 484, 743
Spitzer, L. 1962, Physics of Fully Ionized Gases
Stone, J. M., Gardiner, T. A., Teuben, P., Hawley, J. F., & Simon, J. B. 2008, ApJS, 178, 137
Sutherland, R. S. & Dopita, M. A. 1993, ApJS, 88, 253
Swaters, R. A., Sancisi, R., & van der Hulst, J. M. 1997, ApJ, 491, 140
Tosi, M. 1988, A&A, 197, 47
van Uitert, E., Cacciato, M., Hoekstra, H., et al. 2016, MNRAS, 459, 3251
van Woerden, H., Wakker, B. P., Schwarz, U. J., & de Boer, K. S. 2004, *High-Velocity Clouds*, Astrophysics and Space Science Library, Vol. 312
Veilleux, S., Cecil, G., & Bland-Hawthorn, J. 2005, ARA&A, 43, 769

Vieser, W. & Hensler, G. 2007, A&A, 475, 251
Wakker, B. P., Howk, J. C., Savage, B. D., & et al. 1999, Nature, 402, 388
Wakker, B. P. & van Woerden, H. 1997, ARA&A, 35, 217
Werk, J. K., Prochaska, J. X., Thom, C., et al. 2013, ApJS, 204, 17
Werk, J. K., Prochaska, J. X., Tumlinson, J., et al. 2014, ApJ, 792, 8
Yun, M. S., Ho, P. T. P., & Lo, K. Y. 1994, Nature, 372, 530
Zafar, T., Péroux, C., Popping, A., et al. 2013, A&A, 556, A141
Zschaechner, L. K., Rand, R. J., & Walterbos, R. 2015, ApJ, 799, 61

Gas Accretion and Star-Formation Rates with IFUs and Background Quasars

Nicolas F. Bouché

1 Gas Accretion in the Context of Galaxy Evolution

Two major advances in our understanding of the evolution of galaxies have occurred in the past 5 years or so. First, star forming galaxies (SFGs) are seen to be forming stars at a regular pace, forming the so-called main-sequence (MS; Brinchmann et al. 2004; Noeske et al. 2007; Daddi et al. 2007; Elbaz et al. 2007; Peng et al. 2010; Whitaker et al. 2012, 2014; Tomczak et al. 2016) an almost linear relation between star-formation rate (SFR) and stellar mass ($M_\star$). Second, star-forming galaxies have gas consumption time scales (defined as the time scale to consume the gas at the current SFR, i.e., $M_{\rm gas}$/SFR) that are too short compared to the age of the Universe (at all epochs where gas masses can be measured, currently up to redshifts $z = 2$; Leroy et al. 2008; Genzel et al. 2010; Tacconi et al. 2013; Freundlich et al. 2013; Saintonge et al. 2013, 2016). Thus, in order to sustain the observed levels of star-formation over many Gyrs, galaxies must continuously replenish their gas reservoir with fresh gas accreted from the vast amounts available in the intergalactic medium.

Two additional indirect arguments point towards the necessity for gas accretion. One argument comes from the metallicity distribution of sun-like stars (G-type) in the Milky Way, where the metallicity distribution of G-stars is not consistent with what the expectation from chemical evolution models (whose initial condition is that of a pure gaseous reservoir, a.k.a. a "closed-box") unless some fresh gas infall is invoked (Lynden-Bell 1975; Pagel and Patchett 1975), a discrepancy often referred to as the G-dwarf problem (van den Bergh 1962; Schmidt 1963). Another indirect argument for continuous replenishment of galaxy reservoirs comes from the non-evolving (or very weakly evolving) cosmic neutral density $\Omega_{\rm H\,I}$ for damped Lyα

N.F. Bouché (✉)
IRAP, 9 Av. Colonel Roche, F-31400 Toulouse, France
e-mail: nicolas.bouche@irap.omp.eu

A. Fox, R. Davé (eds.), *Gas Accretion onto Galaxies*, Astrophysics and Space Science Library 430, DOI 10.1007/978-3-319-52512-9_15

absorbers (Péroux et al. 2003; Bauermeiester et al. 2010; Noterdaeme et al. 2012; Zafar et al. 2013; Crighton et al. 2015) which contrasts with the rapid evolution of the stellar cosmic density.

In numerical simulations, accretion of intergalactic gas (via the cosmic web) originates from the growth of dark matter halos which pulls the cold baryons along. In galaxies with luminosities less than L^*, this process is expected to be very efficient owing to the short cooling times in these halos (White and Frenk 1991; Birnboim and Dekel 2003; Kereš et al. 2005; Faucher-Giguère 2011). This phenomenon is often referred to as "cold accretion" and this term describes the halo mass regime ($M_{\rm DM} < M_{\rm crit}$) where the accretion is most efficient (White and Frenk 1991; Birnboim and Dekel 2003). At high masses and high redshifts, cold accretion can occur in the form of cold streams (Dekel et al. 2009), i.e., when the halo is larger than the typical filament cross-section.

Once inside the galaxy dark matter halo, the accreted gas is expected to orbit the galaxy, the cold accreted gas should orbit about the halo before falling in to build the central disk, delivering not only fuel for star formation but also angular momentum to shape the outer parts of the galaxy (Stewart et al. 2011b; Shen et al. 2013; Danovich et al. 2015). Thus, accreting material should co-rotate with the central disk in the form of a warped, extended cold gaseous structure (Pichon et al. 2011; Kimm et al. 2011; Danovich et al. 2012, 2015; Shen et al. 2013), sometimes referred to as a "cold-flow disk" (Stewart et al. 2011a, 2013, 2016). This is further discussed in the chapter by K. Stewart.

In the local universe, such large gaseous disks are often seen around galaxies in H I 21 cm surveys, where the H I disk extends two to three times beyond the stellar radius as in the M33 low surface brightness disk (Putman et al. 2009), the more massive M81 (Yun et al. 1994) and M83 galaxies (Huchtmeier and Bohnenstengel 1981; Bigiel et al. 2010), and others (see chapter by F.J. Lockman). The kinematics of this H I gas in the outer parts show that the gas is systematically rotating in the same direction as the central object at those large radii (two to three times beyond the stellar radius or 20–30 kpc).

As discussed in Stewart et al. (2011a), these gaseous disks (or "cold-flow disks") should produce distinct kinematic signatures in absorption systems. In particular, the gas kinematics seen in absorption spectra are expected to be offset by about 100 km s^{-1} from the galaxy's systemic velocity. It should be noted that these cold-flow disks are the end product of cold accretion and these two terms should not be confused. Moreover, cold-flow disks occur on scales of a few tens of $R_{\rm vir}$ whereas cold accretion (or cold streams) occur on scales of Mpc, hence the possibility to observe with background quasars these two distinct phenomena depends strongly on the gas column density. Theoretical expectations for detecting cold accretion with Lyman limit systems are discussed in Fumagalli et al. (2011), Goerdt et al. (2012), and in the chapter by C.A. Faucher-Giguère. Observational signatures of cold accretion with Lyman limit systems are discussed in the chapter by N. Lehner.

This chapter focuses on the distinct absorption kinematics of "cold-flow disks" which should occur in the high column density regime and well inside the halos of galaxies.

Before discussing the most recent observations, it is important to quantify how much accretion is needed in order to sustain the observed levels of star-formation over many Gyrs to the levels of tens to hundreds of solar masses per year at $z = 1$–2 required by the main-sequence. This can be estimated from the following argument. As shown in Bouché et al. (2010) and many others since (Davé et al. 2012; Krumholz et al. 2012; Lilly et al. 2013; Feldmann 2013, 2015; Dekel and Mandelker 2014; Forbes et al. 2014; Peng and Maiolino 2014), galaxies at $z < 4$ can be thought of as a simplified gas regulator where there is a rough balance between the SFR and the gas accretion rate. The equilibrium between gas accretion and SFR under the "regulator" model (Lilly et al. 2013) or "bathtub" scenario (Bouché et al. 2010) indicates that the accretion rate $\dot{M}_{\rm in}$ should be comparable to the star-formation rate (SFR).

In particular, the equilibrium solution can be found from the continuity equation between the gas supply and gas consumption terms, and equilibrium SFR can be expressed as

$$\begin{aligned} \mathrm{SFR} &\approx \frac{\dot{M}_{\rm in}}{1+R+\eta} \\ &= \frac{\epsilon_{\rm in} f_{\rm B}}{1+R+\eta}\dot{M}_{\rm h} \end{aligned} \tag{1}$$

where R is the gas returned fraction (from massive stars) under the instantaneous recycling approximation, η the loading factor from galactic winds, f_B the universal baryonic fraction, $\epsilon_{\rm in}$ the efficiency of accretion (with $\dot{M}_{\rm in} \equiv \epsilon_{\rm in}\, f_B\, \dot{M}_{\rm h}$), and $\dot{M}_{\rm h}$ the dark matter growth rate (Genel et al. 2008; McBride et al. 2009; Dekel et al. 2013). Equation (1) has several implications. First, it shows that SFR and the accretion rate $\dot{M}_{\rm in}$ are comparable to a factor of order unity given that η is found to be $\approx$1 () and that $R \sim 0.5$ with its exact value depending on the initial mass function. Second, it shows that the evolution of the accretion rate $\dot{M}_{\rm in}(z)$ sets the evolution of the SFR (Bouché et al. 2010). Lastly, it is ultimately the dark matter growth rate that limits the growth rate of galaxies and the evolution of SFR.

Equation (1) is important because it shows that SFR is regulated by cosmological quantities (in other words it is supply limited) and not necessarily the very local conditions (such as gas mass or gas surface density). The gas mass will adjust itself such that there is a close balance between SFR (consumption) and accretion (supply). However, as discussed in Bouché et al. (2010) and Lilly et al. (2013), beyond $z > 4$ or $z > 6$, the dark matter growth rate is so large that the SFR does not have the time to adjust itself towards the (quasi)-equilibrium solution, and thus this solution does not apply.

2 Detecting Gas Accretion

Contrary to gas outflows where observational evidence abounds, direct evidence of gas inflows is notoriously difficult to obtain. As described in this book, this situation has begun to change. One can study gas accretion via Lyman Limit Systems (see chapter by N. Lehner), studies of galaxy spectra (see chapter by K. Rubin), UV spectroscopy of quasars (see chapter by P. Richter), or Lyman alpha emission (see chapter by S. Cantalupo). In this chapter, I will focus on the recent breakthroughs on detecting gas accretion which came from recent technological advancements with integral field units (IFUs).

2.1 Observational and Technological Breakthroughs

To study gas accretion, one can take advantage of one recent key observational result established only about 5–6 years ago, namely that the gas distribution as traced by low-ionization lines (such as Mg II, [Na D]) is highly anisotropic (see chapter by G. Kacprzak). The first evidence which showed that the circumgalactic gas was anisotropic came from a study of [Na D] absorption in stacked spectra from a large sample of ~1000 SDSS galaxies (Chen et al. 2010). Their results showed a strong inclination dependence of the blue-shifted absorption lines. Had the gas been isotropic, this inclination dependence would not have been found.

Another important result came soon after with the investigation of the radial and azimuthal dependence of the low ion Mg II in a sample of 4000 intermediate redshift galaxies with 5000 background galaxies in the COSMOS survey (Bordoloi et al. 2011). They found that—for blue star-forming galaxies—there is a strong azimuthal dependence of the Mg II absorption within 50 kpc of inclined disk-dominated galaxies, indicating the presence of a strongly bipolar outflows. This result was soon confirmed with two key studies at $z = 0.1$–1 using background quasars which have the advantage that they can probe the properties of gaseous halos around individual galaxies (Bouché et al. 2012; Kacprzak 2012). While the presence of strong bipolar outflows leaving galaxy disks more or less perpendicularly, i.e., in the direction of the minor-axis, these studies with background quasars revealed that the low-ionization gas traced by Mg II do also trace the outskirts of gaseous disks when the quasar's location appears to be aligned with the minor-axis. Indeed, the azimuthal angle α[1] distribution of background quasars with Mg II appears to be bi-modal with a main peak at $\alpha \sim 90°$ (aligned with the minor-axis) and at $\alpha \sim 0°$ (aligned with major-axis) (Bouché et al. 2012; Schroetter et al. 2015, 2016).

Thus, a picture is emerging from these observations where outflows occur preferentially along the minor-axis of galaxies, with also large gaseous structures (as

[1] α is defined as the angle between the galaxy major-axis and the apparent quasar location.

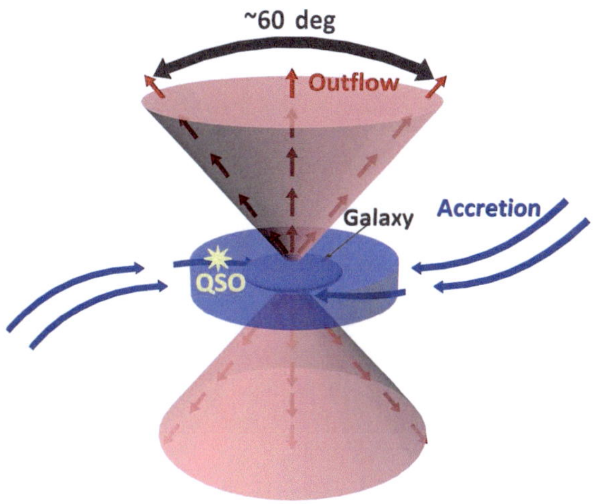

Fig. 1 Schematic representation of gas flows around galaxies [reproduced from Bouché et al. (2013)]

described in the previous section) that are roughly co-planar with the star-forming galaxy. Figure 1 represents schematically the situation with outflows leaving the galaxy along the minor-axis and gas inflows forming the so-called cold-flow disk, the gaseous structure discussed in the previous section and in the chapter by K. Stewart. The cold gas in outflows as traced by Mg II extends to at least 80–100 kpc, but outflows could extend to the virial radius. The cold-flow disk on the other hand is typically 20–30 kpc, i.e., extends to a few tens of the virial radius.

Quasar absorption lines (such as the low-ionization Mg II 2796,2803 doublet) have had the potential to yield important information on gas flows for decades, but the challenge has always been in finding the galaxy associated with a particular absorber. Indeed, in order to connect the properties of the circum-galactic medium (CGM) with those of galaxies, such as SFR and mass, it is necessary to identify the galaxy associated with each metal line absorption system. This identification process has been a challenge more decades because it is inherently expensive and inefficient with traditional imaging and spectrographic instruments. As described in this book, one can take deep imaging, but one will need to perform extensive multi-object spectroscopy of all of the galaxies within a certain impact parameter around the quasar.

But in the past 10 years, there has been a technological breakthrough, with the advancements of IFUs on 10-m class telescopes. Indeed, with an IFU, one can simply point-and-shoot at a quasar field and identify line emitters at the expected redshift of the absorption system (Fig. 2). However, the IFU must have a sufficiently large field of view (FOV). Two IFUs, both on the VLT, have played a major role in this area: the near-infrared IFU SINFONI (with an FOV of 8 arcsec×8 arcsec; Bonnet et al. 2004) and the new optical IFU MUSE (with an FOV of $1' \times 1'$; Bacon et al. 2010).

This technique was pioneered by Bouché et al. (2007) on a sample of 21 Mg II absorbers selected from the SDSS survey using the SINFONI IFU as illustrated in Fig. 2. Given that SINFONI has a small FOV of only 8 arcsec×8 arcsec, one

Fig. 2 Illustration of the advantage in observing quasar fields with IFUs. A line emitter can easily be detected at any position in the field of view by selecting the wavelength slice of the hyper-spectral cube corresponding to the expected redshift of the absorber

has to select the quasar sight-lines with absorbers that have the highest chance to be associated with a galaxy at small impact parameter, i.e., to lie within the FOV. Thanks to the well-known anti-correlation between the Mg II rest-equivalent width and impact parameter (Lanzetta and Bowen 1990; Steidel 1995), the strongest absorbers are the most likely to be associated with a galaxy at a small (of a few arcsec) impact parameter. This selection criteria became the basis for the SINFONI Mg II Program for Line Emitters (SIMPLE; Bouché et al. 2007, 2012; Schroetter et al. 2015).

2.2 *Measuring the Gas Accretion Rate*

Suppose one finds a galaxy with a background quasar within 15–30 kpc of a galaxy, but more importantly, a quasar whose apparent location is aligned with the galaxy's major-axis as illustrated in Fig. 1. This geometrical configuration is ideal to study the properties of "cold-flow-disks" which are created by the global accretion process (or processes) as discussed earlier.

The gas accretion rate can be estimated if one approximates the gaseous structure as a short cylinder of radius b and height h_z which has mass density $\rho(b)$ at the location of the quasar sight-lines, i.e., at the impact parameter b. As the cylinder is intercepted by the quasar line-of-sight at the impact parameter b, the accretion flux $\dot{M}_{\rm in}$ through the inner surface of area $A = 2\pi b\, h_z$ can be estimated from mass conservation argument, yielding the following equation:

$$\begin{aligned}\dot{M}_{\rm in}(b) &= 2\pi b\, h_z\, V_\perp\, \rho(b) \\ &= 2\pi b\, V_\perp\, \cos(i)\, m_{\rm p}\mu\, N_{\rm H}, \end{aligned} \quad (2)$$

where $V_\perp$ is the radial flow velocity (perpendicular to the surface A), i is the inclination of the structure, $N_{\rm H}$ is the total gas column, μ is the mean molecular weight, $m_{\rm p}$ is the proton mass, and the second line uses the identity $m_{\rm p}\mu N_{\rm H} = \int {\rm d}z\rho(b) = \rho(b)h_z/\cos i$. This equation can be used to estimate accretion rates of gaseous inflows in the close vicinity of galaxies as probed by quasar lines-of-sight.

3 Gas Accretion from IFU Surveys

In this section, I discuss example results from the recent IFU surveys that have led to the first constraints on the accretion rates of fresh gas around star-forming galaxies. Before discussing detailed results, I list here several of the key surveys of quasar sight-lines with IFUs. These are

- the SINFONI "SIMPLE" survey of 21 sight-lines with strong (rest-equivalent width > 2 Å) $z \sim 1$ Mg II systems (Bouché et al. 2007; Schroetter et al. 2015), which led to a sample of 14 galaxy-quasar pairs, including the J1422 system discussed in Sect. 3.2 (Bouché et al. 2016);
- the $z = 2$ extension of SIMPLE survey around 18 strong (> 2 Å) $z \sim 2$ Mg II systems (Bouché et al. 2012), which led to only 4 quasar-galaxy pairs, including the HE2243–60 system discussed in Sect. 3.1 (Bouché et al. 2013);
- the SINFONI survey of quasar sight-lines with strong H I absorbers, with column densities log $N_{\rm H\,I}$ greater than 19–20 (Péroux et al. 2011, 2012, 2013);
- the MUSE Gas Flow and Wind (MEGAFLOW) survey around Mg II systems with rest-equivalent width > 0.8 Å, on 22 quasar sight-lines with multiple ($N = 3, 4$ or 5) Mg II systems (Schroetter et al. 2016).

In contrast to the SINFONI surveys where one can only target a single system at a time and have yielded about two dozens galaxy-quasar pairs, the MUSE survey will yield about 80 to 100 galaxy-quasar pairs, thanks to the large wavelength range of the MUSE instrument ($4800 < \lambda < 9300$ Å).

3.1 *Case Study 1: HE2243 at $z = 2.32$*

From a sample of galaxy-quasar pairs, one needs to select suitable pairs for the study of accretion. From Fig. 1 one sees that the galaxy should not be seen face-on and that the quasar apparent location should be aligned with the galaxy major-axis. From the z2SIMPLE survey (Bouché et al. 2012), one star-forming galaxy associated with a $z = 2.32$ Mg II system is located just 26 kpc from the quasar line-of-sight of HE

2243–60.[2] The background quasar whose apparent location on the sky is fortuitously aligned with the galaxy's projected major-axis makes this galaxy-quasar pair an excellent candidate to test the predictions from cosmological simulations (Stewart et al. 2011a, 2016).

Not only IFUs are excellent tools to identify the host galaxy, but also, such observations directly yield the galaxy star-formation rate, the galaxy kinematics and the galaxy's orientation with respect to the quasar sight-line. In this case, the galaxy has a star-formation rate (SFR) of 33^{+40}_{-11} $M_{\odot}$ yr^{-1} (where $M_{\odot}$ is the mass of the Sun), typical for that redshift. The kinematics of the $z = 2.3285$ star-forming galaxy obtained with adaptive optics (AO) assisted SINFONI (i.e., obtained with ~1 kpc or 0.25 arcsec resolution) reveal that this galaxy has physical properties typical for rotationally supported disks seen at that redshift (Förster Schreiber et al. 2006, 2009), with a maximum rotation velocity of $V_{\max} = 150 \pm 15\,\mathrm{km\,s^{-1}}$ determined from 3D fitting. And the apparent quasar location on the sky is only 10° to 20° from the galaxy's projected major-axis.

The kinematics of the absorbing gas along the line-of-sight give clues about the nature of the gas, which were observed with the VLT spectrograph UVES. A direct comparison between the absorbing gas kinematics with respect to the galaxy emission kinematics reveals that the gas seen in absorption 26 kpc from the galaxy—corresponding to seven times the half-light radius $R_{1/2} = 3.6$ kpc or one third of the virial size of the halo $R_{\rm vir}$—is moving in the same direction as the galaxy rotation; i.e., the gas appears to be co-rotating. The comparison is shown in Fig. 3 where panel (a) shows the galaxy emission kinematics as recovered from the IFU data, and panel (b) shows absorbing gas kinematics measured from the absorption lines in the UVES spectra. One can see that all of the low-ionization ions Zn II, Cr II, Fe II and Si II tracing the cold gas show redshifted velocities and the observed velocity field of this galaxy also shows redshifted velocities in the direction of the quasar location. Hence, the absorbing gas appears to be "co-rotating" with the host, meaning that it rotates in the same direction.

However, while the sign of the kinematic shifts agrees, the amplitude of the kinematic differences does not agree entirely. Only a fraction of the velocities seen in absorption are in agreement with those seen in emission from the galaxy (extrapolated at the location of the quasar). This would indicate that a simple rotating disk with circular orbits and a normal (flat) rotation curve can account only for some fraction of the absorbing gas kinematics. The other components have a line-of-sight velocity less than the rotation speed, indicating that this gas is not rotationally supported. In other words, the second group of components is offset by about $100\,\mathrm{km\,s^{-1}}$ from the rotation pattern; hence cannot be gravitationally supported and therefore should be flowing in.

In order to estimate the accretion rate from these observations, one needs estimates of the gas column density and of the inflow speed. A detailed analysis

[2]This system is also a damped Lyman absorber (DLA; Lopez et al. 2002) with a column density of $\log N_{HI} = 20.6$.

of the metal column densities in each of the components (Bouché et al. 2013) indicates that each of these two groups contains about half of the total H I column $\log N_{\rm HI} \simeq 20.6$. Using a simple geometrical toy model of accretion—where "particles" representing gas clouds or absorption components are distributed in an extended disk configuration as in Fig. 1—with cloud kinematics reflecting an inflowing (radial) gas flow, one can generate simulated absorption profiles which can then be compared to the observed absorption profile. Because the relative galaxy orientation is well determined by the VLT/SINFONI data, the only free parameter is the inflow speed which is found to be $\sim$200 km s^{-1}. Figure 3c shows the modeled kinematics with the disk component shown in red, and the inflow component shown in black.

Using the constraints on $V_{\rm in}$ and on $N_{\rm HI}$ from the line-of-sight kinematic shift between the SINFONI and UVES datasets, the estimated accretion rates estimated from Eq. (2) is 30 to 60 $M_\odot$ yr^{-1} given the uncertainties in the column density and inflow speed. The range of accretion rates, 30 to 60 $M_\odot$ yr^{-1}, is found to be similar to the galaxy's SFR of $\sim 33^{+40}_{-11}$ $M_\odot$ yr^{-1}, in agreement with the simplest arguments for galaxy growth given in Sect. 1. Furthermore, for this galaxy's halo mass, $M_{\rm h} \sim 4 \times 10^{11}$ $M_\odot$ (assuming $V_{\rm max} \sim V_{\rm circ}$), the maximum baryonic accretion rate is $f_B\,\dot{M}_{\rm h} \approx 40$ $M_\odot$ yr^{-1} (Genel et al. 2008; Dekel et al. 2013), which implies an accretion efficiency $\epsilon_{\rm in}$ in Eq. (1) of $\sim$100% regardless of the value for the wind loading factor η.

3.2 *Case Study 2: J1422 at $z = 0.91$*

From the SIMPLE survey (Bouché et al. 2007) consisting of 14 galaxies found around strong $z \sim 0.8$–1.0 Mg II absorbers selected from the Sloan Digital Sky Survey (SDSS) database with rest-frame equivalent widths $W_{\rm r}^{\lambda 2796} > 2$ Å, two have the galaxy's major-axis roughly aligned with the quasar apparent location among the galaxies that are not seen face-on (with an inclination >30°) as shown in (Schroetter et al. 2015). One of these two is the $z = 0.9096$ quasar–galaxy pair towards the quasar SDSS J142253.31–000149 (hereafter SDSS J1422–00), where the galaxy's major-axis is about 15° from quasar-galaxy orientation. This is—as already mentioned—the most favorable configuration to look for the kinematic signatures of gas inflows (Fig. 1).

The J1422–00 quasar field was also observed with the MUSE IFU (Bouché et al. 2016), yielding [O II] and Hβ fluxes in addition to the Hα, and [N II] fluxes from SINFONI. Incidentally, the MUSE observations show that there are no companion down to 0.22$M_\odot$ yr^{-1} (5 σ) within 240 kpc (30 arcsec). Combining Hα, [O II] fluxes from SINFONI and MUSE observations, the galaxy has an SFR of 4.7 $\pm$ 2.0 $M_\odot$ yr^{-1} (Bouché et al. 2016), and a metallicity of $12 + \log {\rm O/H} = 8.7 \pm 0.2$. Figure 3 shows the galaxy kinematics which yield a maximum velocity $V_{\rm max} = 110 \pm 10$ km s^{-1}, corresponding to a halo mass of $M_{\rm h} \approx 2 \times 10^{11}\,M_\odot$,

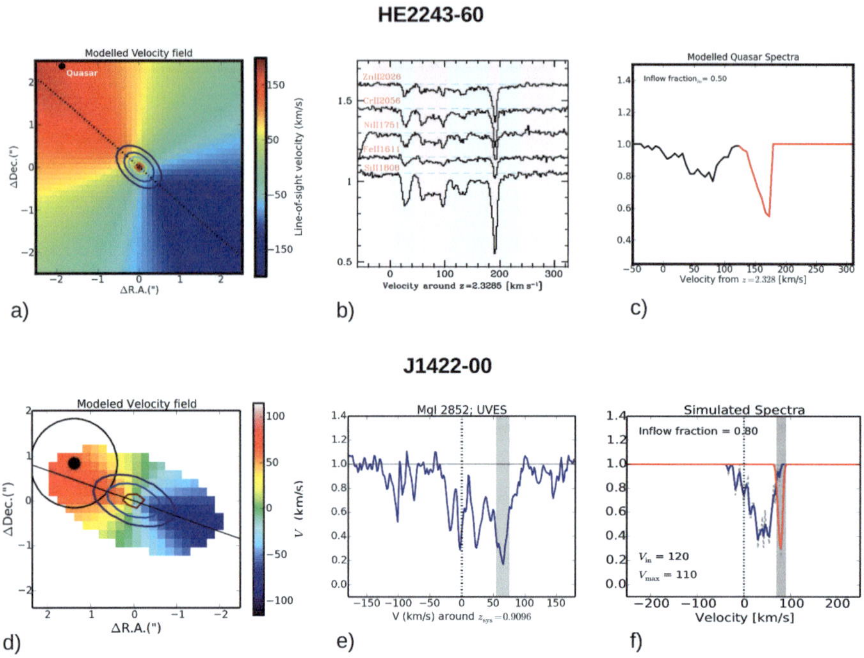

Fig. 3 Panels (**a**), (**b**), and (**c**) show, for the HE2243–60 galaxy-quasar pair, the galaxy kinematics, the gas kinematics at the quasar location, and the modeled kinematics, respectively. The disk kinematics (with circular orbits) is shown in *red* while the gas inflow model is shown in *black*. Panels (**d**), (**e**), and (**f**) are for the J1422–00 pair.

corresponding to a 0.1 $L_\star$ galaxy. For such a halo mass, the average baryonic accretion rate is $f_B\,\dot{M}_{\rm h} \approx 4\,{\rm M}_\odot\,{\rm yr}^{-1}$.

The H I gas column density associated with this 3.2 Å Mg II absorber is $\log N$(H I)$\simeq$ 20.4, measured from a *Hubble Space Telescope*/COS G230L spectra. Using a high-resolution spectrum (UVES) of the background quasar, Bouché et al. (2016) found that the gas kinematics traced by low-ionization lines ($M_{\rm gas}$, Zn, Fe, Al) show distinct signatures consistent with those expected for a "cold-flow disk." Indeed, the absorption profile consists of a component that follows the expected (projected) rotation speed, at $+60\,{\rm km\,s^{-1}}$, with several sub-components that are offset kinematically as shown in Fig. 3b. Using a similar geometrical model as in Sect. 3.1, the absorption kinematics data are consistent with a radial inflow speed of $V_{\rm in}$ $\approx$100 km s^{-1}, when comparing the data (panel e) and the model (panel f) in Fig. 3. With such an inflow velocity, one can estimate the mass flux rate $\dot{M}_{\rm in}$ to be at least two to three times larger than the SFR using Eq. (2).

Interestingly, this galaxy also shows evidence for a galactic wind from the blue-shifted (by $v_{\rm out} = -80 \pm 15\,{\rm km\,s^{-1}}$) low-ionization absorption lines (Mg II and Fe II) in the galaxy spectrum. The estimated mass outflow rate is about 0.5–5 ${\rm M}_\odot\,{\rm yr}^{-1}$ (i.e., a loading factor $\eta \leq 1$), and currently, this is the only galaxy with simultaneous constraints on the gas outflow and inflow rates. With a constraint on the outflow rate and the loading factor, one can better estimate the efficiency of baryonic accretion $\epsilon_{\rm in}$ using Eq. (1). Together these numbers imply that the factor $(1+R+\eta)$ in Eq. (1) is $(1 + R + \eta)\times$SFR that is 5 to 8 $M_\odot$ yr^{-1}. As mentioned, for this galaxy halo mass, the *average* baryonic accretion rate is estimated to be $f_B\ \dot{M}_{\rm h} \approx 4\ M_\odot\ {\rm yr}^{-1}$ using the theoretical expectation for the DM accretion rate from Genel et al. (2008) or Birnboim et al. (2007). Putting it together, the accretion efficiency is high at $\epsilon_{\rm in} \approx 1.0$ (regardless of the loading factor η) in agreement with the theoretical expectation of Dekel and Birnboim (2006). It is important to keep in mind that our measurements on inflow and outflow rates are instantaneous quantities, while this baryonic estimate is time averaged, or averaged for populations of galaxies.

3.2.1 Implications for Angular Momentum

A critical question for these gaseous disks around star-forming galaxies is how much angular momentum they carry as discussed in Stewart et al. (2013) and Danovich et al. (2015). Here, we recall the common relation between the specific angular momentum $j \equiv J/M$ and the spin parameter taken from Bullock et al. (2001), which is

$$\lambda \equiv \frac{j}{\sqrt{2}\,R_{\rm vir}\,V_{\rm vir}}, \tag{3}$$

where $R_{\rm vir}$ and $V_{\rm vir}$ are the halo virial radius and virial circular velocity.

The host galaxy is estimated by Bouché et al. (2016) to have $\lambda_{\rm gal}$ to be 0.04–0.05, using the relation between disk scale length $R_{\rm d}$ and the disk spin λ parameter from Mo et al. (1998). But the angular momentum of the "cold-flow disk" is harder to estimate without a direct size constraint, but one can place useful limits using the minimum radius given by the impact parameter. Doing so, the spin parameter of the cold-flow disk $\lambda_{\rm cfd}$ is estimated to be $\lambda_{\rm cfd} > 0.06$, i.e., 50% larger than that of the galaxy. This limit on the cold-flow disk angular momentum is consistent with the theoretical expectation of Danovich et al. (2015), where the baryons within 0.3 $R_{\rm vir}$ have two to three times the galaxy angular momentum.

3.2.2 Metallicity Aspects

When discussing gas inflows around galaxies, one naturally wonders whether the accreting material has a metallicity of "pristine" IGM gas (i.e., $\log Z/Z_\odot = -2$). Naturally, estimates of the gas metallicity require a H I column density measurement in addition to gas phase column densities for the metallic ions. For both of the cases discussed in this section, the H I column density could be determined directly from either a UVES optical spectrum or a COS UV-spectrum.

In both cases, the metallicity of the absorbing gas is estimated to be much higher than the IGM metallicity of $\log Z/Z_\odot = -2$ for fresh infall, implying a significant amount of recycling. Indeed, the absorbing gas metallicity is estimated to be -0.72 ± 0.04 and -0.4 ± 0.4 for HE2243–60 and J1422–00, respectively. Though, in both cases, the host galaxy ISM metallicity—measured from its nebular lines—is somewhat higher than the absorbing gas, at 1/2 $Z_\odot$ and $Z_\odot$, respectively. Hence, one can conclude from these studies that the gas accretion is already enriched from significant recycling, but is lower than the metallicity of the ISM by about a factor of two.

3.3 Case Study 3: H I Selection

Because a selection using metal lines inevitably leads to a bias in the selection of absorption line systems, C. Péroux and her collaborators have led several studies using H I-selected absorbers, i.e., Damped Lyα absorbers (DLAs) with H I column densities greater than $2 \times 10^{20}\,\mathrm{cm}^{-2}$, and sub-DLAs with H I column densities between 10^{19} and $10^{20}\,\mathrm{cm}^{-2}$. The advantage of the H I-selection is that it bypasses the biases present in metal line selections. The disadvantage of this technique is that the success rate in detecting the host galaxy is much lower.

Using the IFU SINFONI, out of a large survey of 30 H I systems, only 6 were detected (Péroux et al. 2011, 2012, 2013, 2014). Out of these 6 detections they find evidence for the presence of outflows in two systems, while three are consistent with gas accretion and the remaining case is poorly constrained.

Such sample will prove important in order to compare the gas metallicity seen along the quasar line-of-sight with the metallicity of the host as described in Péroux et al. (2016). At the moment, the situation is unclear and large samples are required.

4 Future Perspectives

The case studies described in Sect. 3 are interesting, but the natural question is how common are the features shown in Fig. 3. While SINFONI has played a major role in this area, the wide-field IFU MUSE, one of the second generation instruments for the VLTs which started operation in the Fall of 2014, is changing our view of the CGM and will allow for a major leap forward in our knowledge of gas inflows and outflows. Indeed, thanks to its large $1' \times 1'$ field of view sampled at 0.2 arcsec/spaxel, large wavelength range from 480 to 930 nm sampled at 1.25 Å, and to its unprecedented sensitivity, MUSE (Bacon et al. 2010) will allow us to study gas flows around galaxies with samples ten times larger (i.e., $\approx$100) than current surveys. In particular, we started the MusE Gas FLO and Wind (MEGAFLOW) survey which designed to yield $\approx$100 galaxy-quasar pairs in about 22 quasar sight-lines selected from the SDSS database (DR12) which have multiple (4 or more) Mg II $\lambda\lambda$2796,2803 absorbers at redshifts $0.3 < z < 1.4$ suitable for [O II] $\lambda\lambda$3726,3729 emitters. Results on outflows from the first two fields appeared in Schroetter et al. (2016), At the time of writing, the last data are being acquired.

References

Bacon, R., Accardo, M., Adjali, L., et al., SPIE, 7735, 8
Bauermeister, A., Blitz, L., & Ma, C. 2010, ApJ, 717, 323
Bigiel, F., Leroy, A., Seibert, M., et al. 2010, ApJ, 720, L31
Birnboim, Y., & Dekel, A. 2003, MNRAS, 345, 349
Birnboim, Y., Dekel, A., & Neistein, E. 2007 MNRAS, 380, 339
Bonnet, H., Conzelmann, R., Delabre, B., et al. 2004, SPIE, 5490, 130
Bordoloi, R., Lilly, S.J., Knobel, C., et al. 2011, ApJ, 743, 10
Bouché, N., Dekel, A., Genzel, R., et al. 2010, ApJ, 718, 1001
Bouché, N., Finley, H., Schroetter, I., et al. 2016, ApJ, 820, 121
Bouché, N., Hohensee, W., Vargas, R., et al. 2012, MNRAS, 426, 801
Bouché, N., Murphy, M.T., Kacprzak, G.G., et al. 2013, Science, 341, 50
Bouché, N., Murphy, M.T., Péroux, C., et al. 2012, MNRAS, 419, 2
Bouché, N., Murphy, M.T., Péroux, C., et al. 2007, ApJ, 669, L5
Brinchmann, J., Charlot, S., White, S.D.M., et al. 2004, MNRAS, 351, 1151
Bullock, J.S., Dekel, A., Kolatt, T.S., et al. 2001, ApJ, 555, 240
Chen, Y., Tremonti, C.A., Heckman, T.M., et al. 2010, AJ, 140, 445
Crighton, N.H.M., Murphy, M.T., Prochaska, J.X., et al. 2015, MNRAS, 452, 217
Daddi, E., Dickinson, M., Morrison, G., et al. 2007, ApJ, 670, 156
Danovich, M., Dekel, A., Hahn, O., Ceverino, D., Primack, J. 2015, MNRAS, 449, 2087
Danovich, M., Dekel, A., Hahn, O., Teyssier, R. 2012, MNRAS, 442, 1732
Davé, R., Finlator, K., Oppenheimer, B.D.: 2012, MNRAS, 421, 98
Dekel, A., Birnboim, Y. 2006, MNRAS, 368, 2
Dekel, A., Birnboim, Y., Engel, G., et al. 2009, Nature, 457, 451
Dekel, A., & Mandelker, N. 2014, MNRAS, 444, 2071
Dekel, A., Zolotov, A., Tweed, D., et al. 2013, MNRAS, 435, 999
Elbaz, D., Daddi, E., Le Borgne, D., et al. 2007, A&A, 468, 33
Faucher-Giguère, C.A., Kereš, D., & Ma, C.P. 2011, MNRAS, 417, 2982
Feldmann, R. 2013, MNRAS, 433, 1910
Feldmann, R. 2015, MNRAS, 449, 3274
Forbes, J.C., Krumholz, M.R., Burkert, A., Dekel, A. 2014, MNRAS, 438, 1552
Förster Schreiber, N.M., Genzel, R., Bouché, N., et al. 2009, ApJ, 706, 1364
Förster Schreiber, N.M., Genzel, R., Lehnert, M.D., et al. 2006, ApJ, 645, 1062
Freundlich, J., Combes, F., Tacconi, L. J., et al. 2013, A&A, 553, A130
Fumagalli, M., Prochaska, J.X., Kasen, D., et al. 2011, MNRAS, 418, 1796
Genel, S., Genzel, R., Bouché, N., et al. 2008, ApJ, 688, 789
Genzel, R., Tacconi, L.J., Gracia-Carpio, J., et al. 2010, MNRAS, 407, 2091
Goerdt, T., Dekel, A., Sternberg, A., Gnat, O., & Ceverino, D. 2012, MNRAS, 424, 2292
Huchtmeier, W. K., Bohnenstengel, H. D. 1981, A&A, 100, 72
Kacprzak, G.G., Churchill, C.W., Nielsen, N.M.: et al. 2012, ApJ, 760, L7
Kereš, D., Katz, N., Weinberg, D.H., & Davé, R. 2005, MNRAS, 363, 2
Kimm, T., Devriendt, J., Slyz, A., et al. 2011, preprint, arXiv:1106.0538
Krumholz, M.R., Dekel, A., & McKee, C.F. 2012, ApJ, 745,
Lanzetta, K.M., Bowen, D.: 1990, ApJ, 357, 321
Leroy, A.K., Walter, F., Brinks, E., et al. 2008, AJ, 136, 2782

Lilly, S.J., Carollo, C.M., Pipino, A., Renzini, A., & Peng, Y. 2013, ApJ, 772, 119
Lopez, S., Reimers, D., D'Odorico, S., & Prochaska, J.X. 2002, A&A, 385, 778
Lynden-Bell, D. 1975, *Vistas in Astronomy*, 19, 299
McBride, J., Fakhouri, O., & Ma, C. 2009, MNRAS, 398, 1858
Mo, H.J., Mao, S., & White, S.D.M. 1998, MNRAS, 295, 319
Noeske, K.G., Faber, S.M., Weiner, B.J., et al. 2007, ApJ, 660, L47
Noterdaeme, P., Petitjean, P., Carithers, W.C., 2012, A&A, 547, L1
Pagel, B.E.J., & Patchett, B.E. 1975, MNRAS, 172, 13
Peng, Y.j., Lilly, S.J., Kovač, K., 2010, ApJ, 721, 193
Peng, Y.J., & Maiolino, R. 2014, MNRAS, 443, 3643
Péroux, C., McMahon, R.G., Storrie-Lombardi, L.J., & Irwin, M.J. 2003, MNRAS, 346, 1103
Péroux, C., Bouché, N., Kulkarni, V.P., & York, D.G. 2013, MNRAS, 436, 2650
Péroux, C., Bouché, N., Kulkarni, V.P., York, D.G., & Vladilo, G. 2011, MNRAS, 410, 2237
Péroux, C., Bouché, N., Kulkarni, V.P., York, D.G., & Vladilo, G. 2012, MNRAS, 419, 3060
Péroux, C., Kulkarni, V.P., & York, D.G. 2014, MNRAS, 437, 3144
Péroux, C., Quiret, S., Rahmani, H., et al. 2016, MNRAS, 457, 903
Peroux, C., Rahmani, H., Quiret, S., et al. 2017, MNRAS, 464, 2053
Pichon, C., Pogosyan, D., Kimm, T., et al. 2011, MNRAS, 418, 2493
Putman, M.E., Peek, J.E.G., Muratov, A., 2009, ApJ, 703, 1486
Saintonge, A., Catinella, B., Cortese, L., et al. 2016, MNRAS, 462, 1749
Saintonge, A., Lutz, D., Genzel, R., et al. 2013, ApJ, 778, 2
Schmidt, M. 1963, ApJ, 137, 758
Schroetter, I., Bouché, N., Péroux, C., et al. 2015, ApJ, 804, 83
Schroetter, I., Bouché, N., Wendt, M. 2016, ApJ, 2016 APJ, 833, 39
Shen, S., Madau, P., Guedes, J., et al. 2013, ApJ, 765, 89
Steidel, C.C., in *QSO Absorption Lines*, ESO Astrophysics Symposia, p. 139. Springer-Verlag, Berlin, Germany (1995)
Stewart, K., Maller, A., Oñorbe, J., et al. 2016, ApJ, submitted, arXiv:1606.08524
Stewart, K.R., Brooks, A.M., Bullock, J.S., et al. 2013, ApJ, 769, 74
Stewart, K.R., Kaufmann, T., Bullock, J.S., et al. 2011a, ApJ, 735, L1
Stewart, K.R., Kaufmann, T., Bullock, J.S., et al. 2011b, ApJ, 738, 39
Tacconi, L.J., Neri, R., Genzel, R., et al. 2013, ApJ, 768, 74
Tomczak, A.R., Quadri, R.F., Tran, K.V.H., et al. 2016, ApJ, 817, 118
van den Bergh, S. 1962, AJ, 67, 486
Whitaker, K.E., van Dokkum, P.G., Brammer, G., Franx, M. 2012, ApJ, 754, L29
Whitaker, K.E., Franx, M., Leja, J., et al. 2014, ApJ, 795, 104
White, S.D.M. & Frenk, C.S. 1991, ApJ, 379, 52
Yun, M.S., Ho, P.T.P., & Lo, K.Y. 1994, Nature, 372, 530
Zafar, T., Péroux, C., Popping, A., et al. 2013, A&A, 556, A141

Index

A. Fox, R. Davé (eds.), *Gas Accretion onto Galaxies*, Astrophysics and Space Science Library 430, DOI 10.1007/978-3-319-52512-9

MIX
Papier aus verantwortungsvollen Quellen
Paper from responsible sources
FSC® C105338

If you have any concerns about our products,
you can contact us on
ProductSafety@springernature.com

In case Publisher is established outside the EU,
the EU authorized representative is:
Springer Nature Customer Service Center GmbH
Europaplatz 3, 69115 Heidelberg, Germany

Printed by Libri Plureos GmbH
in Hamburg, Germany